Binary star discoveries
in the URAT1 catalog

by

Martin P Nicholson

Martin Nicholson
Church Stretton
Shropshire SY6 7DQ
United Kingdom

Email – newbinaries@yahoo.co.uk

My New Year resolution for 2015 – "The results of all the solo astronomical data mining projects I have done in the past, or that I will do in the future, must be published and made available to the wider astronomical community."

Looking back, almost every year I have found myself standing in silence in memory of another hobby personality who had died. It didn't matter if the deceased was an astronomer, a "grave hunter" or a postal historian - so often the legacy they left was greatly reduced because so much of their knowledge and experience died with them.

Only a small proportion of the astronomical data mining and astronomical imaging projects I carried out over the last 20 years went through the lengthy – and sometimes controversial – process of third-party publication. This is particularly the case with my work on binary and double stars because, like all the amateur astronomers with an interest in this topic, I found myself communicating with a small number of professional astronomers who seemed to lack any genuine desire for a collaborative relationship with amateur enthusiasts.

The system for providing feedback on discoveries or measurements and for then including these results in the standard catalog has some unusual features. Direct submission to professional colleagues is apparently frowned upon and results "must" be submitted via one of the journals that welcome material of this type. I have found that the requirement that discoveries must appear in a peer reviewed journal as a pre-condition for inclusion in the catalogues is applied arbitrarily and inconsistently – there are many examples of them being included following "personal communication" and I know of other examples where peer reviewed results have not appeared at all. The "two tier system" that United States Naval Observatory (USNO) staff created caused significant ill-feeling and this was compounded when those responsible failed to respond to expressions of concern.

Much the same story can be told about the UK based Webb Society. Unsubstantiated claims of a "special relationship" between this group and the USNO raised false hopes and when the hype didn't match the reality some observers felt somewhat cheated.

Finally there was the Journal of Double Star Observations. Much of the material I submitted over the years to JDSO has either yet to appear or has appeared but was never included in the Washington catalog.

Any astronomer attempting to confirm a discovery claim or wanting to identify possible targets for a research project will require a reliable, easily accessible and up-to-date catalogue that includes all the most recent measurements. What currently exists is an on-line summary of results containing an eclectic mixture of binary stars, double stars and systems that don't exist at all. The problems faced by users are further compounded by the lengthy delay between the making of observations and inclusion in the catalog. The Court of Public Opinion must be wondering why discovery claims made by USNO astronomers have been included in the USNO maintained catalog even before their article was published, but it took from the spring of 2006 until October 2009 to do exactly the same task for other researchers' peer reviewed results. This is particularly strange given that a member of the USNO staff was identified as being an editor of the journal where those 2006 results appeared.

The Washington Double Star Catalog reports the "discovery code" for each entry. A discovery code is exactly that - it records who carried out the work, the painstaking and time-consuming work, resulting in a new discovery. Discovery codes consist of 1, 2 or 3 letters followed by a number and they have been used in the same way since the earliest days of organised astronomical observation. Any astronomer using the catalog in 2015 would assume that the discovery code UC meant that the listed object had been discovered by USNO staff – this is certainly not always the case. Discovery codes have never previously been used to identify the source of the raw data used and this new practice that does not credit the individual researcher is, in my opinion, both highly questionable ethically and seriously misleading to the wider astronomical community.

Over the years I have found a lack of common sense and impartiality from those entrusted with the management and organisation of binary and double star astronomy and it is my belief that there are some serious conflicts of interest in the current system.

Sadly I no longer feel able to share any of my discoveries with fellow enthusiasts prior to formal publication. I have become thoroughly disheartened by the endemic bad practice in the hobby that allows the shameless theft (and I used this word after a suitable pause for thought!) of amateur results by professional astronomers. It has become almost routine for articles to appear that duplicate amateur results from years earlier, either without any acknowledgement at all or with an acknowledgement that is so generic as to be meaningless. Add to that the increasingly frequent failure to honour the long established principle that "first to publish gets the credit" and you have a hobby that no longer has a collaborative or friendly and supportive component.

Martin Nicholson – Shropshire, UK.

May 2015

Binary star discoveries in the URAT1 catalog

Part 1 – Separations of less than 60 arc seconds

Martin P. Nicholson

Ticklerton Barn, Ticklerton, Church Stretton,
Shropshire, England, SY6 7DQ

e-mail: newbinaries@yahoo.co.uk

Abstract: Data mining using the recently published First U.S. Naval Observatory Astrometric Robotic Telescope Catalog (URAT1) has allowed the identification over 9400 common proper motion binary star systems many of which appear to be new discoveries.

Introduction: The First U.S. Naval Observatory Astrometric Robotic Telescope Catalog (URAT1) was made publically available in April 2015. It can be accessed from the Vizier site at

http://vizier.u-strasbg.fr/viz-bin/VizieR?-source=I%2F329

Data mining the catalog was a multi-stage operation.

Stage 1 - Decide on the minimum proper motion to be shown by each component of any previously unreported binary star system and the maximum separation between them. For the purposes of this study the minimum proper motion was 60mas/yr and the maximum separation was 60 arc seconds.

Stage 2 – Download the positional and photometric information from the Vizier site and process it through the purpose built software to identify all double stars with a separation of less than 60 arc seconds. The software takes approximately 20 minutes to process each batch of 1 million stars.

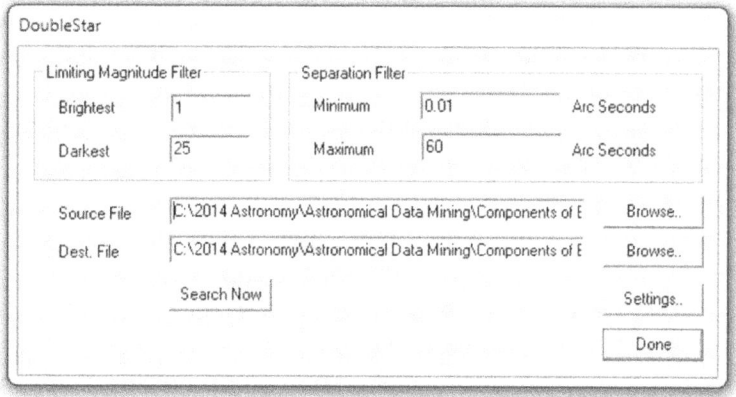

Fig.1 – The input selection screen

Stage 3 – Delete all pairs where the difference in proper motion in either declination or in right ascension is greater than the quoted error in the results.

Stage 4 – Identify the primary component of each system based on the URAT1 model fit magnitude and hence extract the position angle and separation of each pair from the output file generated by the software.

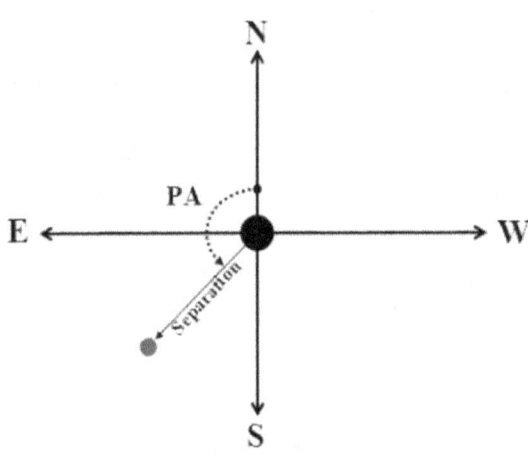

Fig. 2 The key features of a double star

Stage 5 - Cross reference the results obtained with the latest on-line version of the WDS catalogue so that all the known double stars could be removed from the output file leaving only the new discoveries to be presented in this paper.

Acknowledgments

Thanks are due to the anonymous peer reviewer who made a number of helpful suggestions and who processed the manuscript within 72 hours.

Results:

Col 1	Right ascension of the primary star in decimal degrees
Col 2	Declination of the primary star in decimal degrees
Col 3	Mean epoch of the primary and secondary star positions
Col 4	Magnitude of the primary star (The mean magnitude in the 680-750 nm URAT bandpass. This bandpass is between R and I, thus further into the red than UCAC.
Col 5	Magnitude of the secondary star (The mean magnitude in the 680-750 nm URAT bandpass. This bandpass is between R and I, thus further into the red than UCAC.
Col 6	Separation between the two stars in arc sec
Col 7	Position angle between the stars
Col 8	Proper motion of the primary star in right ascension (mas/yr)
Col 9	Proper motion of the primary star in declination (mas/yr)
Col 10	Error in the primary star proper motion (mas/yr)
Col 11	Proper motion of the secondary star in right ascension (mas/yr)
Col 12	Proper motion of the secondary star in declination (mas/yr)
Col 13	Error in the secondary star proper motion (mas/yr)

	Col 1	Col 2	Col 3	Col 4	Col 5	Col 6	Col 7	Col 8	Col 9	Col 10	Col 11	Col 12	Col 13
1	0.0344	77.4436	2013.3895	13.685	16.112	8.16	248.8	86.6	38.7	6.2	86.4	42.7	6.3
2	0.0981	9.3526	2012.9855	14.406	17.578	20.24	131	115.5	19	6.6	113.1	14.5	7
3	0.1467	8.9456	2012.9405	14.787	17.342	6.20	289.1	-16	-71.9	6.6	-12.1	-75.8	6.7
4	0.1892	8.4542	2013.3585	10.83	12.074	4.79	270.7	97.8	21.1	6.5	102.5	13.2	6.6
5	0.2353	74.7091	2013.278	14.23	16.556	6.63	59.9	79.4	35.7	6.2	78.2	30.1	6.5
6	0.2778	19.8735	2013.3005	15.235	16.677	8.48	158.4	-38.4	-64.5	5.6	-36.8	-68.8	5.6
7	0.2895	70.4309	2013.337	16.231	17.533	15.65	37	81.1	-43.5	5.9	83.7	-41.7	6.2
8	0.6077	36.1649	2013.233	18.253	18.542	30.68	26.6	68.2	-18.3	6.6	67.7	-15.1	6.6
9	0.6788	45.0363	2013.5075	17.221	17.508	3.83	330.6	67.7	-12.1	6.5	68.8	-9.2	6.6
10	0.6883	66.8050	2013.414	13.096	14.855	7.40	224.7	61.8	14.9	5.9	61	17.7	5.9
11	0.7755	58.3981	2013.4095	12.248	13.432	5.79	246.4	74.1	-13.2	5.9	72.1	-13.8	5.9
12	0.7952	8.9035	2013.3145	12.038	14.839	6.00	297.8	120.1	14	6.5	124.2	15.5	6.6
13	0.8130	74.7805	2013.49	14.694	18.559	17.88	4.3	158.9	26.6	6.3	160.1	32.7	6.5
14	0.8731	30.0640	2013.4255	14.4	15.094	4.87	40.8	36.5	-108.1	6.2	30.6	-106.6	6.5
15	0.8917	43.1321	2013.6205	13.272	15.716	6.24	96.9	14.4	-61.2	5.5	17.4	-61.2	5.4
16	0.9017	48.5321	2013.4885	14.77	15.539	50.88	47.4	73.2	4	5.5	69.6	7.4	5.5
17	1.0434	67.4219	2013.462	11.619	16.097	7.80	132	108.8	-36.8	5.9	109.2	-38.5	5.9
18	1.1650	60.6795	2013.4355	15.155	16.426	5.03	83.5	69.3	-3.9	6.2	70	3.2	6.4
19	1.2744	64.3745	2013.3415	17.861	18.501	40.29	36.5	-14.1	60.3	6.3	-6.8	60.1	6.7
20	1.4703	3.9967	2013.1965	13.129	14.711	21.39	143.9	69.4	-17.1	6.5	69.6	-16.3	6.5
21	1.5966	49.4554	2013.516	14.683	14.69	8.55	174.4	68.2	-23.4	5.5	68.8	-19.8	5.5
22	1.6469	73.9398	2013.4495	9.638	13.425	7.69	4.9	33.5	135.4	6.2	31.5	132.4	6.3
23	1.6704	28.8285	2013.604	8.786	13.488	16.05	287.9	63.1	-65.6	5.4	59.9	-68.6	5.5
24	1.9041	21.0258	2013.3415	15.539	16.365	13.95	233.7	-13	-65.9	6.4	-12.1	-65.7	6.5
25	1.9530	58.0140	2013.5355	13.482	15.735	8.66	320.3	204.1	4.7	5.5	202.3	8	5.6
26	2.0092	3.7931	2013.0115	14.977	15.074	8.17	357.9	89.9	-30.3	6.7	88.8	-26.9	6.5
27	2.0400	-0.8523	2013.956	11.18	15.456	45.77	35.2	96.9	22.4	5.4	98.1	24.4	5.7
28	2.1027	44.4506	2013.4025	16.932	17.14	31.87	303.6	64.3	10.5	5.6	62.8	9	5.6
29	2.1244	28.4553	2013.517	14.172	14.818	6.06	185.1	-46	-104.4	6.3	-44	-104.6	6.3
30	2.2448	54.5342	2013.5305	16.875	17.178	4.19	220.2	79.8	-59.6	6.2	85.3	-55.3	6.5
31	2.4435	58.1421	2013.4295	16.138	16.38	14.19	285.6	62.9	-41.5	6.2	68.1	-38.6	6.2
32	2.5110	24.0868	2013.124	17.855	18.228	5.00	223.1	76.7	-9	5.3	80.1	-4.7	6.1
33	2.5655	5.4271	2012.997	14.817	15.102	9.43	107.9	58.1	-169.2	6.8	56.4	-169.4	6.8
34	2.5668	39.5069	2013.4035	13.099	13.503	4.97	328.2	86.2	-63	5.5	86.1	-63.9	5.9
35	2.7726	18.7544	2013.6315	11.812	16.871	7.71	270.7	77.3	-76	5.5	77.5	-74.2	5.6
36	2.9449	3.9369	2012.971	15.814	16.381	5.75	311.1	71.2	-69.9	6	71.7	-69.5	5.9
37	2.9502	4.1436	2012.9605	11.538	16.015	59.24	86.9	71.9	-16.1	5.8	72.2	-14	5.8
38	2.9605	39.7256	2013.378	11.321	16.084	41.59	113.7	-35.6	-76.5	5.5	-38.1	-76.5	5.6
39	2.9795	25.3928	2013.3355	17.777	18.476	5.59	55.7	94.6	100.2	5.4	93.6	105.3	6.5
40	3.0383	29.5957	2013.444	11.836	14.95	12.48	150.9	84	-42.3	5.2	84.2	-42.2	5.2
41	3.1327	2.7958	2012.973	15.457	15.737	6.25	323.7	161	5.5	5.8	156.4	1.8	5.9
42	3.1875	48.7150	2013.3195	17.52	17.522	23.34	354.9	-67.9	-50.9	5.7	-69.9	-44	5.6
43	3.2088	65.8741	2013.498	11.141	14.438	6.19	106.9	29.2	-81.4	5.6	28.3	-82	5.6
44	3.2435	46.1063	2013.4915	13.692	14.009	7.03	280.7	129.4	-56.1	5.5	129	-53.7	5.5
45	3.2442	41.6546	2013.4195	13.399	16.554	10.43	55.9	78.5	-16.7	5.5	80.5	-17.8	5.7

46	3.3453	11.4408	2013.559	12.126	15.424	6.85	240.9	91.8	4.2	6.4	90.5	8	6.6
47	3.3730	38.1252	2013.4615	12.698	14.997	10.30	78.7	6.3	-156.3	5.5	5.4	-154.4	5.5
48	3.3962	66.5963	2013.4835	14.186	14.663	4.01	308.5	69.9	41.1	5.9	71.6	38.5	6
49	3.3976	24.9662	2013.5785	16.495	16.763	3.26	0.1	61.4	13.1	5.2	62.6	2.7	5.4
50	3.4363	66.9603	2013.532	15.614	18.132	43.00	19.9	65.3	1.7	5.9	62.7	10.5	5.9
51	3.6037	46.9672	2013.4205	15.112	18.392	57.76	281	98.3	-40	5.5	93.4	-38.6	5.9
52	3.7004	62.1961	2013.5185	8.279	12.573	10.80	212.2	105	-44.1	5.5	102.5	-40.8	5.6
53	3.7618	63.5544	2013.2405	18.299	18.504	31.22	320.9	-68.6	5.1	5.8	-65.5	16	6.1
54	3.8252	20.2955	2013.7065	10.174	12.56	5.13	200.5	38.7	-67.9	5.5	41.3	-66.6	5.4
55	3.8599	61.3904	2013.423	9.905	14.116	10.58	53.2	80.5	-17.4	5.6	81.8	-22.9	5.6
56	3.9164	57.6929	2013.4435	12.622	16.285	59.17	310.1	146.3	-12.6	6	150	-14.7	6
57	3.9232	5.7980	2012.9855	15.652	15.923	6.70	133.7	24.9	-84	6.9	23.8	-81.8	7
58	3.9458	18.8023	2013.507	14.73	15.49	4.46	166.5	79.9	-24.5	5.5	76.8	-26	5.8
59	4.2077	45.2615	2013.4735	13.356	15.362	31.32	263.5	63.5	-19.1	5.5	63.6	-22.3	5.5
60	4.2540	69.4902	2013.4365	10.567	14.595	30.54	17.7	90.6	10.7	5.9	87	9.3	5.9
61	4.2892	31.7038	2013.261	17.084	17.414	5.06	178.5	94.2	-34.2	6.1	97.2	-38.9	6.2
62	4.3316	74.1059	2013.2545	16.246	17.005	3.97	278.6	88.5	3.6	6.3	87.5	0.7	6.7
63	4.5386	29.0820	2013.4625	11.423	18.283	18.26	80.6	-99.2	-34.5	5.2	-100.2	-27.9	5.5
64	4.6145	41.4045	2013.2605	16.042	16.671	4.50	22	69.6	2.8	5.6	66.5	4.7	5.6
65	4.7358	59.8004	2013.453	9.933	13.176	17.47	57.2	83.5	8.6	5.6	81.6	9	5.6
66	4.9130	64.9498	2013.44	12.213	12.873	24.69	12.5	79.6	18.3	6.2	79.7	15.6	6.2
67	4.9386	54.6471	2013.4555	16.364	16.731	3.60	33	-77.3	-35.5	6.3	-79.9	-33.5	6.7
68	5.0366	59.5994	2013.257	18.349	18.942	57.61	82.5	-37.4	68.8	6.8	-36.2	73	7.5
69	5.0606	20.3208	2013.2855	17.44	18.377	10.39	289.2	68.9	-71.3	5.8	68.8	-71.4	6.1
70	5.0910	74.7616	2013.33	17.592	17.657	43.64	221.7	61.7	4.5	6.7	60.7	17.3	6.4
71	5.1783	31.5311	2013.441	16.137	16.475	9.05	138.4	2.6	-87.6	5.2	3.2	-88.2	5.3
72	5.2937	18.2361	2013.349	14.678	16.367	17.97	35.9	126.6	13.3	5.6	126.3	12.2	5.6
73	5.4091	72.7236	2013.4595	15.473	17.658	30.08	259.6	73.3	3.6	6.3	69.9	13.9	6.5
74	5.4476	24.7005	2013.3515	11.897	16.581	7.25	111.1	78.6	-45.6	5.2	79.6	-38.9	5.6
75	5.4553	48.7042	2013.4415	15.471	15.711	5.13	104.1	125.7	-1.7	5.5	125.5	-2.6	5.5
76	5.5232	12.6532	2013.3185	14.262	14.289	29.38	100.3	61.3	3.8	5.5	60.4	-1.3	5.6
77	5.6762	23.0935	2013.209	11.543	15.824	30.85	39.6	87.7	41.4	6.2	86.7	32.2	5.7
78	5.7258	75.7980	2013.3545	13.28	13.848	4.14	184	67.1	13.3	6.3	64.3	10.4	6.7
79	5.7292	50.6909	2013.4155	15.293	17.413	5.92	109.6	-14.5	-118	5.5	-20.4	-117.6	5.7
80	5.9360	55.8882	2013.367	17.452	17.611	44.43	22.9	61.3	-13.4	6.5	62.4	-17	6.9
81	6.0050	26.8566	2013.3105	12.797	14.204	7.00	311.7	-123.1	-76.5	5.3	-118.2	-75.8	5.2
82	6.0297	65.0835	2013.451	15.614	16.363	6.06	184	69.6	15	6.2	69.1	16.7	6.2
83	6.1339	10.6247	2013.137	11.023	16.871	13.82	30	52.1	-80.9	6.4	52.8	-83.5	6.8
84	6.1409	30.2084	2013.5335	10.275	15.117	11.56	97	95.8	65	5.2	89.6	71.4	5.2
85	6.1420	52.8950	2013.186	13.278	17.607	55.84	217.1	-9.7	-62.3	7.7	-10	-64.4	6.1
86	6.1630	10.8873	2013.18	10.353	17.921	26.08	57.8	64.6	-18.2	6.3	65.8	-9.9	7.6
87	6.3337	27.5527	2013.404	10.947	14.868	12.74	137.4	80	-53.6	5.6	80.1	-55.9	5.7
88	6.4549	45.0254	2013.493	13.606	15.17	8.37	48.7	102.9	-12.7	5.5	101.4	-14.1	5.5
89	6.4718	22.0927	2013.2475	13.384	17.648	57.21	209.3	81.5	-5.1	5.2	80.2	-4.7	5.5
90	6.5161	25.5088	2013.4325	15.72	16.132	3.97	268.7	79	-13.3	5.3	81.3	-15.7	5.2
91	6.5796	5.0160	2012.967	13.507	16.624	28.12	212.3	-70.9	-25.5	6.5	-68.3	-22.3	7.1

92	6.7080	46.1929	2013.3995	16.535	17.959	23.67	140.1	81	-14.5	5.6	83.4	-19.2	5.6
93	6.7695	25.0420	2013.6105	6.744	12.003	30.01	265.7	95.2	-21.4	5.1	98.9	-14.4	5.2
94	6.7782	54.0077	2013.373	17.666	18.07	12.91	126	-65.3	-21	5.6	-63.8	-30.3	5.8
95	6.8088	61.8440	2013.2555	17.995	18.392	41.36	259.8	4.1	-68.7	7.4	5.3	-67.9	6.9
96	6.8320	49.7728	2013.3875	16.301	17.024	5.28	287.7	70.3	-23	5.5	69.9	-14.8	5.6
97	6.8706	46.1339	2013.4315	13.757	16.723	14.97	304.7	-64.3	-28	5.5	-64.6	-24.3	5.5
98	6.9547	29.6414	2013.5005	13.911	17.247	17.73	214.5	-72.6	-61.9	5.1	-77.1	-61.9	5.2
99	6.9614	62.3137	2013.425	16.147	16.623	29.32	308.7	74.9	-6	6.4	71.7	-2.6	6.6
100	6.9956	32.7693	2013.401	11.907	13.646	6.20	13.3	-83.4	-7.6	5.2	-85.6	-7	5.2
101	7.0792	18.9983	2013.515	15.445	18.251	6.75	346	100.6	-71.8	5.2	101.5	-68.3	5.4
102	7.0893	16.6994	2013.3415	14.661	14.667	7.11	331.7	62.9	-12.7	6.5	64.4	-12.4	6.5
103	7.1222	12.7796	2013.1435	16.639	16.877	7.88	3.2	-44.9	-60.2	6.6	-42.9	-62.2	6.9
104	7.2648	60.2504	2013.4165	17.09	18.192	59.46	113.6	73.6	23.3	6.5	80.2	11.6	6.2
105	7.2899	4.1083	2012.974	16.614	16.712	56.73	182.2	68.6	-15.7	7.1	66	-17.6	7.5
106	7.2946	57.4773	2013.6015	17.649	18.284	58.56	324.4	-61.9	13.1	5.7	-60.5	5.7	6.1
107	7.3160	62.9191	2013.516	10.624	14.197	8.16	359.4	20.6	-90.2	5.9	24.2	-88.7	5.9
108	7.3229	57.8410	2013.4705	17.534	18.09	47.72	83.4	60.4	-12.3	5.7	60.2	-12.8	5.8
109	7.5225	15.4621	2013.25	15.342	15.565	4.07	198	66.8	20.5	6.7	67.7	20.2	6.7
110	7.6179	31.5132	2013.4195	13.159	15.879	25.34	246.4	65.4	-12.4	5.1	64.6	-15	5.2
111	7.7718	74.1768	2013.419	16.067	16.241	8.06	71.3	64.5	2.8	5.9	61.4	5.1	5.9
112	7.8915	48.2552	2013.4735	14.233	18.042	30.95	290	68	-14.8	5.5	70.3	-11.2	5.6
113	7.9268	74.0574	2013.4695	18.21	18.349	56.62	27.4	-1.5	61	6.1	-4.6	61.9	6.2
114	7.9802	8.9555	2013.108	12.786	12.898	7.45	137.4	92.7	-42.3	6.6	93	-39.5	6.5
115	8.0820	9.3393	2013.582	9.985	11.075	4.73	345.4	-2.9	-72.2	6.3	7	-72.8	6.6
116	8.0891	3.8909	2012.974	11.555	16.167	9.58	311.8	-5	-62.8	6.6	-7.2	-62.2	6.7
117	8.1026	50.3203	2013.478	11.12	13.787	12.04	98.3	-165.6	-51.3	5.5	-163.8	-55	5.5
118	8.1573	36.7061	2013.46	10.17	15.702	10.87	250.9	0.3	-84.9	5.5	-5.5	-88.3	5.6
119	8.4398	58.2893	2013.386	15.299	17.189	7.24	264.7	62.9	7.1	5.9	60.1	14.6	6
120	8.4884	65.0966	2013.494	13.639	17.462	21.41	348.4	124.1	27.2	5.9	125.7	26.5	6
121	8.5343	57.1599	2013.358	15.92	16.949	15.41	43.4	73.9	3.9	6	73.4	3.4	6
122	8.5368	59.8993	2013.5085	7.859	14.918	17.99	139.2	-98.1	-13.3	5.9	-94.1	-6.4	6
123	8.5568	22.6849	2013.126	14.562	15.824	8.79	262	72.4	-49.7	5.3	70.9	-47.8	5.3
124	8.5732	78.8153	2013.625	14.529	17.233	5.26	228.4	98.8	61.1	5.9	98.1	59.7	6.1
125	8.6255	76.1154	2013.5615	16.109	16.505	3.52	232.9	75.8	-56.1	5.9	80.1	-57.6	6
126	8.6390	6.1352	2013.1685	12.997	14.84	8.25	354	-85.8	-24.7	6.6	-86	-24.8	6.6
127	8.6418	39.7440	2013.218	16.717	17.153	16.56	50.8	89.2	-19.9	5.7	87.6	-18.9	5.7
128	8.7750	7.4232	2013.084	14.445	15.575	4.78	66.9	6.9	-79.9	6.6	-5.4	-83.5	7.3
129	8.7946	42.3292	2013.301	12.736	15.639	5.32	123.5	145.7	9.3	5.5	140	11.8	6.2
130	8.8150	82.4218	2013.715	14.397	16.492	7.27	313.3	78.7	1.2	5.8	78.9	-1.9	5.8
131	8.8233	37.3379	2013.1995	17.152	18.664	53.62	260.3	-19.5	-64.7	5.7	-21.7	-65	6
132	8.8614	38.2427	2013.3645	11.948	16.508	10.34	103.9	137.6	-8.1	5.6	137.2	-9.1	5.7
133	8.8835	28.2043	2013.4325	12.581	17.657	11.74	55	80.9	8.1	5.6	79.1	4.5	5.7
134	8.9208	65.6728	2013.5785	15.145	15.33	3.91	75.1	121	85.6	5.9	115.8	84.1	6
135	8.9493	46.5604	2013.487	14.603	14.623	6.26	273.5	72.2	7.6	5.5	72.5	6.1	5.5
136	8.9922	24.5307	2013.341	12.591	13.19	16.86	32.1	74.9	-23.7	6	76.4	-18.7	6
137	9.0451	56.8160	2013.4	13.222	16.648	29.58	25.7	65.1	-28.5	6	62.5	-28	6.1

138	9.1971	11.6980	2013.0875	12.81	13.51	4.76	8.5	-35.6	-62	6.5	-32.9	-66.8	6.7
139	9.3252	75.6064	2013.451	9.686	15.573	34.08	148.1	-126.7	-76.9	5.9	-126	-73.3	6
140	9.4363	22.6786	2013.182	14.468	15.695	5.19	249.7	86	-77.9	5.3	83.6	-82.1	5.6
141	9.4398	40.3024	2013.371	13.098	16.11	8.74	304.4	148.8	-15.9	6.4	147.6	-15.6	5.6
142	9.4398	42.9080	2013.4605	11.219	13.187	13.17	296.3	74.4	-0.8	5.5	74.6	-0.8	5.5
143	9.4431	39.1935	2013.372	13.476	15.574	11.98	354.9	-62.7	-9.2	6.4	-64	-6.2	6.4
144	9.4678	18.8571	2013.184	16.503	16.741	8.33	209.9	76.1	-86.9	5.3	78.2	-90.6	5.3
145	9.5020	24.8317	2013.208	16.2	16.3	16.85	339	63.2	-21.8	5.3	61.5	-22.2	5.3
146	9.5164	40.7760	2013.36	12.132	16.539	8.62	134	106.1	-10.7	6.4	105.9	0	6.4
147	9.6612	20.4098	2013.1955	13.315	17.029	26.66	16.8	65.5	1.2	5.2	65.1	0.3	5.4
148	9.6634	71.8329	2013.441	14.46	14.943	11.02	287	65.6	4.9	5.9	63.4	3.8	5.9
149	9.6996	38.2215	2013.5485	12.456	15.107	5.98	133.7	68.9	-28.1	6.3	69.5	-29.7	6.9
150	9.7642	15.7145	2013.544	12.558	17.675	10.84	39.6	66.7	-0.3	5.2	66.9	-3.2	5.7
151	9.7652	21.6716	2013.309	14.854	17.661	8.05	8.3	112.3	39.7	5.3	106.2	45.8	5.7
152	9.8132	63.6957	2013.5015	18.301	18.925	54.67	223.2	76.7	-5	6.7	78.4	-14.3	7.3
153	9.9327	70.7819	2013.241	12.407	14.179	4.33	324.1	111.9	56.1	5.9	111.8	59	6.5
154	9.9684	46.7729	2013.4545	13.447	15.175	6.71	128.8	-18.5	-69.1	5.5	-23.3	-68.4	5.5
155	9.9839	6.4798	2013.004	15.255	16.295	7.38	304.8	80.2	-57.8	6.7	77.8	-56.3	6.8
156	10.1784	48.8215	2013.4305	13.632	16.53	8.92	87.5	138.4	49.9	5.6	138.1	50.6	5.6
157	10.1987	4.5618	2012.9675	12.906	13.533	5.59	236.7	22.9	-63.8	6.8	25.8	-62.3	6.8
158	10.2632	49.4151	2013.449	13.467	17.854	15.72	56.2	-62.6	-66.9	5.5	-66.1	-65.6	5.7
159	10.4069	58.9235	2013.445	12.186	14.218	12.01	75.2	79.6	10.4	5.6	77.5	15	5.5
160	10.4253	20.1369	2013.4485	7.685	12.9	10.43	286.7	89.3	-73.7	5.1	93.1	-69.3	5.5
161	10.5646	5.6934	2013.1235	9.175	11.985	7.96	214.1	89.8	-3.8	6.4	92.9	-4.2	6.9
162	10.6008	35.7455	2013.46	10.312	14.627	9.85	25.2	113.8	-66.4	5.2	110.3	-68.3	5.2
163	10.6334	33.6658	2013.3925	11.937	13.764	6.15	337.6	90.7	17.9	5.2	91.8	15.6	5.3
164	10.8543	36.6727	2013.1005	17.25	17.592	49.93	213.9	74.9	-15.6	5.7	72.7	-16	5.3
165	10.9908	2.6681	2012.947	14.477	14.629	20.26	279.7	52.7	-77.4	6.8	51.1	-74.9	6.7
166	11.0114	57.6050	2013.4325	10.451	11.873	5.73	162.5	69.6	8.5	5.5	66.3	17.3	6.4
167	11.0550	53.7223	2013.4845	11.434	13.102	8.50	351.9	-16.3	-91.3	5.5	-21.8	-92	5.5
168	11.1074	48.0598	2013.471	13.914	14.642	4.82	337.7	-23.4	85	5.5	-25.3	86.7	5.5
169	11.1125	25.2391	2013.3935	11.711	17.047	8.68	324.2	61.2	-31.1	5.2	60.4	-28.2	5.3
170	11.1879	26.8667	2013.467	16.347	17.024	5.45	95.8	68	47.2	5.4	63.3	47.2	5.2
171	11.1958	16.7788	2013.281	14.844	16.161	8.55	49.1	98.5	4.1	5.9	99.9	2.7	6
172	11.1975	27.9992	2013.474	12.973	14.55	6.34	319.7	71.5	-25	5.2	68.9	-23.7	5.2
173	11.2668	36.5577	2013.256	17.99	18.035	17.01	151.4	76.5	34.8	5.8	76.2	38.7	5.9
174	11.2783	21.6609	2013.4405	16.555	18.917	25.12	94	87	36.9	5.4	87.2	33.7	6
175	11.3802	51.8071	2013.487	11.837	13.985	4.93	165.7	78.8	-0.6	5.5	78.2	2.2	5.6
176	11.4797	50.7932	2013.4845	14.626	15.187	5.97	302.4	84.3	10.1	5.5	84.8	9.1	5.5
177	11.5139	40.8087	2013.4995	7.632	15.675	27.79	235.9	90.1	-55.3	5.5	85	-58.3	5.6
178	11.5173	42.2917	2013.362	15.085	17.481	23.22	269.2	93.5	-3.3	5.5	93.9	-3.1	5.6
179	11.5304	20.5521	2013.2745	14.95	15.925	5.43	37.3	-65.5	-27.2	5.2	-67.8	-29.7	5.2
180	11.5716	50.3000	2013.4685	16.45	17.016	35.41	148.8	66.2	-14	5.6	64	-21.7	5.5
181	11.5954	22.2112	2013.277	14.25	17.697	6.43	207.9	69.7	-21	5.2	69.4	-13	6.8
182	11.6767	16.9256	2013.257	16.317	18.109	34.24	190.4	-13.8	-70	6	-23.4	-69.7	6.6
183	11.6789	60.1447	2013.383	17.352	18.032	59.93	100.6	-4.5	62	6	-3.6	62.5	6

184	11.6801	58.0096	2013.4715	12.042	12.375	6.08	226.8	85.1	-7.5	6	83.6	-9.6	6
185	11.7563	11.9737	2013.618	5.036	12.106	14.33	57.2	76	-13.2	5.7	71.2	-18	5.9
186	11.8245	36.7461	2013.1205	11.016	13.89	5.75	5.5	62.9	-6.2	5.5	64.2	-8.6	5.9
187	11.8247	31.4679	2013.4085	15.006	16.86	10.99	210.4	66.6	14.7	5.2	68.8	13.7	5.2
188	11.8628	59.9598	2013.359	16.335	17.627	5.45	95.7	151.2	-66.6	6.1	147.8	-67.8	6.3
189	11.9117	65.9774	2013.4015	15.178	19.173	13.77	139.1	-65.3	-5.9	6.3	-66.7	4.3	6.7
190	11.9375	27.1099	2013.2915	12.772	16.166	23.29	279.1	12.7	-69.6	5.2	9.4	-67.9	5.2
191	11.9791	20.9254	2013.5805	6.64	13.831	30.17	197.9	158.8	9.7	5	156.3	7.8	5.2
192	12.0047	54.9134	2013.4675	17.425	17.639	55.67	96.4	-25.8	70.4	6.2	-30.6	68.8	6
193	12.1962	72.1365	2013.446	12.967	15.532	9.35	8.5	78.5	-1.2	5.9	76.2	4.4	5.9
194	12.2841	22.2228	2013.3115	14.728	17.181	5.78	175.7	106	-52.7	5.3	104.7	-49.2	5.5
195	12.3047	52.6775	2013.286	18.684	18.868	58.48	114.5	1.5	66.3	6.1	5	64.2	6.9
196	12.3190	36.0484	2013.4295	13.93	14.919	6.16	22.6	67	-39.4	5.5	63.6	-37.9	5.6
197	12.3437	71.8823	2013.435	18.186	18.583	21.82	120.8	19.5	-68.4	6.2	17.3	-72	6.3
198	12.3553	53.2347	2013.5155	10.917	13.834	9.30	20.2	-141	-45.8	5.5	-142.3	-45.6	5.5
199	12.6753	23.4725	2013.3675	11.366	15.603	11.94	96.3	84.2	-17.8	5.2	83.4	-19	5.3
200	12.7353	11.9801	2013.1835	11.666	17.69	14.94	77.6	20.5	-71.7	6.2	16.3	-72	6.8
201	12.7535	35.7101	2013.38	11.102	13.558	5.44	51.9	-62.4	-45.2	5.1	-66.2	-44	5.7
202	12.8510	26.5926	2013.4805	14.897	17.247	5.69	116.7	183	60	5.3	182.9	62.9	6.1
203	12.8783	18.7111	2013.4885	15.323	16.001	5.12	160.1	131.5	-28.6	5.2	130.8	-31	5.2
204	12.9196	58.1492	2013.404	17.489	18.162	36.96	9.6	-61.4	-15.7	5.7	-64.3	-9.5	5.7
205	12.9246	23.8119	2013.4175	11.244	14.884	9.71	106.1	101.7	12.6	5.2	99	11.2	5.2
206	12.9518	23.3101	2013.413	17.063	17.496	53.41	127.2	69.4	-0.8	5.3	67.6	-4.3	5.3
207	12.9694	49.7379	2013.4195	11.755	16.106	33.35	181.4	63.3	2.2	5.5	62.7	-5.3	5.5
208	13.1610	28.5268	2013.404	14.741	16.192	6.05	108.1	96.1	61.9	5.2	98.6	62	5.3
209	13.1611	73.7694	2013.4005	12.841	16.538	6.02	171.4	111.6	6.9	5.9	107.4	13.9	6.1
210	13.3418	3.8702	2013.274	12.004	14.465	10.58	171	75.6	30.1	6.5	73.7	27.9	6.4
211	13.4683	16.6447	2013.2955	11.698	13.375	7.53	154.1	64.2	-35	6.5	61.1	-38.7	6.6
212	13.4830	71.0981	2013.3675	11.185	14.73	6.14	94.6	116.6	1	5.9	115.5	6.9	6.1
213	13.4922	61.6299	2013.318	15.362	18.093	10.32	140.7	66	-16.1	6	65.2	-19.1	6.2
214	13.6618	40.2571	2013.402	13.068	14.69	8.67	126.8	80.2	0.1	5.9	83.2	0.1	5.9
215	13.6741	7.5456	2013.264	9.143	13.683	8.95	168	61.7	15.6	6.4	63.3	11.6	6.8
216	13.6902	68.5228	2013.466	10.22	12.78	23.99	121.7	-75.4	-2.6	5.9	-72.2	-5.4	5.9
217	13.7208	28.7166	2013.378	15.147	15.201	4.15	17.9	80.4	-40.2	5.2	83.2	-41.1	5.7
218	13.7306	34.7986	2013.4825	10.295	15.981	16.68	84.4	94.2	-40.3	5.1	96.1	-37.2	5.3
219	13.7365	19.4738	2013.4165	8.771	13.868	9.80	286.7	89.5	42.3	5.1	93.1	39.3	5.3
220	13.7470	22.1462	2013.2455	12.288	16.222	10.64	23.2	84.3	-8.6	5.2	84.4	-3.5	5.3
221	13.8741	42.1654	2013.1825	13.617	14.747	4.41	352.9	-72	-38.1	5.5	-74.1	-38.7	5.7
222	13.9894	55.3151	2013.493	14.491	18.196	55.88	285.9	75.6	-59.8	6	69.2	-69.8	6.2
223	14.0036	57.4311	2013.357	16.947	18.623	57.42	13.4	82.1	-41.9	6.1	80.6	-37	6.6
224	14.0959	33.0524	2013.379	14.855	16.679	6.63	334.2	67.6	-46.2	5.2	67.7	-41.4	5.3
225	14.1047	47.2761	2013.474	15.085	15.16	14.16	214.1	75.7	12.7	5.5	74.3	8.3	5.5
226	14.2146	8.2316	2013.2915	10.752	16.655	46.33	205.8	-63.5	-35.5	6.4	-61.6	-36.8	6.5
227	14.2322	10.5967	2013.0785	15.006	17.391	16.40	263.2	67.1	-20.2	6.4	69.4	-16.9	6.8
228	14.3265	65.4126	2013.461	18.002	18.391	59.76	224.6	-68.7	-11.1	7.4	-70.6	-16	6.3
229	14.3322	16.0492	2013.365	12.434	17.379	13.02	258.3	77.7	-5.4	6.5	80.5	-7.4	6.6

230	14.3577	38.8948	2013.4305	12.313	15.763	18.89	232	76.9	4.5	5.9	76.8	6.7	5.9
231	14.3720	24.0112	2013.4275	13.37	18.513	16.01	323.1	72.2	1.7	5.2	73.4	-4	5.5
232	14.4002	60.3176	2013.3585	17.764	17.793	41.37	232.9	28.5	72.5	6.6	31.2	71	6.6
233	14.4310	15.2514	2013.3055	11.961	16.958	12.18	311	69.7	6	6.4	71.2	3.7	6.7
234	14.4613	52.6265	2013.2435	18.247	18.38	23.85	265.9	60.9	-8.1	5.7	61.4	-11.4	5.8
235	14.5003	12.0333	2013.575	10.036	15.53	10.31	33.3	69.2	-40	6.4	62.8	-48.7	7.2
236	14.5112	21.6126	2013.0555	15.904	16.076	3.81	151.6	131.3	-22.9	5.3	134.2	-16.1	5.7
237	14.5836	40.4403	2013.369	15.419	16.027	10.07	92.1	-60.5	-15.2	5.9	-60.6	-14.5	5.9
238	14.6021	36.5193	2013.359	15.715	17.535	5.05	231.5	-42.8	-61.7	5.9	-45.4	-60.9	6.1
239	14.6721	16.7346	2013.228	13.922	16.447	6.51	264.4	233.9	-81.8	6.7	236.2	-81.4	7.1
240	14.6783	58.3081	2013.3555	13.967	17.502	12.92	81.5	66.1	-11.1	6	64.8	-11.8	6.2
241	14.6917	32.1057	2013.4915	10.406	15.022	18.65	163	86.4	7.5	5.2	86.1	6.8	5.2
242	14.7230	35.1908	2013.3945	15.506	17.568	7.79	167.3	88.6	12.5	6.4	86.9	8.6	6.5
243	14.7305	32.4406	2013.4015	11.844	12.251	13.55	346.8	12.3	-74.9	6.4	14.9	-73.2	6.4
244	14.7338	16.0886	2013.4815	9.821	15.855	22.58	81	120.7	25.6	6.3	115	30.3	6.5
245	14.7482	4.4801	2013.242	11.4	16.666	10.50	174.1	70.4	-19.9	6.5	73.7	-19.9	6.9
246	14.8676	62.5102	2013.4975	14.952	15.587	3.48	155.6	89.2	40.1	6.5	81.5	51	6.7
247	14.9692	37.6286	2013.3445	13.134	17.865	47.64	28.9	65.3	-33.1	5.9	69.3	-30.4	6.1
248	14.9868	7.1542	2013.304	13.76	14.168	26.44	357.3	63	0.8	6.3	65.5	3.5	6.3
249	14.9945	2.6210	2012.878	15.07	16.543	10.79	318.6	66.6	-46.6	6.6	64.2	-46.4	6.7
250	15.0065	68.1933	2013.4355	12.352	13.758	11.85	244.8	107.9	-3.6	5.9	106.3	-1	5.9
251	15.0292	22.3750	2013.3595	13.43	16.908	18.86	54.5	50.2	-63.8	6.4	46.8	-63.6	6.5
252	15.0995	52.5153	2013.4615	15.011	15.617	27.52	222.5	114	11.5	5.5	115.4	11.1	5.5
253	15.1342	8.6630	2013.036	14.317	16.019	6.18	299.1	43.9	-65.4	6.5	46.6	-65.4	6.8
254	15.1430	17.9284	2013.4275	13.54	14.468	5.58	86.8	86.7	-18.8	5.2	85.2	-16.8	5.2
255	15.2257	16.4389	2013.2785	15.905	16.758	6.42	81.7	75.3	-11.8	6.6	77.4	-19.1	6.8
256	15.3932	28.3172	2013.5975	8.312	15.622	21.33	102.9	-36.5	-80.3	5.1	-41.1	-78.7	5.2
257	15.4130	53.3634	2013.584	13.782	15.633	4.56	146.2	88.3	-11.2	5.5	84.3	-11.6	5.7
258	15.4380	31.3706	2013.243	11.828	15.988	8.21	188.2	18.2	-66.9	5.5	16.2	-68.5	5.9
259	15.4517	42.6703	2013.459	11.127	11.941	5.35	334	121	-31.2	5.5	119.3	-32	5.5
260	15.5849	65.3355	2013.6125	18.576	18.917	54.53	315.7	-79.8	-4.2	6.7	-79.9	9.9	7.8
261	15.5913	17.8478	2013.427	10.03	14.209	7.94	262.9	98.7	93.1	6.5	105.5	90.8	7.1
262	15.6680	83.5726	2013.7265	11.843	18.616	18.81	211.3	85.9	54	6.2	87.8	48.7	6.9
263	15.8007	42.3828	2013.284	15.913	16.198	4.41	284.3	64.8	-10.5	5.6	64.9	-11.4	5.7
264	15.8136	37.2870	2013.5835	13.524	14.745	5.38	241.8	63.8	-15.1	6.4	62.6	-18	6.6
265	15.8164	39.9349	2013.252	16.254	16.572	7.65	111	125.8	-39.1	6.6	122.9	-35.8	6.6
266	16.0016	62.9168	2012.928	19.068	19.19	43.05	271.2	-14.2	-67.5	7.9	-8.3	-66.2	7.8
267	16.1058	43.8349	2013.331	15.585	17.005	8.56	167.4	60.5	-13.4	5.5	61.2	-13.8	5.6
268	16.1373	55.5580	2013.576	9.899	11.398	5.43	36.5	-59	-74.5	5.5	-57.6	-75.1	5.5
269	16.1592	50.3461	2013.309	17.429	19.073	10.88	99	37.5	-83.3	5.6	28.8	-87.2	6.6
270	16.1613	20.5353	2013.3375	10.987	13.741	11.09	304.3	-84.2	25.7	5.2	-81.7	22.6	5.2
271	16.1991	55.9441	2013.468	14.308	17.963	10.23	307.4	-47.2	-73.3	5.5	-44.5	-72	5.7
272	16.2160	42.1985	2013.292	14.939	17.341	38.99	94	61.8	1.1	5.6	63.1	8.1	5.7
273	16.2528	56.8279	2013.3995	17.556	18.275	29.85	69.4	69.6	27.8	5.6	64.7	31.3	5.9
274	16.3128	68.8735	2013.438	16.982	18.064	59.52	259.7	63.9	8.1	6.4	60.8	9.6	6.6
275	16.3521	5.8244	2013.2345	15.994	17.701	55.93	134.7	9.2	-75.7	6.5	2	-75.6	6.7

276	16.4447	22.9012	2013.326	15.281	15.91	6.26	175.4	64.1	-10.3	5.2	64.7	-10.9	5.2
277	16.4573	31.7864	2013.4635	14.683	15.007	7.97	36.3	132.8	-58.8	5.9	134.9	-60.3	6
278	16.5257	20.0802	2013.3435	13.723	16.056	21.52	279.5	94.7	28	5.2	95.9	31.3	5.4
279	16.5290	61.9501	2013.225	15.615	17.644	43.99	99.7	37.9	-79.1	5.9	45.6	-70.7	5.7
280	16.5639	33.8164	2013.252	12.257	18.589	26.66	168.9	95	56.3	5.5	94.6	62.8	6.6
281	16.5726	43.0458	2013.3635	12.848	15.615	19.62	190.3	62.6	-4.8	5.5	63.7	-5.5	5.5
282	16.6279	63.6236	2013.4325	9.388	18.221	32.65	325	76.5	-29.9	5.5	71.8	-39.2	5.7
283	16.7503	2.7652	2012.8635	11.136	14.331	6.26	292.5	85	30.7	6.7	82	31	6.8
284	16.7697	59.5739	2013.3665	16.165	17.648	29.06	119.9	69.3	-23.5	5.6	73.3	-20.3	5.7
285	16.8261	71.8498	2013.399	13.489	16.951	17.54	35.1	29.7	-62.6	5.9	20.7	-63.7	6
286	16.8963	12.7256	2013.0955	12.466	14.278	5.30	74.8	-34.9	-61.2	5.1	-31.6	-65.6	5.4
287	16.9181	-0.7903	2012.87	15.34	15.598	7.16	185.9	-19.9	-61.7	5.8	-21.1	-64.5	5.9
288	16.9511	16.9658	2013.488	13.663	15.887	4.91	341.8	-19.9	-123.7	5.2	-25.4	-119.9	5.6
289	16.9837	23.9920	2013.3735	10.662	14.891	11.48	47	86.5	4.2	5.2	87.3	2.8	5.2
290	17.1223	53.6898	2013.449	18.154	18.174	46.18	275.9	32.7	60.5	5.7	24.3	61.6	5.7
291	17.1271	16.2867	2013.2875	10.148	14.469	8.20	286.3	-20.8	-69.8	5.2	-17.9	-69.2	5.3
292	17.1711	33.4066	2013.396	15.343	16.383	5.13	16.6	78.3	-15.7	5.5	80.9	-16.6	5.6
293	17.3177	23.1638	2013.3455	12.946	16.019	7.86	228.3	85.1	10.2	5.2	84.8	9.6	5.2
294	17.3287	40.7112	2013.3265	12.461	17.286	31.28	28.6	64.7	-4.1	5.5	66.4	-15	5.6
295	17.3357	26.1153	2013.3375	14.104	14.872	5.76	57.1	132	-27.1	5.3	131.8	-28.1	5.2
296	17.3553	61.6707	2013.4375	12.595	14.077	6.81	2.7	78.2	-29.2	5.6	77.3	-31	5.6
297	17.3932	33.8941	2013.487	14.939	16.061	7.61	113	147.1	-49.7	5.5	146.4	-51.1	5.6
298	17.4069	5.5190	2013.0225	15.322	17.333	5.80	57.3	64.6	-1.2	6.6	61.9	3.1	6.7
299	17.4442	26.0534	2013.252	11.424	12.86	10.45	243.8	66	-28.6	5.2	64.1	-27.3	5.2
300	17.4545	28.5290	2013.372	13.469	17.013	15.63	85.5	79	-19.3	5.2	78.3	-17.1	5.3
301	17.5914	37.0415	2013.3475	13.735	16.314	7.46	43.5	60.2	-50.7	5.5	62.7	-50.6	5.6
302	17.5924	-0.1495	2012.87	13.337	16.099	37.02	177.2	63.9	7.8	5.8	64	7.9	6.1
303	17.5971	53.9739	2013.44	13.154	15.861	39.04	248.2	98.8	1.6	5.8	97.7	4.8	5.9
304	17.6138	40.8285	2013.3605	13.154	16.74	18.80	66.9	71.1	-32.1	5.5	71.3	-33.7	5.6
305	17.6955	35.3504	2013.4005	15.279	16.12	9.54	348.5	63.8	-17.6	5.9	65.4	-21.3	5.9
306	17.7044	16.3681	2013.415	15.799	16.46	4.64	331.5	64.6	-0.1	5.2	67.2	-1.8	5.2
307	17.8033	32.4081	2013.321	15.606	16.289	4.49	219.5	21.2	-91.5	5.9	27.7	-91.3	6.1
308	17.9193	44.2085	2013.3585	16.092	16.894	5.99	200.5	74.9	-43.8	5.6	73.8	-39.1	5.6
309	17.9405	71.5519	2013.3895	14.828	16.198	24.24	91.6	-3.5	-95.3	5.9	-4.3	-92.4	6
310	18.0483	66.8499	2013.4415	15.054	17.339	9.10	21.9	91.8	-11	5.9	92.1	-0.3	6.5
311	18.1004	70.1629	2013.4155	18.026	18.467	41.35	254.9	-13.6	-66.2	6	-2.4	-69.1	6.8
312	18.1027	33.2145	2013.4185	16.902	17.2	18.11	84.3	-78	-67.4	6	-81.7	-67.9	6
313	18.2380	62.3107	2013.4675	12.883	14.479	46.81	325.1	65.6	-11.5	5.6	62.9	-8.4	5.6
314	18.2574	18.9879	2013.4405	12.299	14.607	7.96	346.9	79.3	0	5.2	77.9	-0.7	5.2
315	18.3505	37.9085	2013.4065	17.3	19.198	50.45	311.1	-81.3	11.8	5.5	-78.7	12.2	7.6
316	18.3559	66.3827	2013.496	18.558	18.984	51.28	11.9	-113.7	-64.4	6.5	-108.4	-65.5	6.8
317	18.4440	20.1612	2013.271	13.748	18.048	16.90	26.9	-4.7	-121.5	6.3	-2.5	-120.5	7.2
318	18.4639	58.0986	2013.869	13.26	13.354	3.36	252.2	-24.3	-78.8	7.2	-17.2	-81.5	7.7
319	18.6381	68.8314	2013.1615	18.165	18.994	36.92	339.1	-8.8	-63.9	6.3	-2.6	-67.2	6.8
320	18.7393	3.3304	2012.867	15.774	16.992	4.86	1.2	26.8	-70.1	6.8	29.5	-66.3	7.2
321	18.7442	5.9384	2013.2025	12.549	15.993	17.65	118.9	62.2	-27.1	6.5	63.5	-26.3	6.7

13

322	18.7782	13.8941	2013.2695	15.089	16.631	7.85	304	73	5.3	6	71	5.7	6.1
323	18.8281	56.7458	2013.3865	12.24	17.714	16.92	302.9	-69.2	23	5.8	-70.2	23.1	6
324	19.0024	25.4950	2013.2555	13.21	14.479	9.78	115.6	67.3	-15.9	5.2	70.2	-18	5.2
325	19.1310	56.9564	2013.428	14.72	15.087	7.56	59.3	102.7	-3.6	5.9	107	-3.3	5.9
326	19.2307	58.1587	2013.185	18.205	19.082	24.39	123.3	-43.4	-63.9	6	-47.1	-64.9	6.7
327	19.2911	63.2266	2013.3075	18.299	18.342	39.88	248	-65.6	8.7	6	-61.4	16.7	5.8
328	19.3180	56.6137	2013.3905	16.812	17.25	20.19	139.7	89.7	-11.3	5.9	85.8	-9.8	5.9
329	19.4063	59.6049	2013.2165	17.909	18.667	31.93	273.1	-68.6	-14.6	6	-69.8	-21.2	6.4
330	19.4174	58.4973	2013.4535	11.59	13.163	6.78	19.7	102.5	-43	5.8	107.4	-42.2	5.9
331	19.4189	34.0295	2013.2975	15.55	16.594	24.62	341.7	60.9	-15	6	61.6	-10.3	6.1
332	19.4736	6.3299	2013.634	10.209	11.247	5.11	26	-88.4	-132.1	6.4	-91.3	-133.6	7.4
333	19.4951	58.4452	2013.41	13.779	16.435	8.19	136.7	67.4	-12.7	5.8	67.9	-12.8	5.9
334	19.5791	73.9415	2013.3305	18.042	18.389	18.28	45.2	-4.9	-66.9	6	-3.9	-69.4	6.3
335	19.7596	72.7107	2013.508	11.264	11.512	13.62	233.2	-5.7	84.2	5.9	-6.7	80.2	5.9
336	19.8011	22.5089	2013.2295	15.84	17.223	30.99	199.3	66.9	-10.5	5.2	63.4	-15.5	5.5
337	19.8551	5.5252	2012.929	15.342	16.723	5.52	205.3	88.1	19.6	6.7	85.2	30	7.2
338	19.8838	34.7350	2013.5515	11.157	14.512	9.62	345.6	72	-16	5.1	70.1	-12.2	5.2
339	19.9484	43.2332	2013.403	11.856	13.869	12.73	123.4	106.8	7.3	5.5	105.5	10	5.5
340	20.0495	14.0776	2013.189	14.304	14.591	4.88	79.1	69.2	35.9	5.3	68.6	39	5.3
341	20.0512	41.8557	2013.336	15.853	16.453	5.26	54.5	66.8	-36.7	5.5	64.1	-37.7	5.6
342	20.0518	41.7153	2013.2755	17.464	17.805	44.46	236.2	-71.2	-56.6	5.6	-68.1	-56.3	5.6
343	20.0575	66.9202	2013.7305	18.376	18.835	42.13	221.9	65.7	14.2	7.9	64.8	2.7	6.2
344	20.0795	66.7415	2013.421	14.832	17.8	6.96	48.1	119.9	-61.5	6	118.2	-57.4	6.1
345	20.1554	55.1762	2013.491	17.869	18.759	25.92	251.1	79.5	-59.6	6	70.6	-69.3	6.5
346	20.2230	53.5657	2013.2895	17.644	18.335	57.38	3.2	63.3	-1.1	5.6	61.7	-7.2	5.9
347	20.2642	51.6272	2013.0355	18.22	19.004	47.25	72.6	62	-4.7	5.7	64.3	-0.1	6.6
348	20.2801	55.3491	2013.472	8.677	16.319	18.63	310.9	28.7	-160.7	6.2	25.1	-165.6	6.3
349	20.3736	50.9480	2013.3625	15.73	16.401	6.38	292.8	60.7	5.6	5.6	61.2	2.1	5.6
350	20.5067	57.7890	2013.4235	18.265	18.294	40.69	86.3	-62.8	-19	6.8	-60.5	-21.4	6.7
351	20.5231	47.0517	2013.412	12.432	16.472	6.66	328.6	116.6	-12.1	5.5	115.6	-12.1	5.6
352	20.5432	26.2580	2013.4035	11.921	12.659	4.93	147.4	107.9	-17.1	5.2	104.9	-11.5	5.3
353	20.5753	55.0759	2013.4125	15.902	16.949	25.65	293.3	129.6	1.6	6	131.7	5.3	6
354	20.6588	16.5676	2013.406	9.498	17.086	14.44	147.8	63.6	-8	5.5	66.1	-1.3	6
355	20.7768	58.8509	2013.3815	16.443	17.647	4.99	28.7	66.4	-6	6	63.8	-2.7	6.1
356	20.7843	5.7307	2012.9765	13.267	13.885	7.42	341.8	64	4.9	6.5	66.9	8.4	6.7
357	20.9340	31.1185	2013.2975	14.291	15.405	4.53	259.3	99.1	-29.8	5.2	102.9	-26.1	6.5
358	20.9698	64.6806	2013.37	18.449	18.679	35.70	109.7	61.9	14.7	5.9	64	16.8	6.5
359	21.0196	31.7108	2013.354	13.379	17.637	59.79	141.7	12.5	-77	5.2	15.5	-74.3	5.4
360	21.1583	51.0197	2013.5055	10.937	12.206	6.67	113.9	66	-16	5.9	65.1	-15	5.9
361	21.1884	21.3338	2013.5075	13.267	15.271	5.78	167.6	26	-112.1	6	27.6	-107.3	5.9
362	21.2356	56.5424	2013.487	13.554	15.018	24.36	82.9	88.2	70	6.4	82.4	72.3	5.9
363	21.3041	57.0840	2013.376	11.743	17.679	43.59	282.1	89.7	-2.5	5.9	86.6	4.7	6
364	21.4448	27.1554	2013.4135	12.13	13.7	10.21	197.6	-15.9	-72.9	5.2	-15.2	-74.4	5.2
365	21.4627	23.8077	2013.388	15.107	15.398	12.18	167	-35.5	-88.5	6	-36.8	-87.4	6
366	21.4982	28.3424	2013.3595	12.254	16.448	24.55	43.5	67	-28.9	5.2	67.3	-30.5	5.2
367	21.6216	22.5487	2013.4115	10.527	12.401	13.50	241.4	80.8	-19.8	5.9	79.6	-17.1	6

368	21.6700	58.5925	2013.404	13.181	13.447	35.24	174.4	139.1	6.4	6	139.3	5.6	6
369	21.6736	50.0516	2013.294	15.744	17.621	54.30	198.5	117.3	64.5	5.6	122.2	64.8	5.7
370	21.6837	57.7703	2013.4905	8.801	13.291	7.84	179.6	65	-34.2	5.9	67.7	-33.3	6
371	21.9977	63.9611	2013.0125	18.423	18.611	26.25	254.1	66.8	-23.7	6.2	67.9	-12.2	6.6
372	22.1080	23.3860	2013.3335	12.681	17.571	58.71	147.8	61.7	-61.6	5.9	62.4	-64.3	6.1
373	22.1764	14.5035	2013.32	12.552	14.71	7.55	329.9	81.5	8.8	6.4	80.4	9.9	6.5
374	22.2188	63.8143	2013.2205	17.742	18.702	39.59	306.3	72.5	-18.9	6.5	72.1	-8.2	7.7
375	22.2457	58.5835	2013.4445	9.949	15.52	18.81	165.3	78.7	-38.2	5.9	76.4	-34	6
376	22.2787	67.5984	2013.4395	15.944	18.85	20.26	113	-8.5	-61.1	6.2	-3.5	-60.1	6.6
377	22.3603	10.0917	2013.3275	12.251	13.584	10.53	76	64.1	-4.3	6.4	65.4	-7.7	6.4
378	22.3748	48.8132	2013.381	17.565	17.849	36.88	53.9	-21.2	-62.9	5.6	-18.4	-65.5	5.6
379	22.5097	55.5822	2013.3365	18.203	18.956	29.62	29.2	-66.2	-35.3	6.1	-64	-40.1	6.5
380	22.6369	50.2104	2013.452	11.973	16.323	6.72	327.6	-65.4	-50.3	5.5	-67.3	-43.3	5.6
381	22.6417	9.5992	2013.144	12.528	14.827	50.37	294.4	-44.9	-140.8	6.7	-45	-145.3	6.7
382	22.7131	2.9081	2012.968	12.589	14.153	10.81	64	-58.3	-67.2	6.5	-58.6	-67.1	6.7
383	22.7501	34.0209	2013.138	11.626	12.76	4.09	230.5	74.3	-59.8	5.2	72.1	-55.3	5.9
384	22.9033	45.6706	2013.4165	10.545	13.269	7.67	346.3	113.3	-15.8	5.9	112.1	-15.4	5.9
385	22.9328	64.6241	2013.3605	13.309	16.759	20.53	6.4	68.1	23.6	5.9	65.6	25.3	6
386	22.9544	44.5183	2013.396	13.007	15.528	6.63	273.1	64.3	-10.2	5.9	64.3	-5	5.9
387	23.1007	11.3205	2013.1475	10.208	12.537	5.57	200.1	14.1	-60.6	6.4	16.8	-61.2	6.8
388	23.1703	68.1601	2013.4425	10.826	14.6	8.74	232.8	96.7	-114.9	6.3	93.5	-118.9	6.4
389	23.2184	55.5277	2013.404	17.227	17.442	59.68	42.3	22.6	-64.9	5.9	17.9	-69.8	6
390	23.2361	57.8148	2013.3945	16.97	18.917	38.94	177.2	63.7	19.7	6.1	63.8	14	7.7
391	23.3174	13.9621	2013.364	14.818	16.641	6.87	343.2	65.9	-22.9	5.5	64.5	-20.6	5.7
392	23.4258	66.7766	2013.423	13.52	13.827	7.29	169.8	-130.3	-46.5	6.4	-132.7	-51.6	6.4
393	23.4520	30.3624	2013.4475	13.203	14.234	7.68	209	-19.8	-68.4	5.2	-18.1	-67.9	5.2
394	23.4906	53.3413	2013.739	8.025	12.716	5.95	191.5	129.6	-82.6	5.5	134.8	-82	7.6
395	23.6605	30.3539	2013.3775	15.795	16.791	6.26	109	95.8	-44.1	5.2	96.1	-41.9	5.3
396	23.7082	43.5643	2013.3615	16.715	18.035	58.76	315.1	65.3	-7.3	5.9	66.3	-9.6	6
397	23.7183	35.3660	2013.5465	12.508	15.135	6.11	341.8	-72.8	-14.2	5.1	-75.6	-9.3	5.2
398	23.7223	78.1884	2013.3585	13.636	15.774	14.26	16.2	65.4	-22.8	6	62.3	-23.6	6
399	23.7815	20.5137	2013.4675	13.432	14.029	8.24	320.2	10	-71.9	5.5	7.2	-73.6	5.5
400	23.8273	43.0451	2013.4845	8.562	16.424	18.06	231.6	111.6	-90	5.8	106.8	-90.9	6
401	23.9095	-13.2669	2013.943	13.059	14.846	6.63	356.6	99.1	-11.7	5.3	98.9	-11.3	6.8
402	23.9130	28.6956	2013.593	15.23	15.407	4.50	87	129.4	-27.4	6	128.4	-25.6	6.1
403	24.0136	33.2353	2013.3815	11.465	17.498	35.64	114.9	58.9	-77.5	5.2	55.3	-77.5	5.3
404	24.0437	59.1409	2013.4125	13.469	16.289	8.40	145.9	79.4	-21	5.9	75.7	-22.8	6
405	24.0543	15.9985	2013.3555	11.298	15.405	7.70	32.9	64.1	-3.7	6.4	65.8	-0.6	6.7
406	24.0815	23.9183	2013.5565	11.225	17.755	39.12	125.7	-15.2	-73.2	5.9	-17.1	-74.6	6.3
407	24.1013	15.1540	2013.337	17.304	18.253	29.37	182.2	101.2	-22.3	6.5	99.7	-25.4	7
408	24.1276	14.5903	2013.1605	14.232	16.281	10.78	266.6	67.7	-14	6.5	67	-7.9	6.6
409	24.1711	8.4186	2012.9875	10.926	16.029	8.22	221.1	-60.1	8.5	6.6	-63.3	5.8	7.9
410	24.1981	15.7957	2013.4065	13.24	17.539	33.24	275.4	90	-12.2	5.5	92.3	-21.2	6.7
411	24.2870	50.9556	2013.4145	14.031	14.228	15.58	278.4	64.7	0.6	5.5	67.5	0.6	5.6
412	24.3039	79.2362	2013.359	18.077	18.161	15.01	224.7	61.2	13.6	6.6	63.2	10.3	6.7
413	24.3312	61.6937	2013.4115	13.342	14.213	51.98	184.7	86	-12.1	5.9	87.5	-16.4	5.9

414	24.3355	19.6702	2013.322	17.951	18.319	9.92	163	72.6	-1.2	5.8	72.1	6.6	6.1
415	24.3558	80.7996	2013.4435	17.78	18.15	15.69	39.2	72.1	-25	6.5	72.6	-24.3	6.5
416	24.3931	22.5748	2013.6435	8.704	13.487	47.43	249.2	124	-16.5	5.4	124.7	-18.2	5.5
417	24.4271	33.8650	2013.5135	14.343	14.422	13.31	153.5	82.6	-8.8	5.2	83.6	-7.8	5.2
418	24.5186	9.2939	2013.057	12.016	15.097	17.44	245.3	63.8	14.1	6.5	61.8	23.8	6.6
419	24.6080	8.4781	2012.869	16.102	17.205	4.90	232.6	14.9	-125.4	6.8	19.5	-128.1	7.3
420	24.6118	9.6228	2013.1685	10.9	13.816	12.76	148.1	-21.5	-74	6.6	-26.9	-75.1	6.6
421	24.6439	20.8161	2013.4915	11.128	16.464	12.22	356.3	67	-56.4	5.5	66.9	-56	5.6
422	24.7084	5.7763	2013.055	12.418	13.579	6.66	174.4	-47.2	-78.4	6.6	-50	-76.3	6.6
423	24.7102	60.3385	2013.333	16.356	18.413	49.78	322	60.1	-33.1	6	66.1	-23	6.1
424	24.7288	52.7175	2013.4245	10.775	16.937	13.19	138.7	84.4	-31.7	5.5	81.5	-32.7	5.6
425	24.7802	55.4817	2013.3345	17.305	17.716	37.09	135.9	-15	-70.5	6.3	-11	-74.9	6.4
426	24.8130	10.7236	2013.154	14.396	15.339	8.26	102.1	-11.5	-156	6.6	-9.3	-155.4	6.7
427	24.8246	29.9851	2013.2915	16.498	17.043	3.99	49.6	98.8	-12	5.3	97.5	-13.6	5.8
428	24.8781	5.6978	2012.9455	12.96	15.718	7.02	164	-106.1	-15.1	6.6	-106.3	-11.3	6.8
429	24.8901	45.6308	2013.3855	11.879	16.83	12.09	358.2	61.6	-3.6	5.9	62.6	-3.7	5.9
430	25.0462	50.7893	2013.4335	10.918	15.983	10.20	278.3	65.4	-1.9	5.6	65.8	-1.2	5.6
431	25.1170	42.1807	2013.378	16.782	17.024	3.97	133.2	79.8	-111.2	6	75.2	-116.8	6
432	25.1858	8.4074	2012.891	15.05	17.069	7.54	260.7	81.7	-11	6.6	80.3	-13.2	6.9
433	25.2063	25.9876	2013.3135	11.017	17.685	18.41	40	62.6	-23.6	5.1	60.4	-24	5.3
434	25.2747	27.4797	2013.4885	13.733	16.897	24.72	39.3	61.7	-27.2	5.1	61	-25	5.3
435	25.3183	66.1985	2013.392	10.797	15.785	15.11	299.1	-24.7	70.3	5.9	-26.4	70.5	6
436	25.3704	44.7420	2013.423	11.387	14.057	17.17	194.8	137.1	-9.3	5.9	134.3	-5.8	5.9
437	25.3743	40.9459	2013.4285	14.9	16.36	6.69	304	127.8	5.9	5.5	127.6	3.5	5.6
438	25.4103	45.5525	2013.48	11.342	14.806	10.07	123.4	12	-72.3	5.8	12.4	-72.1	5.8
439	25.5911	39.2285	2013.3675	14.41	16.195	8.27	2.8	89.3	-7	5.5	90.6	-7.2	5.6
440	25.6672	32.0323	2013.3415	15.337	16.065	4.49	148.4	74.4	-0.2	5.2	71.5	2.3	5.4
441	25.6742	8.8109	2012.972	12.443	18.538	17.06	154.8	135.7	-151	7	132.9	-145.3	7.6
442	25.6800	22.8910	2013.3575	12.983	15.612	6.60	222.6	64.4	-23.4	5.5	64.9	-22.6	5.6
443	25.7098	29.1036	2013.5565	9.608	17.651	26.66	23.4	3.7	-63.1	5.1	3.6	-65	5.2
444	25.7443	80.3982	2013.579	15.127	16.264	9.57	339	67.6	-35.3	6.3	71.1	-36.5	6.2
445	25.7700	30.7551	2013.673	12.459	15.01	3.93	347.6	63.8	-10.2	5.2	64.7	-11.7	5.5
446	25.8107	38.6920	2013.383	13.804	14.915	55.32	194.5	115.1	5.8	5.5	115.6	7.3	5.5
447	25.8111	25.4583	2013.699	13.2	15.987	6.00	288.9	125	-55.4	5.5	128.9	-56.9	5.7
448	25.8550	41.1534	2013.351	14.512	17.711	15.92	30.1	-8.6	-92.8	5.5	-7.5	-96	5.7
449	26.0571	24.3467	2013.4695	13.222	16.54	7.59	281.1	74.3	-60.3	5.2	71.2	-62	5.2
450	26.2148	65.9428	2013.3595	15.625	18.728	55.93	252.1	7.4	-64.1	6.2	20.1	-62.3	7.2
451	26.2480	30.9828	2013.4355	12.249	13.529	7.30	20.5	9.2	-108.9	5.2	8.8	-107.7	5.2
452	26.3691	11.8997	2012.8905	16.958	17.874	11.82	323.3	63.3	-79.2	6.7	60.5	-81	7.4
453	26.4099	16.4824	2013.446	10.986	16.218	8.63	286.5	87.5	-32.2	5.5	84.2	-36.4	5.9
454	26.4835	7.4291	2013.009	9.499	16.561	26.90	60	85.3	-1.4	6.6	82.3	1.1	6.7
455	26.5069	45.0445	2013.426	14.66	15.607	7.70	249.2	95.9	-8.4	5.9	91.7	-5.5	5.9
456	26.5679	21.2415	2013.5025	10.91	16.137	53.93	302.2	-65.8	-18.1	5.9	-65.5	-19.2	6
457	26.5788	30.7512	2013.274	18.323	18.925	38.89	328.1	85.5	32	5.3	92	23.4	5.7
458	26.6112	72.9482	2013.441	17.767	17.849	22.21	158.3	5	-63.2	5.9	-2.1	-63.7	6
459	26.6486	49.1212	2013.3685	16.511	17.332	13.26	137.6	77.2	-2.8	6	79	-2.1	6

460	26.6623	15.7600	2013.1925	14.828	15.103	4.74	284.4	36.9	-91.2	5.7	40.5	-88.9	5.7
461	26.7365	28.5986	2013.4915	10.174	17.094	20.07	102.1	104.2	-27.3	5.1	100.8	-31.7	5.3
462	26.7990	19.3401	2013.607	12.982	16.6	5.58	332.7	66.6	-8.3	5.9	62.9	-11.4	6.1
463	26.9659	19.4239	2013.4485	12.262	14.868	6.69	249.8	-38.6	-65.4	6	-41.6	-63.9	6.1
464	27.0085	44.2294	2013.507	10.632	15.461	12.16	163.9	103.9	-8.2	6	105.8	-2.7	6
465	27.0501	5.0585	2012.875	15.671	17.273	6.05	150.2	72.9	-34.9	6.7	74.3	-35.2	6.9
466	27.0669	60.1079	2013.29	17.274	17.541	35.78	318.2	81.9	-17.5	6.4	84.1	-19.8	6.4
467	27.0847	-6.4115	2013.502	11.503	11.777	15.17	335.7	70.8	20.8	5.7	69.7	21.7	5.6
468	27.2990	-7.1174	2013.724	9.429	15.297	21.79	333.4	48.7	-64.5	5.5	48.7	-67.6	5.8
469	27.3584	8.4949	2012.867	15.066	16.637	5.33	142.1	65.5	-76.2	6.7	61.4	-78.9	6.8
470	27.3819	26.4813	2013.472	17.613	17.821	3.42	173.5	16.4	-63.5	5.5	16	-66	5.9
471	27.3863	26.1553	2013.3305	15.121	17.285	28.56	19.5	-18.8	-69.8	5.2	-18.9	-68.8	5.3
472	27.3940	24.5757	2013.632	15.395	15.84	5.57	7.8	96.1	-45.5	5.2	91.9	-46.4	5.1
473	27.3945	6.1705	2012.8685	13.768	13.952	8.30	8.5	-20	-109.3	6.7	-25.5	-112.5	6.7
474	27.4035	31.2534	2013.526	8.793	14.756	8.92	273.9	-11.9	-68.7	5.1	-14.9	-64.7	5.3
475	27.4641	26.5516	2013.373	14.093	16.836	7.56	217.7	75.3	-5.9	5.1	73.7	-8.7	5.3
476	27.7180	68.7245	2013.2955	15.509	18.368	6.86	344.1	62.2	-20.3	6.3	62.6	-13.8	7.1
477	27.7994	63.7159	2013.3895	14.468	15.586	10.69	153.3	12	-69.9	6.2	8.4	-72.1	6.3
478	27.9389	22.9318	2013.547	11.024	13.656	15.75	67.8	96.4	-139.9	5.5	99.5	-143.7	5.5
479	28.0501	46.1052	2013.4525	14.547	17.149	9.82	235.9	-20.2	-65.2	6	-22.1	-65.4	6
480	28.2145	-7.3643	2013.83	13.481	14.851	13.00	55.7	67	47.5	5.5	63.9	46.9	6.1
481	28.2159	43.1835	2013.541	14.82	15.08	6.92	264.2	62.9	-34.9	6.3	63.5	-30.3	6.3
482	28.3328	16.5651	2013.204	11.668	18.248	14.74	47	86	-47.3	5.2	83.7	-47.7	5.5
483	28.3405	31.4717	2013.4705	13.354	16.111	18.77	123.7	70.1	-34.4	5.1	70.5	-34.6	5.2
484	28.3497	15.4557	2012.882	16.856	17.099	3.70	147.2	95.8	28.2	5.5	93.9	34.3	6
485	28.4280	50.2765	2013.5175	17.562	17.662	55.53	54.2	-97.8	-19.3	6	-103.4	-8.2	6
486	28.4399	44.5898	2013.4785	12.519	13.654	10.13	47.2	71.5	16.2	6	72.7	15.2	6
487	28.5662	76.5865	2013.4995	12.359	13.022	24.87	68.8	30.9	-73.8	5.6	32.4	-74.2	5.6
488	28.6028	10.3611	2013.4905	7.852	14.845	19.99	247.1	99.9	-41.8	6.2	97.5	-44.8	6.7
489	28.6284	58.7330	2013.5125	12.071	13.843	8.22	8.4	-27.8	76.5	5.9	-31.8	76.8	5.9
490	28.7284	27.8946	2013.4505	13.786	15.465	34.68	18.5	93.8	-10.8	5.1	94.4	-14.6	5.2
491	28.7647	61.0222	2013.342	15.711	18.289	19.86	326.3	94.1	-54.3	5.7	97.1	-49	6
492	28.8412	23.0158	2013.345	10.874	16.394	24.39	12	-69.9	-38.9	5.5	-71.3	-41.6	5.6
493	28.9050	-6.3299	2013.9145	15.405	15.914	9.15	84.8	65.9	1.4	5.9	67.3	3.3	6.4
494	29.1021	17.0002	2013.297	11.292	15.955	15.73	118.4	83.4	-28.1	5.1	83.3	-25.8	5.3
495	29.1489	62.5853	2013.4175	12.898	14.956	8.63	214.9	80.9	22.8	5.6	77.7	20.3	5.6
496	29.1674	42.2520	2013.4995	11.987	16.862	9.17	7.4	150.1	-71.5	6.3	152.3	-70.8	6.5
497	29.1956	24.1491	2013.489	9.782	15.238	15.26	185.3	-39.6	-134.6	5.2	-32	-143.9	5.2
498	29.2214	5.6771	2012.877	16.478	16.872	3.99	66.6	14.9	-78.5	6.7	20.1	-81.4	7.9
499	29.3158	12.9738	2013.805	14.501	14.6	3.93	275.9	-72	-32.6	5.4	-67.3	-35.4	5.4
500	29.3172	47.6313	2013.4	13.667	16.54	16.73	31.6	75.8	-14.1	5.5	76.1	-8.4	5.6
501	29.3590	43.2638	2013.5325	14.493	14.801	4.20	317.7	-53.3	-70	5.5	-51.4	-69.8	5.5
502	29.4092	26.5885	2013.676	13.189	15.409	5.48	323.9	38.6	-129.2	5.2	39.2	-127.8	5.1
503	29.4264	48.1692	2013.434	10.217	16.436	12.67	289.9	62.3	-22.8	5.5	62.4	-22.5	5.6
504	29.5558	63.1898	2013.3475	12.143	16.365	7.11	180.7	129.7	-35.9	5.6	132.4	-28.5	5.9
505	29.5653	62.2861	2013.6645	17.339	18.525	54.10	27.6	6	-63.7	6.8	3.1	-62.7	6

506	29.6349	2.6861	2013.096	13.376	16.124	8.48	336.1	-67.2	-50.7	6.6	-70.9	-50.8	6.8
507	29.6704	62.3505	2013.134	13.05	14.159	3.90	302.2	31.3	-65.8	5.6	31.5	-63.6	6.2
508	29.7868	-10.1859	2013.968	8.576	13.915	12.73	50.6	89	-58.3	5.5	89.5	-53.7	5.4
509	29.8085	41.2428	2013.6045	14.501	14.687	5.29	177.9	72.9	-138.1	5.5	70.9	-135.3	5.4
510	29.8660	-0.2530	2013.319	8.609	13.946	11.38	326.8	16.3	-101.9	5.6	25.3	-98.9	5.7
511	30.0534	61.3810	2013.312	15.965	16.967	4.00	326.3	68.9	-30.1	5.6	72.2	-32	5.9
512	30.1288	47.7283	2013.4425	11.046	16.093	17.01	260.9	90.5	-28.4	5.5	91.4	-30.5	5.5
513	30.1319	59.6526	2013.177	8.712	12.029	5.73	278.3	-161.3	-111.2	6	-160.4	-110.6	6.7
514	30.2050	6.5185	2013.1145	12.875	14.399	7.76	234.1	64.7	1.5	6.7	62.3	-0.2	6.6
515	30.2263	32.1324	2013.324	17.698	18.441	36.89	122.3	80.3	-88.5	5.5	78.9	-95.2	5.5
516	30.2442	79.8679	2013.491	12.805	17.411	8.34	199	124.3	-39.2	6.3	123	-33.2	6.6
517	30.2923	20.0556	2013.2	11.9	16.981	20.48	291.6	81.4	-59.6	5.2	81.9	-60.7	5.4
518	30.3696	14.5579	2013.1575	11.318	16.886	25.84	98.6	104.2	-3	5.2	100.1	-1.1	5.5
519	30.3997	26.2140	2013.562	13.615	18.229	9.01	210.9	100.3	-57.8	5.6	99.4	-64.1	6.2
520	30.4458	1.2848	2013.1665	11.407	11.508	10.44	347.4	66.4	-58.6	6.5	68.6	-60.8	6.4
521	30.5839	-2.8778	2012.9	15.538	16.275	6.04	344.8	76	-119.1	6.7	81.6	-117.8	7.5
522	30.6162	76.8369	2013.5055	11.261	14.129	11.71	5.6	67.4	6.7	5.6	69.5	5.4	5.6
523	30.6453	45.6887	2013.5425	12.445	14.026	29.88	48.8	50.6	-98.5	5.5	51.3	-94.7	5.5
524	30.7154	21.1770	2013.324	13.953	16.935	8.97	189.2	2.2	-62.8	5.2	4	-65.9	5.4
525	30.7493	65.0745	2013.403	15.475	18.379	53.79	300.3	51.8	-70.3	6.3	48.1	-76.1	7.3
526	30.8210	42.7550	2013.477	11.099	11.95	7.16	222.3	41.4	-68.1	5.5	41.6	-66.4	5.5
527	30.9069	62.4587	2013.394	14.885	15.886	8.87	268.4	91	16.7	6.3	92.9	18.6	6.4
528	30.9106	43.9782	2013.487	14.201	14.89	7.18	342.1	-75.2	-29.4	5.5	-75	-30.5	5.5
529	30.9116	52.5682	2013.48	14.067	16.168	7.46	244.8	100.7	-71.8	5.5	101.6	-68.5	5.6
530	30.9716	65.7079	2013.2365	17.617	18.306	45.63	119.6	2.6	61.9	6.4	7.5	60.3	7.1
531	30.9728	63.2516	2013.3425	16.449	18.175	12.63	168.3	113.4	-12.9	6.4	108.7	-15.2	6.8
532	31.0653	67.9499	2013.442	13.403	13.966	23.53	228.2	74.9	-38.3	5.9	75.7	-37.5	5.9
533	31.0997	59.9156	2013.5215	18.255	18.349	52.29	94.4	-60.1	24.9	6.1	-60.3	32.2	6.1
534	31.1548	18.5795	2013.1895	14.052	17.346	10.25	232.4	64.6	-23.2	5.2	65.3	-19.5	5.5
535	31.1849	11.2177	2013.453	12.194	13.491	20.99	78.4	70.9	-73.4	6.3	67	-76.2	6.4
536	31.1950	64.0089	2013.3705	16.509	18.02	16.25	116.9	65.6	-1.3	5.6	62.4	8.4	7.8
537	31.3550	52.7583	2013.4765	15.285	17.394	8.11	109.4	105.6	-45.1	5.5	105.8	-36.9	5.6
538	31.3551	3.2864	2012.993	13.587	17.221	25.87	43.1	60.9	-29.7	6.6	64	-22.4	7.1
539	31.4174	4.8665	2013.236	9.964	16.018	18.33	205.8	119.2	-32.1	6.9	114	-31.4	6.8
540	31.4394	2.9678	2013.337	14.463	14.783	6.28	191.3	0.7	-98.3	6.5	0.4	-100.2	6.6
541	31.5557	49.8384	2013.301	13.605	18.345	7.54	80.5	73.5	-37.5	5.5	78.7	-28.6	7.1
542	31.8035	30.2505	2013.4465	14.358	15.352	58.74	350.9	63.2	5.8	5.2	64.7	3.3	5.2
543	31.8869	57.2680	2013.468	12.496	14.232	8.28	187.3	66.8	1.3	5.8	65.6	1.4	5.8
544	32.1228	50.6883	2013.458	16.295	18.61	10.91	149.8	119.8	52.9	5.6	120.2	62	6
545	32.1281	7.9839	2012.907	16.449	17.546	11.01	46.9	68.4	-8.2	6.7	69.9	-6.6	7.4
546	32.4274	55.4932	2013.507	17.619	17.748	16.00	280.8	-66	21.5	5.9	-67.9	25.1	5.9
547	32.4655	73.9383	2013.4745	12.795	17.172	10.10	167.9	129.2	-7.2	5.9	131	-8.3	6
548	32.5623	18.5798	2013.0775	15.305	18.185	19.37	198.5	-66	-26.2	5.2	-63.6	-24.6	5.6
549	32.5829	3.8821	2013.469	11.628	15.36	6.57	201.2	78.7	-15.7	6.6	82.6	-11	7.2
550	32.5870	-7.1061	2014.0285	15.45	16.604	12.61	318.6	62.4	-36	6.6	67.1	-34.4	7.6
551	32.7802	68.5097	2013.448	13.997	15.04	9.40	70.3	77.8	-37.4	6.3	78.9	-42.9	6.3

552	32.8838	8.2509	2013.2635	10.023	16.154	13.60	255.1	161.4	-69.3	7.1	171.9	-67	7.1
553	32.9160	-7.3725	2013.95	12.047	12.438	11.18	246.6	61.7	21.7	7	62.5	28	6.2
554	32.9177	15.8420	2013.066	15.311	15.658	6.15	210.4	102.6	-3.8	5.7	104	-6	5.7
555	33.0805	15.6545	2013.11	11.508	13.638	5.57	140	67.8	46	5.6	66.4	45.1	5.8
556	33.0961	36.0809	2013.418	14.552	16.129	8.07	247.7	-21.9	-105.7	5.5	-24.4	-104.4	5.6
557	33.0975	3.6679	2013.203	12.498	17.189	12.96	177.9	54.1	-81	6.4	52.2	-78.3	7.7
558	33.1040	36.3874	2013.3365	15.88	18.043	23.68	110.9	65.7	-25.3	5.5	65.5	-23	5.7
559	33.4171	32.9814	2013.616	8.262	14.943	15.12	33.2	91.2	14.4	5.1	91.6	12.2	5.2
560	33.4350	54.1184	2013.4675	14.848	16.458	6.21	130.2	-36.4	-80.6	5.9	-33.3	-82	5.9
561	33.5208	55.9236	2013.7005	13.237	19.016	34.96	214.1	3.8	-68.3	5.8	16.6	-65	7.2
562	33.5361	59.6559	2013.489	13.558	14.281	41.53	339.5	108.2	-98.5	5.9	107.7	-98.3	5.9
563	33.5386	51.3728	2013.376	14.295	14.457	3.96	191.1	32.6	-75.9	5.5	33.4	-74.9	5.9
564	33.5795	-5.4156	2014.013	9.942	12.429	7.55	43.8	102.1	-76	5.3	97.8	-76.7	5.3
565	33.6295	5.7226	2013.007	13.231	17.2	19.68	13.1	66.5	44.1	6.6	67.3	48.1	6.9
566	33.7790	39.3921	2013.543	10.26	14.869	8.22	134.3	-7.9	-103.8	5.5	-12.1	-103.1	5.5
567	34.0115	23.1152	2013.0585	17.038	17.665	19.89	321.5	114.9	-46.6	5.5	114.1	-44.7	5.4
568	34.0913	80.6785	2013.4705	17.28	18.55	55.56	150.6	-112.5	-12.4	6.9	-109.1	-2.9	6.1
569	34.1227	37.6497	2013.455	9.872	14.828	9.66	168.6	126	-12.3	5.5	127.9	-19.7	5.6
570	34.1413	27.6596	2013.333	13.398	18.514	19.97	144.7	107.6	-59.8	5.5	101.9	-58.7	6.4
571	34.1474	63.2697	2013.4455	11.354	16.357	7.48	329.4	88.9	-25.8	5.6	86.8	-26.6	5.7
572	34.3670	62.0135	2013.523	15.24	18.301	20.86	1.4	81.8	0.9	6.2	85.5	2.5	6.3
573	34.4084	61.3860	2013.426	14.212	15.404	33.21	346.9	87.6	-16.4	5.6	85.9	-13.8	5.6
574	34.4194	60.4639	2013.4325	13.387	13.999	30.12	192.6	66.6	-23.2	5.6	63.8	-24.7	5.6
575	34.4231	18.5355	2013.2315	14.079	15.733	16.93	192.4	62.2	-12.5	5.2	61.5	-11.2	5.3
576	34.4551	47.8070	2013.431	15.802	17.139	5.28	45.3	75.1	-109.2	6	74.9	-110.7	6.2
577	34.4567	62.2371	2013.444	12.159	16.342	17.17	127.5	85.4	-21.4	5.6	86.7	-20.9	5.6
578	34.4715	7.7216	2013.1385	13.214	14.43	7.80	253	107.1	-32.7	6.6	111.9	-32.1	6.7
579	34.4942	72.3444	2013.568	8.427	12.272	7.75	154.8	-85.8	68.4	5.9	-90.4	68.6	5.9
580	34.5754	-4.9298	2013.951	14.081	14.563	5.88	15.4	70.6	-20.8	5.7	67.6	26.6	5.9
581	34.6211	14.3503	2013.4305	11.884	12.727	4.92	267.4	73	-18.6	5.1	76.9	-17.8	5.1
582	34.6357	36.4219	2013.4375	16.501	16.798	4.93	194.8	82	-54.9	5.6	83.7	-56.2	5.6
583	34.8048	50.6491	2013.4865	13.433	14.996	9.22	50.5	117.5	102.7	5.9	116.8	103.8	5.9
584	35.0128	18.4867	2013.5355	12.216	15.585	6.43	265.9	126.5	-31.7	5.2	131.2	-28.9	5.4
585	35.1162	41.0323	2013.561	14.806	15.845	4.35	315.8	87.1	38.9	5.5	91.3	40.2	5.7
586	35.1846	67.6242	2013.4335	11.519	15.184	8.36	254.9	104.9	4.6	6	105.1	3.7	6
587	35.2185	2.3092	2014.044	10.848	11.621	12.86	213.8	80.4	12	6.1	77.9	9.2	6.1
588	35.2260	52.8499	2013.5155	13.071	17.9	15.43	250.5	118.7	-79.2	5.9	117.2	-80.2	6.1
589	35.3945	16.7176	2013.1095	14.265	16.725	5.89	207.9	-156.8	-137.1	5.3	-148.1	-140.4	6.3
590	35.4910	67.7334	2013.532	13.026	15.984	5.38	192.3	74.9	-27.9	6	74	-32.5	6.1
591	35.4922	32.2302	2013.4285	12.227	14.682	9.00	48.7	97.7	-71.1	5.2	95.5	-70.8	5.3
592	35.5101	77.9271	2013.533	13.678	16.636	25.41	263	79.5	-5.6	5.8	78.9	1.5	5.9
593	35.5534	53.6433	2013.545	16.452	17.806	13.50	228.5	65.4	-14.4	5.9	65.7	-18.7	5.9
594	35.5553	23.9900	2013.3555	12.391	15.698	44.45	251.8	81.1	-29.5	5.2	80.5	-26.9	5.2
595	35.5688	31.9228	2013.2925	16.559	16.999	5.06	100.4	65.6	-13.4	5.2	64.1	-12.9	5.3
596	35.7573	66.0126	2013.4915	14.206	14.779	8.83	266.1	13.2	-108.4	5.9	14.1	-107.9	5.9
597	35.7814	-3.1681	2014.035	13.231	16.221	10.92	300.7	112.8	9.2	5.3	107.9	9.3	6.3

598	35.7860	58.5578	2013.658	15.759	18.754	10.32	269.3	71.1	-37	5.8	74.5	-37	6.1
599	36.1432	34.3901	2013.5315	13.137	14.165	7.78	62.3	100.9	-70.7	5.2	98.7	-70.5	5.2
600	36.2003	72.4544	2013.581	12.524	15.961	5.50	309.9	75.4	-12.1	5.9	73.4	-11.5	6
601	36.2560	51.0788	2013.4915	14.891	15.074	8.51	272.1	76.2	-51.6	5.5	73.5	-50.7	5.5
602	36.3119	63.5011	2013.4745	15.455	15.508	7.64	277.8	85.3	-33	5.6	84.1	-35.6	5.6
603	36.3887	51.3342	2013.4395	14.883	16.161	9.24	99.9	-22.9	-65.8	5.5	-22	-63.4	5.6
604	36.4116	-5.0742	2013.998	13.899	15.543	7.75	298.2	-46	-80.7	5.4	-42.9	-83.4	5.9
605	36.4164	57.5072	2013.519	10.706	14.118	48.43	2.8	72.7	-21.9	6	71.4	-21	6
606	36.4181	28.7658	2013.714	8.531	11.457	6.11	82.3	165.5	-91.8	5.1	171.8	-90.5	6.6
607	36.4182	-10.5868	2014.0425	14.331	14.705	8.44	173.7	69	-14.3	6	69.8	-14.3	5.4
608	36.4806	69.0653	2013.473	15.457	16.263	4.27	270.5	71.7	-36	5.9	73.2	-36.9	6.2
609	36.5141	65.2521	2013.789	11.63	18.455	29.71	266.6	179.7	1.3	5.8	188.5	8.9	6.5
610	36.6259	33.4742	2013.626	11.217	14.932	9.86	117.5	67.1	-46.9	5.1	64.8	-46.7	5.2
611	36.6423	32.4381	2013.0415	16.337	19.025	52.51	294.1	66.7	-14.9	5.3	60.3	-25.5	7.9
612	36.6985	55.0251	2013.567	18.145	18.613	54.78	32.5	-79.5	-34.2	6.5	-77.3	-45	6.6
613	36.7543	49.2514	2013.4695	10.868	12.16	16.13	334	-81.8	-25.2	5.5	-84.8	-23.1	5.5
614	36.8995	40.5574	2013.7295	15.137	15.351	3.98	88.8	70.2	-26.6	5.4	67.6	-30.1	6.5
615	36.9523	20.1607	2013.1775	12.312	15.575	8.58	355.2	44.2	-74	5.2	46.5	-74.2	5.4
616	36.9657	50.5700	2013.5575	6.653	11.909	14.02	150.9	35.2	-92.4	6	30.4	-92.8	6
617	36.9818	41.1282	2013.4615	16.796	17.055	3.62	133.3	96.6	-23.5	5.6	97.8	-18.9	5.7
618	37.0202	19.9967	2013.2785	8.763	14.553	10.68	20.1	-13.3	-161	5.2	-14.2	-160.8	5.5
619	37.0214	0.0031	2014.0455	14.558	15.478	4.34	256.8	88.7	43.3	5.4	84.3	47.9	5.6
620	37.1175	36.9517	2013.641	8.04	14.718	14.63	37.1	28.9	-111.9	5.4	18.7	-114.3	5.5
621	37.2830	24.6148	2013.3665	13.926	14.158	7.13	316.2	50.2	-85.4	5.5	49.8	-86.2	5.5
622	37.3234	21.5936	2013.396	7.868	13.412	10.56	174.3	103.5	-48.5	5.1	100.4	-50.9	5.3
623	37.3769	27.9722	2013.532	11.974	15.052	8.83	127.8	74.3	-132.7	5.5	72.4	-133.9	5.5
624	37.5564	42.0232	2013.5135	10.505	12.322	5.21	71.7	163.7	-19.3	5.5	166.7	-10.1	5.7
625	37.6904	20.1860	2013.2595	10.832	16.313	14.40	48.7	-37	-78.8	5.6	-37.1	-75.8	5.7
626	37.7722	33.5989	2013.5615	13.246	15.737	8.84	276.3	104.3	-34.2	5.2	107.1	-33.2	5.2
627	37.9626	12.3746	2013.169	14.355	16.392	7.81	325.3	78.3	-61.5	5.2	74.3	-61.5	5.4
628	37.9755	40.0627	2013.422	13.255	13.276	5.81	359.7	199.6	-0.4	5.5	200.5	-2.9	5.6
629	37.9987	41.5214	2013.444	10.363	12.558	5.02	320.8	66	-15.7	5.5	65.5	-18.3	5.5
630	38.0515	49.9951	2013.5515	14.556	14.858	4.29	124.1	33.2	-120	5.5	32.7	-118.8	5.5
631	38.0707	22.3975	2013.1475	14.965	16.878	10.62	18.5	-22.2	-67.1	5.6	-21.6	-68.2	5.8
632	38.2282	19.0090	2012.928	17.498	18.119	15.10	75.5	76	-47.4	6	75.8	-46.9	6.2
633	38.3093	79.2100	2013.2725	18.184	18.391	51.60	354.7	61.5	12.7	6.7	60.4	16.9	7
634	38.3211	-1.0415	2013.9025	15.212	15.466	5.20	137.2	113.6	25.7	5.4	111.8	22.5	5.6
635	38.3270	25.5238	2013.349	13.489	17.416	15.84	49.3	67.1	-0.8	5.2	68.4	-1.2	5.3
636	38.3863	49.3184	2013.4685	17.455	18.091	9.43	359.8	64.6	-2.1	5.6	62.8	-9.1	5.7
637	38.4318	72.5495	2013.5045	13.093	14.866	7.08	162.1	73.2	-47.6	5.9	74.4	-50.2	5.9
638	38.5968	84.2991	2013.7625	10.434	15.975	8.13	200.3	121.1	-64.3	6.1	114	-66.1	6.9
639	38.6216	-6.0518	2013.983	11.773	15.687	11.84	150.2	-73.8	-65.2	5.3	-73.3	-66.7	5.8
640	38.7630	54.5706	2013.571	17.499	18.288	51.54	23.3	-41.4	65.2	5.2	-47.4	65.5	5.4
641	38.7741	14.2821	2013.165	11.317	13.729	5.38	247.7	117.4	-9	5.2	119.4	-9.2	7.3
642	38.7858	-6.4509	2013.981	11.022	11.552	7.75	189.4	84.2	-54.6	5.3	86.3	-55.6	5.3
643	38.8185	52.1362	2013.493	16.823	16.971	3.58	216.1	60.1	-102.3	5.7	67.5	-102.4	5.7

644	38.8223	39.1725	2013.4865	8.699	17.16	46.90	115.3	47.4	-76.8	5.4	50.9	-71.5	5.6
645	38.8238	24.4389	2012.9415	17.625	18.299	27.85	343.1	60.9	-39.1	5.4	61.6	-38.2	5.7
646	38.9231	79.2313	2013.5695	9.819	15.342	18.15	41.1	104.4	19.5	6.3	102.6	16.4	6.3
647	39.0143	49.8333	2013.4305	16.911	18.02	12.04	218.8	7.6	65.5	5.6	8.3	62.7	5.6
648	39.0342	29.9476	2013.3575	15.642	15.955	9.08	182.9	63.6	-5.6	5.2	64.2	-2.9	5.2
649	39.0476	54.7935	2013.4715	10.687	14.308	9.56	232	60.6	-23.1	5.1	60.3	-23.2	5.1
650	39.0514	-4.2580	2013.8245	13.315	14.554	7.23	285.2	74.8	-30.4	5.4	73.8	-28.3	5.4
651	39.0963	30.1376	2013.2705	15.714	17.476	4.85	205.8	160.3	-2.7	5.3	163.5	-5.1	7.2
652	39.1250	50.0606	2013.311	17.824	18.562	54.71	321.4	-10.3	-61.2	5.6	-20.4	-60.4	6.2
653	39.1521	27.1582	2013.3465	10.575	17.253	13.26	35.6	61	-12.8	5.2	62.1	-11.6	5.3
654	39.1627	40.1807	2013.62	11.527	15.87	8.70	163.7	73.2	-21.2	5.4	70.9	-20.6	5.4
655	39.3500	22.5352	2013.1275	15.665	18.425	13.51	256.7	66.4	-28.9	5.7	64.2	-28.5	6
656	39.4287	66.0665	2013.473	17.728	18.62	47.49	163.9	-27.6	-63.6	6.9	-27.7	-63.6	7.6
657	39.5167	27.3104	2013.2535	15.672	16.286	17.14	118.3	63.6	-34.4	5.2	64.4	-35.3	5.3
658	39.5254	22.8793	2012.9445	16.312	17.181	7.09	108	81.1	-23	5.7	80.2	-31.2	5.8
659	39.5346	54.4815	2013.4785	16.995	18.446	40.90	306.2	43.9	67.3	5.5	47.3	69.1	5.2
660	39.5611	25.5118	2013.223	15.723	16.2	4.74	213.2	20.5	-65.2	5.3	21.7	-63.4	5.3
661	39.6028	-2.2464	2013.852	14.58	14.879	4.77	273.1	142	-0.6	6.4	141.7	2.3	6.7
662	39.6178	42.2198	2013.457	11.845	14.139	5.38	179.4	96.9	-32.3	6.2	97.2	-28.7	6.3
663	39.7569	58.8190	2013.4945	14.416	18.151	58.86	131.9	-48.5	-68.6	5.6	-45	-74	5.2
664	39.9954	8.3881	2013.1115	12.693	17.079	8.74	194.1	87.7	-7.6	6.5	84.3	-9.1	7.8
665	40.0434	50.8912	2013.513	11.256	14.563	15.09	325.4	17.8	-79.2	5.9	21.2	-79.5	5.9
666	40.1690	43.5735	2013.499	14.726	15.144	36.21	190.8	-123.1	7.5	6.4	-119.5	10.6	6.3
667	40.3326	55.0651	2013.603	18.169	18.621	50.17	210.6	-69	35.9	5.2	-75.6	28.6	6.2
668	40.3632	15.7718	2013.5775	9.931	11.623	6.64	304.5	124.9	-85	5.5	120.1	-93.8	5.6
669	40.5578	4.0544	2013.5595	10.26	14.086	37.49	152.6	97.8	0.4	6.4	94.5	5	6.7
670	40.6053	27.5325	2013.3075	15.835	18.448	27.67	234.3	135.8	-70	5.3	139.3	-73.8	5.7
671	40.6409	37.2287	2013.5435	10.212	15.786	24.44	342.5	75.1	-36.4	5.8	72.1	-35.9	5.9
672	40.7437	-10.1609	2013.8745	12.412	13.131	24.34	103.1	-60.6	-64.2	5.6	-61.7	-60.2	5.4
673	40.8215	51.4971	2013.569	12.816	14.052	6.82	6.2	82	-4.2	5.8	85.3	-8.7	5.8
674	40.8303	22.5205	2013.324	9.612	16.876	40.38	291.2	22.5	-67.8	5.2	21.3	-69.1	5.3
675	40.9626	31.9255	2013.3545	14.949	18.005	7.11	316.8	85.7	-43.8	5.2	87.9	-44.7	5.5
676	41.0311	48.7438	2013.525	13.64	16.607	30.12	13.1	100.2	-75.4	5.9	96.5	-73	5.9
677	41.0576	24.9946	2013.2645	13.078	15.248	19.12	187.8	-6	-102.9	5.2	-5.6	-101.4	5.3
678	41.2900	61.8302	2013.4885	11.357	13.746	54.20	223.2	68.7	-97.8	5.6	71.2	-95.7	5.6
679	41.3223	58.9913	2013.4825	14.258	14.607	11.30	133.1	35.7	-106.7	5.9	37.6	-102.3	5.9
680	41.3764	33.3810	2013.664	14.668	16.167	4.36	23	65.1	-100.6	5.2	64.1	-105.5	5.4
681	41.5396	-6.6105	2013.654	13.03	14.458	6.64	131.6	59.2	-77.9	5.5	49.3	-82.1	5.9
682	41.5770	62.3209	2013.59	12.894	18.218	48.63	334.1	42.1	-68.1	5.6	41.7	-65.9	5.9
683	41.6414	29.7464	2013.527	8.554	17.414	30.39	353.9	-16.6	-68	5.1	-23.4	-63.2	5.3
684	41.7473	9.1295	2013.337	10.465	16.348	55.30	0.5	-74	-34.1	5.9	-77.5	-36.1	6.3
685	41.8009	28.7657	2013.2345	15.509	17.767	13.06	262.2	28.6	-63.3	6	28.7	-60.9	6.2
686	41.8625	44.9135	2013.3825	11.938	17.784	11.55	270.4	89	-24	6.3	91.5	-17.6	6.8
687	41.9053	7.8425	2012.9445	16.84	17.088	28.95	308.2	-5.1	-101.3	6.4	3.2	-98.3	6.8
688	42.2854	53.5715	2013.471	11.212	17.606	59.81	290.9	0.8	-64.9	5.9	9	-62.4	5.9
689	42.3339	70.4943	2013.5515	8.216	16.789	57.29	264.2	75	-97.9	5.8	81.9	-88	6

690	42.3444	45.3379	2013.368	17.511	17.633	42.64	21.3	-64.9	-4	5.6	-67	-8.7	5.6
691	42.3718	-1.4151	2013.317	12.404	16.647	9.15	181	53.7	-72.9	5.6	55.5	-75	6.1
692	42.4358	52.4704	2013.45	11.434	12.361	10.71	261.4	76.2	-76.5	5.9	74.8	-77.1	5.9
693	42.5199	51.3326	2013.3695	17.091	18.441	5.75	198.5	83.8	-18	6	82.2	-22.5	6.3
694	42.6633	-1.7104	2013.765	13.552	14.655	6.23	329	5.1	-79.1	5.5	8.7	-79.7	5.4
695	42.7352	54.7493	2013.5065	10.748	13.751	19.88	317.5	-17.4	-66	6	-14.2	-64.2	6
696	42.7510	9.4431	2013.164	12.974	17.96	28.05	32.3	131.9	16.8	6.1	125.8	14.2	7.1
697	42.8725	28.0889	2013.417	12.683	14.048	50.65	236.5	60.8	-12.3	5.2	61.2	-10.4	5.2
698	42.9711	22.4681	2013.223	13.381	13.981	5.51	308.8	61.6	39.3	6.1	61.5	38.1	6.2
699	43.0244	62.2639	2013.669	16.901	18.536	53.79	144.4	-64.9	-56	5.9	-60.1	-63.8	6.1
700	43.0953	31.4436	2013.5235	10.382	14.19	7.54	354.9	-72.3	-73.8	5.1	-72.4	-75.9	5.3
701	43.1432	1.6510	2013.3735	13.873	16.72	8.20	13.3	49.9	-71.8	6.4	56.1	-71.7	6.7
702	43.3059	13.8912	2013.5285	13.393	15.5	30.41	163	90.3	-34.1	5.9	88.1	-36.8	5.9
703	43.3108	66.2267	2013.376	18.214	18.748	39.73	225	-76.7	-8	6.4	-78.4	-16.2	7
704	43.3665	57.1367	2013.56	15.184	15.401	4.96	75.2	105.3	-32.2	5.5	105.3	-33.3	5.6
705	43.4290	61.4213	2013.637	18.368	18.672	29.10	240.2	37.5	76.3	6.3	28	79.6	6.4
706	43.4612	51.1221	2013.44	14.649	16.546	40.39	108.6	3.2	-100.7	5.6	4.5	-103.5	5.7
707	43.4837	49.5593	2013.143	18.54	18.647	54.76	199.6	-69.6	-16.8	6.9	-69.1	-20.4	6.6
708	43.5536	-9.9920	2013.968	13.739	16.675	12.39	134.1	60.8	15.6	5.5	61.3	20.8	7.5
709	43.7028	15.6993	2013.509	10.7	14.346	12.96	125.7	63.8	-30.9	5.9	63.8	-32.5	5.9
710	43.7481	32.5362	2013.5505	10.32	11.966	8.79	113.4	84.9	-21.8	5.2	84.3	-22.2	5.2
711	43.7912	84.1348	2013.827	12.041	17.305	53.11	243	74.4	-48.3	5.3	74.5	-49.1	5.4
712	43.8047	54.1335	2013.438	12.929	13.972	12.29	258.6	61.6	-17.6	5.6	61.1	-17.2	5.5
713	43.8164	2.1230	2013.295	13.293	15.988	34.96	67.5	-35.1	-75.3	6.6	-35.7	-73.1	6.6
714	43.8466	27.5309	2013.552	10.213	14.168	14.38	191.7	73.2	-17.1	5.1	72.5	-17.2	5.2
715	43.8664	36.5655	2013.5715	13.11	14.911	5.99	354.4	-12.4	-75.4	5.5	-8.9	-78.8	5.5
716	43.8766	9.7700	2013.1055	16.642	17.028	5.10	35.5	-23	-86.5	6.8	-16.9	-86.8	6.5
717	43.9303	18.0042	2013.2345	16.67	17.484	11.74	73.7	71.2	-28.5	5.7	71.8	-24.4	5.7
718	43.9355	69.4090	2013.484	14.477	15.069	7.04	20.7	92.9	-55	6.2	91.6	-59.3	6.2
719	44.0565	65.3111	2013.562	17.636	18.267	34.06	313.9	24.1	74.1	6.4	24.3	75.3	6.1
720	44.1940	63.4271	2013.5785	10.646	15.43	8.94	234.1	-66.1	36	5.8	-66.8	42.2	5.9
721	44.3076	32.4539	2013.34	16.096	16.58	34.38	217.8	28.1	-84.5	5.3	28.3	-86.5	5.4
722	44.5422	19.2713	2013.3295	12.366	14.333	6.97	8.6	62.2	-40.8	5.5	60.8	-41.5	5.6
723	44.6202	16.7920	2013.387	13.762	16.476	7.12	195.3	183.8	-3.4	6.1	183.3	-2.9	6.5
724	44.7645	63.3883	2013.4155	17.853	18.199	22.25	278.3	35.8	-61.3	6	40.4	-61.3	6.2
725	44.7782	13.3270	2013.258	17.814	18.153	46.46	68	63.3	0.1	6.5	61.2	-4.5	6.3
726	44.8554	63.6549	2013.4615	14.098	17.525	8.11	85.6	-24.1	-104.8	6.3	-29.1	-99.5	6.5
727	45.0284	60.3570	2013.472	13.021	16.146	27.58	240.6	41.2	-61.5	6.3	36.6	-65.1	6.3
728	45.2390	42.3900	2013.5055	13.052	16.626	5.92	167.3	49.6	-98.4	5.5	47.3	-103.8	7.8
729	45.3275	57.1689	2013.512	12.03	14.209	13.76	94.5	93.5	-18.4	5.5	95.5	-18.6	5.5
730	45.3449	42.1402	2013.396	14.784	15.438	38.60	144.1	114.2	-53.6	5.6	112.8	-53	5.6
731	45.3461	34.1770	2013.575	11.433	15.068	45.65	124.3	35	-66	5.2	32.7	-67.5	5.2
732	45.3900	16.3047	2013.3375	13.18	17.455	22.81	309.7	97.9	44	5.2	100.2	44.1	5.4
733	45.4142	33.6367	2013.3375	16.329	17.4	11.97	156.8	50.5	-67.6	5.3	51.6	-70.3	5.4
734	45.4251	23.1042	2013.2655	11.832	13.781	5.38	255	124.1	-19.1	6.6	119.8	-14.8	6.6
735	45.5588	51.5476	2013.395	14.495	17.822	12.73	203.7	81	-93.8	5.6	76.5	-92.1	5.7

736	45.5987	8.4352	2013.526	12.525	13.626	7.43	157.2	85.2	-45.4	6	85.4	-43.9	6
737	45.6090	8.4170	2013.606	10.189	15.304	14.72	335.3	91.5	-10.3	5.9	87.8	-10.8	6
738	45.6144	61.3591	2013.3855	14.271	16.61	8.59	51.8	59.8	-132	6	59.6	-128.4	6
739	45.6407	64.4747	2013.495	11.849	18.126	31.11	218.8	81.4	-1.4	5.9	82.9	1.3	6.2
740	45.6779	46.5281	2013.6485	17.397	18.773	36.33	269.2	15.8	-66.4	5.5	12.5	-63.9	6.2
741	45.7176	35.3386	2013.41	15.158	16.859	5.98	204.3	56.4	-66.3	5.2	58.9	-66.4	5.3
742	45.7675	50.0665	2013.377	15.947	16.36	14.62	229.9	68.5	0.8	5.6	69.8	1.4	5.6
743	45.8288	50.2664	2013.474	14.594	15.755	6.97	140.2	150.5	-78.3	5.6	150.3	-77.1	5.6
744	45.8293	46.7394	2013.5095	14.766	15.858	5.49	181.3	67.6	-10.9	6.3	70.9	-11.3	6.3
745	45.9436	12.1827	2013.1575	16.389	18.614	10.91	296.2	79.8	17.6	5.2	82.1	17.5	6.9
746	45.9663	23.7772	2013.2015	10.171	16.722	21.57	348.8	68.4	-23.4	5.2	68.4	-18.9	5.3
747	45.9890	17.4629	2013.4625	14.692	15.013	4.44	103.1	2.9	-78.8	5.1	-1.4	-79.6	5.4
748	46.0298	-2.2906	2013.612	11.699	13.171	8.49	280.4	130.2	-5.5	5.6	129.3	-10.3	5.6
749	46.0433	13.0059	2012.917	18.203	18.215	33.23	127.5	-27.2	-89.2	6.1	-19.3	-86.8	5.7
750	46.1107	-8.4622	2014.0005	10.552	14.988	9.31	51.2	122.8	7.6	5.3	122.9	10	5.5
751	46.1687	57.2576	2013.511	14.666	16.416	5.23	132.6	69.7	-79.2	5.5	69.7	-79.7	5.6
752	46.2842	58.2776	2013.534	13.305	16.591	6.99	350	55.6	-78.5	5.5	55.3	-78.3	5.6
753	46.3071	28.6270	2013.3935	12.886	14.589	16.40	279.7	64.5	32	5.2	63.5	34.4	5.2
754	46.3858	1.9712	2013.3375	14.915	16.738	6.86	290.2	56.7	-84.7	6.6	59.9	-85.6	6.7
755	46.6225	22.1553	2013.203	17.361	17.497	7.44	265.9	103.9	-48.4	5.5	104.4	-45.3	5.4
756	46.6259	19.4528	2013.308	13.237	14.357	9.07	345.6	63.8	-9.6	5.2	62.7	-8.1	5.2
757	46.7289	35.8097	2013.428	13.533	13.988	4.42	342.7	-68.9	-20.6	5.2	-67.1	-25.8	5.3
758	46.9254	34.8957	2013.433	13.684	16.716	6.02	139.6	115.1	-103.4	5.3	115.7	-99.9	5.6
759	46.9489	-4.8068	2013.259	15.695	15.874	4.83	80.5	99.4	-68.8	6	100	-71	6.6
760	47.0426	26.3000	2013.5335	12.96	13.855	5.17	215.5	64.3	-38.3	5.2	64.7	-36.4	5.2
761	47.1094	9.2979	2013.341	12.909	14.782	6.35	266.8	-81.7	-105.7	6.1	-79.5	-109.1	6.1
762	47.1384	-8.7204	2013.798	10.984	15.774	8.48	230.8	-18.6	-72.1	5.5	-18.9	-69.5	5.9
763	47.2249	29.2177	2013.497	14.121	14.968	29.71	197.8	-62.2	-64.7	5.2	-64.7	-66.4	5.2
764	47.2949	87.7815	2013.7725	16.116	16.751	5.73	147.1	122.6	-43.8	5.5	121	-47.6	5.4
765	47.3984	67.3736	2013.5045	18.33	18.553	59.30	211.6	19.4	-70.1	6.2	9.2	-71.2	6.9
766	47.4006	-10.0258	2013.937	12.11	14.323	8.46	205	103.7	33.2	5.5	103	27.7	5.6
767	47.4696	43.9165	2013.51	16.982	17.081	33.71	77.6	82.8	-36.4	5.5	83.2	-44.4	5.5
768	47.5947	56.4197	2013.585	11.238	17.59	19.27	195.6	50	-67.3	5.9	48.5	-65.7	5.9
769	47.5969	33.6309	2013.3955	15.706	17.588	15.06	36.6	65.2	-20.3	5.2	65.2	-21.5	5.4
770	47.6450	29.5635	2013.236	15.604	17.559	6.71	297.7	70.3	-16.3	5.2	69.2	-19.1	5.4
771	47.7529	75.9272	2013.575	18.24	19.096	54.05	211	65.1	11	6.6	62.9	-1	7.8
772	47.7596	22.5216	2013.262	14.722	15.058	10.04	206.5	16.8	-63.7	5.5	17.4	-66.3	5.6
773	47.9269	49.3850	2013.4525	14.621	17.373	10.19	121.2	12.4	-86	5.5	13.6	-85.4	5.6
774	47.9518	60.8277	2013.37	14.197	17.054	5.38	290.8	83.6	-93.5	5.6	82.9	-91.8	5.8
775	47.9630	27.8129	2013.2665	13.775	18.071	8.23	297.3	-20.3	-73.5	5.2	-24.9	-71.9	5.4
776	48.1484	43.4416	2013.494	12.81	17.112	9.22	233.4	85.9	-35.3	5.5	83.8	-34.5	5.6
777	48.3241	54.8115	2013.537	18.66	18.724	53.52	171.9	48.3	-61	6.3	37.9	-70.1	6.1
778	48.3325	74.8055	2013.56	8.776	11.949	9.91	63.1	65.1	-52.7	6.3	64.8	-48	6.3
779	48.4423	59.0295	2013.4745	17.639	17.881	9.31	83	58.9	-61.2	6.1	60.7	-60.4	6.1
780	48.5388	54.2785	2013.416	14.84	15.269	3.77	139.8	-108.7	58.8	6	-110.2	61.1	6.2
781	48.5691	28.3302	2013.2155	12.496	17.745	39.54	183	67.1	-35.6	5.2	63.7	-36.8	5.5

782	48.6276	35.9451	2013.4045	16.651	17.338	6.44	185.4	64.7	-46.9	5.6	61.9	-48.5	5.7
783	48.7418	34.0955	2013.5155	15.15	16.378	5.27	147.6	21.9	-88.2	6.4	16.6	-92.1	6.5
784	48.8050	23.4462	2013.0835	15.944	16.299	17.94	32.9	94	-17.1	6.2	93.6	-19.3	6.2
785	48.8520	-0.7476	2013.4175	12.458	14.532	10.74	290.8	99.3	-82.4	5.6	98.7	-82.8	5.7
786	48.9125	13.8882	2013.028	17.111	18.668	49.43	249.7	11.1	-60.7	6.1	13.9	-62.3	7.8
787	48.9602	44.0194	2013.6275	17.274	18.314	36.72	107.7	14.9	-66.7	6	4.9	-65.9	7.9
788	48.9702	3.1503	2013.3295	14.182	14.603	5.57	85.7	20.5	95.1	6.4	20.2	94.8	6.5
789	49.1371	38.6339	2013.498	6.693	19.053	43.40	57.4	54.7	-60.7	5.5	45.2	-67.3	5.9
790	49.2958	47.4790	2013.3595	14.922	17.861	8.17	206.1	92.2	-111.1	6.1	91.8	-108.4	6.3
791	49.4525	35.5873	2013.371	15.106	16.966	5.26	323	45.4	-105.1	5.5	41.3	-102.1	5.7
792	49.4863	66.4079	2013.511	15.259	16.15	23.88	200.5	63.6	-25.6	5.9	60.1	-27	5.9
793	49.5114	-2.7499	2013.2355	13.831	16.376	8.52	85.8	39.1	-62.4	5.5	40.6	-60.8	5.9
794	49.5713	17.0808	2013.431	12.204	13.742	7.24	53.5	75.1	4.1	5.9	75	4.7	5.9
795	49.5758	53.5812	2013.4965	14.504	14.545	52.66	326.7	24.1	-95.3	5.5	28.5	-95.4	5.5
796	49.6606	30.6853	2013.233	16.292	18.486	33.61	129.6	108.1	-117.3	5.7	111.4	-120.8	6.3
797	49.7238	7.1655	2013.321	15.906	17.878	7.26	272.1	44.2	-62	6.1	35.3	-63.4	7.8
798	49.8484	16.7067	2013.247	11.324	14.736	6.25	350.5	66.6	-68.1	5.9	70.2	-71	6.3
799	49.9033	2.6718	2013.351	10.936	12.378	7.12	248.1	19.8	-74	6.2	17.3	-75.3	6.2
800	50.0682	71.8843	2013.5855	10.872	13.558	15.33	20.5	63.4	-27.2	5.8	62.1	-27.4	5.8
801	50.1034	13.4888	2013.163	15.3	15.71	4.41	89	81.4	-21.6	5.6	80.3	-22.3	5.8
802	50.1140	70.7133	2013.5955	15.026	17.682	6.64	214.2	82.9	-60.8	5.9	82.6	-57.9	5.9
803	50.1265	51.3429	2013.513	12.08	15.062	45.40	246.3	67.4	-38.6	6.3	65.4	-42.2	6.3
804	50.1327	28.6452	2013.6045	14.391	14.487	6.40	137.7	49.8	-85.1	5.3	48.7	-87.1	5.1
805	50.2178	46.8947	2013.53	11.934	16.063	14.58	131.5	-2	-78.1	5.9	1.7	-78.7	5.9
806	50.3403	48.5602	2013.4715	12.405	17.179	46.15	161.8	16.5	-73.3	6.3	20.1	-70.9	6.4
807	50.4683	-10.0989	2014.047	12.815	13.913	6.53	341.6	67.4	-13.3	5.4	70.2	-13	5.4
808	50.4872	48.8429	2013.413	16.385	18.37	57.87	191.5	16.4	-61.7	6.4	4.1	-62.9	6.5
809	50.4938	13.3918	2013.5605	12.653	13.931	38.82	299.2	63	-14.6	5.4	63.8	-15.5	5.4
810	50.5548	-2.2849	2013.238	13.307	16.905	27.38	129.3	79.7	-16.2	5.5	77.7	-19.3	6.2
811	50.5589	27.0889	2013.0435	12.048	16.902	7.63	292.3	66.2	-49.3	5.3	69.2	-50.3	5.5
812	50.6432	16.9313	2013.4875	15.642	15.674	3.91	196.2	66.7	19.6	5.6	68.2	19.2	5.5
813	50.6606	0.4178	2012.936	17.551	17.835	4.41	315.8	60.3	-87.9	6.6	63.5	-81.8	7.3
814	50.7793	-5.5059	2013.446	14.35	15.188	6.20	249.7	-65.4	6.6	5.6	-65.5	6.8	5.6
815	50.8538	48.4232	2013.3605	16.507	17.94	19.25	334.2	29.9	-124.1	5.6	26.7	-124.1	5.7
816	50.9142	31.3134	2013.4585	12.199	14.276	6.50	237.4	-4.5	-70.8	5.2	-3.3	-73.2	5.3
817	51.0183	26.7355	2013.135	15.998	18.901	11.47	15.2	64.7	-46.5	5.3	69	-44.6	6.9
818	51.1618	64.6610	2013.4405	14.584	14.917	4.84	322.1	-101.9	53.5	5.6	-102.3	53.4	5.6
819	51.1830	1.4030	2013.0585	15.6	17.324	32.62	54	85.4	-34.7	6.6	86.1	-31.2	7.3
820	51.1946	46.5365	2013.4455	16.773	18.067	45.16	155.3	91.1	-88	6	87	-86.4	6.1
821	51.2983	18.0989	2013.2375	13.242	14.684	4.78	106.9	101.9	-30.5	6.1	100.1	-28.6	6.5
822	51.6214	36.1899	2013.4325	12.236	12.331	7.80	109.5	-42.4	-103.8	5.5	-41.7	-104.6	5.5
823	51.6471	-7.0953	2013.531	10.08	15.368	14.97	315.3	74.7	36.9	5.4	77	31.6	5.7
824	51.6670	-6.3051	2013.305	14.582	16.151	21.09	338.5	44	-70.1	5.6	41	-69.2	6
825	51.6696	20.6068	2013.483	12.229	18.167	9.43	265.9	50.9	-107.9	5.2	53.5	-108.1	5.8
826	51.6869	-9.1136	2013.066	15.919	17.194	7.20	272.4	77.1	-3.8	5.9	79.2	-8.9	6.3
827	51.6999	55.1241	2013.545	13.328	17.329	37.53	241.7	30.4	-80.1	5.5	30.2	-82.6	5.6

828	51.7480	21.1076	2013.311	11.468	14.862	21.30	124.1	73.7	-19.5	6	73.1	-19	6
829	51.8007	22.8022	2013.255	13.349	17.77	19.49	275	46.6	-69	6	49.6	-71.9	6.2
830	51.8835	19.1871	2013.6485	8.905	14.849	12.62	204.7	-64.8	-74	5.2	-64.6	-71.8	5.3
831	51.9205	77.3279	2013.5855	17.885	18.406	19.31	318.7	47	61	5.7	38.9	65.9	5.9
832	52.0460	20.8974	2013.1785	14.713	15.096	54.75	301.6	-35.3	-74.5	5.3	-31.6	-79.1	5.3
833	52.0543	7.3022	2013.695	7.824	12.098	32.67	341.3	-15.4	-84.9	5.8	-13.8	-84.7	5.9
834	52.0702	17.5653	2013.4555	12.195	14.219	7.31	251.2	85.9	-63	5.2	90.1	-64.7	5.2
835	52.1075	35.9727	2013.3985	16.539	18.619	36.23	211.7	42	-75.3	5.6	45.9	-72.2	6.1
836	52.1130	39.8140	2013.4865	13.858	16.333	17.96	178.8	-21.3	-75.4	5.5	-21.2	-74.8	5.5
837	52.1452	50.2647	2013.5485	16.288	18.554	59.46	333.7	11.7	-88	6.5	5.5	-84.8	7
838	52.1645	71.6482	2013.54	12.944	17.415	12.82	121.5	69.2	-10.6	5.9	69.7	-7.1	6
839	52.1755	13.1324	2013.4785	9.906	15.026	10.14	192.8	87.4	-30	5.1	84.2	-30.4	5.4
840	52.2547	62.8388	2013.4445	13.438	14.179	48.42	261.2	42.9	-74.5	5.9	42.1	-74.1	5.9
841	52.6279	-6.5911	2013.2995	10.491	17.244	23.56	124.7	78	-69.4	5.6	72.2	-69	6.1
842	52.7050	10.4891	2013.572	12.178	14.29	8.07	335.8	7.6	-116.6	5.9	11	-121	6
843	53.0054	21.9883	2013.4495	13.297	13.508	4.70	57.1	64.5	25	5.3	63.5	18.4	5.7
844	53.0321	43.1359	2013.538	13.654	14.538	6.43	346.5	-47.7	-74.2	5.5	-48.3	-76.5	5.5
845	53.0940	26.4400	2013.0955	15.546	18.663	10.52	70.4	-87.3	-63.6	5.4	-86.2	-63.1	6.1
846	53.1436	-4.8557	2012.944	17.234	17.659	55.71	23.6	73.3	-88.8	7	72.5	-85.4	6.1
847	53.1984	41.8540	2013.5075	17.146	18.683	58.46	323.1	7.2	70.3	6.4	10.4	70.1	6.7
848	53.3993	30.6854	2013.5155	15.449	15.963	4.33	306.1	-95.7	-101.6	5.4	-92.3	-99.3	5.3
849	53.4597	-0.7255	2013.6505	15.66	15.741	8.62	343.5	97.3	-7.1	5.7	97.3	-7.9	5.5
850	53.4625	12.8304	2013.276	16.041	17.269	27.84	68.6	13.8	-62.5	5.2	6.6	-61	5.3
851	53.5240	6.6194	2013.557	13.417	16.801	8.34	17	45.5	-68.8	6	43.6	-69.8	6.1
852	53.5571	53.7596	2013.4825	15.94	18.605	11.14	15.9	63.3	-47.1	5.9	64.6	-39.8	6.3
853	53.7385	-2.2881	2013.5485	10.758	12.972	7.95	38.3	81.3	7.7	5.5	81.3	7	5.5
854	53.7578	43.6874	2013.5395	9.562	11.408	11.62	150.2	62.6	-53.4	5.8	63	-51.6	5.9
855	53.7742	48.2606	2013.503	14.477	16.017	7.32	135.7	-16.1	-68.2	5.9	-10	-67.3	5.9
856	53.8983	40.7779	2013.392	17.379	18.018	59.95	68.9	60	-62.6	5.5	67 8	-60.9	5.6
857	53.9073	25.8908	2013.0745	16.009	16.529	4.10	52	72.3	-24.9	5.4	71.6	-22.8	5.4
858	53.9285	18.0289	2013.315	11.555	14.796	7.22	207	5	-93.5	5.2	5.1	-92.4	5.3
859	54.0798	36.2976	2013.186	18.049	18.31	45.25	311.1	6.9	66.2	5.8	-2.5	64.5	6
860	54.2883	61.9911	2013.46	12.11	14.645	10.17	32.7	46.7	75.5	6	45.5	73.6	6
861	54.4326	57.4967	2013.586	16.979	17.179	4.07	285.3	71.5	-85.3	6.4	75.9	-87.8	6.6
862	54.6140	65.4178	2013.362	16.411	16.527	5.04	62.4	72	-54.7	6	69.4	-58.2	6
863	54.6449	48.2216	2013.473	18.038	18.637	30.80	198	-61.7	-3.3	5.7	-61.8	0.2	6
864	54.6694	44.3148	2013.495	13.582	17.853	26.65	47.5	103.2	-58.1	5.5	106.2	-55.8	5.7
865	54.8491	48.9775	2013.6645	14.352	18.089	6.30	53.2	65.8	-34.5	5.8	69.4	-22.8	6.2
866	54.8504	45.9058	2013.367	18.013	18.052	55.73	64	10	-69	5.6	18.4	-65.7	5.7
867	54.9471	64.8035	2013.509	11.275	15.281	34.88	87	89.3	-33.4	5.9	88.3	-36	5.9
868	55.0452	20.5748	2013.2925	12.075	15.1	17.76	194.8	31.8	-92.9	5.6	29.5	-97.5	5.6
869	55.1055	22.2779	2013.478	16.44	18.658	4.94	227.5	78.4	-22.3	5.6	73	-26.7	5.7
870	55.2115	-3.7791	2013.2725	15.915	15.929	31.79	83.4	83.1	5.9	5.9	83.3	3.9	5.7
871	55.2267	29.8315	2013.2345	12.999	15.933	8.36	332.9	103.1	-63.8	6.2	105.5	-66.1	6.2
872	55.2339	45.6525	2013.59	11.124	12.274	5.30	91.7	82.5	-99.5	5.5	79.4	-101.4	5.5
873	55.3086	1.7266	2013.487	10.482	14.159	10.26	51.9	85.6	16.2	6	86.6	15.5	6.1

874	55.3289	7.6441	2013.4675	15.522	15.611	5.61	5.9	-4.3	-60.1	6.1	-2.2	-61.5	6.1
875	55.3524	49.8483	2013.513	14.526	16.112	49.17	135.9	28.8	72.8	5.8	23.6	76.6	5.9
876	55.3708	37.7950	2013.3935	16.039	17.02	6.77	353.6	0.8	-61.8	5.5	2.5	-61.7	5.5
877	55.4845	34.0459	2013.378	15.608	16.973	8.08	1.7	38.6	-73.7	5.6	37.4	-74	5.6
878	55.6464	65.7535	2013.643	18.359	18.672	20.49	343.4	-35.4	-69.1	6.9	-45.7	-64.5	6.2
879	55.6854	53.3753	2013.487	13.246	17.058	12.18	190.1	89.2	-45.2	5.8	87.4	-40.5	5.9
880	55.7158	-7.6241	2013.362	15.348	15.859	37.81	23.4	60.7	55	5.8	61.2	55	5.8
881	55.7462	36.4843	2013.344	14.605	17.747	7.31	122.6	52.5	-86.9	5.5	53.5	-84.1	5.8
882	55.8300	23.8596	2013.345	13.041	14.207	37.97	169.2	13.9	-62.2	5.5	14.1	-64.8	5.6
883	55.8362	0.7430	2013.366	13.614	16.699	7.43	241.9	-4.4	-98	6.1	-2.6	-97.2	6.5
884	55.9355	53.7492	2013.3985	16.769	18.129	9.58	170.3	-5.3	-73.6	5.9	-5.8	-75.5	6
885	55.9642	-10.7281	2013.9275	14.848	16.21	6.10	26.6	123.9	70	5.6	123.9	64.9	5.8
886	56.1405	75.7150	2013.527	8.54	14.762	14.75	122.5	-95.3	-51.1	5.9	-98	-55.5	5.9
887	56.2231	46.9178	2013.4955	10.247	15.13	11.03	291.5	57.9	-90.4	5.9	58	-88.3	5.9
888	56.2992	82.5674	2013.873	12.187	15.102	12.70	313.2	68.2	-72	5.9	68.1	-69.3	5.9
889	56.3439	37.9778	2013.3825	13.439	13.732	13.39	92.6	29.3	-60.9	5.5	27.1	-61.4	5.5
890	56.3609	14.0540	2013.391	9.189	14.712	9.45	261.1	66	-81	5.1	66.2	-76	6.2
891	56.3950	30.0930	2013.167	16.344	16.586	3.89	213.6	71.8	-24.5	5.5	72.6	-21.1	6
892	56.6242	37.2884	2013.1685	17.903	18.133	42.28	131	69.5	-15	5.8	70.3	-6.4	5.8
893	56.6730	11.6674	2013.4855	13.134	15.624	9.36	27.3	75.5	-34.1	6.4	77.6	-35.3	6.5
894	56.8200	-4.7382	2013.7585	12.646	15.121	5.99	339.5	113.8	-34.1	5.5	111.7	-32.8	5.6
895	56.8460	37.5377	2013.4595	15.287	15.879	4.95	160.6	23.6	-99.7	5.5	23.6	-95.4	5.5
896	56.8563	15.3273	2013.157	16.854	18.163	37.68	195.8	-19.1	-114.5	5.3	-17	-110.1	6.2
897	56.8622	14.5793	2013.557	12.751	13.694	11.92	307.8	63.8	-54.9	5.1	63.8	-53.5	5.1
898	56.8676	-7.0782	2013.3195	16.624	16.651	11.90	131.7	28.2	-71.4	6	35.7	-66.9	5.9
899	57.0146	23.7384	2013.0355	12.445	17.537	7.17	162.8	33.7	-128.9	5.3	34.2	-125.9	5.6
900	57.0562	22.3140	2013.235	11.004	15.967	35.29	62	65.9	-24.2	5.2	65	-24.4	5.3
901	57.1408	-3.0594	2013.3175	15.751	16.887	7.64	68	63.2	-28.8	5.5	65.8	-31.6	6.2
902	57.2756	18.8366	2013.408	11.801	13.868	9.75	295.4	105.2	-23.1	5.3	105.8	-24.3	5.3
903	57.2851	59.6281	2013.448	13.621	14.259	6.43	345.9	140.5	-83.9	5.6	136.6	-81.6	5.6
904	57.4166	34.4182	2013.296	15.157	15.969	12.23	77.2	-67.4	-40.7	5.3	-67.2	-41.4	5.3
905	57.4542	34.7570	2013.265	16.681	17.129	6.78	157.4	65.8	23.4	5.3	66.6	15.6	5.4
906	57.4900	0.1468	2013.7235	13.343	14.149	4.55	152.9	74.2	11.4	6.1	72.3	16	7.9
907	57.5993	32.3684	2013.392	16.771	17.588	6.31	21.9	41.8	-60.2	5.4	45.4	-61	5.4
908	57.6741	7.3593	2013.368	16.443	17.377	25.77	299.9	70.3	-25.6	6.2	69.1	-28	6.2
909	57.6953	-2.9100	2013.5585	10.251	15.186	11.31	160.8	76	-49.6	6.3	80.9	-45	6.5
910	57.7473	-9.0730	2013.3125	14.114	16.765	10.81	0.7	1.9	-87.6	5.5	2.2	-88.6	6.3
911	57.8750	43.5221	2013.4185	14.052	14.414	16.93	15.6	-54.4	-81.2	5.9	-54.3	-80.9	5.9
912	57.8928	48.3411	2013.5465	13.183	15.329	7.83	116.6	83.8	-12.7	5.8	83.3	-13.2	5.8
913	57.9763	20.2780	2013.127	12.603	13.841	6.32	65.5	74.7	-13.8	5.3	71.6	-15.7	5.4
914	58.2525	74.6408	2013.581	13.794	18.297	45.98	194.8	10.2	-67.8	6	9.6	-64.6	6
915	58.3140	14.9105	2013.486	11.221	14.814	7.58	94.9	-39.6	-101.4	5.1	-41.5	-103.9	5.3
916	58.3899	16.6931	2013.5765	9.653	17.068	28.91	346.1	66.1	-53.4	5.1	66.2	-57.7	5.2
917	58.4325	61.0141	2013.2655	18.612	18.872	48.36	314.9	67.6	-15.3	6.5	67.4	-3.7	7.1
918	58.4762	-1.1098	2013.5435	14.897	17.007	8.39	173.2	79.9	-138.9	6.4	80.6	-142.2	7.5
919	58.5383	39.6805	2013.22	18.134	18.863	16.91	83	36.7	-60.1	6.6	33.4	-64.7	6.9

920	58.5677	29.4456	2013.3005	12.815	13.374	6.85	115.3	116.4	-49.8	6.1	113.6	-51.3	6.1
921	58.6218	44.2640	2013.495	14.443	15.941	6.43	5.9	-31.5	-94	6	-27	-94.7	6
922	58.7118	-8.3131	2013.908	14.818	15.359	3.93	331	61.1	-86.8	6.5	60.9	-87.2	5.6
923	58.7151	32.9575	2013.7425	15.41	15.475	3.66	126.1	-117.5	-11	6.9	-122.6	-4.4	5.3
924	58.8172	12.9820	2013.393	12.037	15.119	9.14	58.8	-14.9	-153.1	5.2	-13.4	-151.7	5.3
925	58.8776	38.3137	2013.665	8.652	15.902	12.77	245.1	103.7	-53.7	5.5	99.1	-52.5	5.5
926	58.9025	6.5558	2013.4785	13.76	16.61	24.48	336.2	-20	-62.5	6	-18.9	-62.1	6.3
927	59.0214	-7.5270	2013.607	13.092	13.429	7.28	228.2	75.4	37.1	5.6	77.3	40.3	5.6
928	59.0231	15.5373	2013.365	15.605	18.138	14.37	118.5	75.6	-8.1	5.2	73.2	-3.6	6.2
929	59.0868	36.6477	2013.341	14.036	14.438	4.38	74.6	-10	-80.6	5.6	-9.1	-79	5.8
930	59.1024	45.6419	2013.5225	10.239	17.348	19.79	304.9	37.4	-89	5.9	35.2	-91	6
931	59.1896	39.3812	2013.48	10.602	12.908	7.26	134.3	-113.4	1.1	5.5	-111.6	4.7	5.5
932	59.2235	-1.7612	2013.509	12.814	13.439	6.38	116.5	-2.4	-126.1	5.5	-5.8	-132.1	5.6
933	59.2381	52.3126	2013.488	14.567	16.681	35.59	28.9	32.9	-118.5	5.9	34	-118.9	5.9
934	59.2463	24.6538	2013.242	11.243	17.579	50.53	54.4	186.6	-110.9	5.8	187.2	-111.8	6
935	59.2466	28.0055	2013.401	12.456	14.698	6.44	270.3	37.5	-84.6	5.6	36.2	-84.7	5.6
936	59.4970	67.8836	2013.5315	17.622	17.879	59.71	247.5	-64.1	-6.9	5.6	-62.9	-13.1	5.6
937	59.4972	48.1955	2013.1845	16.321	19.098	47.57	172.9	28	-73.8	6.4	40.9	-68.2	7.1
938	59.5710	0.4805	2013.579	16.572	16.6	4.43	332.1	70.4	-52.4	6	73.2	-53.3	6.2
939	59.7650	17.4283	2013.603	10.993	14.035	6.61	332.8	-81.1	-109.5	5.1	-83.3	-110.5	5.4
940	59.8140	45.9999	2013.498	14.12	15.822	49.92	59	50.8	-133.7	6.1	51.7	-131.3	6.1
941	59.8248	58.3947	2013.546	14.761	16.5	7.14	40.8	115.6	-183.9	5.6	115.6	-179	5.7
942	59.9687	-9.4873	2014.0445	12.029	13.76	7.15	166.3	15.4	-101	5.4	11.3	-101.1	5.5
943	59.9920	-7.5508	2013.454	10.204	15.842	12.86	193.2	61.2	45.9	5.5	62.2	47	5.8
944	60.0135	20.0367	2013.154	15.622	15.915	10.14	236.5	18.6	-68.3	5.3	20.8	-69.4	5.3
945	60.0192	53.7319	2013.4935	11.034	11.965	25.13	219.4	37.8	-64.3	5.9	38.4	-66.1	5.9
946	60.1506	69.1309	2013.518	16.263	16.366	24.58	330.7	78.7	-97.4	5.6	81.7	-96.8	5.6
947	60.1555	47.9062	2013.42	18.027	18.315	27.22	250.4	2.5	60.1	6	-2.9	60.3	6.8
948	60.2351	77.3856	2013.5365	14.954	15.839	3.92	80.5	23.5	-68.7	5.5	28.5	68.6	6.2
949	60.2661	15.2904	2013.4205	12.226	14.055	5.74	222.7	85.3	-18.8	5.1	86.7	-20.6	5.3
950	60.3386	43.2082	2013.463	14.666	15.249	6.85	81.9	39.8	-70.7	5.9	42.3	-67.9	5.9
951	60.3752	15.6251	2013.4585	12.004	16.247	8.15	219.2	65.9	21.8	5.1	66.1	19.3	5.3
952	60.5725	35.8907	2013.507	12.225	14.768	33.54	199.4	67	-80.6	6.4	67.7	-83.7	6.4
953	60.5836	66.3978	2013.5025	10.04	12.2	13.99	304.9	6.2	-63.9	5.6	5.4	-64.1	5.5
954	60.6471	48.2680	2013.4975	12.418	16.412	17.62	251.9	-71.4	-65.3	5.9	-68.8	-64.5	5.9
955	60.7804	39.3911	2013.304	16.507	16.947	5.75	210.2	11.5	-107.3	5.6	12.6	-104.8	5.6
956	60.9513	48.0383	2013.15	18.241	18.686	52.71	219	41.7	66.7	6.2	48.8	64.1	6.6
957	61.0451	51.4712	2013.502	10.418	14.239	7.03	328.3	48.1	-60.7	5.9	49	-61.6	5.9
958	61.1493	53.6993	2013.5125	17.1	18.491	6.10	211.7	60.7	-4.5	6.6	60.3	-3.6	6.7
959	61.2728	-5.2901	2013.483	13.275	14.67	6.36	303.5	19.1	-67.4	5.4	16.1	-68.1	5.5
960	61.3104	32.5347	2013.426	12.653	13.034	8.43	256.8	61.8	-27.8	6.4	60.9	-28.8	6.4
961	61.4238	29.6892	2013.2665	11.071	17.353	42.40	0.3	57.9	-74.8	6	56.2	-75.5	6.2
962	61.4637	21.6520	2013.4875	10.557	13.57	14.16	84.4	25.9	-111.1	5.2	27.9	-106.6	5.2
963	61.4752	66.2386	2013.447	14.645	15.256	10.37	160.7	20.2	-72.9	5.6	17	-73.5	5.6
964	61.7421	23.7345	2013.3365	14.916	16.112	4.85	33	0.5	-83.7	5.2	-1.5	-82.5	5.3
965	61.8216	-8.7661	2013.7625	8.791	13.15	18.47	251.7	110.1	-75.5	5.4	103.3	-76.8	5.5

966	61.9462	29.5239	2013.386	11.552	15.941	7.77	298.6	6.6	-182.5	6	3.4	-184.3	6.2
967	61.9467	34.3357	2013.4305	15.685	16.375	50.79	50.7	84.2	-34.3	6.4	84.9	-33.8	6.4
968	61.9886	11.8640	2013.541	9.876	14.95	9.21	263	29.2	-71.3	5.9	36.9	-67.8	6.2
969	62.0585	44.3068	2013.442	16.405	16.574	5.30	264.3	60.7	-27.2	5.9	62.8	-23.8	6
970	62.3070	39.2102	2013.386	11.809	16.671	39.25	84.5	-26.5	-64.8	5.9	-28.1	-66.7	6
971	62.4045	31.8361	2013.3655	15.208	17.274	39.19	203.3	99.3	-74.3	6.5	98.7	-72.9	6.6
972	62.4148	1.6786	2013.6375	11.055	13.434	9.56	266.9	-39.9	-65.3	6	-41	-63.1	6.1
973	62.4481	2.3708	2013.685	13.27	17.948	25.30	172.3	110.9	70.1	6.1	118.7	62.2	7.1
974	62.5568	6.5391	2013.429	17.189	17.531	5.14	221.9	1.7	-80.8	6.3	-1.5	-82.3	6.1
975	62.6720	2.7229	2013.3565	11.771	15.847	8.21	286.5	80.4	11.8	6.1	80.3	9.3	6.3
976	62.7114	11.4958	2013.51	15.201	15.285	8.09	142.8	61.3	16.4	6.4	60.7	15.3	6.5
977	62.7230	38.1595	2013.507	10.719	14.53	19.41	223	-24.9	-66.7	6	-29	-68.1	6
978	62.7357	-0.4459	2013.84	11.042	14.094	8.34	149	64.4	29	5.5	64.8	31.2	5.4
979	62.8197	22.2562	2013.359	13.656	16.227	7.93	28.4	63.8	-30.8	5.2	62.6	-25.9	5.2
980	62.8238	-3.1016	2013.7145	13.512	14.259	9.84	183.9	76.1	-44.9	5.4	80.1	-44.2	5.4
981	62.8480	45.7997	2013.464	10.754	17.703	30.43	48.2	57.8	-63.6	6.4	53.1	-68.3	6.1
982	62.9809	51.4829	2013.4925	14.715	14.837	13.59	145.1	32.5	-69.1	5.9	32.9	-72.4	5.9
983	63.0677	2.4190	2013.5005	13.64	15.294	28.55	15.9	132.3	34.8	6.1	130.1	36.6	6.1
984	63.1098	7.5708	2013.663	13.137	15.369	4.93	40.1	91.1	-114	6	92.6	-119.4	6.9
985	63.3020	-10.9295	2013.9835	11.073	11.414	5.04	302.4	40.4	-176.7	5.3	34.5	-179.3	5.5
986	63.3924	59.2082	2013.4315	15.376	15.43	6.41	309.3	93.8	-0.9	5.6	90.7	-1	5.7
987	63.4100	38.1910	2013.529	11.579	14.177	8.97	310.3	77.7	-116.6	6.4	76.1	-120.7	6.4
988	63.4133	61.6827	2013.429	17.956	17.996	4.72	349.9	100.1	-23.4	6.3	101	-24.1	6.2
989	63.4511	7.6830	2013.547	13.545	14.59	7.76	165.7	0.1	-62.7	5.9	0.3	-62.1	6
990	63.5234	77.1264	2013.5625	12.539	16.455	8.44	142.4	-73.7	8	5.6	-71.1	16.1	5.6
991	63.5277	25.9854	2013.5345	11.475	12.256	4.73	302.1	32.1	-68.5	5.2	31.4	-68.5	5.2
992	63.5309	-9.8262	2013.882	14.776	14.808	6.89	156	84.4	23.8	5.5	86	25.2	5.5
993	63.5541	65.8001	2013.399	18.476	18.788	44.56	16.8	-68.9	-34.4	5.9	-63	-38.7	6.4
994	63.6034	78.3503	2013.647	14.726	17.729	17.37	183.9	-5.9	-64.6	6.2	-4.2	-63	6.3
995	63.6087	27.8880	2013.2205	16.115	16.724	6.19	10.8	102.5	-4.5	5.3	105.2	-3.6	5.4
996	63.6692	27.9091	2013.425	15.173	15.506	48.60	128.7	87.5	-27.5	5.2	89.4	-26.8	5.2
997	63.6799	43.5522	2013.4785	10.451	15.967	21.19	343.5	48.4	-68.7	5.9	48.1	-69.3	5.9
998	63.7957	35.0103	2013.4685	14.387	15.242	51.99	210	100.8	-35.1	5.2	98.9	-34.9	5.2
999	63.8031	33.4292	2013.575	13.419	13.936	7.16	52.6	64.9	-92.2	5.2	65.7	-92.9	5.2
1000	63.8667	65.1446	2013.4785	9.319	15.362	16.60	149.4	-32.7	-72.5	5.6	-37.7	-66	5.6
1001	63.8725	13.8097	2013.4505	16.17	16.902	4.80	204.8	-22.9	-108.6	5.3	-25.1	-107.4	5.5
1002	63.9241	36.6565	2013.484	13.08	14.492	7.97	206.9	41.9	-64.6	5.3	40.9	-64.3	5.2
1003	63.9768	1.0061	2013.4465	15.696	16.615	7.50	157.3	-4.4	-75.3	6.1	-5.2	-76.9	6.7
1004	63.9963	51.4525	2013.483	10.754	13.883	21.23	276.2	15.2	61.1	5.6	18.8	62.1	5.6
1005	64.0517	59.6704	2013.485	14.122	15.279	5.28	279.2	33.1	-63	5.9	35.2	-61.3	6
1006	64.1354	-6.0409	2013.3185	11.596	17.249	25.39	35.6	104.2	-28	5.5	106	-38.2	6.6
1007	64.2233	-5.1350	2013.68	11.638	13.695	7.22	51.6	90.8	-2.1	5.4	88.1	-3.8	5.5
1008	64.2948	36.5376	2013.339	17.551	17.84	46.90	140.3	-76.1	-20.9	6.5	-80.2	-8.8	6.6
1009	64.2971	52.9418	2013.3695	10.29	17.165	53.28	311.5	15.7	-94	5.6	9.2	-90.7	5.8
1010	64.4510	17.4180	2013.5545	16.912	17.69	37.69	193.1	-11.7	-60.2	5.1	-5.9	-63.2	5.3
1011	64.4797	39.3668	2013.4485	11.98	14.105	33.23	179.8	-2.2	-122.3	5.3	-1.5	-124.8	5.3

1012	64.6255	59.6321	2013.3615	16.318	16.975	56.79	198	-29.6	-72.2	6	-24.8	-74.6	6
1013	64.6437	51.0253	2013.546	17.944	18.167	8.13	119.7	-25.1	-62.6	5.9	-20.6	-61.9	6.1
1014	64.6843	3.1489	2013.781	7.864	14.334	16.57	323.6	62.7	-36.1	5.8	61.3	-35.4	6
1015	64.7689	60.4628	2013.4915	15.101	16.461	6.10	264.4	15.8	-62	5.6	16.9	-63.7	5.6
1016	64.8021	-6.4651	2013.8875	11.863	14.615	6.99	114.5	42.6	-97	5.4	43.3	-98.8	5.5
1017	64.8908	33.2417	2013.488	14.115	15.646	52.46	251	3.3	-80.1	5.5	3.7	-77.3	5.5
1018	64.9332	30.3298	2013.432	11.741	14.469	5.58	320.3	49.1	-64.7	5.5	48.4	-67.3	5.6
1019	64.9892	55.4684	2013.4785	12.073	16.078	17.65	18.3	69.3	-17.9	6.3	67.9	-11.4	6.3
1020	65.0469	27.7993	2013.426	13	15.396	10.21	333	-10.8	-82.7	5.2	-12.7	-81.2	5.2
1021	65.1065	46.6241	2013.469	17.093	17.63	34.30	223.2	39.8	-67.2	6	42.6	-67.4	6.1
1022	65.1241	7.6566	2013.448	16.097	17.207	5.27	52.3	34	-61.3	6.1	33.4	-62.8	6
1023	65.1356	29.5699	2013.479	11.604	14.095	14.61	318.4	72	-41.7	5.2	69.9	-41.2	5.2
1024	65.1766	-2.5175	2013.9335	14.199	15.261	5.68	7.8	34.3	-69.5	5.3	31.2	-69.6	5.4
1025	65.1930	63.6493	2013.263	14.77	19.291	25.69	310	147.4	-114.5	5.6	148.4	-118.9	7.3
1026	65.3201	39.0611	2013.5125	18.417	19.075	30.69	126.3	71.8	-26.8	6.5	69.3	-36.9	6.7
1027	65.3621	60.8717	2013.378	9.923	17.697	19.16	53.3	11	-177.5	5.6	8.9	-181.8	5.9
1028	65.4186	36.9590	2013.36	11.794	16.503	10.87	42.4	49.6	-73	6	50.5	-70.9	6
1029	65.4353	20.4028	2013.408	17.164	17.751	33.49	140.7	99.8	-39.7	5.3	100.8	-36.6	5.4
1030	65.6575	60.4474	2013.4745	18.104	18.808	49.22	334.4	79	-2.6	5.7	81.6	-12.6	6.2
1031	65.7030	22.8316	2013.662	9.872	15.46	14.00	16	-59.2	-74.7	5.4	-56.8	-77.5	5.6
1032	65.7741	69.9403	2013.6365	14.799	19.167	24.72	314.6	13.9	-92.6	6.3	20.9	-93.5	7.6
1033	65.8158	65.5101	2013.5095	15.548	15.662	25.32	20.5	21	-61.8	5.6	20.6	-61.3	5.6
1034	65.8567	-3.4984	2014.009	9.014	15.07	9.36	160.2	61.8	31.6	5.3	64.3	24.5	5.5
1035	65.9836	29.2565	2013.3325	15.993	16.665	4.70	56.3	89.2	1	5.3	86.6	-1.4	5.3
1036	66.0063	12.6530	2013.287	12.038	18.036	15.80	55.8	-52.3	-76	5.1	-51.3	-75.6	7.4
1037	66.0391	33.7828	2013.4795	13.179	18.194	30.99	121.9	-24.1	-101.6	5.9	-23.9	-103.8	6.5
1038	66.0398	30.3207	2013.562	13.113	13.442	9.09	276.6	-16.1	-99	6	-16.9	-97.3	6
1039	66.2551	46.1596	2013.5105	14.573	14.699	6.91	13.6	53.5	-76.9	5.5	56.2	-74.5	5.5
1040	66.2600	50.4759	2013.435	12.089	18.215	12.65	211.8	64	-67.8	5.8	63.2	-71 8	6.3
1041	66.5131	31.4114	2013.387	14.96	17.81	32.62	274	20	-83.8	5.2	21	-85.3	5.4
1042	66.5181	69.3224	2013.294	12.12	14.185	4.15	189	35.9	-84.5	6.3	37	-82.1	6.8
1043	66.6510	-1.2502	2013.7155	15.119	16.343	8.88	147.6	-13.4	-65.6	5.4	-14.2	-63.4	6
1044	66.7144	29.3860	2013.211	16.981	17.738	53.09	344.1	73.9	-22.8	5.3	76.7	-21.4	5.4
1045	66.7896	39.3198	2013.4375	11.278	16.497	13.25	240.4	62.7	-12.7	5.8	63.1	-13.9	6
1046	66.8426	29.0059	2013.2545	14.234	18.513	22.73	76.6	81	-74.3	5.2	84.7	-74	6.7
1047	66.9613	-1.2433	2013.976	15.24	15.279	5.96	201.8	35.6	-76	5.3	35.7	-73.9	5.7
1048	66.9814	-4.8346	2013.9745	16.026	16.474	17.26	86.5	93.1	37.7	5.8	98.1	36.2	6.8
1049	67.0118	40.1826	2013.572	10.107	16.166	15.03	53.8	18.9	-76.6	5.8	12	-80.2	5.9
1050	67.0260	25.4412	2013.484	16.184	18.512	7.08	108.6	0.3	-67.4	5.3	4.8	-64.4	6.2
1051	67.2300	11.3587	2013.6375	10.222	16.399	11.34	35.1	-17.8	-70.8	5.8	-15.9	-71.3	6.1
1052	67.3455	12.7935	2013.5095	12.814	15.262	7.99	213.5	64.4	-33.2	5.8	68.4	-33.2	6
1053	67.4541	47.1441	2013.558	14.877	15.913	4.56	90.9	28.6	-108.6	5.5	27.1	-107.6	5.5
1054	67.4884	56.1545	2013.473	13.98	15.238	10.88	9.9	50.4	-68	5.6	45	-73.3	5.6
1055	67.6801	44.1170	2013.006	18.712	18.981	26.79	137.4	69.3	19.2	6.6	72.3	16.1	6.3
1056	67.8553	55.5402	2013.407	16.953	18.833	23.13	252.2	77.7	16.5	6.5	77.5	22.7	7.5
1057	67.9386	19.8653	2013.7515	16.184	17.882	22.50	168.8	76.8	90.6	5.8	76.2	92.7	6

1058	68.1411	57.8138	2013.57	16.939	18.495	26.84	302.4	-64.7	8.5	5.6	-60.4	17.9	5.6
1059	68.4959	6.2108	2012.943	17.004	17.162	11.30	29.2	114.5	-74.3	6.6	115.7	-81.2	7.1
1060	68.5502	26.0748	2013.531	14.334	16.035	7.70	353.2	40.6	-60.5	5.2	39.7	-62.7	5.2
1061	68.5639	41.2170	2013.2655	18.086	18.311	49.95	140.8	-61.7	-27.4	5.7	-64.9	-16.6	7
1062	68.7205	43.9391	2013.4225	17.104	18.006	16.40	213.1	124.8	-23.4	5.6	121.9	-34.1	5.7
1063	68.7417	86.5020	2013.8985	14.713	16.13	27.36	196.9	41.1	-68.6	5.9	42.8	-68.7	5.9
1064	68.8279	74.6918	2013.5315	9.874	13.239	9.53	128.2	93.9	-88	5.6	88.1	-89.9	5.6
1065	68.8820	63.7889	2013.4995	11.979	13.304	6.43	305.6	-78.5	6.3	5.5	-80.4	10	5.6
1066	68.8849	53.4249	2013.4485	17.177	17.562	57.24	350.7	40.4	-73	5.9	33.2	-78	7.8
1067	68.9737	-13.7799	2014.0485	9.636	14.217	8.66	316.2	14.2	-94.5	5.4	20.5	-91.6	5.4
1068	69.1032	6.7959	2013.32	11.602	13.191	5.14	182.7	61.2	-49.2	5.9	62.4	-47.9	7.2
1069	69.1614	79.1166	2013.6165	11.177	14.728	9.04	4.8	31.9	-87.5	5.9	30	-89.6	5.9
1070	69.2553	5.7902	2013.7495	11.062	13.019	9.33	30.7	-88.7	-6.2	6	-87.5	-9.7	6.1
1071	69.2956	56.6012	2013.515	14.954	15.632	4.55	58	63.2	-40.2	5.5	62	-42.8	5.6
1072	69.3100	30.3999	2013.85	7.65	10.793	6.04	57.1	32.6	-120.7	5.8	41.7	-115.3	6
1073	69.3608	29.2208	2013.5615	14.059	14.456	8.63	12.8	79.2	-122	5.5	76.9	-122	5.5
1074	69.4654	29.8027	2013.492	15.308	16.784	28.47	203.1	66.3	-1.8	5.5	64	-1.4	5.5
1075	69.5051	44.4564	2013.5355	8.743	16.402	41.47	306.5	39.2	-63.1	5.5	40.4	-62.5	5.5
1076	69.5230	2.7464	2013.3175	15.47	15.861	5.22	238.9	16.8	-73.3	6.2	20.9	-71.6	6.5
1077	69.5322	29.0721	2013.4715	18.172	18.498	44.06	185.8	63	9.4	5.6	63.4	-1	5.9
1078	69.6207	35.9501	2013.459	16.088	17.261	10.11	24.6	46.2	-68.1	5.5	48.9	-68.4	5.5
1079	69.8123	44.2766	2013.581	9.168	13.935	14.17	144.6	100.2	-184.3	5.5	100.5	-175.9	5.5
1080	69.8182	39.1102	2013.5225	15.577	17.184	9.43	235.6	200	15.7	5.5	202.9	16.5	5.6
1081	69.8493	63.3878	2013.479	10.985	15.169	7.25	251.5	40.6	-90.3	5.6	39.3	-87	5.6
1082	69.9799	62.6881	2013.387	15.479	15.482	4.77	168.8	-6.9	-86.6	5.6	-4	-84.8	5.6
1083	70.2490	41.2712	2013.2495	17.447	18.189	29.65	182.7	-4.5	-75.8	6.1	-3.6	-74.8	7.1
1084	70.2929	48.0145	2013.47	11.459	16.68	53.80	62.9	78.1	-22.8	6	76.2	-23.4	6
1085	70.3801	39.6899	2013.5135	15.724	16.941	9.27	67	-0.3	-88.7	5.5	-0.1	-88.9	5.5
1086	70.4324	24.2275	2013.5045	17.016	18.299	9.51	23.1	48.4	-93.4	5.7	47.1	-96.3	6.4
1087	70.4732	54.7219	2013.45	15.576	16.398	4.99	279.4	-3.4	-76.1	5.6	-4.8	-75.2	5.6
1088	70.5324	-10.8097	2013.8845	14.848	15.734	5.83	355.7	75.3	-18.8	5.5	80.3	-13.2	5.5
1089	70.5495	35.5614	2013.488	18.469	18.693	5.98	282.4	-70.7	2	7.6	-66.5	-11.4	6.2
1090	70.5579	11.6949	2013.2445	16.161	17.753	5.94	323.4	96.3	-19.1	6.1	101	-15.6	6.8
1091	70.5758	16.5090	2013.4385	12.217	18.223	8.88	93.2	89.2	-134.8	5.1	87.9	-130.3	7.5
1092	70.5762	43.8556	2013.4985	14.145	15.41	54.51	153.7	8.3	-63.8	5.8	10.5	-66	5.9
1093	70.6176	9.7687	2013.5525	14.977	16.396	7.64	216.5	-21.3	-100.7	6	-18.2	-103.4	6.3
1094	70.6298	70.0026	2013.6485	17.73	18.672	43.50	223.2	8.1	65.6	6.3	11.9	68.4	6
1095	70.7857	22.4438	2013.736	14.359	18.393	16.09	27.4	210.1	-109.2	6.3	202	-112.2	7.5
1096	70.8125	30.2506	2013.7525	18.065	18.636	23.45	268.5	33.7	-129.8	7.1	45.8	-119.1	6.9
1097	70.9032	33.2573	2013.536	15.163	15.5	6.90	328.7	34.8	-83.1	6	35.4	-81.7	6
1098	70.9671	46.4250	2013.6215	14.367	18.951	30.28	82.4	-3.8	73.3	5.8	5.3	76.9	7.6
1099	70.9791	24.1212	2013.656	14.718	14.802	59.97	188.7	30.6	-92.2	5.5	28	-91.4	5.5
1100	71.0282	29.5909	2013.621	14.2	16.718	25.01	59.2	39.9	-93.9	5.5	39.9	-96.3	5.5
1101	71.3554	7.5916	2013.7485	8.451	13.732	21.06	56.1	14.2	107.6	5.8	7.6	103.7	6.1
1102	71.4175	56.7216	2013.5235	12.99	15.467	6.69	118.3	75.4	-93.2	6	78.6	-95.7	6
1103	71.5081	43.6266	2013.4595	11.934	17.915	17.21	289.3	89.4	-33.3	5.8	89.2	-30.4	6.1

1104	71.6492	27.2890	2013.4895	14.165	17.209	12.74	131.3	59	-93.5	5.2	62.5	-91.5	5.6
1105	71.8910	40.6564	2013.594	11.836	15.449	58.58	26.4	49.7	-113.6	5.4	50.7	-112.9	5.5
1106	72.0320	68.1574	2013.468	12.529	16.106	11.38	210.7	76.3	-28.9	5.9	76.2	-27.2	5.9
1107	72.1002	30.2719	2013.58	14.903	15.525	5.98	188.2	-0.2	-81.5	6	-4.1	-81.4	6
1108	72.1720	16.6343	2013.7245	7.49	13.403	13.48	68.4	22.5	-78.5	5.1	18.8	-82.7	5.2
1109	72.2069	39.5782	2013.498	15.996	16.33	32.49	316.4	37.7	-86.6	5.5	32	-87.5	5.5
1110	72.2099	75.9412	2013.543	6.177	12.248	13.26	83.1	30.6	-135	5.6	28.7	-132.3	5.6
1111	72.4252	56.6294	2013.495	13.388	17.951	24.25	19	5.1	-67.9	5.5	7.5	-69.7	5.7
1112	72.4471	48.4072	2013.535	14.386	16.862	40.23	310.9	68.9	-75.4	5.9	66.7	-71	5.9
1113	72.4723	49.7184	2013.525	15	16.987	5.60	149.6	73	-25	5.9	66.9	-32.3	6
1114	72.5316	43.4273	2013.525	15.409	16.339	8.53	242.3	12.3	-64.9	5.5	14.1	-66.4	5.5
1115	72.6109	43.2769	2013.4735	16.134	16.615	30.98	3.2	27.8	-120.8	5.5	27.6	-118.2	5.5
1116	72.6198	64.4901	2013.4615	12.54	16.15	15.14	28.5	74.3	-114.3	5.6	71.5	-112	5.6
1117	72.7184	22.0566	2013.506	13.662	16.288	14.55	195.6	-22.1	-134	5.2	-23.2	-132.1	5.3
1118	72.7883	75.9497	2013.4995	12.38	14.957	13.32	271.1	-20.6	94.5	5.6	-17.3	94.4	5.6
1119	72.7952	46.7324	2013.538	11.833	14.757	7.92	222.2	159.2	-75.5	5.5	160.5	-76.5	5.5
1120	73.0175	56.4233	2013.6385	7.976	15.573	16.91	297.5	-8.8	-63.8	5.5	-4.6	-61.6	5.5
1121	73.0379	7.8676	2013.4595	13.954	16.831	8.99	34.4	32.1	-66.6	5.9	32.4	-67.1	6.3
1122	73.0511	72.3497	2013.4835	16.272	17.963	13.20	97	43.3	-159.7	5.7	43.8	-154.8	5.8
1123	73.0757	32.3334	2013.789	11.055	14.622	10.01	105.6	87.2	-72.1	5.1	90.5	-72.5	5.1
1124	73.0775	-6.2801	2013.952	11.787	15.345	13.30	307.7	-11.7	-142	5.3	-12	-140.6	5.4
1125	73.1656	48.8602	2013.541	8.559	12.174	27.26	266.6	46.1	-62.2	5.6	45.1	-61.5	5.6
1126	73.2004	-9.9599	2014.047	14.217	14.421	3.30	333.4	-43.8	-88.2	5.5	-41.3	-88.1	5.9
1127	73.2721	14.8963	2013.788	9.465	14.355	17.47	177.4	62.2	-86.1	5	56.9	-94.7	5.1
1128	73.3031	28.7555	2013.687	13.782	15.008	6.51	177.7	61.9	-42.5	5.1	63.1	-43.8	5.2
1129	73.3101	-4.5445	2013.8945	11.161	11.512	7.09	119.9	79.3	19.2	5.3	79.7	18.8	5.3
1130	73.3170	28.6174	2013.5945	17.103	17.307	12.98	232.3	-6.3	-94	5.2	-4.3	-93.3	5.3
1131	73.4506	48.2567	2013.605	15.894	17.847	57.72	295.5	45.5	-70.8	6.2	34.6	-78.2	6
1132	73.4629	47.3158	2013.4015	11.889	15.931	6.91	354.5	65.5	-86.5	5.8	65.9	-87.9	6.2
1133	73.5091	8.4427	2013.4895	14.067	16.017	6.51	314.3	12.1	-73.6	5.9	4.2	-76.3	6.4
1134	73.5443	40.7850	2013.551	10.524	13.622	12.09	201.3	-74.5	-52.1	5.4	-77.5	-51.5	5.4
1135	73.5798	64.2212	2013.5565	10.34	14.568	16.15	194.6	77.4	-16.6	5.5	77.9	-16.4	5.5
1136	73.6247	41.9628	2013.565	17.807	18.287	26.15	349.7	-68.9	6.7	6.5	-67.6	-2.6	6.6
1137	73.6734	31.3766	2013.5335	14.65	15.736	4.25	275.4	-21	-90.1	5.1	-16.7	-88.6	5.5
1138	73.7183	31.7198	2013.7715	16.677	18.217	32.27	274.2	58	73.6	6.1	59.3	68.2	5.4
1139	73.8363	32.2006	2013.69	11.406	13.42	41.85	280.3	62.8	-76	5.2	61	-76.7	5.2
1140	73.8498	19.8305	2013.632	12.726	12.744	5.44	216.3	64	-39.3	5.2	63.3	-43.8	5.1
1141	73.8948	41.5494	2013.5265	12.066	12.877	6.33	153.2	94	-99.1	5.5	93.3	-100.6	5.5
1142	73.9200	52.9390	2013.4495	14.73	15.939	55.35	5.7	-7.9	-77.2	5.9	-9.9	-75.7	6
1143	74.0527	62.6798	2013.485	15.042	15.986	4.75	230.8	63.6	-69.6	5.6	68.2	-69.5	5.7
1144	74.2499	63.0727	2013.469	14.353	14.4	4.96	258.7	39.5	-66.1	5.6	42.4	-66	5.6
1145	74.2613	15.5125	2013.678	15.99	16.361	4.95	118.9	69.8	38.1	5.2	72.9	37.6	5.4
1146	74.3048	7.1125	2013.573	11.231	13.505	8.80	266.1	-62	-78.1	6	-61.7	-79.1	6
1147	74.4207	15.7193	2013.564	13.95	16.67	7.03	137.6	46.2	-88	5.8	51.9	-84.4	6
1148	74.4377	46.5375	2013.5515	18.11	18.241	52.15	40.5	76.7	28.8	5.6	79.4	33.1	5.8
1149	74.5497	35.7366	2013.556	15.213	18.827	53.19	222.8	7.3	-73	5.2	9.3	-71.8	6.9

1150	74.6301	28.8221	2013.6395	16.055	16.994	6.15	113.3	17.2	-97.7	5.2	16.8	-93.5	5.2
1151	74.6814	10.5210	2013.7465	12.543	13.398	5.90	320.4	79.2	-62.5	6.2	77.5	-61.4	6.4
1152	74.7596	-14.1639	2014.05	13.327	15.027	7.30	202.9	13.7	-100.4	5.5	8.1	-99.3	5.5
1153	74.8723	66.4019	2013.556	9.979	13.111	47.88	58.9	21.6	-82.8	5.9	29.9	-76.8	5.8
1154	75.0380	51.1304	2013.3745	15.235	16.85	4.10	190.9	68.7	-31.4	5.9	73.4	-28.4	6.7
1155	75.1017	45.7385	2013.5795	11.908	12.142	41.28	57.5	62.1	-64.6	5.9	59.2	-62.5	5.8
1156	75.1545	13.4067	2013.3985	17.763	17.961	45.91	242.3	28.3	63.7	6.7	15.9	71.5	7.1
1157	75.1844	13.7862	2013.0455	18.059	18.376	43.55	42.2	-34.7	-68.2	7.1	-28	-68.4	7.9
1158	75.2582	39.5075	2013.4865	11.697	15.48	7.95	134.3	42.2	-99.3	5.4	43.6	-100.3	5.5
1159	75.2668	-0.3611	2014.0365	11.952	15	49.71	331.5	17.3	-63.7	5.3	18.1	-63.2	5.3
1160	75.2945	23.4082	2013.692	12.34	15.948	8.05	53.6	35.4	-82.3	5.1	34	-82.6	5.1
1161	75.3026	62.4673	2013.5525	13.501	13.713	4.66	5.5	80.9	26.7	6	79.8	20.2	6
1162	75.3915	40.5518	2013.325	17.957	18.477	58.45	223.5	-63.5	16.9	5.7	-61.5	21.8	6.1
1163	75.5417	57.1991	2013.653	15.305	17.291	4.84	45.7	66.7	-73.7	5.5	62.7	-76	6.4
1164	75.5568	48.3820	2013.534	13.145	15.855	15.36	93.4	37.3	-75.2	6.3	37.1	-75.6	6.3
1165	75.6730	21.3065	2013.5925	11.617	14.072	6.68	155.2	61.6	-61.8	5.1	62.6	-61.8	5.2
1166	75.8262	36.8909	2013.6525	14.986	15.975	5.58	96.4	67.8	-48.6	5.4	64.2	-48.7	5.4
1167	75.8302	-4.2840	2013.956	11.635	16.059	28.99	304.4	122.4	-11	5.4	124.5	-9.2	5.6
1168	75.9246	38.6000	2013.609	14.983	17.282	14.56	83.5	28.6	-69	5.4	31.5	-71	5.5
1169	75.9711	46.0516	2013.4155	17.307	17.748	42.43	270.6	-65.4	-70.1	6.6	-57.9	-74.1	6
1170	76.2998	54.7692	2013.5765	11.907	15.135	14.52	267.1	63.9	-68.2	5.5	64.4	-66.7	5.5
1171	76.3230	51.0187	2013.5555	9.414	12.385	13.01	328.1	72.2	7.1	5.5	69	8.1	5.5
1172	76.5337	-1.8287	2013.988	15.058	15.149	8.55	173.4	102	-18.1	5.4	103.6	-15.8	5.3
1173	76.6618	45.6134	2013.569	14.587	17.182	8.71	272	-5.8	-100.7	5.9	-6.2	-104.8	5.9
1174	77.0395	33.9777	2013.6885	17.728	18.268	44.66	152.2	-24	70.9	6	-22.7	73.4	6
1175	77.1800	36.2516	2013.823	17.21	18.246	32.68	224.4	-101.9	33.6	6	-100.1	22.2	5.7
1176	77.2144	34.6150	2013.8295	18.096	18.446	52.60	300.2	-12.9	-61	6	-5.2	-60.1	6.3
1177	77.2297	44.5560	2013.5035	13.704	14.543	10.49	271.8	3.7	92.5	5.9	2.2	92.4	5.9
1178	77.2305	20.6477	2013.855	10.43	14.194	5.89	318.5	-56.3	-156.6	5.1	-57.6	-162	5.3
1179	77.2367	-11.6763	2013.8765	10.379	11.142	4.65	168	99.6	-150.7	6.2	98.6	-148.4	6.7
1180	77.2701	3.6700	2013.6965	11.511	14.29	11.20	220.1	8.3	-65.5	5.9	4.6	-65.9	5.9
1181	77.2833	13.3362	2013.675	12.422	14.416	35.51	122.1	65.4	-42.9	6.3	66.4	-38.1	6.4
1182	77.4466	57.7077	2013.593	17.898	18.393	41.48	37.9	75	-32.6	5.6	73.6	-37.3	6.1
1183	77.5395	16.8426	2013.6355	11.043	16.648	55.82	280.7	-39.4	-91.2	6.2	-36.9	-87.7	6.7
1184	77.5941	36.7811	2013.39	16.726	17.895	54.99	25	-4.1	-69.7	5.4	7.1	-69.4	6.4
1185	77.6391	22.4174	2013.7935	15.417	16.764	16.78	98.8	21.7	-69.6	5.1	18.4	-71.9	5.3
1186	77.7302	39.8306	2013.668	13.045	15.323	8.64	234.2	31.5	-103.8	5.4	29.7	-102.3	5.4
1187	77.7492	40.6541	2013.507	14.555	15.618	6.80	352.1	-60.2	-180.8	5.5	-65.2	-182.2	5.5
1188	77.9180	26.3078	2013.5115	14.055	16.116	6.32	340.1	-26	-73.1	5.1	-24.6	-71.2	5.4
1189	78.1172	8.3820	2013.745	11.227	11.243	5.21	14.2	62.8	32.4	5.8	65.8	26.2	6.8
1190	78.1862	56.4735	2013.5835	17.755	17.801	34.84	183.9	13.4	-63.5	6	14.6	-60.6	5.9
1191	78.2513	40.0664	2013.5295	14.885	15.936	13.53	331.9	48.9	-112.9	5.5	48.6	-113.8	5.5
1192	78.3742	39.6351	2013.683	12.706	17.005	8.28	227.8	2.8	-61.4	5.4	1.8	-60.6	5.5
1193	78.5928	43.8546	2013.465	17.876	18.31	43.68	94.5	17.8	-62.2	6.1	25.8	-61.4	6.1
1194	78.6129	58.6560	2013.5785	11.614	14.976	12.58	152.9	-17.4	-75.5	5.5	-23.3	-76.4	5.6
1195	78.9211	42.3892	2013.7885	8.8	11.553	6.12	143.4	36.5	-122.3	5.9	37.4	-121.9	5.9

1196	78.9471	-12.3650	2014.05	7.999	13.457	24.05	132.6	72.1	90.2	5.4	72.2	83	5.6
1197	79.2939	11.2434	2013.657	13.397	14.554	8.43	10.7	75.5	-49.6	6	75	-53.7	5.9
1198	79.3042	21.3555	2013.7095	11.874	14.62	8.25	49.5	15	-76.7	5.1	15.4	-75.8	5.2
1199	79.3264	44.8142	2013.364	16.755	18.765	58.35	1.3	5.8	73.3	6.3	11.9	69.2	7.6
1200	79.3531	65.8998	2013.4655	16.752	16.848	11.29	125.3	0.8	-75.6	5.6	0.5	-76.1	5.6
1201	79.4084	36.2908	2013.701	14.779	18.438	10.25	247.6	-15.3	-70.6	5.9	-12.5	-68.9	6.1
1202	79.6137	56.6673	2013.494	13.033	16.116	6.62	276.6	-65.8	-28.4	5.5	-67.4	-29.8	5.7
1203	79.6549	64.1596	2013.459	10.001	14.634	7.15	134.6	42	-88.6	5.5	43.3	-85.6	5.8
1204	79.8692	31.3368	2013.695	10.915	15.943	10.89	171.2	-23.2	-107.5	5.1	-21.5	-106	5.2
1205	79.9911	40.6621	2013.6315	14.581	15.373	4.05	158.8	-32.4	-66.4	6.1	-34.8	-61.9	6.2
1206	80.0252	60.7723	2013.5415	11.706	13.692	10.93	344.3	2.2	-66	5.9	-0.5	-63.3	6
1207	80.1434	29.0224	2013.6695	9.354	17.121	34.91	333.9	-79.5	-83.5	5.1	-82.2	-88.1	5.3
1208	80.2380	30.6363	2013.72	11.266	15.182	12.78	218.4	37	-88.6	5.1	37.8	-87.9	5.2
1209	80.2857	36.9100	2013.426	17.956	18.052	55.98	193.9	-7	-64.6	6.5	-2.6	-67.5	6.9
1210	80.2907	-5.9628	2013.976	13.059	13.826	6.03	280.6	-31.3	-114.6	5.3	-28.6	-113.3	5.5
1211	80.2975	41.2938	2013.6585	12.241	16.418	7.54	123	5.4	-72.8	6.3	-2	-72.4	6.6
1212	80.3131	2.0226	2013.735	11.852	16.353	21.82	250.6	66.4	23.2	5.8	64.6	26.1	6.1
1213	80.3355	47.4800	2013.501	15.057	15.836	3.69	35.2	53.6	-100.7	5.9	46.6	-100.4	7.1
1214	80.3406	58.4057	2013.533	11.244	13.73	6.29	47.6	8.7	-71.4	5.5	7	-73.3	5.6
1215	80.3456	28.4885	2013.701	11.773	12.434	9.71	87.2	122	-24.7	5.1	121.4	-17.1	5.1
1216	80.4251	20.0541	2013.7345	11.873	14.774	6.51	184.1	24.7	-82.6	5.2	26.2	-82.4	5
1217	80.4767	8.8182	2013.5585	15.47	16.729	52.19	173.1	-6	-73.5	6	-4.7	-74.6	6.2
1218	80.4832	20.9964	2013.5125	17.832	18.837	24.88	220.9	18.1	-69.7	5.5	11.8	-71.2	6.3
1219	80.5201	37.4948	2013.5355	13.319	15.896	6.03	36.6	31.4	-85.8	5.2	30.7	-85.3	5.4
1220	80.6119	59.4936	2013.4765	16.63	17.099	49.05	121.6	18.1	-76.3	5.6	16	-73.3	5.6
1221	80.6658	29.8251	2013.7045	17.778	17.996	39.32	20.8	71.7	19	6.9	74.9	11.6	6.5
1222	80.6738	59.3346	2013.186	13.676	15.709	4.47	304.1	53.5	-94.1	5.5	48.9	-94.3	6.7
1223	80.6808	-8.0317	2014.0395	13.636	15.048	54.03	228.4	36.6	-139.5	5.5	42.3	-136.5	5.4
1224	80.7854	2.8495	2013.886	15.78	16.029	4.44	107.6	69.3	-15.3	5.9	65.7	-16.8	6
1225	80.8249	48.7033	2013.58	15.464	15.819	49.67	151.2	-21.2	-94.9	5.5	-13.7	-96	5.5
1226	80.9892	70.0000	2013.4835	9.667	12.775	7.39	335.5	77.9	-83.6	5.9	78.8	-78.5	6
1227	81.1061	59.3041	2013.5855	17.505	19.039	54.04	290	60.2	-15.9	5.6	60.8	-18.2	7.7
1228	81.2904	34.5213	2013.9665	18.107	18.415	40.41	65.1	-79.6	-33.3	5.8	-79.2	-23.7	7.4
1229	81.3547	11.2182	2013.516	12.785	14.231	6.95	72.5	37.4	-98	6	33.7	-98.5	6.1
1230	81.4557	49.3132	2013.648	15.653	18.426	20.41	56.2	-20.6	-79.7	5.5	-20.1	-79.4	5.9
1231	81.4606	78.3014	2013.607	13.306	14.201	39.90	294.1	-6.9	-69.2	5.6	-5.1	-70	5.6
1232	81.4838	48.2069	2013.5815	13.816	15.995	7.59	6.6	85.2	-103.1	5.5	82.1	-106.7	5.6
1233	81.5097	68.1723	2013.4895	13.792	17.974	14.07	28.3	62.4	-170	6.1	58.3	-167.7	6.3
1234	81.6141	60.9768	2013.486	17.679	18.131	39.09	35.4	-6.8	-70.3	6.1	-0.9	-68.7	6.1
1235	81.6383	14.7270	2013.6125	12.939	15.049	8.97	312.8	38.9	-70.2	5.5	35.5	-68.7	5.4
1236	81.7308	20.0211	2013.8615	14.958	15.352	4.00	238.2	-66.9	-111.4	5.1	-68.7	-115.6	5.6
1237	81.7580	10.0439	2013.574	14.659	18.676	57.14	303.9	14.1	-67.2	6	26.9	-62.2	7.6
1238	81.7949	27.4964	2013.661	13.713	15.112	5.99	238.7	-46.6	-105.6	5.2	-44.8	-108.9	5.1
1239	81.8210	50.3696	2013.518	17.956	18.181	47.77	131.5	0.8	71.7	5.6	-7.5	68	5.7
1240	81.8273	9.9120	2013.6195	10.102	13.343	10.54	308.4	55	-82.3	5.9	59	-80.1	6
1241	81.8982	59.8999	2013.5795	12.955	13.087	10.78	357.3	35	-72.5	5.9	35.4	-73.4	5.9

1242	81.9185	60.5666	2013.6045	10.124	15.171	16.86	13.6	-7.7	-143.3	5.9	0.1	-138.7	6
1243	82.0498	50.3909	2013.5975	10.726	12.837	8.69	117	31.7	-64.3	5.5	32.1	-64.1	5.5
1244	82.2276	40.7313	2013.563	16.999	18.664	48.24	148.5	71.3	46	6.1	67.3	50.2	6.2
1245	82.2936	4.0172	2013.9575	13.592	15.424	4.98	44.2	163.2	-12.1	5.9	158.2	-16.5	6.2
1246	82.3079	19.0465	2013.592	12.118	13.739	17.17	54.3	108.7	-16.4	5.2	110.5	-14.6	5.2
1247	82.4362	-7.7597	2014.0445	15.539	15.819	43.44	170.4	-14.9	-104.1	5.6	-10.5	-107.3	5.8
1248	82.4520	31.6083	2013.6205	11.209	14.574	8.41	341.6	43.2	-65.4	5.2	43.2	-64.6	5.2
1249	82.4791	34.1076	2013.6065	16.606	16.997	7.19	153.4	25.4	-93.7	5.2	20.9	-93.5	5.2
1250	82.6157	39.9376	2013.544	11.781	14.956	8.29	191.5	1.4	-66.4	5.2	9.7	-65.5	5.3
1251	82.6562	57.5132	2013.3195	13.864	15.639	4.80	223.5	30.9	-62.6	5.5	28.8	-62.9	5.8
1252	82.6768	57.6931	2013.5255	13.847	17.53	52.99	312.3	-62.5	-45.3	5.5	-60.7	-44.8	5.6
1253	82.8813	10.1628	2013.661	11.743	14.342	21.92	241.9	-28.1	70.4	5.8	-31.6	68.2	5.9
1254	82.9336	37.4325	2013.862	18.346	19.184	37.81	200.7	-73.8	53.6	5.6	-81.8	42.6	5.9
1255	83.0442	17.1586	2013.5125	10.788	17.285	12.04	156.8	84.6	-62.7	5.5	84.3	-68.8	6
1256	83.1216	53.5074	2013.5215	14.045	15.615	7.59	188.4	-11.3	-88.5	5.5	-10.5	-92.9	5.5
1257	83.2672	8.8046	2013.4185	16.863	17.449	9.05	289.2	11.9	-67.5	6.1	2.8	-68	6.2
1258	83.2752	56.2790	2013.5525	13.865	15.569	24.29	6.2	43.6	-72.2	5.5	43.3	-72.5	5.5
1259	83.2960	22.5632	2013.5605	17.072	17.686	34.35	330	-61.6	-11.6	5.3	-63.6	-11.9	5.4
1260	83.3991	7.4765	2013.196	16.496	18.024	17.88	197.6	-11.5	-68.4	6.6	-4.5	-68.8	7.1
1261	83.4604	52.1992	2013.2315	11.404	12.938	4.54	267.4	6.5	-75.3	5.5	16.7	-75.9	5.9
1262	83.5002	-3.6270	2014.054	10.743	15.29	14.30	296.9	-6.5	99.9	6.2	-11.5	100.3	6.2
1263	83.6015	-1.3685	2014	15.539	15.595	25.79	356.1	-6.6	-67.6	6.3	-4.1	-66.9	6.3
1264	83.6494	51.6035	2013.579	14.51	17.325	26.07	101.8	-22.7	-94.7	5.5	-21.5	-92	5.5
1265	83.6594	-2.5714	2014.042	12.406	13.53	5.80	120.3	-58.5	-79.8	6.1	-59.5	-81.2	6.4
1266	83.6874	18.3572	2013.605	13.876	14.256	7.67	133.4	-50.5	-106	5.2	-50.5	-106	5.2
1267	83.7129	71.0913	2013.504	12.278	14.912	5.57	295.8	26.4	-79.1	6	27	-76.6	6.2
1268	83.7326	36.8900	2013.651	9.424	14.102	16.72	103.4	46.3	-82.7	5.4	47.9	-87	5.5
1269	83.8726	57.1155	2013.5095	13.301	14.514	7.42	89.7	-1.1	-77.7	5.5	-0.8	-78.2	5.5
1270	83.9480	34.4857	2013.571	15.696	16.692	8.10	109.7	-54	-76.3	5.2	-54.4	-73.1	5.2
1271	84.0731	33.1001	2013.6115	12.228	13.27	9.26	124.1	42.2	-73.2	5.2	42.9	-76	5.2
1272	84.4560	55.0363	2013.4965	10.623	18.041	20.26	275	29.7	-82.9	5.5	26.2	-81.4	5.7
1273	84.5421	-15.9062	2013.171	12.825	13.791	6.58	128.8	15.2	-65.1	5.9	14.2	-63.1	6
1274	84.5579	49.3164	2013.5585	11.626	14.116	7.43	86.6	7	-82.1	5.5	7.9	-80.2	5.5
1275	84.6063	48.3362	2013.57	13.493	15.73	6.52	166.2	-8.7	-71.6	5.5	-6.1	-75	5.5
1276	84.6183	35.8919	2013.6435	9.788	14.73	50.81	170.7	-34.1	-63	5.4	-34.6	-61.5	5.5
1277	84.6960	39.4427	2013.5365	13.285	13.688	8.05	224.3	42.8	-84.7	5.5	45.3	-85.2	5.5
1278	84.7622	37.0592	2013.5275	16.449	17.477	19.83	3	7	-84.3	5.5	2.6	-82.5	5.5
1279	84.8493	20.1929	2013.258	12.817	13.195	4.01	204.5	63.6	-29.4	5.2	66.3	-20.8	6.3
1280	84.8744	80.6247	2013.7155	13.104	18.472	13.85	305.6	-70.2	34.9	6.2	-72.9	32	6.7
1281	84.9382	11.9418	2013.334	14.192	16.584	6.83	296	-32.6	-134	5.9	-35	-133.8	6.2
1282	84.9657	34.6182	2013.574	15.611	16.831	23.01	37.3	9.4	-70.4	5.2	9.8	-67.6	5.2
1283	85.0623	36.1895	2013.6075	11.987	12.7	8.52	6	93.9	-36	6.3	92.5	-33.6	6.3
1284	85.1706	36.3466	2013.6335	14.807	15.248	7.61	296.7	-133.4	-48.5	6.4	-133.9	-44.5	6.3
1285	85.1860	58.8224	2013.5645	13.374	16.087	11.56	240.6	-61.5	-38.6	5.5	-63.9	-40.2	5.5
1286	85.1898	19.1318	2013.6045	12.104	14.192	24.98	254.9	-4.7	-85.9	5.1	-4.1	-87.3	5.1
1287	85.2468	3.5499	2013.8775	15.969	18.112	32.57	287.9	27.3	-94.7	6	16.4	-94	7.3

1288	85.2626	43.6296	2013.5175	17.78	17.844	42.46	216	61.7	-16.3	5.9	60.2	-15	6
1289	85.2973	-0.2175	2014.006	14.56	15.918	31.06	75.2	30.2	-79.3	5.3	26.6	-77.1	5.5
1290	85.3048	9.5490	2013.8395	8.327	14.651	24.94	132.6	-28.7	-69.3	5.6	-29.8	-67.2	5.8
1291	85.3360	49.5999	2013.5855	17.193	17.906	57.86	116	60.5	-23	5.5	60.5	-19.2	5.6
1292	85.4187	56.1440	2013.5305	10.859	16.647	26.38	20.3	-15.7	-74.7	5.9	-18.6	-72.1	6
1293	85.5392	62.6190	2013.5945	11.368	14.583	34.62	349	21.4	-62.4	5.8	23.4	-62.8	5.8
1294	85.6595	29.8478	2013.6345	14.135	14.82	5.97	75.6	82.2	-52.4	5.2	80.2	-52.4	5.2
1295	85.8567	55.1747	2013.4915	17.878	19.259	28.05	303	2.4	-60.7	5.6	-3.2	-62.6	7.4
1296	85.9169	23.7485	2013.3915	15.41	16.367	8.34	166.4	138	3.7	5.2	137.1	1.7	5.3
1297	85.9689	42.2377	2013.5145	11.354	14.11	17.17	182.5	-12.7	-65.7	5.9	-11.7	-67	5.9
1298	85.9831	20.5011	2013.5205	12.083	15.608	17.44	239	52	-85.2	5.2	50.2	-83.4	5.2
1299	85.9883	63.2540	2013.3385	12.425	13.893	4.06	170.8	14.7	-94.9	5.8	12.3	-96.2	6.1
1300	86.1031	81.4012	2013.778	11.869	16.181	7.39	302.5	1.9	-76.4	5.5	4.1	-77.6	5.6
1301	86.1719	21.3476	2013.6155	13.804	15.431	5.70	224.7	-113.2	110.3	5.2	-112.1	110.6	5.3
1302	86.3046	27.6532	2013.527	16.2	17.53	6.30	80.1	-46.1	-100.4	5.2	-43.9	-99.2	5.5
1303	86.3114	31.6483	2013.539	12.843	13.941	57.74	191.8	10.9	-133.5	5.2	11.7	-129.5	5.2
1304	86.3133	29.5650	2013.5425	12.535	15.306	21.52	30.9	-3.8	-99.8	5.1	-2.5	-95.8	5.2
1305	86.4006	10.2941	2013.266	17.636	17.792	34.23	268.5	-14.1	-84.7	6.5	-13.5	-85.9	6.5
1306	86.4333	16.6701	2013.586	13.17	13.288	6.45	160.3	83.4	-87.9	5.2	85.2	-88.9	5.2
1307	86.6617	44.6033	2013.5405	17.419	17.783	16.72	2.8	-10.8	-63.2	6.2	-7.7	-61.9	6
1308	86.8792	-3.4891	2014.058	13.318	16.041	8.17	258	-31.4	-146.3	5.3	-31.1	-139.5	5.5
1309	86.9749	13.0882	2013.445	14.171	17.793	19.19	33.9	44.4	-72.3	5.1	38	-74.4	5.3
1310	86.9981	45.6580	2013.6025	11.129	16.192	46.11	230.9	-0.2	-91.3	5.8	-2.1	-91.3	5.9
1311	87.1405	8.8696	2013.598	12.872	16.422	9.16	344.3	9	-91.1	5.8	7.4	-92.4	6
1312	87.3159	18.1892	2013.586	17.273	17.323	3.69	42	66.2	-48	6.5	64.1	-54.5	6.2
1313	87.3453	19.0016	2013.521	14.004	14.063	7.92	222.6	59.3	-111.6	6.1	60.7	-112.5	6.1
1314	87.3466	48.9791	2013.5735	15.055	16.019	10.53	214.2	-12.6	-64	6	-11.9	-65.5	6
1315	87.3735	56.5649	2013.5755	10.488	12.865	15.19	293.5	6.1	-61.4	5.8	7.1	-60.2	5.8
1316	87.3889	61.5900	2013.2055	11.708	15.894	5.81	330.8	1.5	-64.8	5.5	1.9	-64.1	6.7
1317	87.3907	46.0225	2013.634	16.142	17.94	57.88	249.6	-4	-61.1	5.6	-3.5	-60.6	5.9
1318	87.5281	19.0510	2013.388	16.254	17.209	7.01	83.8	4.3	-63.7	6.1	4.9	-66.9	6.2
1319	87.5348	16.8071	2013.422	16.016	16.693	10.15	259.4	36.7	-91.5	5.2	35	-92	5.2
1320	87.7402	20.9474	2013.1005	18.245	18.454	57.77	309.8	35.9	-60.9	6.4	29.1	-61	7
1321	87.7663	59.5472	2013.509	15.321	18.333	12.76	32.9	-17.3	-121	5.8	-12.4	-123.5	6.2
1322	87.7788	37.1527	2013.616	18.127	18.559	36.84	231.4	-17.9	-68.7	6.7	-27.4	-69.4	7.9
1323	87.8651	68.4214	2013.446	10.964	17.642	20.66	294	2.7	-114.2	6.3	-0.4	-110.8	6.5
1324	87.9103	55.4761	2013.5055	14.707	14.849	7.99	354.8	-8.7	-77.5	5.8	-12.1	-76	5.8
1325	88.0329	17.3004	2013.5915	15.177	15.574	50.21	153.5	23.3	-66.8	5.1	23.9	-68.1	5.1
1326	88.1048	35.0265	2013.6795	11.648	13.665	5.98	48.5	16.5	-65.9	6.1	15.4	-67.7	5.9
1327	88.1463	11.0787	2013.4	17.704	17.753	21.09	96.2	64.5	11.3	6.4	64.4	3.1	6.8
1328	88.1621	47.1143	2013.523	8.562	14.913	27.04	121.4	-44.8	-155.3	5.5	-38.1	-156.6	5.5
1329	88.2869	47.7854	2013.6135	12.535	12.656	10.19	207.5	25.3	-68.2	5.4	25.2	-68.3	5.4
1330	88.3019	3.9169	2014.0515	13.27	14.339	5.01	135.3	-16.5	-121	5.7	-7.4	-120.1	5.8
1331	88.3579	46.3960	2013.6275	16.1	16.311	9.63	163.9	52.3	-114	5.5	49.4	-110	5.6
1332	88.3756	63.9064	2013.488	13.936	14.237	10.99	227.7	10.7	-78	5.6	10.1	-76.3	5.6
1333	88.4111	31.6773	2013.49	11.22	13.11	6.67	320.5	16	-84.2	6.1	17.5	-84.3	6.1

1334	88.4916	68.3132	2013.4915	13.479	15.294	7.64	217.8	-21.7	-72.9	6	-20.4	-72.5	6.1
1335	88.5079	60.0008	2013.3015	14.887	15.699	4.34	237.8	90.4	-86.9	6	93.3	-87	6.3
1336	88.5119	15.1111	2013.598	8.448	17.754	21.10	238.4	108.2	26.6	5.3	106.5	24.4	6.2
1337	88.5777	15.9696	2013.6155	18.139	18.231	53.94	321.2	-62.3	-21.4	5.6	-62.4	-18.8	6.4
1338	88.6875	-4.2450	2014.0465	13.67	14.448	13.09	207.7	-30	-121.5	5.3	-32.3	-120.3	5.3
1339	88.7107	16.0912	2013.601	12.18	13.127	6.11	50.4	29.5	-66.3	5.5	29	-66.8	5.4
1340	88.7302	22.3004	2013.2945	16.892	17.309	4.46	23.3	87.6	-79.4	5.6	90.3	-77.9	6.1
1341	88.7363	42.9467	2013.446	16.674	17.584	52.09	128.9	2.5	-61.8	5.6	7	-61.2	5.6
1342	88.8005	37.3146	2013.5675	16.388	17.115	10.44	190	-5.7	-63.8	5.9	-4.3	-65	5.9
1343	88.8068	53.1617	2013.557	14.245	14.757	13.92	148.9	12.8	-89.7	6	14.2	-91	5.9
1344	88.8248	0.4114	2014.0285	15.664	15.997	6.69	136.4	73.3	-115.2	5.7	76.9	-118.5	5.7
1345	88.8906	-13.9695	2014.05	10.383	13.294	15.45	78.3	-18.4	-62.3	5.5	-18.4	-61.3	6
1346	88.9971	59.4502	2013.5955	13.494	14.318	5.80	102.9	2.3	-71.2	5.5	-2.5	-71.4	5.5
1347	89.0835	27.5643	2013.4455	15.481	17.055	29.42	100.5	17	-95.9	5.2	17	-96.5	5.3
1348	89.3307	1.6495	2014.0085	14.978	15.404	9.43	124.5	6.8	-67.2	5.7	5.5	-65.6	5.8
1349	89.3847	14.7288	2013.714	13.348	17.182	6.72	261.8	7.1	-63.4	5.4	6.5	-61.2	5.6
1350	89.3942	66.3935	2013.526	13.803	14.559	6.10	62	-62.5	-4.1	5.6	-64.1	-5.8	5.6
1351	89.5295	4.9872	2014.02	14.633	15.631	4.65	187.6	50.4	-68	5.7	43.9	-70.6	5.8
1352	89.6328	5.0509	2013.9925	11.492	15.078	7.69	101.7	32.1	-79.9	5.7	30.6	-78.6	5.7
1353	89.7899	40.0929	2013.3945	14.229	16.199	21.09	223.9	36.8	-71.6	5.5	35.2	-70	5.6
1354	89.8005	46.5194	2013.26	12.669	14.02	4.46	139.1	-93	-12.9	5.5	-94.8	-16	5.9
1355	89.8358	45.1150	2013.4255	15.94	16.876	8.40	224.9	14.8	-96	5.6	14.3	-96.1	5.6
1356	89.8632	57.2444	2013.488	14.49	17.269	5.75	268.9	19.8	-67.4	5.5	17.5	-64.5	5.9
1357	89.9740	6.3362	2014.038	13.934	15.357	4.55	267.6	13.6	-66.2	5.6	24	-62	7.3
1358	89.9756	-6.6835	2014.104	13.853	15.484	14.23	30.8	69.4	-31.6	5.3	68.2	-31.2	5.6
1359	90.0374	74.8918	2013.5285	12.774	12.91	22.697	227.8	9.2	-65.3	5.6	12.8	-66.2	5.6
1360	90.0973	42.1080	2013.4985	11.288	11.704	11.481	227.3	4.2	-65.2	5.5	5.9	-67.4	5.5
1361	90.2195	38.2894	2013.4445	14.919	15.301	11.683	278.7	36.5	-64.1	5.5	39	-63.5	5.5
1362	90.3236	28.2444	2013.4095	13.549	16.388	10.556	56.2	12.2	-74.4	5.2	11.4	-74.7	5.3
1363	90.3260	27.2809	2013.589	8.891	13.683	38.176	269.2	-10.5	-126.3	5.2	-13.1	-122.2	5.2
1364	90.4008	32.8523	2013.533	13.764	17.151	6.549	16.2	87.9	2.7	6.1	83.8	-4.5	6.2
1365	90.4040	33.2006	2013.7645	12.802	15.921	6.013	322	29.8	-83.2	6.1	32.1	-85.2	6.2
1366	90.4876	60.8787	2013.5615	15.709	17.443	5.869	79.2	49.6	-102.7	5.6	48.7	-102	5.6
1367	90.4949	3.0870	2013.912	17.106	17.916	6.516	150.7	-5.4	-116.6	6	-9.8	-119.3	6.3
1368	90.5619	37.3003	2013.46	15.455	17.269	23.581	72.7	-21.6	-94.9	5.5	-20.5	-94.8	5.9
1369	90.5819	-13.8815	2014.05	13.903	14.929	13.88	280.7	4.1	-64	5.5	-2.5	-61.7	6.3
1370	90.5912	-13.9261	2014.05	9.904	14.815	20.329	324.3	-16.6	-94.5	5.5	-8.1	-94.8	5.5
1371	90.7348	43.4550	2013.518	16.182	17.18	30.615	156.1	-34.6	-61.5	5.5	-35	-64	5.5
1372	90.7998	68.0461	2013.4355	16.395	17.372	5.429	235.5	5.1	-63.9	5.6	4.9	-63.6	5.7
1373	90.8857	6.6733	2013.9015	12.715	17.5	35.628	311.6	145.7	-116.5	5.8	143.7	-125.8	5.9
1374	90.9134	51.7741	2013.6305	14.712	16.407	6.135	38.4	10	-71.5	5.9	12.9	-72.7	5.9
1375	90.9147	20.9738	2013.5275	10.275	15.822	10.442	201.9	45.3	-66.5	5.5	42.9	-64.6	5.6
1376	90.9166	43.4513	2013.415	15.963	16.659	15.354	27.2	-1.6	-83	5.5	-0.8	-85	5.6
1377	90.9409	56.6898	2013.547	14.007	14.941	12.564	335.7	-14.3	-132.1	6	-18.3	-133.9	6
1378	90.9646	45.8791	2013.557	13.909	15.622	25.643	335.9	32.4	-81.6	5.5	30.6	-79.5	5.5
1379	90.9994	14.6997	2013.4985	12.317	14.573	5.372	127.7	-26.1	66.8	5.4	-30.7	67.8	6.4

1380	91.0513	40.6139	2013.492	13.117	17.306	31.286	33.7	16.1	-68.4	5.5	17.6	-69.6	5.5
1381	91.0724	12.2943	2013.694	10.259	13.233	25.161	250.9	36.5	-106.7	5.4	33.6	-109.4	5.5
1382	91.1138	26.1841	2013.402	14.75	16.45	14.606	317.7	-2.6	-92.7	5.2	-0.8	-92.8	5.3
1383	91.1275	13.9644	2013.82	9.062	14.495	22.868	237	-77.2	-84.2	5.3	-76.3	-85.3	5.4
1384	91.2213	-4.7391	2014.0575	12.664	17.044	14.534	15	-3.7	-61.5	5.3	-2	-61.3	5.8
1385	91.2403	9.9946	2013.5355	17.559	17.735	27.168	134	60.8	-20.1	6.2	65.2	-13.3	6.2
1386	91.2640	7.3914	2013.7055	14.403	15.799	8.724	347.4	-186.3	-102.8	6	-184.6	-104.7	5.9
1387	91.3743	50.7572	2013.7415	17.606	17.798	28.965	341.3	-17.1	-65.5	6.9	-9	-64.7	6
1388	91.4273	34.7868	2013.5885	11.442	13.694	8.209	9.7	12.7	-66	5.9	11.9	-65.7	5.9
1389	91.4911	62.4758	2013.5265	9.72	14.131	9.075	142	15.3	-96.2	5.5	11.1	-93.8	5.6
1390	91.5212	52.6220	2013.6365	11.538	15.105	19.409	71.7	30.4	-96.7	5.8	30.2	-96	5.8
1391	91.6431	-3.4938	2014.0405	13.895	14.732	8.642	64.9	7.5	-63.3	5.3	5.9	-65.5	5.3
1392	91.6513	39.0950	2013.516	8.504	12.247	22.844	197.9	-38.4	-156.5	5.5	-37.4	-158.3	5.5
1393	91.6651	12.8712	2013.6125	14.966	15.095	10.463	238.6	-2.8	-82	5.9	0.3	-81.3	5.9
1394	91.7179	52.3594	2013.615	11.405	12.797	6.998	278.3	-55.1	-131.6	5.8	-57.7	-130	5.9
1395	91.7185	61.8770	2013.5465	12.944	16.455	15.098	280.9	15.1	-70.3	5.5	16.8	-70.3	5.6
1396	91.7630	41.7611	2013.5915	8.408	10.376	10.123	238.3	7	-86.3	5.5	8.2	-86.6	5.5
1397	91.8623	48.4817	2013.637	13.292	13.56	40.295	302.8	-84.2	-35.6	5.8	-82.4	-34.6	5.8
1398	91.8992	-5.4976	2014.1245	13.079	16.767	58.73	34.4	-0.9	-72.3	5.3	-2.9	-73.8	6.1
1399	91.9328	12.3802	2013.618	13.638	16.268	15.006	334	29.2	-92.4	5.4	28.1	-92.2	5.5
1400	91.9381	11.5748	2013.5645	17.479	18.504	41.063	320.2	-75	-17.7	7.4	-75.3	-29.6	6.9
1401	91.9734	27.7717	2013.6205	11.434	17.022	39.101	109.3	109.5	-177.6	5.3	112.5	-181.1	5.2
1402	91.9814	42.8940	2013.5705	11.454	16.009	35.155	307.3	25.2	-78.8	5.4	27.9	-76.6	5.5
1403	92.0152	5.3558	2014.009	13.733	15.255	9.783	246.4	128.2	34.3	5.7	129.4	32.6	5.7
1404	92.0639	59.1302	2013.4915	15.675	16.726	6.378	11.5	-59.7	-140.7	6	-62.1	-140.2	6.1
1405	92.0796	34.9767	2013.482	12.846	14.377	13.244	342.6	78.3	-84.1	6	75.4	-82.1	6
1406	92.0889	48.9906	2013.6185	15.313	16.399	9.514	54.9	-2.8	-70.5	5.8	-3.9	-69.6	5.8
1407	92.1034	20.7926	2013.572	13.113	14.463	25.153	27.4	51.2	64.9	5.1	51	66.6	5.1
1408	92.2563	11.1032	2013.667	12.17	13.746	14.791	349	67.2	-20.6	5.9	67.1	-20.6	5.9
1409	92.3348	36.8709	2013.612	9.151	17.155	16.689	315.7	23.3	-65	6.3	11.5	-69.3	6.4
1410	92.3815	-10.0348	2014.106	13.454	14.803	23.389	50.7	20	-186.9	5.4	25.1	-179.6	5.5
1411	92.3984	47.5687	2013.688	10.628	14.253	24.9	212.5	4	-60.6	5.4	3.6	-61.3	5.4
1412	92.4749	25.3065	2013.287	18.054	18.75	35.637	6.5	-66.8	1.1	5.6	-65.4	4.5	5.8
1413	92.5776	-2.3564	2014.0275	13.812	14.678	10.145	122.3	-19.2	-75.1	5.3	-22.9	-72.9	5.3
1414	92.7011	3.2530	2014.04	14.388	16.224	45.5	280.9	-45.9	-150	5.7	-47.6	-148.2	5.9
1415	92.7377	40.7498	2013.434	12.912	14.854	25.698	78.4	-67.4	-40.7	5.5	-66.2	-40.9	5.5
1416	92.7948	35.9575	2013.592	11.981	13.476	27.026	345.9	19.6	-62.4	6	17.7	-62.7	6
1417	92.8624	6.9135	2013.9425	11.846	16.153	6.877	148.9	32.8	-107.7	5.8	31	-111.8	6.1
1418	92.8764	54.3727	2013.583	15.127	16.523	4.595	274.8	-1.9	-128.6	6	-0.3	-128.6	6.1
1419	92.8776	11.5109	2013.6495	12.092	14.804	20.305	99.6	123.2	-47.4	5.8	124.8	-48.2	5.9
1420	92.9602	36.6490	2013.592	13.428	16.751	18.6	84.9	-27.4	-91.5	5.4	-24.8	-89.4	5.5
1421	93.0616	69.5794	2013.5035	10.526	17.996	15.228	237	67.7	-162.6	5.9	70.3	-163.5	6.2
1422	93.1270	8.7919	2013.9095	11.237	13.77	5.679	149.6	84.6	-67.5	5.7	79.4	-65.9	5.8
1423	93.1297	60.5235	2013.515	12.646	15.45	11.596	200.3	7.8	-87.5	5.8	8.4	-83.7	5.9
1424	93.1331	24.7865	2013.5015	14.245	14.493	9.849	324.5	4	-83.2	5.2	0.2	-80.9	5.2
1425	93.1885	47.0124	2013.556	11.543	13.771	32.734	151.8	10.1	-129	5.5	12.8	-127.8	5.5

1426	93.2825	23.9292	2013.569	10.987	16.514	44.681	108.2	-71.1	-47.1	5.2	-70.6	-47.7	5.3
1427	93.2853	5.3767	2013.9205	16.733	17.649	14.757	213.9	29.3	-75.9	6.3	25.3	-76.9	6.4
1428	93.3329	50.6870	2013.7445	17.648	18.255	5.37	333	10.7	-70.4	5.5	2.8	-71.7	5.9
1429	93.3347	46.1666	2013.5315	10.604	13.818	6.801	230.5	-24.8	-86.9	5.5	-21.7	-88.3	5.5
1430	93.3416	-10.1925	2014.071	14.4	14.587	28.917	106.2	-29.7	-77.6	5.4	-27.7	-79.8	5.4
1431	93.3728	17.5361	2013.5925	10.601	15.34	29.158	348.3	-8.9	-98	5.1	-7.6	-97.4	5.2
1432	93.5467	9.1909	2013.908	17.63	18.069	29.477	342.1	-34.4	66.7	6.4	-37	65.8	6.3
1433	93.6211	70.9411	2013.365	13.531	16.523	5.819	207.4	52.2	-116.4	5.9	54	-112.7	6.5
1434	93.6791	40.2655	2013.58	15.288	16.193	4.202	153.4	-21.2	-64.4	6	-22.3	-62.8	6.1
1435	93.7143	73.4443	2013.5635	10.821	14.384	47.032	357.6	42.7	-112.5	5.9	41.6	-117.5	6
1436	93.7362	29.4129	2013.457	15.736	16.214	18.094	261	-8.9	-61.7	5.2	-8.1	-60.4	5.2
1437	93.7940	25.6724	2013.544	11.878	17.681	10.667	129.1	-14.4	-63.9	5.2	-17	-65.9	5.3
1438	93.8441	51.1508	2013.599	11.235	14.862	42.3	150.7	-10	-70.9	5.5	-11.3	-72.9	5.5
1439	93.8985	-10.9721	2014.075	12.186	14.89	45.397	56.2	31.1	-79.1	5.4	28.8	-77.3	6.6
1440	93.9122	36.1487	2013.7215	6.95	7.357	11.393	216.1	-67.8	3.2	6.1	-65.2	-1.8	6
1441	93.9190	34.4505	2013.519	15.087	16.237	7.488	119.1	54	-144.6	5.5	51.3	-142.5	5.6
1442	93.9519	-8.4445	2014.099	13.351	14.063	39.606	84.2	-46.7	-86.3	5.4	-47.1	-89.4	5.4
1443	93.9620	66.1824	2013.552	10.177	16.18	41.32	29.9	4.9	-63.9	5.8	5.8	-65	5.9
1444	93.9769	10.2114	2013.528	12.812	16.207	12.149	316.4	1.7	-64	5.9	0.8	-61.9	6.1
1445	93.9931	51.9299	2013.617	12.767	16.556	13.959	5.9	-2.4	-60.3	5.5	-1.9	-61.7	5.5
1446	94.0543	68.8823	2013.5145	13.354	14.176	19.444	291.4	-23.7	-128.7	5.9	-26.3	-125.9	5.9
1447	94.1321	70.6698	2013.527	14.805	16.756	18.842	83.6	58.5	-81.2	5.9	58.7	-82.9	5.9
1448	94.1398	60.8651	2013.476	15.587	16.06	4.689	251.7	14.9	-63	5.6	16.9	-63.1	5.7
1449	94.2350	13.7266	2013.616	14.564	16.694	55.65	3.5	-10.3	-63.6	5.1	-9.7	-61.8	5.1
1450	94.3222	17.6057	2013.4475	16.191	18.586	35.883	251.3	-6	-66.5	5.1	-2.3	-64.5	7.1
1451	94.3470	57.4145	2013.571	7.498	13.533	15.326	16.1	-39.1	-61.7	6.1	-40.6	-61.7	6
1452	94.3791	11.3259	2013.5325	14.975	15.969	9.926	61.2	31.5	-106.1	5.9	28.8	-105.4	5.9
1453	94.4290	14.5173	2013.426	15.945	18.492	34.257	203	-6.8	-64.4	5.2	-4.5	-63.9	6
1454	94.4303	68.5621	2013.4865	13.563	15.119	30.703	237	78.6	-71	5.9	79.8	-69.7	5.9
1455	94.4802	44.0949	2013.443	17.552	17.618	37.325	202.3	71	-25.7	5.5	72.8	-27.9	5.9
1456	94.5568	28.6876	2013.5025	12.886	13.656	32.763	310.9	0.1	-66.1	5.2	-1.7	-65	5.2
1457	94.5728	-9.4038	2014.0525	12.306	13.487	6.231	181.3	-13.1	-65.6	5.4	-13.4	-63.3	5.4
1458	94.6073	45.4929	2013.5165	16.388	17.366	10.708	31.3	-5.2	-113.4	5.5	-6	-115.8	5.6
1459	94.6631	58.3243	2013.5715	10.796	16.026	31.435	11.1	57.8	-79.7	5.9	55.6	-82	6
1460	94.6883	22.8306	2013.5175	13.521	14.33	29.305	336.6	-22.6	-95.3	6.4	-19.5	-94	6.4
1461	94.7173	57.1375	2013.6245	8.017	13.152	54.294	42.4	53.7	-111.8	5.5	61	-103.9	5.5
1462	94.9061	8.2794	2013.817	10.561	14.747	22.898	239.5	-94.6	-49.3	5.8	-95.4	-56.1	5.9
1463	94.9336	8.7407	2013.8525	14.217	16.629	17.065	43.3	-17.2	-78.6	5.7	-16.1	-79.6	5.9
1464	94.9772	34.7087	2013.769	17.588	19.014	49.575	253.9	-2	-64.9	5.3	-12.1	-61	6
1465	95.0016	34.6968	2013.6365	17.811	17.989	21.502	180.9	18.5	-102.7	5.6	24.3	-100.1	5.8
1466	95.0183	17.5543	2013.436	17.198	18.784	44.512	329.7	-62.1	-11.9	5.7	-60.9	-12.8	6.3
1467	95.1509	85.0299	2013.921	10.787	18.476	31.58	101	15.9	-182.9	5.8	24.1	-174.2	6.8
1468	95.2213	45.9726	2013.1965	12.57	14.233	4.093	62.5	33.3	65.8	5.5	30.4	65.8	6.2
1469	95.2418	-3.7563	2014.0725	8.754	14.038	43.786	183.2	-12	69.6	5.3	-10.4	72.7	5.3
1470	95.2643	36.8083	2013.622	15.924	18.679	11.523	93.9	-27.7	-83.2	5.5	-35.9	-76.8	6.5
1471	95.3472	32.8270	2013.6865	9.697	17.349	15.27	225.5	73.1	-101	5.2	73.3	-100.2	5.5

1472	95.3593	20.6074	2013.542	13.595	16.01	6.356	324.8	-3.8	-95.5	6	-9.1	-97.2	6.2
1473	95.3645	-12.4607	2014.114	12.032	15.3	17.709	352.8	44.9	-63.5	5.4	44.5	-66.8	5.5
1474	95.5247	48.9503	2013.4995	14.334	15.258	4.424	1	54.1	-67	6	55.5	-71.2	6.3
1475	95.6980	20.4478	2013.713	8.744	12.199	42.599	89.5	-43.2	-110.3	5.9	-42.7	-105.8	6
1476	95.7334	13.4096	2013.563	18.509	18.783	22.399	85.9	60.1	30.7	6	66.6	19.6	6.6
1477	95.7590	55.5547	2013.627	12.898	14.272	4.627	277.6	-35	-152.2	5.8	-37.4	-152.2	5.8
1478	95.7920	-5.5130	2014.0675	12.773	14.856	5.907	356.8	45.1	-157.1	5.3	42.9	-160.2	5.5
1479	95.8228	-3.3645	2014.115	13.566	15.811	20.612	180.6	-6.1	-85.5	5.3	-1.9	-85.2	5.3
1480	95.8442	-7.6920	2014.098	15.497	15.524	11.797	143.2	-14.3	-65.6	5.4	-16.3	-68.3	5.5
1481	95.8982	78.0357	2013.505	12.944	17.676	13.27	16.2	45.7	-108.5	5.6	50	-109.7	5.8
1482	95.9359	39.2852	2013.622	14.25	14.259	6.383	174.1	-60.4	-67.7	5.5	-60.6	-65.7	5.4
1483	95.9819	71.0523	2013.4695	15.805	16.09	22.511	274.9	66.8	-4.2	5.9	66.6	-4.4	5.9
1484	96.0028	20.5387	2013.654	13.207	16.318	47.444	323	98.9	71.1	6.4	96.9	79.1	6.8
1485	96.0341	50.9555	2013.765	12.703	13.867	4.711	265.6	14.7	-94.2	5.9	16	-94.7	5.7
1486	96.1730	6.7230	2013.912	14.693	17.578	12.063	298.1	29.4	-61.9	5.7	27.1	-62.7	6
1487	96.2094	60.6995	2013.5535	12.931	14.885	59.092	69.7	2.7	-71.8	5.5	6.5	-74.2	5.5
1488	96.2458	54.7517	2013.5515	14.543	14.868	6.675	214.7	8.8	-117.6	5.8	11.3	-117.2	5.8
1489	96.2868	36.7634	2013.615	15.444	15.523	4.553	337.5	83.1	-53.8	5.9	82.6	-55.1	5.9
1490	96.3339	11.2330	2013.6275	7.214	12.503	9.251	115.3	-135.1	-55.7	5.8	-140.5	-45.1	6.4
1491	96.4383	12.2122	2013.6655	12.645	17.474	8.515	198.6	-5.4	-68.5	5	-6.9	-66	5.3
1492	96.4406	15.3641	2013.632	16.897	17.694	18.141	348.5	43	-60.7	5.2	39.5	-63	5.2
1493	96.4519	36.3827	2013.8195	17.532	18.849	56.294	19	-73.6	35.5	7.2	-70.6	36.1	7.4
1494	96.4801	6.9271	2013.834	9.784	10.277	32.848	176.4	-13	-64.1	5.7	-11.8	-63.3	5.7
1495	96.5534	11.9208	2013.6105	13.756	14.239	54.969	160.2	-0.4	-86.9	5.1	-3.8	-89.1	5.1
1496	96.6474	-14.0656	2014.05	14.446	15.543	4.714	74.7	10.5	-112.9	5.3	5.9	-111.4	6.2
1497	96.6634	13.1008	2013.813	6.977	12.669	32.51	328.6	9.7	-100.9	5.6	5.4	-100.4	5.1
1498	96.6752	23.1120	2013.47	12.701	13.703	23.981	159.7	-21.7	-76.9	6.4	-22.1	-74.5	6.4
1499	96.7066	10.2442	2013.6615	14.08	15.84	16.269	334.1	-32.6	-80.2	5.8	-30	-80.3	5.9
1500	96.7539	40.5609	2013.534	9.416	13.221	8.216	262.5	90.1	-166.4	6	93.9	171.3	6
1501	96.7652	79.3836	2013.7005	13.357	15.915	28.579	162.8	-17.7	-83.7	5.8	-17.7	-85.3	5.8
1502	96.7821	72.8971	2013.5915	14.313	15.746	13.719	187.1	82.2	-120.5	6.3	84.8	-123.4	6.3
1503	96.7974	21.3931	2013.4385	12.659	14.311	7.265	218.9	37	-104.9	6.5	35.8	-104.6	6.4
1504	96.9461	30.0763	2013.4555	13.935	17.7	18.111	160.9	132.9	20	5.2	130	20.7	5.4
1505	97.0329	35.6108	2013.5525	11.055	15.479	37.443	192.2	-72.4	-42	5.2	-71.3	-46.8	5.2
1506	97.0584	-4.5625	2014.104	13.99	14.212	7.223	298.3	7	-79.2	5.3	6.8	-81.5	5.3
1507	97.0622	25.8001	2013.6065	12.522	16.161	6.376	7.4	42.4	-62.1	5.2	42.3	-65.5	5.4
1508	97.1006	61.0758	2013.585	11.522	12.402	15.615	310.9	15.6	-93.7	5.5	14.3	-94.1	5.5
1509	97.1187	61.1033	2013.4735	12.811	16.6	6.389	141.5	-87.3	-62	5.5	-85.9	-62	6.1
1510	97.1298	59.1057	2013.5645	15.269	16.822	36.681	349.4	53.9	-168.1	5.6	52	-168.3	5.6
1511	97.1434	81.6446	2013.812	11.441	14.303	6.504	223.6	17.3	-91.3	5.8	11.7	-88.7	5.9
1512	97.1595	28.6416	2013.5015	17.557	18.128	47.151	308	-61.4	-6.3	5.2	-62.7	-12.2	5.4
1513	97.1616	40.6859	2013.535	14.391	15.643	7.573	321.2	5.6	-70.9	5.3	6	-71.1	5.2
1514	97.1690	51.8096	2013.677	14.614	15.636	16.633	189.3	4.1	-126.7	5.8	5.4	-131.2	5.8
1515	97.2004	15.1163	2013.5725	12.339	16.679	24.825	230.5	13.8	-68.9	5.1	11	-68.4	5.2
1516	97.2283	52.1257	2013.606	7.213	14.013	20.35	202.3	-66.1	-23.2	5.9	-65.2	-27.6	5.8
1517	97.2628	46.0782	2013.617	13.094	13.173	9.064	26.1	-7.6	-71.4	5.4	-8.5	-71.7	5.4

1518	97.3075	7.8458	2013.7755	15.756	16.048	14.019	193.8	22.3	-80.1	5.8	17.2	-78.1	5.8
1519	97.3496	28.0319	2013.527	18.012	18.38	5.626	319.2	17.4	-68.4	5.3	21.2	-70.8	5.5
1520	97.3607	6.1702	2013.1255	16.098	16.246	3.148	348.3	-11.8	-68.5	6.6	-14.1	-67.4	6.4
1521	97.4092	67.6747	2013.4645	13.601	16.762	10.98	69.4	100.2	30	5.9	95.9	31.1	5.9
1522	97.4798	71.2163	2013.5405	11.447	17.068	30.452	289.3	-25	-66.4	5.9	-23.4	-65.4	5.9
1523	97.5463	36.7256	2013.4085	14.602	14.737	43.003	74.3	47.5	-63.9	5.2	47	-68.1	6.7
1524	97.5805	47.1078	2013.6715	15.687	15.909	4.247	353.8	-13.4	-69.6	5.4	-15.4	-65.9	5.5
1525	97.6920	70.6577	2013.4325	14.309	17.386	6.062	116	24.4	-93	5.9	17.8	-91.1	6.2
1526	97.7524	23.8870	2013.598	12.618	14.194	34.165	4.9	-13.3	-67.7	5.1	-9.1	-68.8	5.1
1527	97.7588	4.5078	2014.0385	7.582	13.004	13.743	122.1	49	-84	5.8	44.8	-89.2	5.7
1528	97.8450	42.6404	2013.56	15.741	17.031	21.629	342.7	13.1	-119.2	5.5	15.7	-121.7	5.5
1529	97.8495	66.9253	2013.543	13.048	17.194	34.446	288.3	81.7	-8.6	5.9	82	-3.1	5.9
1530	97.9120	57.8611	2013.6625	13.995	18.155	43.144	231.7	0.6	-63.8	5.9	3.2	-61.2	6.1
1531	97.9882	41.5658	2013.529	11.149	12.764	7.551	318.2	-94.5	-29.1	5.3	-95.7	-26.7	5.2
1532	98.0081	64.6172	2013.5715	15.674	16.51	4.92	339.1	0	-82.8	5.5	0.8	-80.9	5.6
1533	98.0115	39.7421	2013.5205	9.316	16.116	18.147	35.5	-78.1	-58.6	5.3	-84	-51.5	5.3
1534	98.0412	57.0362	2013.593	10.75	14.045	7.436	321.2	-26.5	-97.4	5.5	-27	-98.7	5.5
1535	98.1714	-8.3508	2014.1055	11.844	12.709	36.417	148.7	-16.3	-60.5	5.4	-12.3	-60.9	5.4
1536	98.2460	42.6687	2013.588	11.604	13.501	29.927	138	65	-127.2	5.5	64.6	-127.3	5.5
1537	98.2537	10.1085	2013.567	14.752	16.93	29.922	334.7	-61.7	-115.3	5.9	-61.7	-113.7	6.1
1538	98.2538	19.4111	2013.6185	15.05	15.658	8.499	338.9	34.7	-75.5	5.1	32.5	-73.1	5.1
1539	98.2948	62.0400	2013.6135	14.547	15.707	31.95	84.7	23.3	-136.4	5.5	21.3	-136.4	5.5
1540	98.3030	5.4626	2014.067	7.636	10.762	10.413	309.3	100.4	-87.5	5.6	95.7	-97	5.7
1541	98.3339	43.1195	2013.5715	16.589	18.226	6.067	125.8	23.3	-69.3	5.5	23.9	-68.8	5.8
1542	98.3532	26.4703	2013.5495	15.643	18.161	46.851	172.7	10.8	-62.8	5.2	14.1	-61.2	5.3
1543	98.3620	22.7490	2013.524	11.141	13.406	33.421	311.3	67.5	-77.3	5.2	65.4	-78.4	5.2
1544	98.4451	18.3587	2013.5325	14.03	17.039	48.99	319.1	10.6	-63.8	6.3	15.8	-61.1	6.4
1545	98.4625	10.5466	2013.618	16.503	17.678	42.865	264.5	-63.7	-99.7	5.9	-65.5	-100.5	6.2
1546	98.5531	-14.1017	2014.05	14.245	14.447	11.624	273.7	-60.8	-45	5.3	-61.5	-48.8	5.2
1547	98.5789	3.5521	2014.032	14.492	15.981	8.682	202.1	46.7	-76.4	5.7	44.8	-77.2	5.7
1548	98.5832	52.1419	2013.607	13.038	15.225	12.31	125.1	-35	-81.5	5.5	-33.8	-80	5.5
1549	98.6770	11.5917	2013.436	17.436	18.045	26.348	62.7	-24.4	-74.3	6	-33	-67.8	6.6
1550	98.6997	71.1774	2013.5375	14.452	15.573	7.897	14.1	99.2	19.8	5.9	98.2	15.5	5.9
1551	98.8313	-4.3335	2014.128	13.047	13.244	15.785	112.9	10.7	-70.5	5.3	4.8	-70.6	5.3
1552	98.8638	17.1934	2013.5255	13.961	16.599	15.878	80.2	11.5	-66.8	5.1	9.5	-65.2	5.2
1553	98.9086	-10.2734	2014.0615	14.227	16.717	14.27	25.1	-43.9	61.8	5.4	-40.5	65.6	6.5
1554	98.9392	5.4421	2013.9665	12.714	15.031	12.255	253.3	26.4	-85.3	5.8	26.2	-88.7	5.7
1555	99.0496	74.9365	2013.5685	9.602	13.071	13.006	317.5	-29.7	-105.8	5.6	-35.4	-107.9	5.6
1556	99.1057	33.7395	2013.7125	14.745	16.664	27.802	138.7	-75.3	-32.1	5.2	-77.9	-35	5.7
1557	99.1075	34.9837	2013.6835	13.246	13.382	5.468	104.6	23.7	-92.6	5.2	23.4	-96.7	5.2
1558	99.1232	28.4327	2013.542	11.815	14.496	7.09	28.8	-0.6	-100.6	5.2	-1.1	-100.7	5.2
1559	99.1594	27.5907	2013.558	9.749	12.724	17.327	263.6	26.2	-91.7	5.2	21.5	-93.3	5.2
1560	99.1680	13.5333	2013.5925	13.62	18.357	13.336	227.7	13.4	-128.7	5.1	7	-133.6	6.4
1561	99.1775	13.3020	2013.7215	10.874	11.592	39.537	153	32.3	-70.3	5	33.3	-70	5.1
1562	99.1930	37.8510	2013.5625	9.09	14.004	45.311	303.9	-60.7	-224.8	6.2	-66	-221.6	6.1
1563	99.2102	60.4224	2013.5915	14.355	16.851	34.438	255.1	41.4	-105.9	5.5	41	-104.4	5.6

1564	99.3048	-0.4133	2014.082	14.93	15.121	7.335	283.4	23.8	-75.1	5.5	31.2	-76.1	5.5
1565	99.3258	-6.7388	2014.1165	13.083	13.61	14.258	114.3	55.9	-93.6	5.3	54.6	-92.5	5.3
1566	99.3842	22.9683	2013.5355	13.212	15.222	9.906	156.1	-28.7	-75.7	5.2	-30	-78.6	5.2
1567	99.4052	79.1307	2013.6465	13.763	17.705	13.198	322.4	-47.9	-145.2	5.9	-47.3	-142.5	6
1568	99.4335	-9.3923	2014.078	10.879	12.541	11.448	291	-6.6	-69.3	5.4	-3.5	-71.1	5.4
1569	99.4631	68.8169	2013.2985	16.251	18.619	23.394	140.3	18	-91.1	5.9	22.6	-85.7	6.5
1570	99.4913	39.7527	2013.5205	13.653	15.335	10.601	348	-50.7	-77.9	5.9	-52.4	-77.5	5.9
1571	99.5090	-7.2565	2014.084	9.34	14.254	53.507	112.7	-0.5	-80	5.3	5.7	-83.4	5.4
1572	99.5733	18.4734	2013.6785	10.654	10.838	55.811	63.6	-75.7	-24.8	5.1	-75.9	-23.9	5.1
1573	99.5918	62.2384	2013.6025	9.127	15.675	12.275	84	-12.4	-85.5	5.5	-17.5	-83.8	5.6
1574	99.6876	55.0885	2013.457	14.931	15.384	4.033	2.1	0.6	-63.8	5.5	-1.9	-64.9	5.9
1575	99.7323	30.9463	2013.5115	16.486	16.851	4.238	98.1	14.6	-61.6	5.5	10.7	-61.1	5.6
1576	99.7893	9.4631	2013.49	15.445	15.657	4.257	212.1	56.9	-70.5	6	61	-71.8	6
1577	99.9099	62.3083	2013.6125	12.414	15.496	42.012	146.6	-118.3	-49.4	5.5	-117.6	-51.2	5.5
1578	99.9537	9.7181	2013.636	14.369	17.331	9.525	264.4	-8.3	-85.8	5.8	-13.9	-82.4	6
1579	100.0103	-4.1719	2014.1005	10.418	14.967	16.33	318.9	54.4	-116.2	5.3	50.8	-119.3	5.3
1580	100.1102	11.2962	2013.595	15.068	16.384	5.986	230.4	67.2	-42.9	5.8	68.7	-43.1	6.1
1581	100.1134	64.2536	2013.579	13.696	15.427	14.345	24.8	14.2	-138.4	5.5	16.1	-136.4	5.6
1582	100.1284	29.8097	2013.612	10.109	13.248	36.183	168.8	-22.7	-67.6	5.8	-23.8	-65	5.9
1583	100.2034	31.9893	2013.7325	9.737	9.901	55.681	37.4	58.9	-73.2	5.4	62.6	-65.5	5.4
1584	100.3307	82.2668	2013.928	9.041	13.515	8.562	281.5	48.9	-154.2	6.1	50.2	-155.1	6.2
1585	100.3474	-2.3505	2014.093	7.582	11.42	6.74	276.8	71.2	-142.2	5.4	73.4	-144.6	6
1586	100.4083	56.1583	2013.6745	12.46	17.824	8.738	187.8	-61.6	-27.1	5.5	-60.5	-24.2	6.2
1587	100.4937	20.8392	2013.628	15.819	18.024	56.444	282.8	3.5	-66.2	5.5	-5.9	-66	5.6
1588	100.6005	57.7513	2013.6365	13.135	17.486	12.811	342.4	67.1	-49.1	5.9	61.6	-50.7	6
1589	100.6338	33.0019	2013.4995	18.044	18.606	56.378	251.5	-7.2	-64.8	5.6	-6.3	-65.8	5.9
1590	100.6484	23.2299	2013.3245	16.079	18.026	30.562	281.4	-67.8	37.2	6.7	-73.2	25.9	6.1
1591	100.6820	36.0673	2013.677	10.946	14.433	11.63	57.6	-6.1	-61.7	5.8	-6.8	-61.8	5.8
1592	100.7198	18.0145	2013.5635	12.378	14.502	9.503	99.3	15.3	-71.4	5.9	12.9	-71.5	5.9
1593	100.7406	34.4502	2013.696	14.479	16.248	9.226	106.8	-3.7	-66.2	5.5	-7.7	-66.4	5.4
1594	100.7762	-6.6231	2014.0955	12.977	15.068	5.371	231.6	-71.5	19.3	5.4	-66.5	25.8	6.2
1595	100.7871	39.3454	2013.5405	10.344	12.195	56.298	349.7	6.2	-103.5	6.3	4.8	-106.8	6.3
1596	100.7966	50.5405	2013.62	13.393	13.631	5.738	102.5	-9.8	-70	5.5	-7.6	-70.9	5.5
1597	100.8627	76.8079	2013.5075	13.181	18.488	10.197	44.5	22.6	-166	5.9	24	-163.7	6.4
1598	100.8684	59.0655	2013.6175	14.923	16.6	5.379	178.1	18.6	-88.7	5.5	18.6	-89.2	5.6
1599	100.9026	8.8621	2013.5495	13.289	18.045	10.613	250.4	72.4	-65	5.8	66.8	-66.2	6.9
1600	100.9495	29.1567	2013.469	10.259	18.155	35.962	67.1	5.8	-63	6	0.7	-64.6	6.4
1601	100.9639	60.8124	2013.708	10.244	17.316	15.143	217	-5.4	-68.2	5.5	-13.2	-69.7	5.6
1602	100.9874	-10.1397	2014.1	10.088	11.579	49.427	122.7	55.9	-89.1	5.4	52.9	-86.4	5.3
1603	101.0129	30.3411	2013.5105	12.533	14.531	18.287	117.7	-3.6	-61.5	5.5	-4.2	-63.3	5.5
1604	101.0482	53.2952	2013.652	6.133	14.019	41.674	67.5	40.7	-194.2	5.6	39.1	-196	5.5
1605	101.0817	10.3336	2013.7065	15.322	15.823	14.48	54.5	-17.9	-110.4	5.8	-19.6	-111.3	5.9
1606	101.1627	28.9240	2013.4825	12.633	14.338	5.643	234.8	191.7	-13.5	5.5	189.8	-13.8	5.6
1607	101.2935	6.2416	2013.774	15.213	16.076	5.974	346.1	60.3	-76	5.8	59.8	-78.6	6
1608	101.4105	9.5414	2013.4495	13.31	18.078	14.095	65.7	-6	-81.8	5.8	1	-81.2	6.4
1609	101.4116	13.4447	2013.573	13.101	14.441	4.35	182.7	70.5	-21.8	5.9	73.8	-17.3	6.7

1610	101.4179	28.9172	2013.4505	12.018	16.518	29.028	258.8	-8.8	-63	5.4	-12.3	-63.5	5.5
1611	101.4761	26.4060	2013.367	16.19	17.292	14.228	99.4	-73.4	-112.9	5.6	-71	-111.3	5.6
1612	101.5286	9.0790	2013.6225	12.539	16.122	7.126	300.2	-98.3	-24.8	5.8	-98.7	-28.3	6.1
1613	101.5825	77.1933	2013.567	11.212	12.632	16.581	107.8	24.1	-81.9	5.9	24.7	-82.6	5.9
1614	101.6047	84.1033	2013.958	14.542	16.023	23.742	219.1	-129	-41.8	5.7	-129.3	-45.6	5.7
1615	101.6158	19.4846	2013.607	12.575	14.754	20.415	197.1	79.3	-102.9	5.2	78.5	-106.1	5.2
1616	101.6182	-6.6060	2014.0935	9.654	16.652	16.233	214.7	4.8	-79	5.3	-0.3	-81.3	6.2
1617	101.7070	36.6387	2013.5985	11.569	13.307	5.605	177.5	-43.7	-73.8	5.2	-43.7	-72.4	5.3
1618	101.7610	64.5207	2013.6215	16.216	16.313	7.641	330.4	27.8	71.1	5.5	25	72.4	5.5
1619	101.8486	6.1288	2013.804	12.5	15.12	25.304	356.2	68.1	-171.8	5.9	71.1	-170.8	5.9
1620	101.9149	5.9494	2013.526	14.65	16.809	23.062	211	101.4	-136.8	6	102.5	-138.6	6.7
1621	101.9714	58.5003	2013.5705	12.662	13.844	20.399	41.1	-50.6	-124.3	5.5	-53.7	-125.1	5.6
1622	102.0055	8.8128	2013.2925	16.385	17.916	52.021	63.4	-12.5	-62.4	5.9	-3.2	-63.3	7.1
1623	102.0195	14.0579	2013.668	11.291	15.504	10.186	158.4	66.7	-70.4	5.9	64.7	-71.3	6.1
1624	102.0477	0.3051	2014.0695	7.419	11.082	7.29	146.2	-72.1	39	5.7	-75.7	34.9	5.7
1625	102.1871	-6.8990	2014.116	14.407	14.416	29.461	36.1	2	-65.8	5.3	-1.1	-64.1	5.3
1626	102.1971	66.2790	2013.6055	13.503	16.194	28.857	293.1	103	43.4	5.8	107.8	44.7	5.9
1627	102.2015	31.8195	2013.629	16.233	18.217	46.928	87.1	-63.4	-1.2	5.5	-64	4.4	5.9
1628	102.2221	49.8843	2013.6955	13.923	14.017	6.658	252.9	66.2	29.3	5.8	68.7	30.7	5.8
1629	102.2528	50.2834	2013.638	13.41	16.856	12.812	310.5	-0.2	-70.6	5.8	-1.9	-71.5	5.9
1630	102.3077	37.4295	2013.6135	15.775	17.547	16.345	86.6	-22.6	-85.9	5.3	-15.5	-88.1	5.4
1631	102.4123	53.0310	2013.56	7.883	9.769	21.835	68.3	-7.1	-60.5	5.9	-8.2	-62.1	5.9
1632	102.4193	3.3756	2014.123	12.596	14.077	4.361	243.4	26.4	-192.8	5.7	33.6	-191	7.9
1633	102.4229	66.9127	2013.5475	11.608	14.598	6.098	265.4	3.5	-66.2	5.8	7.3	-66.1	6
1634	102.5611	-4.9847	2014.103	10.448	15.655	21.902	90.6	-22	-61.5	5.7	-24.5	-63.5	5.7
1635	102.5803	54.9092	2013.331	11.583	12.041	5.624	220.4	-17.4	-79.9	5.9	-20.6	-75.4	6.2
1636	102.6362	84.8926	2013.985	10.339	12.311	30.314	14.8	14.4	93.4	5.8	13.8	93.2	5.8
1637	102.7083	11.4902	2013.585	15.535	15.656	6.227	246	-20.7	-76.1	5.9	-18.2	-73.1	5.9
1638	102.7272	48.4535	2013.6445	14.686	15.346	33.615	347.1	43.4	-129.2	5.8	44.3	-130.6	5.9
1639	102.7775	29.6749	2013.6	13.387	13.565	34.355	263.5	-44.6	-71.4	5.2	-46.1	-71.8	5.2
1640	102.7847	84.5495	2013.9965	10.227	11.97	6.168	307.5	1.6	71.9	5.8	-1.8	71.3	5.9
1641	102.7907	28.1028	2013.4965	14.387	14.461	13.938	233.1	-95	-62.3	5.2	-95	-61.7	5.2
1642	102.7914	3.0949	2014.1535	8.929	16.627	17.414	184.4	-87	-44.5	5.7	-87.8	-49.3	6
1643	102.8348	17.3960	2013.5825	13.182	16.833	15.522	219.1	66.5	-91.5	5.6	63.7	-92.8	5.7
1644	102.9214	11.5196	2013.2785	18.064	18.18	12.5	88.5	23.2	-69.8	6.1	24.6	-67.9	6.7
1645	103.0700	11.5973	2013.7495	12.496	14.459	18.167	250	-17.6	-112.5	5.8	-21.8	-111.1	5.8
1646	103.1027	-8.0576	2014.1205	14.694	14.988	4.091	68.2	-24.1	-80.7	5.4	-28.1	-80	5.6
1647	103.1302	26.2294	2013.3005	15.188	18.401	56.174	130.9	30.8	-66.3	5.6	41.6	-60.1	6.8
1648	103.1382	58.3344	2013.664	10.78	11.571	6.134	97	-3.5	-80.4	5.9	-4.9	-79	5.9
1649	103.1678	27.7471	2013.4225	13.619	16.977	14.795	269.9	-79.8	61.9	5.2	-83	59.9	5.3
1650	103.2371	-0.4052	2014.1415	13.641	13.657	22.653	4.5	89.5	-60.6	5.7	89.4	-57.9	5.7
1651	103.2732	63.8106	2013.603	11.453	14.141	6.654	127.9	-8.2	-107.5	5.5	-7.8	-105.8	5.6
1652	103.2758	23.4500	2013.4575	11.408	12.319	19.181	205	72.8	-34.7	5.2	73	-37.2	5.2
1653	103.3168	30.4341	2013.544	14.503	14.917	7.287	187.5	-122.3	-85.4	5.5	-124.1	-85.3	5.5
1654	103.3616	4.3254	2014.1315	15.649	16.496	5.453	267.4	83.7	-63.5	5.8	78.6	-62.6	6.3
1655	103.3707	-3.3842	2014.156	14.9	15.763	5.227	100.1	-79.6	-48.6	5.7	-83.4	-40.2	7.3

1656	103.3926	58.2833	2013.632	10.857	11.816	10.854	86.4	-2.5	-81.4	5.5	-1.6	-80.2	5.5
1657	103.4459	47.5674	2013.543	12.638	16.889	18.117	10.676	34.3	-64.7	5.5	35.6	-62.6	5.5
1658	103.4584	82.4470	2013.895	14.903	15.48	3.911	244.8	-15.6	-88	5.5	-5.1	-92.9	6.6
1659	103.4619	-3.8276	2014.0865	10.838	15.504	7.322	322.9	-10.2	111.9	5.8	-2.3	107.4	5.9
1660	103.4805	20.3242	2013.4845	13.669	17.18	18.952	116.8	44.8	-130.4	5.2	43.3	-129.8	5.3
1661	103.4907	3.0067	2014.126	12.679	13.216	20.078	141.9	70.3	-46.1	5.7	70.2	-45.1	5.7
1662	103.5967	78.9503	2013.5715	15.878	16.239	4.904	99.1	-15	-66.7	5.6	-21.6	-62.1	5.7
1663	103.6196	7.7940	2013.402	15.825	18.867	13.604	207.8	14.9	-72.8	5.8	14.2	-73	7.8
1664	103.6389	2.4678	2014.163	16.412	16.629	10.61	1.2	-70.6	-22.8	5.8	-72.6	-26.4	6.6
1665	103.7004	13.9729	2013.74	11.329	14.702	44.847	50.5	35.9	-69.7	5.8	36.8	-71.9	5.9
1666	103.7756	54.3942	2013.529	10.011	11.987	4.883	145.6	-10.6	-62.2	5.5	-10.2	-61.9	5.8
1667	103.8146	10.2431	2013.5255	14.306	17.612	7.114	341.7	-28.9	-129.4	5.9	-32.7	-127	6.6
1668	103.8885	40.3692	2013.667	15.479	16.087	11.256	127	-31.9	-74.7	6	-34.6	-71	6
1669	103.9069	-9.4900	2014.068	8.351	9.467	13.894	300.5	-21.4	-72.4	5.4	-20.9	-71.9	5.4
1670	103.9393	54.5520	2013.676	9.743	11.385	35.607	316.5	0.1	-67.5	5.5	3.3	-66.2	5.5
1671	103.9662	61.4985	2013.6655	9.781	14.349	21.859	293.5	-3.3	-114.7	5.5	-3.4	-112.6	5.5
1672	104.1794	-7.1091	2014.092	14.404	15.229	19.642	27.4	-10	-96.8	5.4	-15	-93.5	5.5
1673	104.2203	38.7597	2013.6825	13.077	17.388	24.591	214.6	29	-110.8	5.4	26.6	-112.2	5.5
1674	104.2736	0.6090	2014.136	10.027	13.269	29.649	122.3	62.8	-99.8	5.7	63.5	-95.8	5.7
1675	104.3537	49.3908	2013.6765	12.115	14.305	7.685	341.2	1.2	-76.1	5.4	3.3	-73.6	5.5
1676	104.3543	52.1762	2013.7295	12.884	13.919	4.593	188	-193.3	-12.1	5.5	-194.9	-15.1	5.6
1677	104.5143	15.1760	2013.429	15.527	17.682	58.418	66.1	28.6	-60.5	5.9	17.6	-67.5	6.3
1678	104.5295	-9.8698	2014.0835	9.412	11.442	9.59	335.5	-76.5	-73.9	5.4	-76.3	-78.2	5.3
1679	104.5328	18.7492	2013.353	16.356	17.021	3.679	338.7	-68.1	-64	5.2	-61.9	-72.9	5.7
1680	104.6078	75.3717	2013.5475	16.437	17.21	7.807	1.2	62.4	-21	5.6	63.1	-18.6	5.6
1681	104.6089	-12.9923	2014.0795	8.4	12.36	50.037	102.8	71.5	-125.2	5.3	80	-120.4	5.4
1682	104.6618	6.2104	2013.856	12.95	14.457	7.086	22.4	-51.9	75.8	5.8	-55.9	75.4	5.8
1683	104.6822	47.6493	2013.664	11.301	11.318	20.547	147.4	37.7	-72.5	5.4	34.9	-70.6	5.4
1684	104.6848	-9.8628	2014.125	12.753	14.063	8.752	190	-31.7	-89.8	5.3	-32.2	-89.7	5.3
1685	104.6960	16.3778	2013.3445	16.365	18.099	49.428	306	-20.2	-70.7	5.7	-13.3	-74	5.7
1686	104.7638	56.5169	2013.719	14.505	16.824	8.599	82.6	-31.1	-138.6	5.5	-32.4	-137.2	5.5
1687	104.7647	-11.7278	2014.0875	12.67	14.265	12.872	52	11.9	-97.8	5.3	13.2	-97.9	5.3
1688	104.7648	35.9677	2013.6725	14.33	15.831	6.512	188.7	13.8	-99.1	5.4	17	-101.5	5.5
1689	104.7816	43.6175	2013.5695	10.218	14.685	24.555	246.8	64.3	-41.3	5.4	61.9	-39.7	5.5
1690	104.7914	-6.6216	2014.1345	15.41	15.57	5.857	316.8	17.3	-108.1	5.4	17.5	-110.5	5.3
1691	104.8095	30.4368	2013.53	15.571	16.07	6.77	221	4.7	-80.3	5.2	3.5	-80.4	5.3
1692	104.8104	40.8698	2013.6335	14.054	14.595	37.274	51.8	-34	-73.7	5.4	-35.4	-73.7	5.4
1693	104.8419	-0.8809	2014.1075	13.083	17.099	13.984	152.4	-58.8	-191.6	5.3	-59.8	-186.1	5.7
1694	104.8529	22.1129	2013.3925	14.303	16.941	15.149	39.9	4.3	-77.2	5.2	7.3	-77.1	5.3
1695	104.8900	18.8667	2013.703	10.645	16.189	23.118	38.9	-66.6	-40.7	5.1	-66.8	-37.4	5.1
1696	104.9101	28.5668	2013.4835	14.835	17.364	6.536	114.2	-33.8	-72.5	5.2	-35.2	-70.3	5.5
1697	104.9419	18.5398	2013.298	17.345	17.775	3.72	134.7	25.1	-62.4	5.3	17.8	-63.5	5.8
1698	104.9900	46.7259	2013.698	13.437	15.753	5.727	349.1	-21	-95.2	5.4	-20.5	-96.3	5.5
1699	105.0610	24.9730	2013.2745	16.379	18.278	10.955	149.4	12.6	-69.6	5.3	8.3	-69.5	5.6
1700	105.1476	30.9160	2013.58	16.269	16.375	4.655	349.9	38.8	-72	5.3	33.7	-72.3	5.3
1701	105.2180	-1.4424	2014.1585	13.671	15.982	13.497	171.1	-68.7	12	6.1	-68	8.1	6.1

1702	105.2729	54.5680	2013.6005	15.791	17.416	9.909	112.5	-5.7	-68.7	5.5	-7.7	-69	5.6
1703	105.3052	-13.6201	2014.0945	13.56	15.051	5.61	169.6	-7.9	-96.3	5.4	-10.3	-94.6	5.3
1704	105.3271	52.0527	2013.685	9.688	11.253	4.828	288.2	-17.4	-65.3	5.5	-21.6	-64.8	5.5
1705	105.4329	21.2381	2013.3055	14.361	14.664	8.655	273.3	-0.5	-61.4	5.3	0.1	-61.3	5.2
1706	105.4340	30.3162	2013.5615	14.702	15.917	58.898	111.1	-1	-63.3	5.2	-2.9	-64.2	5.3
1707	105.5459	-6.0395	2014.139	14.285	15.633	22.04	92.2	112.5	-159.9	5.3	114	-161	5.5
1708	105.5782	16.9525	2013.852	8.554	12.558	9.814	32.3	-36.3	-85.3	5.7	-37.9	-87.3	5.8
1709	105.6380	25.8564	2013.432	16.693	17.059	5.641	353.9	64.2	-53.1	5.3	62.9	-51.9	5.3
1710	105.6678	35.9358	2013.682	10.884	15.101	41.32	4.7	2.6	-148.4	5.2	-0.7	-149	5.2
1711	105.7127	14.9380	2013.6485	12.427	13.695	15.497	101.4	-41.4	-109.5	5.5	-44.4	-111.3	5.5
1712	105.7915	73.3633	2013.5475	17.699	17.941	3.816	232.6	-26.4	-60.7	5.7	-17.9	-65.3	5.7
1713	105.8498	1.5384	2014.0175	13.365	16.065	5.669	47.8	121.3	-130.3	5.9	126.9	-131.9	6.6
1714	105.8502	8.9448	2013.6665	12.221	12.979	6.363	27.1	-66.4	5.8	6	-65.6	2.6	6
1715	105.9423	5.9899	2013.7515	13.463	14.001	10.969	140.1	35.7	-69.1	5.8	42.9	-67.7	5.8
1716	105.9953	47.0500	2013.538	7.173	15.126	18.396	145.3	-83.9	8.7	5.4	-83.4	19.2	6
1717	106.0309	20.3935	2013.425	12.539	18.34	15.361	75	69.7	-53.8	5.2	68.1	-51.6	5.5
1718	106.0542	60.9802	2013.5825	13.562	18.314	9.74	102.8	-7	-62.1	5.5	-8.8	-63.8	6.6
1719	106.0812	-6.5920	2013.836	16.246	16.68	30.401	334.9	25	-106.1	5.6	28.9	-110.2	6
1720	106.1384	25.6198	2013.496	11.898	16.245	7.188	336.5	-6	-61.6	5.2	-11.3	-62	5.3
1721	106.1756	17.3665	2013.4545	12.226	12.866	4.586	201.8	34.5	-111	5.6	33.9	-107.2	5.9
1722	106.1884	-6.0358	2014.08	13.215	14.086	12.21	179	40.2	-147.2	5.5	38.1	-143.7	5.5
1723	106.1932	59.0130	2013.637	11.858	12.494	13.221	103.4	7.7	-69.9	5.5	6.7	-69.4	5.5
1724	106.2275	28.1167	2013.759	8.875	9.454	12.801	168	74.4	-92.1	5.1	73.5	-93.5	5.1
1725	106.2892	39.0576	2013.677	12.089	17.3	47.188	279.6	3.6	-64.7	5.2	0.6	-62.9	5.3
1726	106.5986	40.7868	2013.7005	14.618	15.06	4.081	0.2	4.2	-63.6	5.2	7.3	-64.3	5.4
1727	106.6848	10.0780	2013.44	17.066	18.641	18.626	19.2	60.5	30.6	6.1	63.2	25.2	6.8
1728	106.6870	40.8664	2013.605	13.12	16.019	7.918	15.5	-32.7	-77.7	5.2	-30.5	-77.7	5.3
1729	106.6961	42.3676	2013.642	14.836	16.01	15.152	349	-13.5	-62.9	5.4	-7.4	-63.6	5.4
1730	106.7130	25.5328	2013.429	14.199	16.129	4.828	231.8	65.3	-105.9	5.2	65.7	-101.2	7.2
1731	106.7421	13.2920	2013.713	10.278	15.037	7.409	165.3	15.8	-65.6	5	19.4	-66.9	5.4
1732	106.7751	82.2807	2013.974	13.767	16.748	6.588	137.5	5.5	-63.6	5.4	2.6	-61.4	5.7
1733	106.8300	52.6365	2013.573	14.833	14.909	5.061	254	-31.7	-90	5.6	-28.4	-94.7	5.5
1734	106.8733	18.5422	2013.7045	10.419	14.075	13.829	166	54.1	-65.9	5.1	50.2	-67.9	5.1
1735	107.0076	58.2708	2013.6815	10.384	11.382	4.485	79.5	-24.3	-151.5	5.9	-34.8	-152.3	7.5
1736	107.0138	50.7666	2013.5405	12.272	15.39	6.814	158.6	33.2	-71.9	5.5	31.7	-69.4	5.6
1737	107.1134	65.5915	2013.5675	13.677	18.487	47.823	342.8	-5.5	-67.7	5.6	-6.8	-71	5.8
1738	107.1479	50.8952	2013.605	14.156	15.026	8.007	14.4	-73.5	-27.4	5.5	-72.7	-27.7	5.5
1739	107.2412	23.1679	2013.4185	14.91	16.843	7.85	333.7	-11.2	-92	5.9	-13.1	-89.7	6.1
1740	107.2890	41.3038	2013.594	13.85	17.264	55.774	50.3	90.9	-182.8	5.6	94.1	-186.7	5.8
1741	107.3052	63.0531	2013.6615	14.816	15.501	10.173	301.3	65.3	-12.9	5.5	65.8	-6.7	5.5
1742	107.3061	61.1584	2013.667	13.534	14.034	50.631	29.3	-28.9	-105.1	5.5	-29.1	-106.1	5.5
1743	107.3864	5.5667	2013.869	10.391	13.067	8.558	276	106.4	-180.2	5.9	103.9	-187.1	6
1744	107.4108	-4.7425	2013.9115	12.327	13.211	5.642	169.7	20.1	-97.3	5.3	15.8	-97.1	5.5
1745	107.4272	46.0255	2013.654	11.892	12.12	35.093	71.7	-121.7	-66.2	5.9	-121.8	-62.2	5.9
1746	107.4744	44.7527	2013.691	10.981	14.777	12.795	293.1	-67	-51	5.9	-69.6	-51.4	5.9
1747	107.4987	56.5859	2013.7025	11.219	14.594	7.149	25.8	-48.3	-63.6	5.9	-50	-60.7	5.9

1748	107.5871	49.9306	2013.613	12.248	17.223	20.197	192.2	-8	-82.9	5.5	-6.7	-82.8	5.5
1749	107.5953	-4.9753	2012.947	17.805	17.895	41.737	172.8	-14.6	-88.1	6.1	-22.4	-82.5	7.8
1750	107.6018	37.1960	2013.725	11.48	16.298	9.339	281.3	-58.8	-68.6	6	-59.8	-69	6
1751	107.6838	14.3700	2013.369	17.564	17.594	57.143	316.3	-65.7	-3.6	5.2	-63.1	2	5.3
1752	107.7385	57.0579	2013.7125	13.449	14.657	34.643	342.2	93.7	-22.8	6	93.4	-20.3	6
1753	107.7581	-0.7823	2014.0645	15.136	15.387	3.179	315.8	-61.8	-66.6	5.7	-57.1	-70.3	5.4
1754	107.8185	16.3971	2013.5725	12.139	16.712	21.469	340.2	-99.6	-44.4	5.1	-101.3	-45.5	5.3
1755	107.9679	50.9156	2013.6315	10.247	12.527	9.028	147.9	-29.9	-81.5	5.5	-29.6	-84.9	5.5
1756	108.0253	8.8915	2013.753	12.186	14.569	12.502	110.6	-7.9	-99.1	5.8	-6.6	-98.5	5.9
1757	108.0773	27.2194	2013.5895	14.341	14.372	9.086	75.3	-42.4	-162.6	5.5	-39.9	-160.6	5.5
1758	108.1287	40.5071	2013.7685	16.459	16.895	4.592	61.9	7	-68.4	6.3	6.3	-67.2	6.4
1759	108.1347	54.7433	2013.6045	11.735	18.597	52.067	4.7	-18.4	-71.5	6	-25.9	-72.6	6.6
1760	108.2509	-8.1126	2014.144	8.222	15.673	14.986	316.2	49.8	-60.4	5.4	52.7	-61.1	5.6
1761	108.2785	51.8075	2013.5965	13.47	15.309	49.994	312.5	17.9	-76.2	5.5	19.9	-74.1	5.5
1762	108.3165	20.2587	2013.4625	13	17.21	33.96	124.6	-65.1	-11.1	5.1	-64.4	-8.2	5.3
1763	108.4142	-11.2454	2014.09	12.052	14.012	47.701	140.5	37.9	-89.2	5.4	42.2	-88.7	5.4
1764	108.4626	3.2240	2013.7535	13.918	14.322	14.579	107.4	12.6	83	5.8	14.2	82.9	5.9
1765	108.5457	34.0932	2013.6385	15.622	17.46	27.279	128.1	-76.9	-37.6	6	-79.4	-32.8	6
1766	108.6742	31.3521	2013.681	10.646	12.536	15.555	296	-16.6	-104.3	6	-14.3	-105.9	6
1767	108.7429	81.1848	2013.7745	14.928	15.939	13.652	197.4	68.9	22.1	5.5	66.2	27.2	5.5
1768	108.7788	57.2767	2013.6875	8.101	12.61	28.058	140.5	98	-84.7	5.5	98.1	-80.4	5.5
1769	108.8324	7.4351	2013.7605	12.981	13.879	10.358	131.7	-133.3	45.6	5.9	-131.9	44.8	5.9
1770	108.8381	3.0238	2013.599	12.781	16.333	19.361	215.1	43.2	-150.4	5.9	35.7	-151.3	6.4
1771	108.8876	10.5047	2013.36	16.699	17.342	13.426	50.9	38.6	-74.3	6.1	39.5	-74.6	6.3
1772	108.8957	74.0453	2013.6095	12.191	15.845	12.815	267.6	29.7	-135.1	5.6	34.9	-133.4	5.6
1773	108.9549	17.5910	2013.643	11.613	12.461	6.178	166.3	-141.8	8.2	5.1	-140.7	7.5	5.2
1774	108.9640	45.5739	2013.685	12.43	16.239	10.611	282.1	-9.3	-69.9	5.6	-8.8	-73.1	5.6
1775	109.0348	79.6099	2013.686	12.723	14.118	24.432	342.1	154.9	18.3	5.5	158.8	15.4	5.6
1776	109.0353	10.6656	2013.4915	10.84	13.443	6.719	204.1	-6.2	-78.1	5.8	-5.8	-77.2	6.1
1777	109.0433	-2.2629	2013.9005	9.954	13.96	45.659	30.3	-43.1	-63.8	5.4	-43.2	-63	5.5
1778	109.0546	11.3623	2013.6475	12.449	13.399	31.925	30.5	-6.1	-91	5.9	-10.6	-89.1	5.9
1779	109.1589	-8.6817	2014.138	13.808	15.958	52.141	337.6	-2.3	-61.1	5.4	-5.2	-62.2	5.8
1780	109.1859	-1.0331	2014.002	15.008	15.279	4.926	236.8	-61.6	-34.6	5.4	-61	-33.1	5.3
1781	109.2485	65.2620	2013.5855	13.961	14.472	11.144	289.6	-8.7	-63.1	5.5	-5.9	-61.5	5.5
1782	109.2548	41.4050	2013.613	14.582	16.947	11.918	71.7	15.2	-65.7	5.2	14.7	-64.9	5.3
1783	109.2685	2.5577	2013.585	15.969	16.222	7.53	253.3	19.9	-104.5	6	23.1	-109.4	6.1
1784	109.2851	-4.6156	2013.3565	14.976	18.012	24.184	33	-126.2	-18.9	5.5	-131	-24.9	6.2
1785	109.2877	47.6547	2013.689	9.538	9.762	12.99	169.5	61.6	-102.9	5.6	61.8	-106.6	5.6
1786	109.4463	-8.9021	2014.1405	10.356	12.468	57.132	293.6	-47.3	89	5.4	-54.3	89.1	5.4
1787	109.7010	62.6534	2013.668	14.153	14.492	13.982	248.8	-19	-71.6	5.5	-16.5	-73	5.6
1788	109.7078	46.6661	2013.7315	17.185	18.456	25.283	346.3	22.9	-61	6	21.3	-61.4	7.1
1789	109.7156	17.8672	2013.7005	13.643	15.506	14.282	120.1	64.2	-65	5.1	66.2	-65.2	5.1
1790	109.8016	20.9595	2013.813	8.413	13.649	31.993	333.8	-40	-120.8	5.3	-39.3	-116.3	5.5
1791	109.8189	84.2326	2014.0005	13.215	18.586	36.686	259.7	8.3	-65.6	5.8	1.9	-63.8	6.2
1792	109.9454	-4.2669	2012.947	17.272	17.696	28.381	41.9	39.2	-69	6.4	45	-61.1	7.4
1793	109.9560	71.8771	2013.6055	13.092	13.844	18.242	241.1	-35.9	-131.5	5.6	-33.5	-134.1	5.6

1794	109.9813	33.5927	2013.749	12.68	13.904	4.032	275.3	-36.7	-78.7	5.8	-30.1	-82.9	5.9
1795	110.1586	36.9626	2013.6925	11.158	16.327	12.531	5.5	-1.8	-163.7	5.8	-1.7	-165.7	5.9
1796	110.3049	-11.6634	2014.1235	11.279	15.454	7.938	152.8	-199.4	-23.7	5.5	-198.5	-24.4	5.4
1797	110.3586	52.0076	2013.6115	14.46	15.956	7.645	98.7	12.1	-82.5	5.6	7.8	-82.1	5.6
1798	110.3754	12.5299	2013.7285	10.37	13.508	48.79	152.6	-114.9	-14.7	6.3	-115.9	-13.5	6.4
1799	110.4087	4.9945	2013.3985	16.961	17.155	48.563	7.7	-69.1	16.2	6.2	-67.9	24	6.1
1800	110.4190	35.8507	2013.695	16.111	17.511	26.137	184.8	17.7	-110.4	5.8	15	-109.4	5.9
1801	110.4298	-1.1105	2013.424	15.867	16.017	4.546	101.2	-0.7	-79.2	5.6	1.2	-76.4	7.2
1802	110.5023	47.2039	2013.6625	12.054	15.085	15.865	339.4	-38.1	-89.1	6	-39	-87.5	6
1803	110.5574	6.3857	2013.5925	11.959	15.537	44.302	20.9	43.5	-83.2	5.8	46	-83.7	6
1804	110.6357	35.8922	2013.7695	9.802	10.408	30.447	323	10.9	-139	5.8	21.7	-143.2	5.7
1805	110.6410	5.7861	2013.409	17.493	17.859	46.538	249.3	40.7	-70.8	6.2	46.9	-66.2	6.4
1806	110.6731	3.4547	2013.531	15.646	16.19	8.792	267.7	16.1	-77.2	6	20.8	-78.1	6
1807	110.7249	-5.3254	2013.425	11.322	17.454	17.658	244.5	-92.3	-60.4	5.5	-95.1	-61.6	6.2
1808	110.7463	53.8277	2013.704	10.607	13.538	7.354	0.2	17.8	-80.3	5.8	15.6	-78.2	5.8
1809	110.7559	31.7657	2013.677	9.959	14.208	19.276	247.1	-43.8	-134	5.8	-42.9	-134.7	5.8
1810	110.7781	14.1110	2013.376	16.399	17.745	9.335	121.5	0.4	-65.6	5.2	1.5	-68.4	5.4
1811	110.8454	32.9902	2013.6885	12.514	12.881	7.995	286.8	2.4	-66.2	5.4	1.6	-66	5.4
1812	110.8467	79.1970	2013.73	11.058	13.606	21.539	102.8	-31.4	87.5	5.8	-29.1	88	5.8
1813	110.9333	13.8629	2013.6265	12.143	14.694	7.618	197	1.8	-60.3	5.1	0	-61.7	5.1
1814	110.9364	73.4397	2013.5785	14.212	15.599	9.543	286.3	25.4	-85.8	5.6	22.9	-87.3	5.6
1815	110.9573	47.8849	2013.654	12.311	14.44	7.86	236.9	-6.2	-92.3	6.3	-2.4	-90.6	6.3
1816	110.9660	-7.9593	2013.872	9.312	12.743	20.313	14	-37.5	138.9	5.5	-32.1	135.3	5.7
1817	110.9747	9.9208	2013.4955	11.929	16.365	8.016	164.4	50.7	-89.6	5.8	53.9	-88.5	6.2
1818	111.0017	15.6406	2013.662	10.824	12.783	10.025	206	-60.5	-60.7	5.1	-62.8	-58.1	5.1
1819	111.1105	33.5660	2013.5925	16.629	16.69	4.225	275.1	-20.2	-78	5.5	-14.2	-79.7	5.7
1820	111.1397	50.7964	2013.665	10.204	12.793	19.776	130	74.8	-115.8	5.8	71	-113.2	5.8
1821	111.2499	3.7485	2014.043	9.14	12.333	5.212	10.6	-13.8	-99	5.7	-7.9	-100.6	6.6
1822	111.2810	16.0294	2013.874	10.953	11.488	4.21	14.4	-69.7	7.9	5.1	-68.4	5.4	5
1823	111.4238	29.5379	2013.63	13.072	16.624	6.909	355.2	-46.9	-84	6	-47.3	-88.1	6.1
1824	111.4890	-0.0216	2013.1245	17.325	17.56	47.442	90.3	5.4	-65.1	6.5	5.5	-63.5	6.6
1825	111.5904	12.8432	2013.449	15.16	17.206	7.077	253.3	7	-146.9	5.1	7.9	-145.4	5.4
1826	111.6389	66.3372	2013.5865	11.557	15.587	19.734	31.3	-46.6	-113.3	5.6	-46.3	-112.2	5.6
1827	111.6451	-3.4226	2013.892	14.764	15.36	8.927	278.7	21.3	-67.7	5.4	17.9	-69	5.4
1828	111.6671	26.9793	2013.624	10.626	12.518	5.395	331.9	-126.2	-97.4	6	-127.4	-99.3	6.1
1829	111.7409	38.7464	2013.7085	11.038	16.882	40.25	122.9	-14.3	-115.9	5.2	-12.3	-117	5.3
1830	111.7714	21.4223	2013.6145	13.875	14.577	39.004	100.3	-6.2	-66.1	5.1	-5.7	-64.6	5.2
1831	111.8110	4.4943	2013.6745	14.671	15.255	10.244	182.9	75.6	-38.2	5.9	75	-37.4	5.9
1832	111.8202	22.9240	2013.318	16.904	18.486	7.719	53	24.3	-67.1	5.2	21.9	-67.2	5.7
1833	111.8831	10.9133	2013.658	15.719	16.129	28.213	237.8	-21	-146.8	6.1	-22.2	-144.1	6.1
1834	111.8854	33.9448	2013.6875	14.451	17.178	14.536	210.3	-55.4	-124.6	5.4	-58.9	-124.1	5.6
1835	111.9164	24.3362	2013.865	10.663	12.283	6.934	320.6	-7.7	-106.6	5.1	-8.9	-101.7	5.1
1836	111.9229	25.6611	2013.563	15.295	16.724	8.226	322.5	41.6	-62	5.2	39.6	-63.4	5.3
1837	111.9619	42.4716	2013.673	14.257	15.941	32.074	172.8	48.9	-147.9	5.2	47.2	-148.9	5.3
1838	112.0169	11.1470	2013.765	13.95	14.722	40.192	51.9	8.9	-91.2	6	12.1	-92.5	6
1839	112.0177	-13.3739	2014.05	10.534	15.5	13.438	222	-65.6	-48.3	5.5	-63.9	-49	5.5

46

1840	112.0205	27.5675	2013.6785	10.002	15.044	37.023	242.2	-6.6	-64.7	5.1	-6.4	-66.6	5.2
1841	112.0654	43.7589	2013.68	14.177	14.866	53.744	343.5	4.5	-67.1	5.9	9.2	-68.3	5.8
1842	112.0664	53.3001	2013.646	12.511	14.474	22.415	350.8	61.4	-26.2	5.8	62.6	-23	5.8
1843	112.0867	31.0030	2013.659	13.769	16.116	9.409	296.2	22.7	-152	5.5	22.2	-149.7	5.5
1844	112.1128	-1.7896	2013.6585	12.501	15.061	11.256	13.9	-92.5	20.3	5.5	-95.3	20.7	5.5
1845	112.1245	89.6838	2013.101	14.777	15.918	17.542	307.3	202.3	-146.7	6.7	199	-142.5	6.8
1846	112.1249	-12.5835	2014.1285	10.779	12.474	24.57	237.7	-76.9	54	5.5	-78.3	53.5	5.6
1847	112.1295	34.6177	2013.7705	11.14	12.896	34.146	86.4	94.3	-55.2	5.4	88.1	-55.2	5.4
1848	112.1364	29.0525	2013.6725	12.507	15.46	26.043	228.5	-63	-25.2	5.1	-65	-25.3	5.2
1849	112.2071	71.4665	2013.6835	13.081	15.507	5.891	119.7	4	-84	5.6	4.9	-82	5.8
1850	112.2094	19.0504	2013.7175	10.365	13.736	28.943	141.5	24.7	-73.9	5.1	24.8	-76.1	5.1
1851	112.2159	73.2563	2013.59	13.564	16.25	44.254	311.2	-26.7	-74.4	5.6	-28.8	-75.2	5.6
1852	112.2667	5.7670	2013.593	14.392	15.562	42.199	56.1	62.6	-66.4	5.9	60.3	-68.7	6
1853	112.2805	4.7739	2013.3945	13.177	17.37	16.273	21.9	63.1	-148.7	5.9	58.3	-158.5	7.2
1854	112.2808	39.1631	2013.6655	14.585	17.505	12.757	24.9	56.9	-83.5	5.2	57.1	-89.5	5.3
1855	112.4078	20.7262	2013.369	17.687	17.872	44.655	256.3	-77.8	-1.5	5.3	-79.9	-7.3	5.7
1856	112.5078	31.0549	2013.676	10.579	14.681	48.193	223.6	-27.3	-110.6	5.4	-32.7	-108.6	5.5
1857	112.7015	53.9625	2013.6425	14.917	16.259	9.548	331.3	-42.2	-122.9	5.5	-40.4	-122.8	5.5
1858	112.7034	17.7580	2013.6965	13.631	14.278	38.457	280	-18.6	-68.3	5.1	-19.5	-66.9	5.1
1859	112.7419	55.9632	2013.7135	16.853	17.592	6.114	111.1	-6.2	-69.8	5.9	-9	-71.2	5.9
1860	112.7489	32.3632	2013.5145	18.119	18.645	48.338	46	-17.3	-65.9	5.6	-14.9	-63.6	7
1861	112.8236	-7.4032	2013.8405	10.156	11.853	10.304	283.8	53.6	-111.8	5.4	52.1	-106.9	5.5
1862	112.8578	26.0941	2013.6245	10.946	16.142	10.756	145	-110	24.2	5.9	-110.9	31.3	6
1863	112.8596	-2.6193	2013.823	9.655	14.63	10.522	337.3	-146.5	109.6	5.4	-149.4	106.4	5.7
1864	112.9009	62.0193	2013.71	10.773	11.995	22.818	74.7	34.9	-152.8	5.5	29.4	-159.4	5.5
1865	112.9075	78.9570	2013.724	15.045	15.091	3.993	285.6	-1.7	-74.4	5.7	-7.9	-70.5	5.7
1866	112.9375	48.7822	2013.6845	16.183	17.807	12.672	203.2	-99.2	-10.8	5.8	-99	-12.7	6
1867	112.9595	8.8156	2013.8685	10.14	13.52	16.579	183.2	-24.2	62.5	5.7	-22.1	61.2	5.8
1868	112.9897	18.2790	2013.6555	12.691	15.867	10.116	146.8	-176.2	8.6	5.1	-175.2	4.4	5.2
1869	113.0217	80.8570	2013.937	13.479	18.159	8.779	275.2	-19.9	-88.6	5.5	-25.3	-85.9	5.8
1870	113.0237	-8.8820	2014.1125	6.568	9.523	23.797	112.8	-98.8	-183.5	5.3	-96.8	-178.4	5.4
1871	113.0644	7.1770	2013.591	13.982	14.506	13.621	243.1	83.6	-81.1	5.9	83.1	-80.4	6
1872	113.1340	68.8548	2013.6545	13.618	14.839	41.196	133.6	-0.1	-100.9	5.5	-0.3	-103.2	5.5
1873	113.1410	29.0243	2013.633	13.061	15.287	34.033	13.3	-15.4	-67.3	5.1	-16.4	-64.7	5.2
1874	113.1509	23.7069	2013.5385	12.967	16.771	9.424	30.1	9	-97.1	5.1	12.8	-96.7	5.3
1875	113.1601	48.0607	2013.599	17.705	18.068	9.331	310.8	66	-16.6	5.9	64.3	-17.3	6
1876	113.2043	17.9651	2013.792	9.391	10.635	23.815	56.2	-31.8	-80.4	5.8	-36.2	-81.8	5.9
1877	113.3889	-7.4763	2013.6305	14.191	15.217	6.821	256.1	-19.1	-83.1	5.6	-19.7	-81.8	5.5
1878	113.4160	0.4990	2013.702	12.498	13.818	31.129	2.1	-6	-166.7	5.9	-2.2	-167.6	6
1879	113.5852	6.5093	2013.6915	12.102	16.396	7.39	244.7	-12.7	72.6	5.8	-6.2	71.9	6.2
1880	113.6728	65.8155	2013.6355	16.475	16.674	51.983	142.8	-11.9	-70.5	5.5	-3	-70.6	5.5
1881	113.6765	31.7619	2013.678	15.01	16.075	12.083	25.4	29	-68.5	5.4	29.8	-69.5	5.5
1882	113.6837	48.9808	2013.6775	10.364	13.689	12.298	60.7	-27.4	-106.8	5.8	-26.2	-107.7	5.8
1883	113.7194	22.8709	2013.4285	15.053	19.02	14.928	323.4	22.1	-72.8	5.1	11.1	-76.8	6.1
1884	113.7481	9.9125	2013.917	12.082	13.067	55.714	22.3	-36.1	65	5.7	-42.5	63.9	5.7
1885	113.7551	51.4407	2013.6485	11.908	13.773	11.449	120.9	-41.6	-62.6	5.8	-40.4	-61.8	5.8

1886	113.8159	-9.9946	2014.1275	10.169	12.905	20.235	281.9	26.9	-62	5.3	27.8	-63.4	5.3
1887	113.9043	23.1674	2013.609	17.855	19.502	51.316	324.9	2.9	-62.7	5.3	-5	-62.1	6.3
1888	113.9179	27.8235	2013.7155	14.098	15.183	17.574	102.5	-91.8	-60.5	5.1	-92	-59.5	5.2
1889	113.9248	55.8672	2013.6865	14.693	15.727	5.848	28.2	-15.7	-65.6	6	-19	-67.8	6
1890	113.9694	-5.9678	2012.9465	16.883	17.049	31.383	192.6	0.1	71.8	6.3	-6.1	73.5	6.7
1891	114.0194	35.2857	2013.7385	11.432	12.828	22.11	286.8	21	-79.9	5.4	19.6	-83.5	5.4
1892	114.0209	36.8440	2013.7755	10.257	14.714	13.077	179.5	-51.1	-99.7	6.2	-57.8	-98.1	6.3
1893	114.0666	48.2005	2013.688	15.41	17.99	8.654	192	-12.2	-85.1	5.8	-16.4	-83	6.1
1894	114.1145	51.3055	2013.666	15.467	15.992	16.427	358.5	-28.8	-69.4	5.8	-28.4	-71.7	5.8
1895	114.1401	61.2840	2013.7035	13.594	14.981	8.262	221.2	20.2	-64.7	5.5	18.5	-62.6	5.5
1896	114.1639	21.4107	2013.5215	15.899	17.753	5.195	202	-37.4	-96.8	5.2	-33.1	-93.9	5.8
1897	114.2744	2.7203	2013.1315	17.671	18.411	25.829	1.9	65.6	-12.4	6.3	68.7	-11.9	6.6
1898	114.2829	48.8149	2013.6505	12.984	16.148	43.227	202.6	11.7	-84.2	5.8	11.3	-86.4	5.8
1899	114.2965	-4.5860	2013.6365	10.62	15.558	26.063	92.1	-12.3	-82.7	5.4	-16.2	-83.2	5.6
1900	114.4567	38.6707	2013.845	16.976	18.237	5.312	174.5	-19.6	-97.9	5.2	-21.3	-96.9	6.1
1901	114.5575	-2.2932	2013.802	11.085	15.211	19.414	234.6	-62.6	13.4	5.4	-66	12.4	5.5
1902	114.6245	-3.1919	2014.0515	10.162	10.317	7.145	304.3	-85.6	-104	5.4	-85.2	-109.4	5.3
1903	114.6548	68.4069	2013.68	14.179	15.954	6.823	120.8	7.9	-77	5.5	7.6	-77.4	5.5
1904	114.6609	47.8349	2013.6585	12.387	15.359	48.646	10.4	-72.8	-102.8	5.5	-72.2	-99.3	5.5
1905	114.6864	40.0057	2013.7	11.271	14.331	35.931	294.3	40	-73.6	5.2	39.4	-71.4	5.2
1906	114.7988	69.4325	2013.5095	15.141	18.409	7.194	221.2	-107.3	-91.7	5.5	-99.5	-101	6.8
1907	114.8041	11.3461	2013.5145	14.885	18.529	19.309	5.1	17.7	-105.8	5.8	11.4	-107.8	6.9
1908	114.8359	71.1379	2013.7175	15.928	17.381	7.709	139.9	16.7	-73.5	5.5	13.3	-72.7	5.6
1909	114.8829	24.5950	2013.4315	11.35	13.617	5.581	125.7	-61.3	-39.3	5.1	-64.3	-36.3	5.4
1910	114.8981	34.0354	2013.836	17.3	18.504	11.644	168	-38	-87.4	5.2	-37.4	-89.2	5.5
1911	114.9239	7.9241	2013.509	16.244	18.11	54.715	217.6	38.2	-88.3	5.9	34.7	-86.7	6.8
1912	114.9531	-6.2442	2014.0705	7.701	9.678	15.726	168.4	-85.6	14.8	5.3	-83.3	12.8	5.4
1913	114.9539	12.4345	2014.03	10.47	15.353	6.94	102.1	35.6	-94.9	5	37	-96.6	6.2
1914	114.9782	4.7857	2013.4215	14.692	16.394	5.498	13.7	19	-109.1	5.8	16.1	-108.6	6.5
1915	115.0526	46.3187	2013.7545	11.939	12.833	10.083	324.1	84.2	-98.8	5.4	81.2	-101.8	5.4
1916	115.0577	-7.5619	2013.8745	15.251	15.334	4.697	281.7	-2.5	-173.9	5.8	0.9	-175.9	5.4
1917	115.2399	6.6845	2013.652	14.182	14.872	6.969	352	179.7	-41.5	5.9	171.7	-38.7	6
1918	115.2535	18.0935	2013.5605	14.134	17.803	5.379	189.4	110.5	-64.9	5.2	116.6	-64	6.2
1919	115.3070	49.9640	2013.628	17.436	17.587	12.273	237.9	-19.7	-76.7	6.1	-10.3	-76.6	6.1
1920	115.3788	31.4372	2013.756	13.026	17.156	39.96	357	86.1	-89.3	5.2	86.8	-85.4	5.3
1921	115.4114	46.5166	2013.803	13.715	14.224	6.028	178.8	8.8	-61.7	5.4	8.6	-61.2	5.4
1922	115.4598	33.9087	2013.7405	13.658	14.474	4.533	358.8	-113.4	5.5	5.1	-111.2	1.8	5.3
1923	115.5276	-10.0088	2014.1345	12.598	15.872	18.293	268.7	-79.9	-1.7	5.4	-80.6	-1.8	5.6
1924	115.5750	4.3392	2013.6065	16.602	17.596	9.026	158.8	-13.2	-76.1	6.2	-13.1	-79.9	6.3
1925	115.6474	32.9122	2013.8395	14.003	16.072	5.966	300.1	-4.3	-135.6	5.2	-3.6	-136.5	5.2
1926	115.6691	51.1770	2013.672	14.133	16.127	44.674	265	-64.6	1.2	5.9	-66.8	-1.4	5.9
1927	115.6766	41.5050	2013.7835	9.521	10.459	21.155	356.4	-38.7	-147.4	5.2	-44	-144.4	5.2
1928	115.7844	6.1310	2013.341	16.444	16.792	4.038	144.3	33.9	-89.6	6.2	33.8	-91.9	6.2
1929	115.7854	-8.7765	2013.605	12.816	16.481	14.479	209.3	-43.7	67.3	5.5	-52.1	61.5	6.1
1930	115.7946	4.4000	2013.77	11.032	13.796	28.528	235.9	-60.1	-80.9	5.8	-62	-81.3	5.9
1931	115.8560	-8.1680	2013.7035	12.501	15.956	28.521	168.9	20	-62.3	5.5	18	-61.7	5.6

1932	115.8683	36.7689	2013.981	14.796	15.443	4.673	169.4	-97.3	-90.7	5.2	-98.8	-91.9	5.3
1933	115.8797	3.4855	2014.1275	7.103	9.817	9.544	146.8	14.8	-84	5.7	19.9	-79.9	5.7
1934	115.8993	61.8776	2013.69	12.123	16.27	11.124	294.5	-63.8	-114.8	6	-65.3	-115.8	6
1935	115.9462	43.7946	2013.756	6.986	14.489	24.037	60.4	-91.8	-14.5	5.5	-92.7	-15.9	5.4
1936	115.9519	7.5887	2013.666	14.563	15.636	10.073	16.6	16.1	-66.3	5.9	15.6	-66.3	5.9
1937	115.9639	1.1234	2013.3655	15.828	16.777	29.604	357.9	-9.9	-69.6	6.3	0.3	-71	6.2
1938	116.0033	55.3788	2013.8435	16.003	17.733	4.9	137.2	78.3	-12.8	5.5	72.6	-22.5	7
1939	116.0471	21.7079	2013.5765	14.133	17.251	8.889	241.4	-84	-62.6	5.1	-88	-64.2	5.3
1940	116.0925	6.0174	2013.639	12.508	13.776	5.62	154	67.1	-62.7	5.9	71	-63.1	6
1941	116.2313	71.6409	2013.6645	10.097	15.735	13.673	272.8	49.4	-63	5.5	47.2	-60.1	5.5
1942	116.2691	18.8000	2013.5725	13.444	16.208	7.002	125.9	-83.1	5.8	5.1	-82.9	9.2	5.2
1943	116.3321	53.5370	2013.663	12.599	14.371	5.316	118.5	-65	-20.4	5.5	-66	-18.9	5.5
1944	116.3380	13.1969	2013.956	14.714	15.145	4.086	45.8	-24.1	-68.3	5	-26.1	-69.6	5.1
1945	116.3418	64.1761	2013.488	12.817	16.006	5.38	348.2	-59.9	-153.1	5.8	-60.2	-154.3	6.5
1946	116.3719	-5.4532	2013.7015	12.968	15.517	7.249	252	46.2	139.4	5.5	43.7	142.5	5.7
1947	116.3814	24.7336	2013.7435	8.76	17.231	46.443	312.4	-10	-73.7	5.9	-2.5	-72.4	6.1
1948	116.4177	49.5296	2013.678	14.59	15.419	15.227	129.1	-101.2	-161.4	5.9	-105.6	-159.6	6
1949	116.4437	81.4010	2013.8655	11.968	16.655	15.094	88.7	5	-102.2	5.4	-0.6	-100.3	5.5
1950	116.5408	62.6195	2013.506	13.206	18.682	20.146	200.5	-60.9	-25.8	5.8	-62.7	-18.6	6.3
1951	116.7202	11.8191	2013.5835	14.931	17.433	59.406	359.3	21	-64.7	5.9	28.3	-65.2	6.1
1952	116.7327	42.6737	2013.746	16.375	17.399	5.295	280.1	-44.8	-73.4	5.5	-34.4	-75.6	5.8
1953	116.7617	27.9445	2013.7465	12.48	15.229	53.051	186	15.3	-77.8	6	14.4	-80.5	6
1954	116.7922	10.4775	2013.6425	15.118	15.494	4.6	145.8	11.8	-101.9	6	12.6	-100.3	6.1
1955	116.8025	12.4411	2013.789	12.997	14.3	11.456	61.6	35	-78.5	5	32.9	-78.4	5.1
1956	116.8230	62.3224	2013.594	16.613	16.682	4.121	193	-67.2	-66.1	5.9	-66.6	-62.5	6
1957	116.8600	1.7149	2013.4755	16.332	16.613	41.249	30.6	38.5	-93.5	6.6	41.3	-91.4	6.6
1958	116.8886	-4.9750	2013.3315	14.534	17.244	9.925	74.6	88.4	-72.5	5.5	86.7	-70.4	6.7
1959	116.9365	25.0201	2013.5845	15.924	16.328	6.763	77.9	15.7	-65.9	6.3	11.5	-65.7	6.4
1960	116.9540	52.1058	2013.6585	16.139	17.401	10.42	241	146.5	-33.5	6	148.8	-37.7	6.1
1961	117.0310	50.2170	2013.6665	10.734	10.765	31.202	341.9	-31.9	-153	6	-31.8	-154.2	6
1962	117.0660	33.3803	2013.7095	13.243	17.076	34.982	82.9	-16.7	-70.2	5.1	-16.5	-73.2	5.2
1963	117.0908	40.2568	2013.7055	15.813	15.934	18.324	338.2	-67.7	-49.3	5.2	-68.9	-49.5	5.2
1964	117.2048	46.1962	2013.7885	14.206	16.609	5.606	117.7	-23.5	-81.6	5.5	-24.9	-82	5.7
1965	117.3528	49.2587	2013.93	18.058	18.845	21.794	85.9	-97.5	-44.8	6.3	-103.9	-33.5	7.3
1966	117.3618	24.0856	2013.5795	15.764	17.4	33.383	339.6	-68.3	-6.6	5.4	-68.1	-7.6	5.5
1967	117.3779	58.6474	2013.6465	11.815	13.624	26.688	187.1	12.4	-116.3	5.5	13.6	-115.1	5.5
1968	117.3827	30.9946	2013.608	17.987	18.048	29.486	207	-124	-98.2	5.4	-123.7	-98	5.4
1969	117.4141	34.9942	2013.8945	14.11	15.764	4.655	229.6	80.5	-74.3	5.2	82.9	-72.2	5.4
1970	117.4685	3.6846	2013.898	14.653	15.126	4.804	184	148.2	-87.9	6	151.9	-87.3	5.8
1971	117.5232	0.0128	2013.386	14.392	17.996	49.349	258.9	20.3	-72.5	5.5	25.3	-67.4	6.9
1972	117.5372	-8.1534	2013.823	10.895	11.177	23.745	200.2	-79.5	-126.1	5.6	-80.7	-126.7	5.7
1973	117.5422	70.9835	2013.618	15.64	16.022	6.062	338.5	-10.7	-129.7	5.8	-12.5	-125.4	5.9
1974	117.5569	27.2263	2013.772	12.289	14.33	8.499	344.6	64.8	-4.1	5.3	62.2	-7.8	5.4
1975	117.6314	14.8415	2013.931	7.593	14.381	27.413	12	-65.2	-30.4	5	-64.2	-27.7	5.1
1976	117.6805	6.9995	2013.336	16.815	17.927	13.401	344.5	-29.1	-75.7	6.1	-31.9	-76	6.2
1977	117.7411	4.4832	2013.751	14.596	14.834	5.049	190.5	-156.3	-59.6	6	-160.4	-58.4	5.9

1978	117.7420	13.3170	2013.5985	16.704	17.745	21.341	15.9	-81.5	-20.3	5.1	-80.7	-19.1	5.2
1979	117.7577	40.1019	2013.8275	14.862	16.747	7.352	185.7	-133	-157.9	5.2	-128.2	-153	5.3
1980	117.7624	47.1202	2013.7995	15.548	17.49	15.463	72.6	-18.4	-94	5.2	-18.6	-97	5.3
1981	117.8403	28.9716	2013.816	10.904	11.994	8.636	28.1	-158.1	-12.4	5.4	-156	-9.7	5.4
1982	117.8808	-9.2843	2013.7815	12.826	13.691	15.892	209.6	-52.8	-176	5.6	-55.7	-174.4	5.7
1983	117.9755	24.6017	2013.6725	13.505	13.873	5.076	226.1	-86.6	39.1	5.4	-86.1	39.5	5.5
1984	118.0342	-11.9967	2014.133	10.689	14.469	11.171	70.6	-55.6	-68.5	5.4	-57.9	-72.2	5.5
1985	118.0376	-8.5166	2013.615	15.074	15.442	26.083	244.2	70.5	34.9	5.7	69	36.1	5.7
1986	118.1290	77.8876	2013.7955	14.361	15.047	25.961	0.5	-141.5	-63.9	5.6	-139.6	-60.1	5.5
1987	118.3639	8.1353	2013.5275	11.591	16.852	9.746	34.1	95.1	-117.4	6.3	93.3	-120.6	6.8
1988	118.4518	-5.9386	2013.3315	14.101	17.018	20.938	88.4	-11.4	-81.8	5.5	-11.8	-80.7	5.8
1989	118.5291	36.0133	2013.828	11.353	13.014	13.358	147.1	65.8	-30.5	5.2	66.4	-36	5.2
1990	118.5715	16.0419	2013.697	14.343	16.611	9.169	74.5	-13.9	-94.7	5.1	-11.3	-93.4	5.2
1991	118.5835	11.7337	2013.7065	15.067	15.324	5.682	184.2	49.8	-121.7	5.9	48.5	-120.3	6
1992	118.5963	70.1117	2013.6755	11.326	17.391	11.539	199.6	-61.8	-55.8	5.8	-64.7	-53.5	5.9
1993	118.6127	43.7435	2013.7725	15.139	16.243	6.461	124.1	-33.6	-94.2	5.2	-34.6	-94.8	5.2
1994	118.6235	31.3235	2013.751	14.784	15.87	10.579	323.1	-24.3	-89.2	5.2	-25.2	-89.2	5.2
1995	118.6273	78.1123	2013.7615	10.686	15.108	44.971	211.2	0.6	-80.4	5.8	2.7	-79	5.8
1996	118.7491	4.7616	2013.558	12.904	15.988	12.471	100.8	-19.1	-82.4	5.9	-15.7	-82.6	6
1997	118.7775	25.8345	2013.6535	13.325	16.223	5.789	274.6	-59	-121.9	5.1	-57.5	-120.7	5.5
1998	118.8638	29.1583	2013.889	8.055	11.538	14.902	182.3	32.6	-63.8	5.1	37.2	-63.7	5.1
1999	118.8985	44.1279	2013.6945	12.387	13.05	8.521	196.7	-100.1	-49.2	6	-98.6	-50.5	6
2000	118.9200	77.5503	2013.705	10.778	13.532	7.91	222.3	-48.5	-70.4	5.5	-48.3	-67.4	5.5
2001	118.9245	35.9694	2013.882	16.984	18.959	25.006	287.5	16.1	-61.9	5.2	10	-65.9	6.2
2002	118.9264	-0.5539	2013.5365	11.604	17.047	15.353	249.9	-1.6	-68.9	5.4	-3.2	-69.7	6.4
2003	118.9666	5.5236	2013.681	15.864	15.901	3.992	25.4	11.4	-63.7	6.1	10	-61.5	6
2004	118.9918	53.4855	2013.7235	14.789	15.211	5.002	107.5	10.4	-112.3	5.9	8.9	-109.2	5.9
2005	119.0932	64.7771	2013.693	13.979	16.897	23.986	123.6	-40.5	-61.5	6	-39.8	-60.7	6
2006	119.1492	25.8490	2013.6675	15.951	16.375	34.224	293.5	-22.8	-81.4	5.2	-19.5	-79.3	5.2
2007	119.1520	-1.2304	2013.469	15.081	17.746	11.038	129.1	-79.6	-22	5.5	-80.2	-24	6.7
2008	119.1734	18.8726	2013.6505	13.245	16.79	7.942	355.4	-81.7	-12.1	5.4	-83.6	-16.3	5.6
2009	119.1866	46.3321	2013.789	15.529	15.586	20.285	120.7	-10.4	-136.2	5.8	-9.5	-137.8	5.8
2010	119.1870	6.9531	2013.8145	9.114	15.382	25.719	285.9	-71.6	-27.6	5.7	-69.4	-32.6	5.9
2011	119.2838	-5.8658	2013.691	10.393	14.61	45.922	11.6	26.6	-75.5	5.5	18.9	-76.9	5.6
2012	119.3555	45.3490	2013.7505	11.427	12.553	4.138	284.9	-46.4	-81.9	5.5	-36.9	-86.7	6.6
2013	119.4533	27.6595	2013.7615	12.869	16.517	59.24	20.7	1.2	-64.2	5.1	-0.8	-63.3	5.2
2014	119.4620	69.8846	2013.717	9.656	13.246	13.859	72.4	-185.8	-35.8	6.2	-188.6	-28	6.3
2015	119.6049	64.8703	2013.617	14.102	17.566	12.057	347.1	-58.5	-78.1	5.9	-52.1	-80.5	6
2016	119.6288	15.5036	2013.8055	12.571	13.204	16.092	208	-78.7	-112.7	5.1	-77.5	-112.7	5.1
2017	119.6340	72.9151	2013.719	12.833	16.622	17.023	351.9	-1.2	-63.6	5.5	-2.4	-62.4	5.5
2018	119.6535	7.9939	2013.788	10.41	13.081	46.596	178.4	62.9	-102.4	5.8	60.4	-106.1	5.9
2019	119.7082	46.4449	2013.8105	14.18	15.919	12.975	326.9	-93.2	4.9	5.5	-93.6	0.2	5.5
2020	119.7101	3.8537	2013.7205	14.643	15.845	6.728	240	-75.1	34.1	5.9	-73.7	33.2	5.9
2021	119.7514	73.5871	2013.677	11.219	12.852	8.717	210.2	-25.9	-69.9	5.5	-27	-70.6	5.5
2022	119.9084	29.4319	2013.683	15.668	16.691	4.818	214.4	-81.2	-23.6	5.2	-84.2	-26.5	5.3
2023	119.9492	22.8314	2013.7035	11.532	16.053	8.26	322.6	16.4	-62.1	5.8	14.5	-62.8	5.9

2024	120.0199	50.5540	2013.661	14.447	15.096	6.934	237.1	-37.9	-84.1	5.5	-37.9	-84.7	5.5
2025	120.0699	43.8240	2013.73	12.378	16.752	35.772	116.9	-16.1	-72.2	5.5	-13.7	-71.2	5.6
2026	120.0964	40.1089	2013.897	13.249	14.289	4.252	298.9	-77.7	-89.7	6.4	-78.3	-81.6	6.1
2027	120.0965	59.8962	2013.6855	15.022	15.764	27.904	206.6	-13.7	-109.2	6	-14.4	-105.7	6
2028	120.1125	-3.6621	2013.736	11.413	16.113	33.832	236.7	27.3	-73	5.4	26.3	-71.7	5.5
2029	120.1260	51.5317	2013.641	12.493	16.096	17.654	164.6	-9.6	-65.2	5.5	-12.4	-62.4	5.5
2030	120.1869	52.2780	2013.787	10.578	14.478	6.44	52.8	35.2	-81.4	5.5	28.9	-85.3	5.5
2031	120.1968	38.9663	2013.9095	13.387	15.405	6.885	271.8	-47.6	-63.7	5.9	-50.6	-64	5.8
2032	120.2023	-6.2910	2013.367	14.421	16.29	5.576	296.8	40.1	-176.4	5.6	51.8	-181.6	6.3
2033	120.3563	56.9432	2013.621	8.459	10.668	6.13	32.7	-9.7	-78.3	5.5	-12.4	-78.3	5.5
2034	120.3725	19.6775	2013.6145	12.91	17.242	57.369	131.4	37.7	-76.8	5.9	38.7	-76	6
2035	120.4705	-7.3920	2012.947	17.585	18.208	26.404	65.5	89.6	-46.6	6.4	98	-33.7	7.2
2036	120.5427	67.6973	2013.61	15.029	16.427	5.169	358	21.6	-98.5	5.5	25.6	-100.5	5.8
2037	120.5640	62.3718	2013.794	15.214	17.001	5.969	335.5	3.3	-105	5.8	0.8	-105.4	5.9
2038	120.6122	22.0858	2013.6145	15.794	15.927	7.716	35.9	-49	-141.1	5.2	-48.1	-137.2	5.2
2039	120.6139	59.5311	2013.721	13.806	14.765	22.008	9.4	-36.4	-81.1	5.5	-37.2	-81.4	5.5
2040	120.6379	7.7965	2013.7135	13.975	14.438	21.932	5.9	-65.6	0.8	5.9	-63	-0.7	5.9
2041	120.6691	8.6874	2013.523	11.951	14.362	6.055	85.3	-2.2	-131	5.9	-4.1	-130.5	6.3
2042	120.6992	19.6179	2013.4205	15.857	18.484	9.814	358.1	78.4	-106.2	5.2	78	-103.1	6
2043	120.8211	52.2678	2013.669	14.549	15.711	9.216	62.7	18.2	-141	5.5	17.1	-139.1	5.5
2044	120.8359	7.1237	2013.9265	8.494	12.206	36.756	12.9	-16.1	-121.8	5.8	-23.1	-125	5.9
2045	120.9199	-7.1025	2013.801	15.521	15.935	6.406	239.7	-60.6	-29	5.7	-61.1	-25.1	6.3
2046	121.0899	44.7465	2013.6515	16.583	17.045	12.72	36.2	-3.5	-67	5.3	-3.4	-66.4	5.3
2047	121.0955	43.9616	2013.824	9.008	14.528	13.484	52.8	136.1	-15.4	5.2	132.6	-18	5.2
2048	121.1364	13.0318	2013.6045	14.793	17.011	5.284	95.8	-24.1	-86.7	5.9	-27.9	-83.7	6.3
2049	121.2043	19.3334	2013.729	11.795	14.818	39.221	301	-27.1	-74.7	5.9	-28.3	-74.7	5.1
2050	121.2250	-2.3872	2013.787	11.859	12.768	15.546	162	58.9	-77.9	5.5	58.6	-77.2	5.5
2051	121.2474	4.1573	2013.4515	14.932	16.777	11.119	86.6	-58.7	-94.7	6	-61.1	-95.5	6.2
2052	121.3085	29.4408	2013.6895	12.4	15.996	9.513	75.9	-46.7	-89.9	5.9	-49.7	-92.3	6.1
2053	121.3294	11.1122	2013.7575	13.429	16.037	14.905	296.4	37.5	-79.4	5.9	38.8	-80	6
2054	121.3367	24.4024	2013.6685	15.59	17.427	54.883	249.5	-51.1	-110.5	6	-53	-113.6	6.1
2055	121.3383	65.3543	2013.734	12.69	15.02	44.522	202.3	-131.6	-73.3	5.5	-134.7	-75	5.5
2056	121.3844	31.4140	2013.739	11.959	16.828	25.019	120.9	-43.4	-97.7	5.2	-43.3	-100	5.2
2057	121.4288	11.6067	2013.2825	15.275	16.196	3.787	109.2	13.6	-159.8	6.1	10.3	-156.2	6.9
2058	121.4546	76.7478	2013.6385	11.12	14.606	10.726	347.3	-22.3	-74.4	5.5	-32.4	-72.4	5.5
2059	121.5669	62.8771	2013.6905	15.808	16.084	23.734	293	71.4	18.4	5.5	72.3	20.9	5.5
2060	121.6143	0.3518	2013.521	15.387	15.704	8.173	188.6	-60.2	-32.6	6	-60.1	-37.9	6.1
2061	121.6152	42.5299	2013.73	17.439	18.252	7.922	180.7	-49	-60.2	5.6	-54.6	-60.6	5.8
2062	121.6338	67.9271	2013.776	14.519	17.159	12.715	198	-26.7	-87.5	5.5	-28.7	-84.3	5.6
2063	121.6597	-5.6369	2014.0685	9.263	15.197	11.922	327.6	-44.4	-83.5	5.4	-42.8	-80	5.5
2064	121.7420	11.8762	2013.6355	15.329	15.571	12.298	180.5	-33.4	-81.6	6	-32.9	-80.6	6
2065	121.7524	57.5173	2013.7505	13.773	14.471	5.969	255.7	-24.3	-135.8	5.5	-23.8	-136.6	5.5
2066	121.8353	61.2026	2013.8375	15.153	18.408	10.401	302	-70.7	-91.3	5.5	-76.7	-92.8	6.2
2067	121.9497	12.2832	2013.744	14.518	15.341	5.538	81	7.1	-60.9	5.1	11	-61.5	5.1
2068	121.9645	30.4362	2013.7905	8.481	17.217	22.385	299.9	-8.7	-108.9	5.1	-9.3	-111.9	5.3
2069	122.1034	29.8765	2013.821	13.074	18.197	13.912	185.6	-32	-62.6	5.1	-31.7	-62.4	5.3

2070	122.2275	-6.4167	2013.882	12.215	14.764	37.424	247.7	-65.2	38.9	5.4	-65.6	36.9	5.5
2071	122.2850	-4.4542	2013.5075	13.283	16.426	9.263	55.7	-7.6	-60.9	5.4	-11.2	-62.1	5.8
2072	122.3324	19.5107	2013.787	10.347	13.751	8.294	300	172.7	-120.8	5.1	171.4	-119.5	5.2
2073	122.3927	35.1315	2013.698	11.246	14.599	15.275	72.4	71.8	-7.7	5.2	72.9	-6.1	5.2
2074	122.5544	33.7748	2013.954	10.644	11.174	5.347	294.7	-2.1	-69.2	5.1	-1.6	-69.4	5.1
2075	122.5821	21.4143	2013.739	14.551	14.741	9.033	343.6	-86.4	-42	5.9	-84.1	-38.5	5.8
2076	122.6096	53.7035	2013.615	11.472	16.093	10.569	34.1	-9.3	-108.8	5.5	-11.6	-105.4	5.5
2077	122.6148	70.6551	2013.7025	17.025	17.871	12.665	148.6	-37	-62.3	5.5	-32.3	-65.3	5.6
2078	122.6525	19.3782	2013.857	13.111	15.134	4.412	50.6	-81.9	-16.4	5.2	-84.2	-18.4	5.2
2079	122.7338	77.6418	2013.6865	7.907	12.702	15.812	274.2	-130.2	-96.6	5.6	-135.3	-86.3	5.5
2080	122.7343	48.2538	2013.8755	10.662	16.155	8.948	329	58.1	-66.9	5.4	58.7	-69.7	5.5
2081	122.7661	30.3144	2013.9795	10.319	10.853	6.457	229.4	-69.6	-8	5.1	-70	-11.6	5.1
2082	122.8048	31.2167	2013.6525	15.12	16.826	6.047	34.9	14.2	-95.6	5.2	16	-98.5	5.3
2083	122.8344	57.7316	2013.6465	14.164	15.51	4.158	29	-119.3	-59.7	5.5	-120.7	-66.9	5.6
2084	122.8641	62.7294	2013.7065	13.531	14.147	4.394	11.2	-35.1	-73.7	5.5	-29.9	-78.4	5.6
2085	122.8947	6.7455	2013.668	11.688	15.538	10.711	44.4	52.5	-65.8	5.9	49.1	-69.1	6.1
2086	122.9511	-7.2759	2013.955	9.172	10.417	14.744	146.2	-97.2	16.7	5.4	-96	12	5.4
2087	122.9545	27.6847	2013.727	15.874	17.374	49.125	118.2	-65	-2.6	5.1	-65.1	-5.6	5.2
2088	122.9554	13.4453	2013.9835	7.19	15.115	42.542	195.8	-124.1	-86.6	5	-126.9	-85	5.1
2089	122.9612	25.6020	2013.7745	12.909	13.311	4.638	13.7	-89.2	-22.9	5.2	-90.6	-18.5	5.1
2090	123.0026	38.5370	2013.8565	12.096	13.656	6.392	270.1	-75.4	-62.4	5.2	-75.6	-61.7	5.1
2091	123.0919	74.2342	2013.6935	10.012	16.223	40.55	104.3	-10.2	-62.6	5.5	-11.6	-63.7	5.5
2092	123.0976	58.1541	2013.687	11.49	16.658	9.158	214.3	16.5	-112.2	5.5	22.7	-112.2	5.6
2093	123.1150	27.6949	2013.7665	17.852	18.635	12.998	170.1	-69.6	-64.7	5.3	-71.1	-62.4	5.5
2094	123.1179	34.2439	2013.6045	16.088	18.78	55.849	186	50.5	-65.1	5.2	58.6	-61.8	5.5
2095	123.1251	20.7231	2013.5495	17.515	17.63	9.117	30.3	-66.4	-54.1	5.3	-67.1	-57.6	5.3
2096	123.1594	14.4012	2013.7455	14.585	15.77	5.685	291.9	97.2	-8.2	5.1	100.2	-8.6	5.1
2097	123.1717	40.0887	2013.889	13.582	16.898	7.352	200.8	3.5	-90.6	5.1	-1.4	-90.6	5.2
2098	123.2102	56.6586	2013.64	14.739	16.452	7.02	90.7	-12.2	-100.4	5.5	-12.2	-100	5.5
2099	123.2299	7.9702	2013.8595	9.87	14.859	18.778	2.3	-44.2	62.3	5.8	-45.5	62.9	6
2100	123.2497	55.5821	2013.757	14.493	15.369	6.006	32.3	6	-130	5.5	8.7	-130.5	5.5
2101	123.3497	14.7382	2013.7855	12.393	15.843	27.403	266	87.6	-22.2	5	88.1	-20.7	5.1
2102	123.4082	29.5492	2013.7175	13.96	15.952	14.206	72	8.9	-64.8	5.1	10.3	-64.8	5.2
2103	123.5427	2.5694	2013.6	11.64	16.599	13.438	300.3	-59.3	-196.3	6	-62.4	-196.2	6.5
2104	123.5533	14.1469	2013.6815	14.44	16.717	22.082	36.1	-65	-69.6	5.1	-63.5	-66.6	5.2
2105	123.5573	-5.4554	2014.03	10.312	11.052	5.117	320.6	-95.7	37.9	5.4	-94.4	32.2	5.3
2106	123.5740	25.2136	2013.668	15.589	16.836	14.757	83.6	65.6	-49	5.2	61.5	-51	5.2
2107	123.5999	22.6978	2013.6475	12.613	13.683	5.396	100.6	-14.7	-127.6	5.2	-15.8	-124.4	5.2
2108	123.6101	12.7022	2014.07	10.575	11.063	5.484	199.1	-92.7	4	5	-89.3	2.1	5
2109	123.6338	12.2827	2013.6905	12.662	13.783	32.077	345.4	33.4	-70.4	5.1	31.6	-72.1	5.1
2110	123.6525	63.4264	2013.4835	14.421	15.54	4.658	256.6	6.8	-108.6	5.9	13.2	-108.1	6.5
2111	123.6817	46.8432	2013.744	13.139	16.262	15.931	8.4	-51	-70.7	5.5	-47.4	-71.4	5.5
2112	123.6825	21.9357	2013.4685	13.344	13.719	4.701	282.3	83.7	-71.9	5.2	85	-72.4	5.3
2113	123.8404	63.8412	2013.774	11.596	16.246	35.003	341.5	2.8	-71.5	5.4	6.3	-67.8	5.5
2114	123.8404	78.7037	2013.735	17.248	17.774	7.268	119.2	61.3	-89.4	5.6	61.6	-82.7	5.7
2115	123.8717	50.1951	2013.716	13.35	14.838	6.322	121.8	34.4	-60.9	5.7	31.7	-61.1	5.8

2116	123.8914	29.8490	2013.7575	14.31	15.041	17.706	288.1	46.9	-78.2	5.1	48.8	-82.2	5.2
2117	123.9233	23.4492	2013.729	13.854	16.029	6.421	79.8	27.2	-69.9	5.1	27.5	-70	5.2
2118	123.9378	-10.4487	2014.1495	8.674	13.847	26.738	142.4	-61	-46.7	5.4	-62.2	-40.8	5.4
2119	124.0083	80.6701	2013.8475	14.47	15.638	29.374	18.7	-38.4	90.8	5.5	-38.2	87.2	5.5
2120	124.0656	27.1460	2013.8	10.401	15.368	35.24	257	24.6	-94.7	5.1	25.3	-94.7	5.1
2121	124.1580	15.0004	2013.6975	13.508	16.304	11.15	100.9	-7.8	-73	5.9	-3.9	-72.4	5.9
2122	124.1672	68.1458	2013.7465	14.273	16.445	5.388	103.9	26.3	-169.4	5.5	18.7	-166.5	5.8
2123	124.1862	37.1135	2013.709	13.903	18.737	53.915	303.1	9.5	-105	5.2	12.5	-100.4	6.8
2124	124.2436	27.7333	2013.707	15.364	16.675	5.336	334.6	-46.8	-130	5.2	-47.1	-133.4	5.3
2125	124.2609	55.0620	2013.687	15.082	15.418	31	109.6	-16.9	-105.4	5.5	-20	-105.1	5.5
2126	124.3064	27.3353	2013.897	11.759	14.538	8.415	31.3	-2.7	-84.3	5.1	-0.8	-85.3	5.1
2127	124.5930	47.0992	2013.7365	11.777	15.474	8.117	33.6	-46.4	-79	5.2	-49.8	-80	5.3
2128	124.6500	-5.9083	2013.89	12.742	13.58	22.107	149.5	150.3	-155.2	5.5	145.5	-152.8	5.5
2129	124.6578	17.6049	2013.9395	10.205	12.529	14.199	84.7	35.8	-90.7	5.3	34.6	-92.7	5.3
2130	124.6987	68.1078	2013.832	11.521	12.628	6.222	49.1	64	-149.2	5.5	64.3	-145.4	5.5
2131	124.8205	28.7787	2013.8385	10.864	14.461	9.814	114.6	-74	-56.3	5.5	-72.5	-55.7	5.5
2132	124.8374	47.5334	2013.713	14.822	15.031	53.731	74	-11.2	64.5	5.9	-7.7	65.7	5.8
2133	124.9306	21.1128	2013.756	12.003	16.797	11.546	272.7	68.7	-9.2	5.1	69.5	-11.8	5.2
2134	124.9499	42.4566	2013.798	12.579	16.995	29.863	9.1	0.3	-99.3	5.8	1.3	-99.7	5.9
2135	124.9662	14.0683	2013.6435	14.415	17.048	27.203	13.1	85.4	8.1	5.4	85.6	11.5	5.6
2136	124.9665	65.6790	2013.757	10.956	15.451	22.708	143.6	-11.3	-93.5	5.5	-13.9	-92.1	5.5
2137	125.0047	52.4618	2013.6555	12.657	13.712	9.087	91.7	-52.9	-106.3	5.8	-55.1	-108.5	5.8
2138	125.0092	48.3886	2013.798	13.9	14.912	45.186	291.5	-101.5	-86.1	5.8	-103.3	-85.7	5.8
2139	125.0154	56.1763	2013.6815	13.82	17.29	8.008	157.7	-92.1	-76.1	5.5	-89.2	-73.1	5.6
2140	125.0182	77.0543	2013.8625	17.612	18.035	7.713	285.1	-56	-82	6	-51.5	-84.4	6
2141	125.0237	44.2416	2013.683	11.47	15.271	7.691	168.2	-29.1	-66.8	5.9	-26.6	-68.3	5.9
2142	125.0506	-2.5205	2013.769	14.804	15.437	13.876	45.8	37.5	-176	5.5	39.4	-178.6	5.6
2143	125.1797	-0.0158	2013.134	16.281	17.186	55.702	89.6	-60.2	-75.9	5.8	-55.4	-75.3	6
2144	125.1803	11.5766	2013.516	16.964	17.6	4.346	265.5	-88.7	18.7	6	-85.1	23.1	6.5
2145	125.1930	62.6035	2013.742	11.882	15.374	13.275	75.2	39.2	-78.5	5.5	40.5	-73.9	5.5
2146	125.2440	14.6557	2013.566	17.161	17.293	9.249	196.7	-22.9	-108.3	5.7	-18.5	-110.8	6.1
2147	125.2478	17.0467	2013.53	16.501	17.674	30.62	283.3	56.4	-74.5	5.5	57.6	-77.7	5.7
2148	125.2709	17.1549	2013.771	11.524	15.507	32.319	29.8	32.6	-135.7	5	30.5	-140.3	5.1
2149	125.3172	1.6781	2013.3275	12.336	17.961	19.802	235.1	-10.1	-64.5	5.9	-21.3	-64.9	7.7
2150	125.3264	10.3743	2013.761	15.268	15.772	13.242	71.2	72.7	17.8	5.9	68.9	20.8	5.9
2151	125.3608	67.6950	2013.699	9.533	14.596	14.997	112.7	29.1	153.5	5.5	34.8	147.3	5.5
2152	125.4241	4.6394	2013.118	16.834	18.388	34.979	25.5	67	-4.9	6.7	64.9	1.8	6.7
2153	125.4954	20.0620	2013.7525	12.547	14.994	21.519	46.3	31.9	-61.5	5.1	35.7	-61.6	5.1
2154	125.6322	37.4628	2013.688	12.083	17.164	32.941	50.9	-40.3	-127.9	5.2	-42	-129.9	5.4
2155	125.6374	11.7515	2013.724	12.128	15.328	7.63	148.5	-192.5	-23.3	6	-194.4	-27.6	6.2
2156	125.7293	27.3933	2013.733	14.546	16.612	6.924	30.3	-36.4	-60.1	5.8	-37.7	-61.8	5.9
2157	125.8133	41.6875	2013.77	12.348	15.893	9.021	245.2	-53.8	-131.7	5.5	-55	-129.9	5.6
2158	125.9101	2.1787	2014.0085	9.019	12.383	6.854	116.3	21.9	-177.3	5.9	13.8	-178.2	5.8
2159	125.9203	14.5364	2013.7315	13.394	15.569	28.552	321.6	33.3	-138.4	5.1	34.2	-138.7	5.1
2160	125.9533	57.4173	2013.646	7.721	9.055	57.739	180.8	-80.2	41.7	5.5	-81	32.1	5.5
2161	125.9886	12.7980	2013.7665	11.104	16.488	18.953	209.8	7.1	-95.3	5	6.1	-95.3	5.2

2162	126.0066	50.1222	2013.7605	16.144	16.671	5.138	201.9	-5.5	-98	5.8	-6.6	-101.1	5.8
2163	126.0427	-1.8337	2013.9065	12.412	14.782	42.84	123.6	-85.9	-8.5	5.4	-84	-8.9	5.4
2164	126.0649	44.9490	2013.832	7.526	8.709	29.129	242.1	-63.7	-173.5	5.2	-61.8	-176.3	5.2
2165	126.0725	-4.4040	2013.309	15.115	16.884	24.675	300.7	119.9	-99.8	5.8	120.6	-99.8	6
2166	126.1939	14.6923	2013.326	15.47	15.579	3.758	70.6	-87.8	84.6	5.1	-90.5	79.3	7.6
2167	126.2646	3.7799	2013.7155	13.03	13.933	6.01	182	40	-65	5.9	41.8	-65	5.9
2168	126.2931	18.3487	2013.8385	11.687	15.166	37.816	168.6	-65.3	-56.9	5.1	-65.2	-56.4	5.1
2169	126.3555	7.9501	2013.844	13.498	15.463	7.435	276.8	-36.4	-78.7	5.9	-39.7	-78.1	5.9
2170	126.3880	48.6898	2013.7215	12.578	12.91	8.506	198.7	-9.9	-62.4	5.8	-5.8	-62.6	5.8
2171	126.3952	21.0996	2013.717	15.892	16.271	6.633	130.7	11.5	-69.4	5.1	3.8	-69.7	5.2
2172	126.3995	45.6558	2013.642	15.675	18.834	22.489	238.4	-2.3	-78.9	5.2	4.4	-75.3	5.7
2173	126.4694	29.4211	2013.7025	16.836	17.252	16.302	288.3	-27.1	-75.1	5.9	-31.6	-75.4	5.9
2174	126.4775	65.6608	2013.7615	13.436	15.298	38.137	194.3	49.6	86.1	5.5	49.6	86.4	5.5
2175	126.4847	55.3946	2013.7685	7.988	16.501	32.352	8.2	-31.6	-76.7	5.5	-31.4	-79.5	5.5
2176	126.4942	40.3474	2013.7905	11.641	15.275	9.892	202.6	-40.7	-116.6	5.5	-43	-115.6	5.5
2177	126.7193	52.1763	2013.862	9.09	9.394	16.03	290.5	-25.4	107.3	5.4	-20	107.5	5.4
2178	126.7408	13.1112	2013.732	11.503	13.266	5.836	149.4	-62.2	-5.9	5	-61	-4	5.1
2179	126.7447	85.8125	2013.9985	14.132	16.945	43.141	88.5	-28.1	-69.4	5.8	-36.8	-62	5.8
2180	126.7571	65.6107	2013.7445	13.85	14.013	15.563	41	-63.1	-32.9	5.5	-61.8	-35.2	5.5
2181	126.7664	3.5143	2013.7185	9.979	15.811	16.596	144.2	10.6	-62	5.8	16.8	-63.6	6
2182	126.7683	29.5738	2013.9425	8.393	12.28	25.409	183.5	-125.4	-38.5	5.1	-129.8	-37.8	5.1
2183	126.8421	-2.4979	2013.652	11.436	16.208	31.49	144.3	75.5	-41.4	5.4	72.6	-41.9	5.8
2184	126.8656	23.2693	2013.548	16.637	16.764	3.693	7.2	-28.7	-67.4	5.2	-31.7	-69	5.6
2185	126.9114	15.7806	2013.658	16.506	18.931	7.896	322.6	46.3	-82.2	5.3	47.8	-81.5	7
2186	126.9615	10.0353	2013.7405	11.993	16.227	21.827	127.5	-61.8	-67	5.9	-63.3	-63	6
2187	126.9707	53.7804	2013.7315	12.659	13.656	9.532	82.3	-13.9	-113.3	5.5	-12.2	-114.8	5.5
2188	127.0745	49.0454	2013.7665	13.886	15.183	27.334	14.2	-70.6	-79.2	5.5	-71.1	-81.7	5.5
2189	127.0938	0.4356	2013.376	16.324	16.772	27.685	265.3	0.6	-115.7	6.1	3.1	-118	6.5
2190	127.0989	53.8098	2013.873	12.088	13.876	4.653	145.4	-46.5	-97.7	5.5	-48.4	-97.6	5.6
2191	127.1133	30.6305	2013.961	14.363	15.015	4.367	150.5	-12.8	-108.8	5.8	-12	-107.3	5.7
2192	127.2316	9.6728	2013.805	13.958	14.08	27.472	95.1	-39.4	-124.3	5.9	-43.8	-116.5	5.9
2193	127.2793	-11.3495	2014.117	10.1	13.981	13.644	224.6	-61.6	-62.7	5.4	-62.1	-60.5	5.5
2194	127.2910	35.1663	2013.7225	14.476	17.759	16.223	211.2	63.1	-8.3	6	65.1	-11.9	6.1
2195	127.3241	27.1738	2013.7125	12.972	17.695	10.179	251.4	-28.6	-61.7	5.1	-31.9	-62.7	5.2
2196	127.3273	75.9363	2013.8185	12.771	17.217	14.22	139.9	-99.7	-41.7	5.5	-102.3	-41.7	5.5
2197	127.3570	30.5625	2013.7685	12.543	15.193	16.903	312.9	-56.2	-64.1	5.2	-57.1	-62.1	5.2
2198	127.4850	47.1119	2013.7125	14.983	16.182	23.319	336.6	-69.2	-15.1	5.2	-67.9	-17.1	5.2
2199	127.4905	2.7399	2013.981	10.701	11.729	6.022	58.1	-56.6	-113	5.9	-53.6	-107.7	5.7
2200	127.5022	28.2391	2013.7225	13.948	16.379	37.234	307	-104.7	-53.8	5.1	-106.2	-52.3	5.2
2201	127.5163	51.3490	2013.813	12.397	14.058	42.109	236.7	-48.2	-107	5.4	-42.3	-108.5	5.4
2202	127.5490	38.4581	2013.7075	16.042	16.813	20.578	278.7	-42.6	-61	5.5	-45.6	-62.2	5.5
2203	127.6251	23.6099	2013.824	11.868	13.097	6	21.5	-59.1	-110.1	5.4	-58.6	-109.4	5.4
2204	127.6623	59.6029	2013.8245	10.309	12.78	12.969	33.2	-36.9	-90	5.4	-35.8	-91.1	5.4
2205	127.8361	44.4015	2013.734	11.886	16.73	34.856	27.7	-123.4	-123.9	5.2	-120	-119.4	5.3
2206	127.9376	-6.9663	2013.7475	12.193	15.696	16.451	273.2	-91.4	-23.6	5.5	-91.5	-28.4	5.6
2207	127.9538	11.1745	2013.823	10.523	12.766	10.442	150.5	-60.1	-75.8	5.9	-59.2	-75.7	6

2208	127.9851	18.5760	2013.832	14.296	15.946	11.448	259.4	-58.4	-99.6	5.4	-60.3	-98.7	5.4
2209	127.9882	17.8352	2013.9475	8.737	10.275	46.087	352.5	24.1	-100	5	20.7	-101.4	5
2210	128.2334	30.5128	2013.646	14.825	17.242	11.227	296.7	5.9	-72.6	5.2	3.5	-71.6	5.3
2211	128.2628	-3.5982	2012.999	16.943	17.595	53.885	2.7	-66.9	-22.8	5.9	-60.3	-32.3	6.3
2212	128.2903	32.2933	2013.8285	11.377	12.601	25.828	167.9	-65.2	-42.5	5.1	-66.4	-37.2	5.2
2213	128.3001	26.7643	2013.6935	13.516	16.966	43.017	207.8	-84.9	-31.9	5.1	-87.5	-30.6	5.2
2214	128.3748	17.7485	2013.754	13.851	14.132	8.178	96.9	-109.5	-14.8	5.1	-108.4	-15.1	5.1
2215	128.4651	-0.9494	2013.7495	10.818	12.552	8.601	319.2	-64.4	28.5	5.4	-64.2	29.6	5.5
2216	128.4659	24.4372	2013.6295	15.876	17.164	42.631	69.5	-63.4	116.8	5.2	-67.3	121.9	5.3
2217	128.4778	12.3300	2013.7945	12.385	12.8	38.666	295.8	62.2	-44.8	5.1	63.2	-42	5
2218	128.5808	10.6701	2013.667	12.105	17.14	28.23	232.4	45.6	-62.3	6.2	44.8	-60.2	6.5
2219	128.5848	64.2472	2013.7975	11.068	12.557	7.691	216.1	-98.8	-105.6	5.5	-100.1	-105.6	5.5
2220	128.5974	65.3151	2013.791	9.696	13.796	50.615	255.4	-30.1	-85	5.5	-24.8	-88.1	5.5
2221	128.6485	37.9407	2013.8005	17.678	18.389	10.753	74.8	34.6	-111	5.9	37.4	-104.9	6.1
2222	128.6541	33.8636	2013.749	13.083	16.04	15.281	145.4	-15.5	-61.6	5.2	-15.6	-62	5.2
2223	128.6798	17.4519	2013.8265	9.948	13.001	17.23	140.7	-71.1	-50.3	5	-76	-45.2	5.1
2224	128.6914	26.0634	2013.77	11.118	17.92	15.899	32.5	-38.3	-95.6	5.1	-41.1	-97.4	5.5
2225	128.7025	22.9642	2013.797	13.966	16.41	27.693	300.8	-54.4	-96.6	5.4	-55	-93.2	5.5
2226	128.7266	4.8465	2013.865	9.594	12.157	27.213	265.4	37.9	-85.6	6.2	43.9	-81.3	6.3
2227	128.7459	8.3349	2013.955	12.2	15.26	5.995	61.9	-77.1	-10.4	6.3	-73.3	-12.2	6.3
2228	128.7943	-2.3934	2013.246	14.794	17.095	37.257	183.7	-29	-60.9	5.6	-28.2	-63	5.9
2229	128.8619	24.2606	2013.8905	10.5	11.876	33.014	92	-9.2	-104.2	5.1	-3.9	-102.6	5.1
2230	128.8887	-6.8051	2013.476	14.479	16.117	29.906	23.5	30.9	-124.6	5.6	31.7	-130.3	5.8
2231	128.9388	9.9697	2013.8245	12.253	15.021	12.893	142.7	-98.7	-76.5	6.3	-99.1	-73.9	6.3
2232	128.9433	80.6606	2013.8275	11.311	15.306	6.928	50.2	18	-79.4	5.5	12.3	-80.9	6.2
2233	128.9631	15.6952	2013.7135	14.415	14.789	12.387	68.2	84	-59.8	5.1	85.5	-61.3	5.1
2234	128.9948	-10.4507	2014.115	11.777	15.687	27.625	135.5	-98.7	-57.9	5.4	-97.4	-65.3	5.7
2235	128.9981	14.7967	2013.714	15.363	15.706	3.938	40.6	-22.5	-124.1	5	-26.8	-121.6	5.5
2236	129.0302	0.2666	2013.036	16.868	17.657	10.875	352.4	30.1	-92	6.6	39.8	86.1	7.5
2237	129.0491	69.5129	2013.7575	11.366	15.438	35.476	125	-51.9	-93.5	5.5	-51.6	-90.1	5.5
2238	129.0866	46.6968	2013.6215	16.214	17.311	7.43	229.5	-8.8	-84.7	5.9	-8.4	-83.1	5.9
2239	129.1752	63.9724	2013.7935	12.55	15.407	22.517	176.1	9.6	-61.8	5.4	9.4	-60.8	5.5
2240	129.2319	14.9344	2013.8685	11.919	15.442	12.157	165	-69.7	-10.8	5	-73.4	-8.5	5.1
2241	129.2322	39.1254	2013.9295	10.382	15.38	7.664	288.9	-9	-88.6	5.2	-4	-86.8	5.4
2242	129.3807	35.2736	2013.6345	15.918	16.005	36.184	82.9	-10.2	-79.8	5.2	-0.6	-78.6	5.2
2243	129.3850	-7.3979	2014.0095	12.474	14.934	6.909	287.8	-76.6	18.5	5.4	-75.2	17	5.5
2244	129.4728	31.7040	2013.7865	12.766	16.973	14.899	72.5	-86.1	-88.6	5.2	-83	-90.6	5.2
2245	129.5897	46.6702	2013.6385	14.923	17.717	38.094	320.7	-64.3	-22.4	5.5	-66.7	-22.3	5.6
2246	129.6030	-6.4149	2013.998	14.335	14.908	5.095	221.1	2.1	-102	5.5	-1.4	-100.6	5.4
2247	129.6920	13.4093	2013.596	16.422	18.661	18.207	72.2	62.9	-62.4	5.2	69.6	-55.1	6.9
2248	129.7006	15.9593	2013.78	11.334	13.664	6.234	7.6	47.2	-94.7	5	45.2	-95.4	5.2
2249	129.8019	41.5966	2013.686	11.604	17.936	42.134	346	-40.5	-67	5.2	-45.6	-67.4	5.4
2250	129.8539	49.5505	2013.7005	16.852	17.903	5.724	325.7	-89.2	-62.7	5.6	-85.2	-59.4	5.6
2251	129.8877	19.1186	2013.868	16.826	18.695	32.886	30.1	-61.2	-77	5.5	-64.8	-68.7	6.4
2252	129.8926	22.8225	2013.7355	11.815	15.493	8.7	315.6	12.4	-64.3	5.3	13.8	-65	5.4
2253	129.9278	18.0238	2013.871	13.355	15.104	28.303	206.1	-88.8	8.1	5.3	-87.5	7.8	5.4

2254	129.9493	8.9391	2013.7965	11.622	13.231	6.654	88.8	-208.7	186.2	6.2	-213.4	182.4	6.4
2255	129.9675	44.9798	2013.8205	11.987	12.123	7.006	182	-184.8	-177	6	-185.3	-178.2	6
2256	129.9929	57.3466	2013.8135	13.199	13.709	12.932	316.3	18	-120	5.8	19.7	-118.3	5.8
2257	130.0047	37.8907	2013.5695	16.072	17.206	7.246	78.4	-65.5	-48.7	5.3	-68	-45.8	5.3
2258	130.0176	79.8669	2013.8405	16.002	17.787	14.971	46.1	-66.1	-61.4	5.5	-66.1	-62.2	5.5
2259	130.0371	66.3107	2013.7725	10.79	13.99	7.921	256.8	-52	-66	5.9	-51.3	-66	5.9
2260	130.0407	66.4120	2013.7385	12.128	15.652	22.222	72.1	-90.1	-37.7	5.9	-92.5	-38.6	5.9
2261	130.1030	48.9859	2013.691	14.799	17.042	19.662	47.9	149.1	-142.1	5.6	152.5	-141.4	5.7
2262	130.1073	-6.6801	2013.6545	12.758	16.663	9.533	143	-114.4	-81.9	5.5	-117.1	-82.7	6.2
2263	130.1571	22.5563	2013.8525	12.181	14.907	7.74	34.9	81.3	-49.2	5.3	77.5	-50.4	5.4
2264	130.2334	79.5094	2013.8325	12.36	13.737	48.99	136.4	69.2	61	5.5	74	58.3	5.5
2265	130.2672	33.7590	2013.923	15.345	15.385	5.038	222.8	-64.9	17.6	5.9	-67	14.4	5.9
2266	130.2822	72.0820	2013.783	15.979	16.561	10.581	169.5	-37.9	-85.3	5.5	-36.9	-81.6	5.6
2267	130.3368	1.1574	2013.764	15.295	15.643	22.733	342.7	73.8	-118	6.3	70.6	-118.9	6.2
2268	130.3976	34.0746	2013.831	9.554	14.36	8.342	229.5	-120.4	-83.2	5.9	-126.1	-72.8	6
2269	130.4475	38.5840	2013.9985	9.89	14.728	8.409	84.8	-120.4	-93.2	5.2	-128	-95.5	6.1
2270	130.4559	-6.1974	2014.067	14.025	14.582	3.933	49.5	-15.9	-77.1	5.5	-16.5	-76.7	5.8
2271	130.6153	-6.3249	2013.449	15.392	16.577	5.808	7.4	71.9	-80	5.5	70.5	-80.5	6.9
2272	130.8555	48.8614	2013.881	7.696	7.799	10.129	330.6	-72.9	-51.2	5.5	-73.3	-52	5.5
2273	130.8613	63.8191	2013.7855	13.458	16.938	7.256	234.5	76	13.4	5.9	79.3	11.5	6
2274	130.8650	13.1295	2013.833	11.129	11.174	7.571	4.6	-68.8	-61.8	6	-64.4	-61.4	6
2275	130.8944	63.1406	2013.7535	14.228	16.432	19.541	199.4	-40.9	-61.1	5.9	-41.8	-63	6
2276	130.9123	-0.5226	2013.743	15.807	16.136	8.463	71.3	16.1	-68.6	5.5	19.8	-69	5.8
2277	130.9339	27.4115	2013.7205	17.26	17.324	9.023	54.7	-75.3	-65.7	5.2	-74.2	-66.5	5.3
2278	130.9706	35.6213	2013.74	12.41	16.927	23.538	93.9	-12.1	-84.8	5.2	-17.3	-81.8	5.2
2279	130.9751	58.4795	2013.7715	14.741	15.452	23.297	347.2	-29.4	-79.4	5.8	-30	-81.2	5.8
2280	131.1934	12.7272	2013.6165	14.414	16.676	5.042	255.6	-53.2	-63	6.3	-58.4	-61.6	7
2281	131.2026	-1.1684	2013.834	9.64	13.095	36.856	344.1	113.6	-81.3	5.4	105.1	-91.6	5.6
2282	131.2332	7.5066	2013.841	16.79	18.514	34.81	68	-0.1	-68.2	6.1	-9.7	-66.1	7.2
2283	131.2654	13.5203	2013.7965	11.547	15.786	20.558	109.6	-114.4	-135.9	5.4	-114.9	-139.3	5.5
2284	131.2935	35.6492	2013.921	12.827	13.807	4.503	145.6	-125.3	-102.7	5.2	-127.9	-100.9	5.2
2285	131.3224	60.7242	2013.873	10.151	12.485	7.982	359.8	-77	5	5.8	-76.2	9.9	5.8
2286	131.3427	47.2790	2013.6695	17.478	18.148	13.017	62.4	12.9	-62.4	5.5	8.6	-65.9	5.5
2287	131.3481	60.0167	2013.863	10.273	11.517	4.197	207.1	-122.8	4.9	5.8	-111.8	-3.7	7
2288	131.3498	46.1708	2013.797	7.916	11.326	32.329	359	7.8	-80.4	5.4	10.7	-80.4	5.4
2289	131.3515	8.2042	2013.741	12.408	15.723	14.323	346	5.8	-138.4	5.9	4.6	-139.9	6
2290	131.3608	10.9128	2013.842	10.496	13.577	13.539	60.6	-214.7	-43.4	6	-218.4	-50.2	6
2291	131.3833	44.4419	2013.6995	11.594	16.2	19.287	85.4	-139.9	-185.3	5.5	-140.4	-178.8	5.6
2292	131.4345	10.8862	2013.652	11.583	17.795	39.317	345.8	-2.1	-86.9	5.9	-10.5	-89	6.2
2293	131.4702	-6.9664	2014.1735	7.676	12.281	18.385	13.2	55.7	-71.6	5.3	54	-78.4	5.3
2294	131.5526	7.7393	2013.809	12.98	15.802	47.849	3.4	165.9	-102.5	5.9	165	-105.4	6
2295	131.5597	27.5945	2014.002	7.065	10.126	41.009	174	10.7	-88	5.1	7.1	-85.8	5.1
2296	131.5802	50.3216	2013.7155	14.237	16.667	17.124	268.9	-26.5	-91.9	5.5	-27.9	-91.8	5.6
2297	131.6645	39.0239	2013.6285	13.72	17.443	25.057	33.7	-66.9	-38.9	5.2	-69.7	-37	5.3
2298	131.7085	30.4205	2013.8675	13.331	14.311	19.596	57.8	22.2	-78.4	5.1	25.7	-77.5	5.1
2299	131.7156	-2.3106	2013.4785	13.82	16.546	10.855	345.3	-46.4	-75.1	5.6	-44.2	-74.2	6.3

2300	131.7406	54.1314	2013.8005	15.574	16.463	11.987	134.2	4.5	-167.7	5.9	2.6	-169.4	5.9
2301	131.7467	60.4775	2013.7585	11.285	17.141	10.714	172.6	9	-125.5	5.9	12.1	-121.1	6.1
2302	131.7632	28.5488	2013.8295	11.421	13.913	10.308	123.8	-51.2	-123.9	5.1	-53.8	-122.3	5.1
2303	131.8068	61.7410	2013.6565	15.573	17.813	14.326	129	58.3	-82.9	6	60	-87.3	6.1
2304	131.8459	8.2026	2013.571	16.059	17.312	15.198	245	-18.1	-77.9	6	-18.3	-81.1	6.2
2305	131.8610	2.4262	2014.169	13.934	16.783	14.083	294.7	176	-166.4	5.7	175	-165.3	6.2
2306	131.9507	36.6555	2013.7125	12.581	16.387	19.515	254.3	-105.3	9.4	5.2	-102.8	10.2	5.2
2307	132.0015	26.1131	2013.5985	16.228	17.078	6.052	54.4	-25.1	-60.4	5.2	-24.4	-60.2	5.2
2308	132.0147	-6.7873	2014.1635	12.639	16.354	17.04	308.1	80.4	7.6	5.3	80.9	13.6	5.6
2309	132.1486	24.8528	2013.9085	10.376	11.675	36.042	222.6	-7.8	-102.9	5.1	-8.9	-101.7	5.1
2310	132.1511	-13.8857	2014.206	10.911	11.673	7.354	92.2	-190.3	0.8	5.4	-196.4	7.9	5.8
2311	132.1988	19.9550	2013.976	13.242	16.551	7.155	176.3	-82.1	-30.1	5.9	-81.1	-28.9	6.2
2312	132.2897	-4.3217	2013.8675	12.174	12.708	7.578	110.1	-119.9	74.2	5.6	-119.9	71.4	5.6
2313	132.3323	1.4750	2014.0715	13.92	14.401	37.521	76.6	54.5	-116.6	5.8	61.3	-116.5	5.8
2314	132.4415	6.6072	2013.997	11.1	11.668	4.98	260.4	45.5	-71.2	5.1	42.6	-71.3	5
2315	132.4467	13.1177	2013.3335	13.727	15.958	5.236	24.1	108.3	-111.7	5.1	105.5	-116.8	5.8
2316	132.4573	11.1581	2013.852	9.452	15.278	21.788	209.5	-73.1	27.1	5	-76	24.4	5.1
2317	132.4721	16.8300	2013.899	10.956	12.547	18.441	304.4	64	-66.9	5	61	-68.4	5
2318	132.4748	27.6162	2013.742	15.045	15.722	30.395	282.3	-42.5	-85.9	5.1	-41	-84.6	5.2
2319	132.4910	17.2903	2013.6565	14.593	16.067	6.79	222.1	-5.9	-65.9	5.1	-2.8	-65.2	5.2
2320	132.5010	-3.3616	2013.966	9.885	12.978	18.593	44.6	-77.8	25.1	5.4	-78.3	26.7	5.5
2321	132.5738	18.8994	2013.972	12.497	14.907	5.607	199.4	-62.6	-95	5.4	-61.6	-93.8	5.4
2322	132.5887	28.8198	2013.7005	14.644	17.11	10.823	32	-39.6	-135.9	5.2	-38	-134.9	5.3
2323	132.6461	-9.9076	2014.178	13.336	13.802	17.572	197.3	-83.1	-25.3	5.3	-87.3	-25.8	5.3
2324	132.7159	69.2202	2013.8275	10.989	13.848	7.283	280.1	-27.4	-173.1	5.9	-26.3	-171.4	5.9
2325	132.7194	25.0381	2013.7475	10.232	16.225	11.224	279.6	-75.4	-9.1	5.1	-72.8	-6.4	5.4
2326	132.7229	57.9588	2013.8155	12.982	15.041	45.651	282.7	-55.9	-76	5.8	-55.2	-77.6	5.9
2327	132.7499	75.9669	2013.951	11.88	13.133	5.321	129.4	-39.4	-66.6	5.9	-36.3	-65.4	5.8
2328	132.7647	-0.9284	2013.8435	10.829	15.351	14.977	141.5	-76.7	-7.3	5.4	-78.8	-4.8	5.6
2329	132.8676	7.5143	2013.8395	13.394	17.373	20.161	129.3	18.4	-63.5	5	17.2	-66.7	5.2
2330	132.8958	35.5291	2013.7195	13.524	16.263	18.726	191.1	-37.8	-97.9	5.8	-36.6	-100.9	5.9
2331	132.9458	41.5709	2013.692	15.32	17.807	32.709	21.8	-115.3	16.2	5.2	-110.6	18.7	5.3
2332	133.0239	32.7810	2013.6675	14.736	15.667	12.109	13.7	-31.4	-86.7	5.8	-35.5	-88.5	5.8
2333	133.0246	19.6337	2013.7265	16.462	16.715	10.279	80.9	-69.7	-67.9	5.5	-69.7	-67.3	5.5
2334	133.0985	28.4752	2013.7455	11.827	12.832	17.883	126.6	-5.2	-105.1	6	-6.1	-101.9	6
2335	133.1213	9.4152	2013.6075	15.058	16.635	19.784	328.2	-1.4	-60.9	5.9	1.7	-63.8	6.1
2336	133.1219	21.2387	2013.895	10.97	13.047	6.341	240.8	40.2	-68.4	5.3	39.9	-66.1	5.4
2337	133.2082	8.9623	2013.7285	11.445	15.489	55.976	255.4	-80	-39.9	5.9	-80.3	-43.3	6
2338	133.2375	15.8005	2013.752	16.419	17.263	33.408	210.5	-3.2	-92.4	5.2	-2.9	-90.1	5.2
2339	133.2388	-8.5654	2014.1725	10.538	15.877	22.173	337.3	68.7	-10.3	5.3	67.1	-14.7	5.5
2340	133.2518	-12.7701	2014.093	10.304	12.859	18.291	275.4	-72.5	14.7	5.1	-75.7	14.6	5.1
2341	133.2618	-8.9742	2014.1585	14.432	14.789	6.64	241	-12.6	-71.7	5.3	-11.4	-71.8	5.4
2342	133.2649	-4.6036	2014.133	11.907	15.565	17.097	112.4	20.4	-176.8	5.3	20.8	-177.1	5.4
2343	133.3308	43.7218	2013.6385	16.403	17.916	10.411	298.8	-41.9	-67.6	6.1	-43.6	-63	6.1
2344	133.3334	48.6000	2013.647	15.943	17.376	9.574	317.9	-34	-89.9	5.5	-32.8	-89	5.5
2345	133.4044	55.5337	2013.7575	15.872	15.901	14.966	270.3	77.6	-30.4	6	74.9	-29.5	6

2346	133.4189	3.3612	2014.0595	15.031	15.782	9.46	50.3	-1.9	-73.5	5.8	-3.7	-71.5	5.8
2347	133.4559	20.4613	2013.891	14.083	14.773	9.884	332.8	74.8	-38.6	5.3	73.1	-35.3	5.4
2348	133.4595	26.3332	2013.674	15.179	17.473	22.603	303.9	-7.4	-117.8	5.4	-5.4	-117.6	5.6
2349	133.5964	78.1720	2013.98	13.939	14.579	4.491	121.2	-17.5	-90.6	5.4	-17.6	-87.8	5.5
2350	133.5995	79.6814	2013.939	10.144	13.76	6.63	2.3	-90.5	-49.3	5.5	-87.9	-52.1	5.5
2351	133.6257	40.1327	2013.6915	7.411	15.553	27.269	254.8	35	-163.9	5.3	36.8	-158.6	5.2
2352	133.7652	44.8759	2013.6505	15.958	17.535	29.355	36.5	22	-78.9	5.4	11.5	-81.3	5.5
2353	133.7661	31.0285	2013.645	15.048	17.848	6.952	301	-78.1	-69.1	5.8	-81.2	-66.3	6.1
2354	133.7961	13.4850	2013.6975	13.33	15.291	21.881	211.1	-91.7	-13.1	5.1	-89.3	-11.2	5.1
2355	133.8850	37.8950	2013.684	12.237	15.619	8.98	163.4	-64.9	-30.5	5.2	-64.1	-28.6	5.2
2356	133.9852	38.2442	2013.6635	11.293	16.128	14.203	32.5	44.9	-116.3	5.2	47.4	-118.6	5.3
2357	134.0572	4.4870	2014.134	15.901	16.297	4.395	169.7	-104.4	87.5	6.2	-108.1	87.2	6.3
2358	134.1029	-9.2749	2014.1115	15.866	15.985	8.597	316.6	36.5	-90.8	5.7	36.5	-89.5	5.6
2359	134.1907	7.7969	2013.93	10.292	13.29	6.833	190.9	70.2	-121.3	5.9	71.6	-116.2	6.1
2360	134.3219	50.2425	2013.6375	13.569	17.11	6.361	43.2	64.5	-19.9	5.8	62.4	-17.4	6.4
2361	134.4126	47.7225	2013.661	13.452	13.767	9.575	272.2	-15.3	-88	5.4	-17.2	-89.4	5.4
2362	134.4265	55.3669	2013.827	12.192	12.453	12.143	258.8	162.3	62.1	5.8	163.6	64.4	5.8
2363	134.4397	-12.8924	2014.1145	11.981	13.453	11.063	164.8	-63.6	31.4	5.2	-65.3	33.5	5.1
2364	134.4500	22.4276	2013.733	15.115	17.775	13.621	267.6	4.3	-81.4	5.4	8.6	-80.3	5.6
2365	134.4611	20.0279	2013.8125	15.614	15.66	12.68	287.2	-100.6	54	5.4	-99.4	57.2	5.4
2366	134.4966	5.5723	2013.9	12.71	13.561	11.681	134.7	-77.3	-23.7	5.9	-77.5	-25.5	5.8
2367	134.5085	11.5047	2013.8595	10.634	13.296	52.67	73.6	172.3	-187.9	6	172.8	-186.7	6
2368	134.5419	52.4541	2013.7075	15.224	15.664	5.756	281	-130	-133.6	5.8	-130.2	-129	5.8
2369	134.6606	4.4860	2014.113	15.911	16.179	33.408	59.1	17.1	-72.1	5.9	17.7	-70.5	5.9
2370	134.7405	9.6165	2013.8525	14.544	14.664	51.633	298.4	-112.3	4.4	5.9	-108.4	5.2	5.9
2371	134.8313	-2.6754	2013.9415	10.001	12.217	23.336	309.1	-22.4	-116.1	5.5	-28.9	-113.9	5.4
2372	134.8488	56.2361	2013.7125	14.949	17.469	6.359	157.8	-33.5	-72.9	5.8	-38.1	-73.5	6
2373	134.9677	60.4496	2013.8145	10.961	11.969	9.492	151.3	74.5	11.7	5.8	71.5	11.7	5.8
2374	135.0327	-1.7419	2013.5825	10.579	11.838	5.098	214.5	77.4	-75.8	5.4	77.3	-71.4	5.8
2375	135.0588	11.4110	2013.892	14.208	14.593	7.35	251	-51.9	-75.1	5.9	-50.3	-74.8	5.8
2376	135.0661	-5.7587	2014.1115	12.867	13.126	7.555	174.1	65.3	-53.4	5.3	67.9	-51.9	5.3
2377	135.0737	34.8624	2013.8255	8.109	12.402	7.022	133.2	-129	-61.6	5.2	-130.1	-72.9	6.2
2378	135.0787	9.2884	2014.0865	10.369	11.186	4.338	262.9	-66.7	-11.5	5.8	-65.1	-10.5	6.1
2379	135.1310	-6.2567	2014.0205	10.874	15.799	20.196	241	-63.3	-51.9	5.4	-66.1	-48.4	5.6
2380	135.2034	4.7935	2013.9995	9.339	11.857	12.332	257	191.1	-115.3	5.8	188.2	-124.2	5.9
2381	135.2054	23.7795	2013.935	8.618	13.732	55.387	79.1	44.9	-141.8	5.3	38.2	-147.1	5.4
2382	135.2092	17.4968	2013.8165	12.883	12.989	30.23	344.7	-24.8	-99.5	5.1	-23.2	-102.3	5.1
2383	135.2164	6.7256	2013.8135	12.474	14.414	44.669	198.6	-162.3	44.8	6	-161.6	47.1	6.1
2384	135.2266	10.5779	2013.903	10.937	13.7	11.535	352.2	-119.4	-51.2	5.9	-118.1	-51.3	5.9
2385	135.3614	62.9287	2013.777	14.258	15.351	33.38	161.6	-125.3	-24.4	5.6	-121.5	-21.8	5.6
2386	135.5085	60.1406	2013.7155	16.223	16.668	35.082	11.9	-79.8	-95.4	5.6	-82.1	-91.7	5.6
2387	135.5208	36.7290	2013.7925	12.512	18.777	38.314	136.8	-85	-19.6	5.2	-86.1	-21.7	6.2
2388	135.5288	11.6435	2013.9465	13.289	15.309	9.3	265.3	-91.9	-58.3	5.9	-94.1	-57.4	5.9
2389	135.5820	61.6606	2013.802	11.449	11.691	15.031	208.8	-43.8	-78.9	5.5	-42.5	-74.6	5.5
2390	135.6332	28.3737	2013.8885	12.13	13.138	20.423	150.1	-43.6	-89.6	5.4	-42	-88	5.4
2391	135.7014	23.1404	2014.0795	11.084	13.315	6.102	41.7	-91.3	-32.7	5.3	-93.9	-32.9	5.3

2392	135.7639	-11.6703	2014.178	14.262	14.733	14.828	53.9	-105.9	-6.8	5.4	-107.8	-4.6	5.5
2393	135.8946	28.8211	2013.91	12.12	16.017	7.206	275.8	-65.8	-41.6	5.4	-67.5	-39.4	5.4
2394	135.9312	53.9669	2013.734	14.023	16.215	55.87	206.4	-36	-74.9	5.8	-42.3	-76.4	5.8
2395	135.9385	7.0121	2013.8525	16.123	17.872	25.236	73.7	-76.9	-49	5.9	-74.9	-53.2	7.3
2396	136.0256	34.5470	2013.7635	12.337	17.618	10.351	129.3	59.3	-137.7	5.2	55	-145.7	5.5
2397	136.0630	3.0260	2014.1545	8.529	8.809	32.226	326.9	56.6	-85	5.8	54	-90.7	5.7
2398	136.1462	-10.0459	2014.1515	13.912	15.429	4.86	181.9	33.5	-114	5.4	31.8	-111.5	7.5
2399	136.1610	28.4040	2013.7715	15.017	17.455	6.569	122.6	-6.9	-81.7	5.2	-15.6	-83	5.3
2400	136.1695	67.1632	2013.744	15.125	15.29	11.306	180.6	74.6	-57.4	5.6	74.8	-56.9	5.6
2401	136.1801	22.0576	2013.853	13.274	17.76	7.033	318.2	78	-46.2	5.3	79.4	-43.1	7.5
2402	136.1918	-7.5045	2014.1845	12.288	16.037	6.926	242.3	-77.4	-85.6	5.3	-71.4	-88.1	6.1
2403	136.2476	-3.9875	2014.15	7.872	8.798	10.565	84.6	-74.5	5.3	5.4	-76	8.7	5.3
2404	136.2479	2.4362	2013.8925	11.749	12.076	10.043	68.3	-208	1.3	5.9	-202.2	8.3	5.9
2405	136.2796	32.8675	2013.7795	11.243	16.231	20.335	231.9	-62.2	-18.6	5.1	-62.2	-16.6	5.2
2406	136.3043	22.3074	2013.9555	13.176	16.875	45.065	256.3	-78.2	-26.8	5.3	-79.3	-23.8	5.4
2407	136.3144	17.9828	2013.638	15.973	16.472	5.95	118	23.1	-61.9	5.5	21.1	-60.2	5.5
2408	136.4397	55.5290	2013.799	7.79	13.963	56.393	52.9	-211.8	-7.2	5.9	-216.1	1.8	5.9
2409	136.5109	55.1588	2013.73	11.517	16.185	21.463	59.9	-96.5	-108.7	5.9	-100.2	-107.2	5.9
2410	136.5375	12.3805	2013.907	14.593	15.076	18.961	96.2	-8.9	-77.9	5.9	-8.7	-78.2	5.9
2411	136.5424	32.8592	2013.732	14.074	15.031	7.731	42.2	-17.3	-70	5.8	-14.3	-73.1	5.8
2412	136.5564	15.4890	2013.69	14.907	15.397	9.701	177.1	-74.3	-29.9	6	-74.4	-29.3	6
2413	136.5649	17.1117	2014.0505	7.324	10.062	56.704	320.4	0.8	-68.1	5.8	-1	-66.4	5.9
2414	136.5685	57.5361	2013.764	10.044	18.44	18.291	222.3	-89.8	-68.5	5.8	-95.2	-65.1	6.2
2415	136.5863	29.7954	2013.783	11.201	14.766	37.827	179.1	-15.2	-92.5	5.1	-15.9	-91.6	5.2
2416	136.5974	-8.0331	2013.9895	13.999	16.346	10.173	72.5	-66.3	32.3	5.4	-67.7	36.3	5.7
2417	136.6271	-2.3659	2013.642	14.059	14.725	21.674	246.5	-12	-77.4	5.6	-14.3	-78.5	5.5
2418	136.6904	-3.7821	2014.174	15.29	16.166	15.19	78	17.2	-68.6	5.6	12.6	-69	5.7
2419	136.7262	13.9795	2013.872	11.724	17.387	11.258	314	-12.7	-70.5	5.9	-10.6	-73.2	6.1
2420	136.7294	65.9711	2013.806	13.296	14.779	18.419	212.9	67.4	-11.5	5.6	66.3	-13.9	5.6
2421	136.7400	14.2483	2013.7875	16.997	18.56	7.329	309.3	-86.6	-33.4	6.1	-85.8	-44.4	6.6
2422	136.8190	21.9364	2014.1275	9.364	11.6	7.785	4.8	-72.7	-40.1	5.8	-69.7	-41.1	5.8
2423	136.8767	36.3786	2013.715	10.001	13.794	8.93	289.6	-34.3	-149.3	5.2	-34.5	-148.8	5.2
2424	136.8979	2.3800	2013.5755	12.026	17.329	14.788	341.3	-87.3	3.5	5.9	-83.5	5.2	6.5
2425	136.9224	10.0059	2014.147	15.473	15.684	6.908	218.2	-67.5	0.6	5.8	-70	-1.6	5.8
2426	136.9552	85.5881	2013.932	14.486	14.591	58.011	45	27.6	-166.3	5.7	27.4	-167.1	5.7
2427	136.9692	-7.1393	2014.173	8.425	10.03	7.129	136.1	75	-54.4	5.3	68	-56.4	5.4
2428	137.0992	27.5357	2014.0125	8.098	8.11	51.769	57.6	-47.2	69.6	5.1	-47.7	69.8	5.1
2429	137.1501	15.4505	2013.583	16.456	16.922	19.543	10.5	-60.6	-12.5	6.4	-61.4	-10.5	6.5
2430	137.2169	51.6045	2013.8595	5.324	13.782	26.276	33.8	-135.1	-34.9	5.6	-129.5	-40.6	5.5
2431	137.2400	46.3757	2013.684	10.5	16.645	34.941	343.5	69.4	35.5	5.2	69.2	35.2	5.3
2432	137.2975	28.9376	2013.7805	13.317	16.212	8.819	261.2	-8.6	-77.5	5.7	-10.6	-76	5.8
2433	137.3341	7.3770	2014.038	12.899	16.271	18.127	155.4	41.5	-146.6	5.8	39.9	-145.6	5.9
2434	137.3530	3.6733	2014.1435	11.324	12.913	8.394	34	-97	-17.1	5.7	-96.4	-14.2	5.9
2435	137.5125	21.5499	2013.8385	14.534	16.293	24.404	189.1	-78	-23.4	5.4	-79.5	-21.2	5.4
2436	137.5391	65.1028	2013.9235	13.269	14.161	4.69	129.2	-83.4	-110.5	5.8	-85.2	-113.1	5.8
2437	137.5673	44.7586	2013.6725	13.162	16.302	6.699	266.8	-62.4	21.8	5.2	-62.4	19.8	5.3

2438	137.6070	12.0513	2013.929	13.548	15.265	24.984	219.3	-74.7	31.4	5.9	-77.1	30.3	5.9
2439	137.6114	14.3937	2013.7875	15.032	15.564	22.797	73.8	-40.5	-64.2	6	-40.7	-65.9	6
2440	137.6472	8.6657	2014.154	14.617	17.979	9.208	28.3	-67.9	-41.7	5.8	-64.7	-48.6	6.1
2441	137.9424	16.2752	2013.696	13.823	14.045	8.999	202.6	29.9	-71.7	5.1	30	-69.8	5.1
2442	137.9722	23.9485	2013.947	13.856	14.936	17.063	47.5	-72.2	10.7	5	-70.8	8.4	5.1
2443	138.0113	37.2265	2013.586	12.173	13.882	5.303	157.5	-63.3	0.1	5.2	-62.6	1.1	5.3
2444	138.0754	-9.4082	2013.9385	10.74	16.06	28.075	64.3	-87.3	11.2	5.3	-85	9.7	6
2445	138.1173	49.2058	2013.68	7.357	12.214	8.155	75.8	-56.7	-176.6	5.8	-63.2	-176.1	6.1
2446	138.2633	1.9427	2013.6225	10.921	16.105	23.74	93.2	-64.4	-62.3	6.4	-66.7	-60.8	6.7
2447	138.3350	56.2357	2013.7635	11.733	13.432	51.838	87.7	-40.5	-71.6	5.5	-40.2	-70.8	5.5
2448	138.3490	47.1521	2013.6785	12.5	17.623	14.824	138.4	48.3	-80.6	5.2	47.4	-81.5	5.3
2449	138.4895	45.2010	2013.712	11.415	16.461	22.605	19.4	-62.6	-23.6	5.2	-61	-23.2	5.3
2450	138.5014	26.1762	2014.072	8.726	11.217	11.524	348.7	-89.8	-22.1	5.7	-90.1	-29.5	5.7
2451	138.5223	48.2147	2013.825	10.06	15.428	9.036	2.3	-47.7	-68.8	5.4	-43	-69.3	5.8
2452	138.5705	12.6271	2013.9605	13.699	15.534	10.521	214.8	-95.5	-12.3	5	-96.4	-11.6	5.1
2453	138.6393	-3.5202	2014.157	13.101	13.23	14.576	358.7	-68.5	8.5	5.3	-66.2	6	5.3
2454	138.6856	18.1043	2013.61	14.413	18.107	10.977	113.1	-130.3	-93.9	5.2	-129	-94	5.6
2455	138.7076	69.7511	2013.7595	15.106	17.285	22.895	28.8	30.6	-126.6	5.9	35.8	-126.4	6
2456	138.7417	43.1853	2013.7615	10.386	13.868	12.21	74.7	-67.6	21.7	5.2	-67.2	24.6	5.2
2457	138.7598	47.0731	2013.7745	14.701	15.033	8.284	151.3	60.7	-13.5	5.2	61.5	-14.7	5.2
2458	138.7886	23.3749	2014.0385	7.849	10.882	59.957	170.7	-126.8	-158	5	-129.6	-155.3	5.1
2459	138.8108	8.4030	2014.1945	10.479	14.763	39.828	7.4	-57.2	-79.3	5.7	-60.9	-79.6	5.8
2460	138.8223	40.8608	2013.6515	12.57	14.746	8.862	116.2	22.6	-94.6	5.2	21.5	-94.2	5.2
2461	138.8822	67.3818	2013.6505	16.409	17.336	14.961	324.6	67	-45.5	6	70.8	-41.3	6
2462	139.0896	4.3841	2014.1465	14.969	15.104	3.196	241	-77.7	-84.7	6.1	-79	-81.2	6.1
2463	139.1068	60.7370	2013.806	9.099	17.559	31.499	24.8	16.3	-89.2	5.4	7.1	-88.5	5.5
2464	139.2023	-6.0900	2013.984	13.363	15.481	8.024	261.2	-73.1	26.4	5.4	-71.4	22.7	5.5
2465	139.2509	7.9522	2014.123	12.367	14.233	4.635	105.1	-17.5	-68.4	5.7	-23.6	-67.8	5.9
2466	139.2594	53.4305	2013.694	12.778	16.86	8.061	2.7	-14.2	-76.4	5.9	-15.3	-78.8	6
2467	139.3236	26.9364	2013.9135	13.103	16.522	33.886	358.9	58.8	-81.5	5.5	55.1	-81.3	5.5
2468	139.3338	67.2788	2013.768	11.049	14.661	8.152	200.5	-57.4	-77.5	5.5	-60.7	-77.5	5.5
2469	139.3818	60.2000	2013.7715	13.339	14.052	42.981	220.7	-24.6	-63.2	5.4	-23.4	-64.1	5.5
2470	139.4393	53.2151	2013.7385	12.978	16.317	26.171	226.5	140.1	-33.4	6	138.5	-34.6	6
2471	139.4990	51.6375	2013.7575	13.922	14.264	24.402	184.8	-88.4	-69	6	-89.1	-72.8	6
2472	139.5561	46.3757	2013.679	11.229	17.154	15.915	214.2	-45.1	-104.9	5.2	-39.7	-110.2	5.3
2473	139.5565	30.5386	2013.791	13.501	14.091	5.623	110.2	-133.1	-43.2	5.2	-132.9	-44	5.2
2474	139.6139	60.6266	2013.808	13.41	13.747	34.78	263.2	-184.4	51.1	5.5	-185.7	49.3	5.5
2475	139.6378	27.5289	2013.87	14.01	15.327	15.992	256	-23.2	-74.5	5.5	-27.8	-73.6	5.5
2476	139.6383	-3.8321	2014.2075	11.622	15.864	10.953	349.1	6.4	-65.9	5.3	6.4	-65.4	5.5
2477	139.6404	15.2703	2013.687	12.53	16.514	8.745	346.9	-101.7	-112.6	5.1	-105.9	-112.9	5.4
2478	139.6848	17.1913	2013.427	17.026	17.29	9.01	333.9	-50	-88.4	5.3	-51.3	-87.3	5.3
2479	139.7241	64.4165	2013.8125	13.983	14.523	17.813	195	119.1	-67.3	5.5	117.6	-71.6	5.5
2480	139.7243	-8.7848	2014.013	13.608	14.109	28.261	72.4	-76	-132.9	5.4	-73.1	-127.1	5.4
2481	139.8399	53.7848	2013.7475	10.419	17.317	15.303	229.9	-91.2	-35.8	5.4	-93.7	-33.7	5.5
2482	139.9107	21.0017	2013.7625	15.368	15.912	6.825	52.4	-44.9	-96.7	5.1	-44.2	-93.5	5.2
2483	139.9278	4.7246	2014.2015	13.793	16.598	12.621	114	-190.5	-12.6	5.7	-189.6	-12.1	5.7

2484	139.9332	31.1957	2013.755	14.72	15.872	4.764	10	52	-65.8	5.2	49.8	-65.9	5.3
2485	139.9630	41.6770	2013.8145	13.563	15.118	4.376	193.6	-127.2	-87.9	5.9	-125.1	-77.7	5.9
2486	139.9679	4.8699	2014.2085	15.633	16.427	4.705	337.3	-19.9	-109.3	5.8	-18.7	-111.8	7.7
2487	140.0017	30.8776	2013.856	10.301	15.119	32.107	221.2	-73.3	-31.5	5.1	-68.7	-33.7	5.2
2488	140.0165	36.4202	2013.6665	14.936	15.489	7.439	183.9	-29.4	-68.6	5.8	-29.1	-69.1	5.9
2489	140.0250	55.9345	2013.721	15.007	17.719	10.871	353.6	-132.7	-42.9	5.5	-133.5	-40.3	5.6
2490	140.0699	20.3512	2013.6265	14.584	16.565	8.253	51.3	-76.7	-75.4	5.1	-76.3	-75	5.3
2491	140.0833	61.7972	2013.668	13.002	15.604	12.007	294.9	-72.1	-20.6	5.5	-71.6	-22.4	5.5
2492	140.1110	5.0353	2014.1735	14.37	14.885	10.741	26.8	61.3	-59.6	5.7	65.5	-59.9	5.8
2493	140.1174	-0.5123	2014.151	13.077	14.27	21.163	161	-112.5	86.3	5.8	-113.3	90.1	5.8
2494	140.1492	18.7093	2013.979	13.632	15.476	5.15	302.1	30.5	-66.1	5.1	26.9	-66.6	5.4
2495	140.1918	29.4645	2013.86	17.757	18.744	7.591	128.5	-65.9	-7.6	5.7	-64	-8.6	7
2496	140.2738	7.2391	2014.1585	14.009	15.034	9.254	260.7	-23.9	64.9	5.8	-29	64.2	5.8
2497	140.2821	42.6337	2013.757	12.893	13.101	23.826	61.3	-94.7	-56.1	5.2	-92.2	-56.7	5.2
2498	140.3070	68.7184	2013.6675	15.48	17.656	19.796	167.4	-55.7	-75.5	5.6	-49	-81.8	5.6
2499	140.3702	28.8639	2013.84	16.406	18.884	44.468	246.5	15.8	-66	5.5	21.1	-66	6.2
2500	140.3948	26.9296	2014.032	13.59	13.754	29.606	59.9	45.3	-64	5.4	47	-65.7	5.4
2501	140.4152	-1.5014	2014.1505	14.549	14.793	3.425	143.5	-76	9.5	6.6	-75.6	12.8	5.9
2502	140.4376	49.2662	2013.6925	14.969	15.566	8.234	152.4	45.3	-65.8	5.5	47	-65.2	5.4
2503	140.4840	44.7459	2013.7305	12.579	17.092	9.222	263.5	-109	20.1	5.2	-111.2	19.3	5.4
2504	140.5383	30.6943	2013.774	17.033	17.539	41.775	116.4	-31.4	-110.4	5.2	-32.6	-106.5	5.3
2505	140.5477	34.0572	2013.7175	14.177	14.199	13.597	76.8	54.7	-88.3	5.2	54.9	-87.8	5.2
2506	140.5959	29.5279	2013.8935	9.982	12.588	6.761	85.2	24.1	-84.2	5.4	18.2	-87.1	5.6
2507	140.6288	4.3840	2014.19	12.495	15.825	9.891	251.9	-77.5	1.8	5.7	-78.1	4.7	5.8
2508	140.6305	-4.9015	2014.2085	13.91	15.947	8.177	237.5	112.3	-186.1	5.8	117.4	-192.6	6.1
2509	140.7561	-0.3398	2014.1755	13.153	14.289	4.533	241	113.3	-147.7	5.9	110.8	-147	6
2510	140.7586	68.9764	2013.6445	14.706	17.684	18.14	84.9	-92.8	-24.6	5.5	-94.7	-27.3	5.7
2511	140.7723	19.1574	2013.7025	11.099	13.551	33.892	267.7	-40.8	-76.7	6.3	-42.1	-74.5	6.3
2512	140.7758	22.3050	2014.132	8.988	10.234	8.269	343.4	-131.8	-174.7	6.1	-139.3	-180.8	6.1
2513	140.8497	26.9946	2013.855	17.17	17.35	21.517	58.9	-26.8	-61.9	5.6	-30.1	-61.8	5.5
2514	140.9164	0.0701	2014.236	16.26	17.159	56.056	33.4	24	-63.5	6.2	12.7	-69.6	6.8
2515	140.9225	17.1328	2013.6005	10.745	18.289	26.303	105.3	-85.3	-34.6	5.1	-88.5	-36.4	5.5
2516	140.9335	51.9965	2013.8015	13.304	15.491	7.875	85	11.2	-109.8	5.8	13.6	-108.3	5.8
2517	141.0667	35.2789	2013.714	14.331	15.815	13.385	303.1	-149.9	2.2	5.5	-151.3	3.2	5.5
2518	141.1141	-1.0325	2014.197	14.638	15.968	6.012	298.7	-63.4	-8.7	5.8	-60.7	-18	6.1
2519	141.1538	42.7504	2013.69	15.102	17.648	8.187	118.2	-15	-65.7	5.5	-13.9	-67.6	5.7
2520	141.2461	4.6068	2014.131	10.527	12.868	12.685	235	-47.4	-73.9	5.7	-43.8	-75.1	5.7
2521	141.2820	63.9714	2013.7605	12.854	13.919	14.133	296.4	-61.5	-115.3	5.5	-59.1	-114.9	5.5
2522	141.2859	5.4695	2014.1685	10.643	14.989	13.424	67.8	-173.9	77	5.7	-175.6	75.7	5.7
2523	141.3494	61.4633	2013.8005	10.986	13.997	8.272	183.7	51.1	-108.2	5.5	51.7	-109.4	5.5
2524	141.3655	21.0420	2013.912	13.421	18.232	17.412	14.1	-135.8	-78	6.3	-139.6	-74.8	6.6
2525	141.4223	44.5664	2013.794	13.165	16.775	24.49	160.6	-8.8	-83.3	5.5	-11	-84.7	5.5
2526	141.4243	46.4308	2013.7185	15.993	16.331	7.845	284.5	-174.2	-198.4	5.6	-175.2	-200.1	5.7
2527	141.4876	9.6132	2014.191	13.411	15.084	15.839	45.8	-76.5	-65.8	5.7	-79.9	-64.1	5.7
2528	141.5803	29.9974	2013.817	12.163	15.18	34.798	337.5	19.3	-71.2	5.4	18.4	-71.5	5.4
2529	141.6238	16.5373	2013.9825	9.16	13.107	13.247	217.2	-73.5	-29.1	5	-72.5	-22.5	5

2530	141.6457	45.7895	2013.812	12.625	15.404	13.381	259.2	5.9	-74	5.4	2.2	-75.8	5.5
2531	141.6488	69.3733	2013.6615	11.167	16.496	10.163	96.6	-69.2	-47.6	5.5	-69.7	-43.3	5.6
2532	141.6685	85.8142	2013.915	14.837	16.176	27.798	12.8	-76.3	-22.6	5.9	-73.1	-33.2	5.9
2533	141.6727	20.9250	2013.7465	15.244	16.091	24.549	310.7	-128.4	-45	5.2	-128.9	-47.3	5.2
2534	141.7092	1.9617	2014.1915	11.627	16.49	8.738	235.7	5.7	-73.5	5.7	0.8	-74.9	6.1
2535	141.7151	41.1759	2013.751	11.126	17.307	13.523	52.9	-97.1	-13.3	5.2	-98.1	-10.8	5.3
2536	141.7390	18.6279	2013.761	14.235	14.248	45.604	234.6	124.4	-99.8	5.2	114.9	-100.7	5.2
2537	141.7638	52.2239	2013.772	12.292	16.739	12.89	215.8	-64.1	-45.5	5.4	-62.3	-49.9	5.4
2538	141.7685	13.2106	2013.596	16.416	17.488	23.894	232.6	-7.4	-105.9	5.5	-7.3	-100.8	5.7
2539	141.8047	12.2443	2013.9175	14.474	17.591	7.671	42.2	-77.9	5.3	5.4	-73.9	8.5	5.6
2540	141.8579	39.4282	2013.7675	10.528	14.154	7.15	202.4	25	-137.4	5.2	27	-136.9	5.2
2541	141.8607	13.0367	2013.739	12.423	17.426	8.41	71.1	11.3	-92	5.4	10.6	-93	6.5
2542	141.8629	53.2860	2013.7675	13.117	15.561	6.957	71.1	-92.1	-18.3	5.4	-93.1	-12.3	5.5
2543	141.8688	39.5044	2013.8485	8.978	12.248	56.154	289.9	109.5	-157.9	5.2	113.9	-153.8	5.2
2544	141.8746	53.6344	2013.6815	14.087	16.152	52.543	147.3	-28.5	-96.2	5.5	-27	-94.4	5.5
2545	141.8988	-2.6896	2014.1735	10.141	14.549	40.519	133	-67.2	6.5	5.3	-70.2	2.8	5.3
2546	141.9592	-5.6633	2014.1905	11.49	15.978	15.937	267.4	-66	37.7	5.3	-65.2	37.6	5.4
2547	141.9753	28.6392	2013.8655	13.619	15.242	45.692	115.5	-114.4	-52	5.5	-107.8	-55.1	5.5
2548	141.9956	43.2774	2013.748	12.288	17.326	47.431	276.8	-20	-67.4	5.5	-22.5	-65.7	5.6
2549	142.0686	51.9315	2013.8	10.734	13.381	31.562	91.8	-139	40.9	5.4	-142.1	35.2	5.4
2550	142.1570	12.6644	2013.6715	18.217	18.365	10.038	9.7	-34.3	-129.6	6	-37.1	-135.1	6.4
2551	142.1683	16.8505	2013.8035	12.271	14.975	7.946	226.2	-79.7	-31.7	5.4	-80.6	-35.2	5.5
2552	142.2127	5.6641	2014.2035	13.378	17.315	8.101	320.8	-74.3	23.5	5.7	-77.7	23.1	6.1
2553	142.2239	27.9289	2013.8085	16.55	17.501	7.452	275.6	-115.7	-35.8	5.6	-120.5	-35.1	5.6
2554	142.2313	54.1451	2013.807	13.913	15.198	4.574	48.9	-23.4	-61.8	5.5	-21.9	-65.7	5.5
2555	142.2548	-9.5224	2014.1765	14.38	14.693	6.139	289.8	-79.2	-14.5	5.4	-78.2	-15.6	5.4
2556	142.2588	2.7620	2014.133	9.816	12.449	12.279	157	-79.3	36.4	5.7	-82.9	33.3	5.7
2557	142.2904	3.8349	2014.167	14.483	14.876	37.851	227.7	-126.8	-154.6	5.7	-127.2	-154.8	5.7
2558	142.3129	14.7647	2013.745	15.218	16.308	4.458	156.7	-66.4	-44	5.5	-70.5	-42.3	5.8
2559	142.3181	56.7144	2013.896	12.532	15.65	5.959	138.6	84	-54.4	5.5	79.5	-60.2	5.8
2560	142.3658	28.8797	2014.0105	13.364	13.89	4.099	209.9	6.1	-94	5.1	7.2	-93.2	5.3
2561	142.3748	-4.0340	2014.201	15.273	15.647	9.793	196.1	-65.3	17.1	5.3	-67.8	17.2	5.3
2562	142.4371	24.5336	2013.8085	13.468	14.155	27.308	84.8	-24.8	-87	5.1	-25	-84.9	5.2
2563	142.4532	-2.7445	2014.1795	11.937	14.483	25.888	47.5	-70.9	-32.3	5.3	-69.4	-34.4	5.3
2564	142.4818	22.4538	2013.7645	14.286	15.087	37.398	357.2	-8.8	-89.6	6.3	-9.2	-90.8	6.2
2565	142.5070	-7.6767	2014.1875	11.009	13.494	22.726	358.2	-68.2	0.8	5.3	-67.7	-0.3	5.3
2566	142.5331	5.7167	2014.1775	13.881	15.159	11.153	203.8	74.9	-96.2	5.7	70.7	-96.5	5.8
2567	142.5758	66.5277	2013.71	15.55	17.891	6.799	194.6	-87.4	7.6	5.5	-84.1	12.2	5.7
2568	142.6201	38.3659	2013.639	16.338	17.214	8.187	242.6	-158.8	-122.6	5.3	-167.1	-121.4	5.4
2569	142.6478	55.9813	2013.6385	17.11	18.526	13.415	189.8	-20.7	-67.7	5.5	-22.1	-69.3	6
2570	142.6485	10.6017	2014.172	8.231	8.266	13.825	242.3	-194.4	-16	5.7	-202.5	-14.6	5.8
2571	142.6993	-7.3856	2014.168	14.003	14.701	32.088	253.4	-59.5	-98.6	5.3	-55.1	-96.8	5.4
2572	142.8208	6.4992	2014.1725	15.098	15.463	5.515	78.3	-17.2	-79.2	5.8	-13	-77.8	5.8
2573	142.8290	65.8751	2013.7285	16.882	17.645	12.258	15.7	-5.1	-78.3	5.5	-4.6	-78.9	6.3
2574	142.8710	14.6577	2013.889	10.881	14.234	12.399	309.9	-73.2	-2	5.3	-72.8	-2.9	5.4
2575	143.0058	12.0559	2014.2545	13.494	13.586	4.601	150.2	-109.3	-23.2	5.2	-107.5	-23.5	5.3

2576	143.0143	88.1109	2013.8135	11.627	15.324	16.901	65.7	-78.2	-19.3	6	-81.8	-15.4	6
2577	143.0590	47.1499	2013.7825	10.762	15.069	16.324	113.6	40.5	-68.3	5.2	35	-68.2	5.2
2578	143.1226	42.0879	2013.849	13.874	14.372	6.259	131.1	-83.2	4.5	5.2	-82.3	2.9	5.2
2579	143.1426	55.1919	2013.7155	17.002	17.182	7.899	207.8	-63.1	-65.4	5.5	-65.9	-66.4	5.5
2580	143.2209	15.1557	2013.9225	12.353	18.632	12.17	26.6	-120.9	-153.2	5.4	-120	-156.6	6.7
2581	143.2304	28.7799	2013.873	13.72	13.848	6.929	73.4	2.8	-61.3	5.1	0.3	-62.4	5.1
2582	143.3527	-5.3077	2014.193	12.051	17.714	15.257	317.2	-77.2	-20.1	5.3	-76.8	-24.8	7.1
2583	143.3537	7.5922	2014.166	14.028	14.407	3.807	76.1	10.5	-131.5	5.9	3.8	-136.8	7
2584	143.4153	25.8730	2013.9315	14.254	14.752	5.01	169.7	-79.4	-110	5.1	-79.6	-107.6	5.1
2585	143.4467	-10.3290	2014.1705	8.679	11.665	29.375	18.2	-88.7	6.3	5.4	-88.3	4.4	5.4
2586	143.4552	24.2627	2013.914	9.401	17.16	28.202	264.1	-168.5	-38.6	5.7	-171.9	-43.3	6
2587	143.4684	15.4916	2013.9345	9.093	13.458	34.212	47.7	-46.9	-86.6	5.3	-43.7	-93.2	5.4
2588	143.5306	12.0253	2014.183	13.902	16.178	5.228	277.6	-69.9	13.2	5.3	-71.8	11.9	5.6
2589	143.6267	22.8122	2013.898	11.532	15.729	10.087	85.7	63.8	-47.9	5.1	60.5	-47	5.1
2590	143.6552	55.4076	2013.7375	12.338	13.07	5.19	167.7	-65.8	-21.7	5.9	-65.8	-21.9	5.9
2591	143.6654	41.7252	2013.768	10.943	15.114	10.11	329.6	27.1	-78.5	5.8	25.8	-78.9	5.9
2592	143.6949	43.8793	2013.7765	13.989	16.384	10.55	308	-83.4	3.2	5.5	-84.2	-0.1	5.5
2593	143.8084	60.8916	2013.885	7.297	10.289	10.793	355.4	-45	-65.9	5.5	-42.7	-68.5	5.4
2594	143.8163	34.9619	2013.837	10.488	10.947	44.041	289.3	-88.7	12.7	5.2	-87.7	13.5	5.2
2595	143.8291	52.5267	2013.7665	13.839	14.55	14.294	194.8	-2	-90.9	5.9	-3.8	-91.1	5.9
2596	143.8629	46.5028	2013.687	14.073	16.912	7.679	242.7	-40.8	-117.9	5.5	-42	-123.5	5.6
2597	143.8816	15.2363	2013.698	15.83	16.191	16.031	85.7	-122.3	9.1	5.6	-123.8	7.6	5.6
2598	143.9193	38.5256	2013.778	13.323	16.967	9.189	146.8	81	-117.9	5.2	79.6	-115	5.2
2599	143.9943	52.6931	2013.7475	12.148	15.436	16.004	253.2	-63.5	-81	5.8	-61.3	-78.8	5.8
2600	144.0067	42.8586	2013.7925	11.137	11.426	12.707	15.3	-116.2	51.9	5.5	-116.2	58.1	5.5
2601	144.0602	47.4315	2013.815	10.565	11.995	9.139	7.5	-60.7	-37.1	5.5	-62.4	-38.2	5.5
2602	144.1301	-6.3711	2014.1895	13.184	16.349	13.884	110.8	-99.3	53.3	5.3	-102.7	50	6.3
2603	144.2579	34.7453	2014.0355	13.008	14.918	5.26	182.1	-70.4	16.8	5.2	-73.4	10.5	5.8
2604	144.2809	0.5048	2014.1405	10.278	12.874	28.495	101.7	140.7	-133.7	5 7	141.1	-135.4	5.7
2605	144.3447	59.9315	2013.709	12.8	15.427	9.19	64.3	138.6	-41.3	5.8	134.6	-42.2	5.9
2606	144.3766	25.6811	2013.919	10.774	14.577	6.81	344	-92.5	-47.3	5.9	-92.9	-46.5	5.9
2607	144.3821	28.6354	2013.8145	11.071	18.507	18.994	48.7	-136.9	-163.3	6	-130.9	-167.1	7.1
2608	144.4599	30.4328	2013.9155	14.453	15.509	8.594	97.7	-28.1	-86.2	5.2	-23.2	-89.3	5.2
2609	144.5050	14.3469	2013.9995	6.855	14.91	24.549	230.2	-81.9	-83.4	5.8	-77.8	-92.7	6
2610	144.5332	1.7137	2014.196	13.129	18.306	9.698	185.5	39.3	-80.8	5.7	27.6	-85.6	7.2
2611	144.5736	71.9471	2013.776	10.157	17.649	17.403	189.8	-63.4	-63.7	5.9	-65.1	-66.6	6
2612	144.5756	68.2836	2013.803	11.923	14.206	6.797	93.5	-103	-62.4	5.5	-101.7	-61	5.5
2613	144.6227	52.7398	2013.6665	16.143	16.43	4.174	313.7	94	-18.2	5.9	92.3	-16.3	6
2614	144.6702	-11.1679	2014.057	10.505	13.333	50.376	328.6	-60.3	42.7	5.7	-60.6	42.7	5.7
2615	144.6775	44.5203	2013.776	14.862	15.978	17.801	120.3	-77.3	-71.2	5.5	-75.6	-72.4	5.5
2616	144.6901	10.7775	2014.1615	7.102	10.311	8.282	99	-61.2	-62	6.1	-63.3	-63	6.1
2617	144.8398	-5.2979	2014.145	10.906	14.741	13.375	87.2	43.1	-71.1	5.3	47.1	-70.6	5.4
2618	144.8691	49.0375	2013.7045	11.404	15.022	6.596	357.5	-70.7	-93.1	5.8	-69.3	-92.8	5.9
2619	144.8695	-8.0230	2014.171	11.787	16.011	46.454	314.2	-121.1	-159.4	5.4	-125.7	-164.6	6.2
2620	144.9814	22.3612	2014.0085	9.673	11.184	7.874	325.5	-166.1	-50.4	5.1	-172.5	-52.8	5.1
2621	145.0808	16.6231	2013.876	17.663	19.077	42.789	29.8	82.9	-82.6	6.3	70.7	-91.4	7.5

2622	145.1879	5.0870	2014.189	11.973	17.077	10.985	260.9	-69.7	-64.5	5.7	-70.5	-60.9	6.1
2623	145.1976	77.3568	2013.772	16.296	16.775	6.507	115	-93.5	-76.7	5.6	-93.7	-77.5	5.6
2624	145.2351	46.1251	2013.702	15.696	16.413	49.357	248	-98.2	37.2	5.2	-97.5	38.7	5.2
2625	145.3154	7.1281	2014.178	14.531	15.103	22.725	244.9	-55.4	-170.5	6.1	-52	-169.1	6.1
2626	145.3525	18.9648	2013.714	16.32	17.425	4.434	346.4	-110.1	-54.9	5.2	-112.2	-54.9	5.3
2627	145.3661	-5.9198	2014.1635	15.526	15.839	5.174	204.1	-26	-64.1	5.4	-22.8	-63.1	5.7
2628	145.3878	22.9202	2013.89	15.945	16.444	6.918	237.7	-85.7	-10.9	5.1	-87.5	-9	5.1
2629	145.4282	32.4973	2013.8305	12.589	13.666	12.373	202.6	-84.5	-42.2	5.8	-81.5	-43	5.8
2630	145.4356	14.0223	2013.87	15.104	16.299	10.513	102.1	-47.4	-66	5.4	-46.3	-69.1	5.4
2631	145.4421	10.6821	2014.208	11.963	14.204	5.945	344	3.4	-62.1	6	4.1	-63.9	6.1
2632	145.4741	12.3075	2014.06	13.433	14.016	8.398	186.1	-133.1	-35	6.1	-131	-34.3	6.1
2633	145.5369	57.3631	2013.694	17.13	18.296	7.811	250.3	-72.2	-19.3	5.5	-67.6	-26.7	6
2634	145.5928	64.3729	2013.672	12.386	16.867	14.789	277.6	-166.9	-70.1	5.5	-165.9	-69.2	5.6
2635	145.6213	40.9732	2013.7795	7.746	11.479	11.951	201.7	-25.3	-78.8	5.2	-29.3	-75.2	5.2
2636	145.6345	71.8900	2013.7865	13.108	13.531	13.81	279.6	-12.5	-65.3	5.5	-11	-64.6	5.5
2637	145.6707	33.5608	2013.8445	13.349	14.856	45.6	242.6	-76.8	-11.7	5.1	-76.8	-11.5	5.1
2638	145.7110	33.1184	2013.8495	11.395	13.779	6.06	170.4	-87.6	-57.3	5.2	-88.9	-58.7	5.1
2639	145.7190	6.5446	2014.16	13.874	15.567	9.32	146.3	-2.4	-82.8	6.1	-3.8	-82.4	6.1
2640	145.8218	27.6779	2013.827	14.136	14.56	5.95	1.9	-3	-94.1	5.6	-6	-96.5	5.6
2641	145.8785	3.2548	2014.1875	15.076	16.164	40.225	214.4	-21.9	-68.8	5.7	-19.7	-69	5.8
2642	145.9302	46.7118	2013.599	12.941	15.168	5.949	106.7	-117.7	-30.5	5.2	-124	-30.8	5.6
2643	146.2094	1.7785	2014.21	14.402	16.96	6.808	307.6	-30.9	-89.9	5.7	-27.7	-91.6	6.1
2644	146.2197	39.3486	2013.699	15.491	17.28	6.397	220.8	-49.2	-99.2	5.2	-49.2	-95.6	5.3
2645	146.2293	23.2966	2013.9005	9.261	15.938	27.73	98.4	-108	-30.3	5	-111.3	-32.8	5.2
2646	146.2546	43.2274	2013.8245	8.676	8.763	5.217	308.6	-5.3	-89.9	6.6	-6.7	-94.3	6
2647	146.2746	33.6994	2014.0255	11.26	17.34	8.982	326	-21.9	-64.9	5.1	-11.3	-70.4	6.9
2648	146.3156	60.4208	2013.7095	13.898	16.13	10.545	252.7	-59.3	-68.3	5.5	-62.3	-68	5.5
2649	146.3180	20.6730	2014.104	12.307	12.649	4.023	205.1	-81.8	-97.6	5	-79.8	-97.7	5.2
2650	146.3406	46.2754	2013.6965	15.415	15.842	21.73	21.8	-190.1	-133.9	6.1	-190.2	-132.6	6.1
2651	146.3548	3.7218	2014.19	14.947	17.321	29.464	253.6	-31.3	-102.2	5.7	-38.9	-100.7	6
2652	146.3582	15.6780	2014.0235	9.646	11.786	35.074	69.1	-102	61.6	5	-101.7	71.3	5
2653	146.4939	28.8606	2013.865	9.535	9.96	37.768	179.7	-28.6	-60.7	5.4	-27.8	-63	5.4
2654	146.5229	25.6022	2013.8525	10.909	14.175	16.059	184.3	-40.2	-80.3	5.4	-46.7	-81.7	5.4
2655	146.5556	0.3517	2014.2395	17.417	17.521	3.968	334	-172.9	-36.1	7.2	-167.9	-37.6	7.7
2656	146.5782	50.5986	2013.7085	13.286	16.859	11.28	192.3	-100.4	-29	5.8	-99.3	-28.9	5.9
2657	146.5968	8.0040	2014.1915	14.091	16.158	8.248	169.5	28.3	-73.6	5.8	24.7	-73.5	6
2658	146.6898	14.1637	2013.954	10.618	14.801	56.264	12.1	-32	-63.1	5	-33.5	-66.3	5
2659	146.7463	-13.1323	2014.1015	13.764	14.89	10.122	338	-182.9	-13	5.6	-181.4	-12.2	5.6
2660	146.7611	40.4314	2013.7505	12.908	18.623	13.481	13.4	-90	-41.9	5.2	-89	-48.4	7.1
2661	146.8186	8.5704	2014.14	7.161	9.338	59.247	190.2	-85.1	-31.2	5.8	-80.7	-34.3	5.8
2662	146.8711	56.0330	2013.817	9.754	14.214	8.969	143.4	-45.3	-70.6	5.8	-48.3	-66.6	5.9
2663	146.8867	18.0319	2013.987	13.548	15.4	4.577	96.6	106.2	-90.8	5.2	108.6	-94.3	6
2664	146.9853	36.8577	2013.8635	12.107	12.864	9.158	129.7	-79.3	-91.7	5.2	-76.9	-94	5.2
2665	147.0050	81.6035	2014.015	14.351	17.982	17.237	147.3	-82.1	-70.8	5.8	-83.5	-69.1	5.9
2666	147.0538	39.1350	2013.853	12.812	14.548	5.71	182.8	34.5	-105.3	5.2	35.2	-105.9	5.2
2667	147.0587	-10.1924	2014.094	8.663	10.026	11.318	226.8	133.9	-199.1	5.7	139	-197.5	5.8

2668	147.1852	41.3973	2013.641	17.711	17.913	4.755	289	-95.1	-30.8	5.3	-91.3	-29.1	5.5
2669	147.2018	-1.8175	2014.2085	15.133	15.344	34.192	243.8	62.9	-24.5	5.4	60.5	-24.9	5.3
2670	147.2028	33.4041	2013.8755	17.1	17.103	8.948	189.8	64.6	-15.7	5.2	64	-15.7	5.2
2671	147.2624	27.0022	2013.772	13.264	13.366	13.468	16.4	34.8	-68.8	5.5	32.9	-68.4	5.5
2672	147.2666	9.8722	2014.2075	13.571	16.347	8.818	23.1	-80.6	-14.4	5.7	-80.6	-17.2	6.1
2673	147.2832	-0.8996	2014.2095	14.358	16.292	9.674	333.4	-72.1	46.5	5.3	-72.9	46.3	5.5
2674	147.3040	32.9959	2013.831	15.445	18.47	11.94	285.3	105.9	96.9	5.1	103.3	96.6	5.5
2675	147.3258	7.8384	2014.1835	13.224	17.107	9.523	331.3	-118.6	-11.7	5.8	-117.6	-16.8	5.9
2676	147.3296	20.7101	2013.8825	14.487	15.065	16.055	78.6	-85.4	27.6	5.1	-84.9	29.1	5.1
2677	147.4261	-4.9420	2014.211	15.944	15.962	6.444	186.3	-87.6	-9.5	5.5	-84.6	-11.8	5.7
2678	147.4520	19.8183	2013.9185	12.841	17.462	23.132	29.6	-95.6	21.8	5.1	-99	23	5.2
2679	147.4755	5.9642	2014.139	9.988	15.187	9.882	47.5	83.2	-8.9	5.7	79.6	-12.5	5.8
2680	147.6204	22.3701	2013.923	15.086	15.131	15.572	25.7	21.7	-65.1	5.1	21.2	-65.7	5.1
2681	147.6213	59.6349	2013.431	15.03	16.119	4.063	248.4	-30.2	-69.9	5.8	-31.6	-68.7	6.3
2682	147.6219	50.7176	2013.6975	11.005	15.861	19.847	188.6	-100.3	-133.1	5.8	-103.6	-134.7	5.9
2683	147.6235	45.0833	2013.783	7.457	11.137	53.062	10.9	-85.8	-94.3	5.2	-82.3	-95.1	5.2
2684	147.6568	30.8675	2013.832	18.446	18.485	4.339	133.2	-77.1	-62.7	5.8	-79.3	-63.2	7.3
2685	147.6569	39.8384	2013.807	11.451	13.964	24.517	143.8	-165.6	-61	5.2	-164.8	-65.3	5.2
2686	147.6890	24.1601	2013.8255	15.979	17.828	53.72	107.1	13.6	-67	5.2	17.1	-63.1	5.3
2687	147.6991	75.2511	2013.72	14.538	15.536	25.672	32.6	-113.4	-57.1	5.6	-113	-56.3	5.6
2688	147.7042	16.1450	2013.9305	10.382	17.215	11.44	38.7	-75.1	76.5	5.4	-73.1	80	6.3
2689	147.7090	-12.9100	2014.051	13.873	14.35	4.294	168.7	-72.6	-59.5	6.1	-79.3	-56.1	6.3
2690	147.7651	1.0713	2014.204	13.628	16.142	26.885	304.8	-69.5	-31.4	5.7	-66.4	-37.2	5.8
2691	147.7836	2.6331	2014.1825	16.549	17.438	5.899	124.8	23.3	-66.1	6.2	24.4	-67.8	6.1
2692	147.8903	52.7705	2013.7205	12.338	18.86	23.444	241.3	43.7	-83.1	5.8	38.6	-82.2	6.8
2693	147.9139	21.2029	2013.912	11.46	17.011	9.353	317.5	-74.1	45	5.1	-72.3	41	5.3
2694	148.0323	30.8276	2013.8045	12.053	17.034	11.694	287.2	-132.4	-129.1	5.2	-133.2	-124.7	5.3
2695	148.2431	22.9008	2013.802	16.626	17.484	17.661	142.1	-90.9	-60	5.2	-89.8	-61.5	5.2
2696	148.3819	31.2436	2013.7895	14.904	18.23	10.875	308.9	74.2	-33.6	5.2	73.4	-32.5	5.6
2697	148.5613	39.1031	2013.8565	10.522	15.386	23.398	119.3	-121.5	-74.1	5.2	-120.5	-75.4	5.2
2698	148.5969	42.3293	2013.7365	14.993	15.234	27.204	229.5	-72.9	21.3	5.5	-72.4	22.9	5.5
2699	148.6291	50.0097	2013.756	12.817	14.161	13.843	85	-101.9	-15.1	5.9	-102	-19.2	5.9
2700	148.7083	34.5384	2013.952	13.668	14.237	10.8	89.4	-64.1	-6.7	5.1	-62.3	-8.1	5.1
2701	148.7389	6.3387	2014.0515	16.449	17.251	11.775	23.2	-64.4	-2	5.8	-62.5	-13	6.5
2702	148.7538	48.5718	2013.7385	11.542	12.082	9.874	303.2	156.5	-2.1	5.9	150.2	0.1	5.9
2703	148.7605	68.9392	2013.838	10.181	10.258	8.933	273	-68.9	-58.2	5.5	-67	-58	5.5
2704	148.8511	25.9849	2013.806	13.912	14.018	10.991	298.1	-41	-64.7	5.4	-40.2	-66.9	5.4
2705	148.8656	35.0022	2013.8625	10.189	12.894	31.287	59.5	-37.4	-70.8	5.1	-36.3	-73.5	5.1
2706	148.8778	34.3571	2013.997	15.196	16.286	5.11	112.6	-28.9	-94.1	5.1	-29.7	-95.2	5.2
2707	148.9183	58.6619	2013.6555	16.863	18.317	12.824	262.9	-8.8	-97	6	-9	-95.4	6.2
2708	148.9199	1.8005	2014.136	14.751	17.017	8.871	146.7	-76.2	-0.5	5.8	-77.8	-8.3	6.1
2709	148.9499	0.3303	2014.1805	13.079	14.859	17.141	359.7	-67	-3.8	5.7	-68.2	-7.3	5.8
2710	149.0388	-7.0032	2014.135	9.682	14.36	10.8	109.9	89.7	-86.9	5.4	86.3	-88.4	5.8
2711	149.0474	46.2803	2013.724	15.455	15.532	8.603	327.2	-74.1	-24.8	5.5	-74.6	-23.3	5.5
2712	149.0827	25.8579	2013.666	16.913	18.264	36.88	171.8	-67.4	18.2	5.5	-69.3	14.5	5.7
2713	149.0945	13.4321	2013.714	13.587	16.595	5.597	308.9	-11.2	-71.1	5.8	-12.3	-67.5	6.3

2714	149.1121	-1.9135	2014.1775	13.567	15.808	13.984	164.1	-100.7	-90.9	5.4	-97.6	-86.2	6.2
2715	149.1205	69.5300	2013.694	12.863	16.301	36.451	76.4	-186.2	-162.8	5.6	-186.9	-162.1	5.7
2716	149.1963	2.8989	2014.1875	9.221	17.342	16.509	54.4	13.4	-70.8	5.7	23.1	-69.5	7.6
2717	149.2882	20.6057	2013.9345	8.928	14.335	8.987	35.3	7.6	-115.1	5.3	7.7	-115.2	5.6
2718	149.2952	16.4329	2013.917	13.945	14.96	21.32	35.6	-26.9	-81	5.8	-28.1	-80.1	5.8
2719	149.3593	30.5332	2013.9595	11.332	15.593	8.403	61.8	-23.2	-69.4	5.1	-25.4	-68	5.1
2720	149.4026	0.4876	2014.2025	14.423	16.888	40.882	190.6	-79.8	-17.5	5.7	-81.1	-20.9	6.1
2721	149.4446	28.2730	2013.702	12.632	16.966	19.756	101.8	-64.7	-36.8	5.4	-61.4	-35.9	5.5
2722	149.4490	43.5889	2013.7955	11.74	12.677	6.537	218.2	65.3	-41.5	5.9	66.1	-38.3	5.9
2723	149.4516	26.9384	2013.6655	15.546	16.557	5.151	97.7	-116.3	-108.6	5.5	-116.5	-111	5.8
2724	149.4817	37.9311	2013.834	16.028	16.077	20.68	62.4	-18.3	-67.3	5.2	-9.5	-66.3	5.2
2725	149.5061	22.1393	2013.8195	13.832	15.105	15.636	163	-97.4	-29.8	5.1	-99.2	-24.9	5.1
2726	149.5170	59.7846	2013.5875	11.476	12.956	4.935	287.2	-63.4	-61	6	-61.5	-58.2	6.2
2727	149.5206	73.1801	2013.6875	15.361	16.411	5.704	274.7	-141	-87.6	5.6	-138.1	-84	5.6
2728	149.5600	24.2011	2013.8785	13.598	15.448	9.595	184	-98	-21.9	5.4	-97	-24.9	5.4
2729	149.5736	13.6427	2014.024	13.463	17.423	7.626	177.6	46.6	-68.8	5.8	45.7	-64.6	6.2
2730	149.5793	23.4009	2013.835	13.885	14.959	6.709	338.9	-11	-111.7	5.1	-11.1	-113.2	5.1
2731	149.6522	31.3498	2013.764	15.285	18.763	14.942	3.6	-18.1	-65.6	5.1	-22.4	-62.9	6.8
2732	149.6871	79.7271	2013.928	15.52	16.016	4.969	173.9	-41.3	-75.3	5.8	-38.5	-78.7	5.9
2733	149.7006	79.8877	2014.054	15.958	16.875	4.612	75	67	-24.9	5.8	64.9	-21.9	6.4
2734	149.7725	53.2637	2013.8035	8.172	15.172	16.97	208.4	-60.8	-10.4	5.4	-61.2	-9.3	5.5
2735	149.7862	59.4033	2013.749	10.621	14.679	46.531	158.6	-123.7	-30.1	6	-128.9	-32.9	6.1
2736	149.7885	16.6429	2013.7525	17.236	18.883	20.598	157.2	-64	-24.3	6	-64.6	-15.2	7.9
2737	149.7960	9.0277	2014.239	16.895	17.906	17.16	90.7	-66.7	-18.8	5.9	-63.3	-20.3	6.5
2738	149.8359	57.0772	2013.8165	11.625	12.672	52.822	176.3	-114.6	-44.8	5.9	-114.6	-51	5.9
2739	149.8621	24.0790	2013.7445	17.769	17.797	56.881	225.7	-5	-63.4	5.5	-6.1	-62.1	5.5
2740	149.9218	3.2403	2014.133	12.985	15.034	5.263	167	-75.8	-45.5	5.8	-78.5	-43	5.7
2741	149.9597	74.1723	2013.7865	13.443	17.178	9.094	245.2	-62.5	-78.9	5.5	-60.2	-76.5	5.6
2742	150.0357	21.6042	2013.9535	9.65	14.651	9.445	270.9	37.2	-90.5	5.1	36.8	-92.2	5.2
2743	150.0544	6.2507	2014.128	7.593	9.946	23.781	21	-97.1	15.7	5.8	-99.8	16.4	5.8
2744	150.0739	68.4841	2013.6425	16.041	16.899	15.509	13.3	-39.9	-86	5.5	-40.2	-83.8	5.6
2745	150.1629	24.2534	2013.94	15.663	16.834	4.691	218.9	-26.1	-129	5.4	-25.8	-127.8	5.6
2746	150.1660	28.1731	2013.895	7.839	12.717	56.712	341.1	-41.7	-140.8	5.4	-44.2	-138.1	5.4
2747	150.1897	18.8759	2013.919	10.825	13.418	14.896	317.8	79.9	-52.8	5.1	79	-53.4	5.1
2748	150.2173	45.7200	2013.7125	14.303	16.464	13.018	13.4	-46.7	-80.3	5.5	-49.5	-80.7	5.6
2749	150.2343	49.6257	2013.7	12.776	13.97	7.387	109.3	150.6	-37.8	5.5	147.6	-38.2	5.5
2750	150.3122	58.3108	2013.6765	17.063	17.396	3.743	19.4	-140.5	25.3	6.1	-148	13.1	6.3
2751	150.3331	36.4563	2014.0115	17.06	17.31	5.367	359.7	98.3	-166	5.2	91.6	-168.9	5.3
2752	150.3401	23.8740	2013.83	14.831	16.027	6.224	171.7	-162.5	-35.4	5.1	-161.4	-33.4	5.2
2753	150.3543	19.3508	2013.987	9.574	12.109	9.98	100.2	-65.6	-50.6	5.1	-65.6	-54.6	5
2754	150.3569	32.6517	2013.876	11.291	13.621	48.541	72.1	-81.5	-51.6	5.1	-84	-48.4	5.1
2755	150.3894	23.0014	2013.8195	12.446	14.603	9.667	36.8	-104.9	4.1	5.1	-105.9	4.9	5.1
2756	150.4183	76.6065	2013.7685	14.751	16.418	42.235	132.7	-158.3	-64	5.5	-163	-61.3	5.6
2757	150.4267	38.3146	2013.906	16.046	16.51	52.06	143.5	-78.3	2.5	5.2	-75.4	3.6	5.1
2758	150.5233	79.7055	2013.986	11.892	14.853	6.015	178.1	166.4	-122.4	5.9	165	-118.5	5.8
2759	150.6452	26.5073	2013.812	14.891	16.093	5.65	18	-75.5	-13.7	5.2	-74.8	-17.6	5.3

2760	150.6463	34.9507	2013.744	15.69	18.375	7.574	282.1	-26.9	-75.4	5.2	-34	-69.8	5.5
2761	150.6582	54.7785	2013.8095	13.453	14.839	6.458	186	-100.1	-148.1	5.5	-101.5	-147.8	5.5
2762	150.8351	43.0671	2013.8225	15.798	18.586	34.776	13	-64.3	-41	5.2	-62.5	-39.1	5.7
2763	150.8755	34.4803	2014.078	13.591	15.015	4.064	261.8	-102.4	-61.5	5.1	-101.4	-58.1	7.2
2764	150.8760	62.5654	2013.846	7.03	15.594	49.262	167.9	-93	-107.4	5.8	-87.7	-116.1	5.8
2765	150.8844	26.3224	2013.839	12.541	15.267	16.028	277.7	114.1	-155.9	5.2	113	-158.7	5.3
2766	150.9161	5.4808	2014.0115	13.894	14.873	4.073	323.2	105.7	-95.8	6	105.5	-93.9	6.5
2767	150.9204	10.8144	2014.144	10.622	12.425	37.607	85.2	-93.2	-30.2	5.8	-85.8	-38.8	5.7
2768	150.9269	48.3310	2013.7365	13.005	13.82	56.371	298.8	-67.6	1.4	5.5	-68.1	-0.3	5.5
2769	150.9325	3.4648	2014.151	15.422	15.81	4.04	232.1	21.4	-65.1	5.9	23.6	-63.3	7.3
2770	150.9849	2.1272	2014.0855	14.843	15.053	20.509	325.4	-77.9	26	5.8	-74.9	28.8	5.8
2771	150.9855	74.7544	2013.7865	12.659	12.707	34.047	12.6	-89	-63.2	5.5	-90.7	-59.7	5.5
2772	151.0687	-1.6927	2014.1465	8.812	9.42	6.43	295.4	-65.9	-19.2	5.4	-66.6	-23.2	5.4
2773	151.0739	47.5310	2013.673	14.634	16.499	26.436	334.9	-71.2	4.8	5.5	-74.1	6.1	5.6
2774	151.0979	36.1997	2013.933	13.732	13.94	20.525	260	66.2	-53.8	5.1	66.8	-53.5	5.1
2775	151.1008	61.6643	2013.725	11.893	14.644	7.25	204.5	74.1	22.2	5.8	71.5	20.6	5.8
2776	151.1455	24.1161	2013.821	13.912	16.759	11.775	96.1	-60.1	20	5.2	-60.7	22.3	5.3
2777	151.1732	43.0040	2013.7005	14.704	16.256	11.02	115	53.9	-93.9	5.5	56.7	-86.2	5.2
2778	151.1975	25.7218	2013.8275	13.646	16.491	11.451	313.4	-102.1	38.9	5.2	-102.9	37	5.3
2779	151.2077	23.5037	2013.8785	9.787	15.6	15.499	209.4	-74.9	1.2	5.1	-76.3	1.8	5.1
2780	151.2117	52.6827	2013.7055	13.432	16.177	11.397	253	-63.4	-25	5.5	-64.9	-25.7	5.5
2781	151.2480	5.5929	2014.1595	11.484	17.838	19.43	58.4	63.1	-26.4	5.8	65.2	-22.3	6.1
2782	151.2497	46.5705	2013.8175	8.92	11.271	8.024	249.5	-111.6	-107.3	5.2	-115.6	-110.7	5.2
2783	151.2861	42.7071	2013.7555	11.148	13.853	15.07	113.3	-67.8	-80.8	5.2	-67.4	-83.4	5.2
2784	151.3194	50.9545	2013.765	17.238	17.357	10.832	4.3	-83.7	-8.3	5.5	-87.1	-2	5.5
2785	151.3451	63.8302	2013.7055	13.552	14.22	4.485	236.6	76.2	-69.7	5.9	75.5	-65.3	5.9
2786	151.3584	1.9845	2014.0215	10.138	11.44	26.853	210.4	37.1	-80.3	5.8	35.3	-81.2	5.8
2787	151.4068	37.0936	2013.887	9.819	15.845	14.267	80	-62.1	-29.4	5.2	-65.2	-28.6	5.2
2788	151.4824	39.5820	2013.8345	7.47	11.291	21.962	251.9	-76.6	-26.9	5.2	-82.8	-20.4	5.2
2789	151.5250	22.9100	2013.9735	11.83	15.072	7.115	104.9	-58.4	-66	5.1	-59.3	-67.2	5.2
2790	151.5614	29.5969	2013.8115	16.088	16.923	6.404	103.5	-86.2	-39.7	5.1	-86.1	-41	5.2
2791	151.5665	23.7822	2013.819	12.761	15.213	6.561	150	-21.8	-75.6	5.1	-23.8	-76.4	5.2
2792	151.5696	41.0239	2013.8195	9.548	16.024	24.966	149.6	-89.2	-35.2	5.2	-92.5	-32.1	5.2
2793	151.5931	-0.7747	2014.142	12.172	14.5	30.818	242.9	-94.7	-15.8	5.4	-98.1	-16.4	5.4
2794	151.5995	27.0476	2013.8255	10.721	11.445	12.465	306.9	-66.9	-9.3	5.1	-66.8	-10.1	5.1
2795	151.6009	39.2676	2013.8405	13.508	15.998	17.889	129.7	-67.1	-4.2	5.2	-66.6	-6	5.2
2796	151.6282	10.0122	2014.2	14.56	16.92	5.127	134.9	-63.9	-20.3	5.7	-61.5	-26.3	6.2
2797	151.6317	16.9473	2013.875	16.665	17.204	5.477	281.1	-90.6	-20.3	6	-93.2	-20.9	6.1
2798	151.7298	4.9178	2014.088	13.868	15.47	25.136	205.7	-16.4	-128.2	5.8	-19.7	-130.1	5.8
2799	151.7716	24.6166	2013.916	12.591	13.676	5.195	51.5	-73.1	12.1	5.1	-76.4	10	5.2
2800	151.7843	77.8641	2013.839	13.05	17.812	12.425	282.5	140.2	11.7	5.5	134.7	18.9	5.7
2801	151.9041	54.2665	2013.795	13.091	15.056	45.491	0.1	-70.5	-30	5.5	-66.9	-31.3	5.4
2802	151.9572	21.8394	2013.742	15.672	16.484	28.302	24.8	-67.4	7.6	5.1	-68.9	9	5.2
2803	151.9715	20.8796	2013.763	12.746	16.967	13.415	165.2	-18	-88.1	5.1	-16.5	-87.3	5.2
2804	152.0690	33.4864	2013.897	10.089	15.912	43.687	104.4	70.4	-48.6	5.7	68.6	-48	5.8
2805	152.1097	29.9196	2013.792	16.9	16.967	5.136	127.4	-25.8	-66.9	5.2	-25.2	-66.5	5.2

2806	152.1491	-0.2381	2014.1065	12.289	15.454	30.287	272	-142.2	-44.2	5.6	-148.6	-43.6	5.7
2807	152.1532	38.3225	2013.8525	15.316	15.604	12.428	44.3	26.1	-67.6	5.2	25.1	-68.3	5.2
2808	152.1719	-9.8475	2014.1535	14.06	15.706	7.735	313.2	74.9	-53.1	5.4	71.7	-53.3	5.9
2809	152.2394	19.8502	2013.927	14.071	18.261	7.356	188.4	-93.6	-6.5	5.1	-97	-8.8	5.5
2810	152.3209	35.6039	2013.921	11.247	13.225	10.677	202.1	-60.8	-114.4	5.1	-62.4	-115.8	5.1
2811	152.3400	11.4504	2014.115	16.102	17.18	4.911	320	43.2	-116.2	5.8	44.1	-118.6	6.4
2812	152.4091	40.6871	2013.855	15.24	15.466	35.305	203.9	-60.1	-76.2	5.8	-58.7	-76.7	5.8
2813	152.4270	55.4799	2013.828	10.617	12.352	6.642	225.5	-87.5	13.5	5.4	-86.6	10.8	5.4
2814	152.4418	-5.0273	2014.0935	9.052	14.968	20.893	338.5	-143.9	-47.2	5.5	-142.7	-46.2	5.9
2815	152.4631	27.1640	2013.8675	14.793	16.043	4.707	150.8	-81.3	-99.7	5.4	-76.4	-103.1	5.5
2816	152.5115	30.1718	2013.788	16.183	18.471	19.951	138	-61.9	-80.3	5.2	-53.4	-82.7	5.5
2817	152.5856	56.4867	2013.811	11.945	13.823	14.475	199.8	-93.8	-66.5	5.5	-97.6	-63.4	5.5
2818	152.6269	74.0671	2013.701	11.754	14.506	10.005	135.6	9.3	-61.9	5.5	13.2	-63.8	5.5
2819	152.6383	25.6525	2013.894	13.059	17.817	26.538	254.6	-84	-30.4	5.4	-87	-33.6	5.6
2820	152.6828	49.8598	2013.7675	11.258	13.461	14.701	124.1	-95.6	-77.7	5.9	-97.7	-74.6	5.9
2821	152.7583	16.5899	2013.955	11.259	15.246	23.105	343.4	-64	18.4	5.8	-63.3	19.5	5.8
2822	152.8206	48.0569	2013.7005	9.839	14.97	30.197	34.9	-71.2	-36.4	5.8	-72.3	-36.7	5.9
2823	152.9038	-4.7839	2014.119	13.785	14.314	11.907	22.5	17.7	-86.5	5.4	16.5	-86.9	5.3
2824	153.0155	10.3395	2014.0795	14.327	14.417	8.182	132.9	-96.3	-10.9	5.8	-97.8	-10.4	5.8
2825	153.0160	47.8463	2013.78	9.305	11.384	29.861	148.1	39.1	-62.8	5.5	37.8	-62	5.5
2826	153.0360	54.0754	2013.768	10.626	17.457	42.757	178.8	-6.2	-77.7	5.6	-10.4	-76	5.5
2827	153.0839	27.1033	2013.967	8.308	16.072	13.059	335.8	-98.3	-78.4	5.4	-94	-88.7	7.1
2828	153.0851	71.8802	2013.739	11.683	17.412	13.738	271.4	81.9	-33.8	5.5	85	-36.5	5.6
2829	153.1623	-9.7546	2014.0785	11.712	16.09	21.043	318.4	-86.7	33.2	5.4	-84.9	38.2	7.2
2830	153.2461	47.1757	2013.7135	9.817	11.034	30.882	154.9	-122	5	5.5	-126	3.4	5.5
2831	153.2886	71.2784	2013.6855	10.84	15.45	28.974	52	-67.7	-41.4	5.5	-68.2	-37.7	5.5
2832	153.2907	-8.7819	2014.073	11.398	12.557	5.089	204	109.6	-129.5	5.4	110.1	-127.2	5.9
2833	153.3726	11.5986	2014.027	14.226	17.11	12.906	309.8	-89	-73.2	5.8	-88.9	-74.1	6
2834	153.4426	3.6217	2014.2005	11.418	13.526	16.775	81.1	-39	75.5	5.7	-42	74.6	5.7
2835	153.4699	15.0151	2013.89	11.475	14.94	12.64	244.9	-65.9	30.3	5	-66.5	28.9	5.1
2836	153.5079	22.4560	2013.858	9.383	13.519	35.838	306.7	60.2	-100.2	5.1	62.3	-92.3	5.1
2837	153.5505	31.3036	2013.7765	13.149	16.602	43.058	262.3	-77.1	-38.4	5.2	-79.2	-37.5	5.2
2838	153.6021	23.3516	2013.7585	13.856	15.719	28.608	317.4	19	-60.2	5.1	18.4	-60.2	5.1
2839	153.6142	57.1077	2013.7965	13.11	14.433	10.279	75.5	-62	87.1	5.5	-64.1	86.8	5.5
2840	153.6326	1.8858	2014.206	14.028	15.527	21.292	226	-130.8	-49.6	5.7	-136.2	-51	5.8
2841	153.7171	16.2759	2013.909	11.001	16.883	41.548	324.5	-61.7	3.8	5	-64.4	3.8	5.1
2842	153.7577	24.3139	2013.8175	11.684	15.778	55.121	5.5	11.8	-67.6	5.4	10.3	-67.5	5.4
2843	153.7939	11.8786	2014.1165	9.229	13.508	17.204	314.4	50.7	-64.4	5.7	49.9	-67.6	5.8
2844	153.8239	47.4295	2013.7385	11.232	12.347	47.876	52.5	-77.5	-120.8	5.2	-80.2	-117.5	5.2
2845	153.8345	70.7818	2013.742	11.547	12.046	6.663	201.1	-9.9	70.1	5.5	-10.2	76.7	5.5
2846	153.9323	17.0864	2013.884	15.769	17.268	5.5	207	-139.7	-113.2	5.1	-135.1	-108.4	5.3
2847	153.9427	53.3499	2013.8785	10.67	12.18	5.805	164.8	-93.1	-128.6	5.5	-98.8	-127.9	5.5
2848	153.9564	51.0316	2013.756	13.938	14.976	15.146	150	138.9	-46.1	5.6	140.6	-46.4	5.6
2849	154.0824	76.2598	2013.8545	15.399	18.354	24.184	326.7	-80.8	-89	5.6	-80.7	-82.7	5.8
2850	154.0854	27.2928	2013.9215	15.503	18.477	20.683	344.9	-65.8	-51.4	5.1	-60.2	-57.8	5.4
2851	154.1192	73.6876	2013.641	12.365	17.185	10.405	53.1	-66	-63.5	5.6	-64.5	-64.7	5.6

2852	154.1428	26.3372	2013.8005	14.452	15.302	8.47	266.6	-15.8	-107.3	5.1	-16.8	-107.3	5.2
2853	154.1575	11.9103	2014.172	13.568	13.949	4.031	319.1	-60.7	-75.8	6.4	-57.2	-75.5	6.1
2854	154.1739	38.9387	2013.7975	14.762	15.44	18.641	155.7	-139.3	14.8	5.2	-139.3	14.3	5.2
2855	154.2127	39.5910	2013.863	11.256	15.693	9.153	185.7	-89	-34.3	5.2	-91.5	-31.6	5.2
2856	154.2247	40.4905	2013.8145	14.081	14.241	20.165	315.4	-97.7	-31.1	5.2	-95.7	-33.3	5.2
2857	154.2871	45.2251	2013.785	11.4	15.879	20.772	128	-89.9	27.8	5.2	-89.6	25.8	5.2
2858	154.3277	57.2315	2013.855	17.846	17.866	6.482	18.7	-67.3	-33.1	5.6	-62.6	-37.8	5.7
2859	154.3562	21.3125	2013.844	11.653	18.192	30.986	102.8	51.6	-126.5	5.1	55.5	-126.3	5.3
2860	154.3675	34.2818	2013.892	16.933	18.562	11.46	247.3	-64.7	-16.2	5.2	-62.5	-22	5.3
2861	154.3844	10.2605	2014.1045	12.191	17.048	26.05	1.2	110.5	-43.7	5.8	108.9	-42.1	5.9
2862	154.4599	17.3185	2013.958	13.407	14.133	6.079	75.2	-97.1	-36.7	5	-97.5	-38.3	5
2863	154.4600	27.2014	2013.8185	10.34	16.23	19.334	238.2	-91	-17.4	5.1	-90	-18	5.2
2864	154.5132	-10.3328	2014.14	9.999	14.686	33.292	258.1	-61.1	37.5	5.4	-63.4	32.6	5.5
2865	154.5317	48.6291	2013.833	15.611	15.749	3.509	312.8	-63.3	-71.4	5.7	-57.1	-80.5	5.8
2866	154.5686	18.9754	2013.947	9.163	17.213	14.666	153	-100.8	10.9	5	-98.8	6.8	5.3
2867	154.6773	45.6237	2013.6715	14.258	16.097	5.94	338.4	-23.9	-63.8	5.2	-20	-63.5	5.3
2868	154.6858	8.8514	2014.107	13.102	15.083	15.584	136.9	-74.3	-35.3	5.8	-72	-36.3	5.8
2869	154.7619	30.6212	2013.8635	12.973	14.087	4.973	233.5	-109.9	-29	5.2	-106.1	-27.2	5.2
2870	154.7671	18.1413	2013.716	15.382	17.558	7.061	51.3	-9.3	-95	5.1	-8.8	-93.7	5.3
2871	154.7744	15.4079	2013.9525	13.584	14.718	6.091	252.8	-73	-21.6	5	-71.1	-22.2	5
2872	154.8027	-3.0978	2014.098	13.627	14.629	5.152	271.3	-150.7	4.5	5.4	-143.8	2.9	5.8
2873	154.8318	20.1978	2013.939	14.459	16.495	4.903	228.5	-62.4	29.4	5.1	-62.6	30.7	5.3
2874	154.8365	-0.0773	2014.1715	13.838	15.84	5.664	327.1	-29.9	-79.2	5.3	-27.9	-81.2	5.5
2875	154.9224	71.9586	2013.665	14.973	16.441	31.967	62.6	-71.1	-48.6	5.5	-72	-51.9	5.5
2876	154.9276	20.4036	2013.891	10.548	13.399	11.993	108.9	-84.9	-34.6	5.1	-86.4	-32	5.1
2877	154.9813	72.3669	2013.6345	10.622	12.957	9.095	7.1	-171.7	-46.7	5.6	-173.9	-49.6	5.6
2878	154.9976	-9.0593	2014.1865	6.445	13.243	23.354	313.7	-64.9	-69.1	5.4	-59.1	-75.5	5.5
2879	155.0133	40.4787	2013.8275	9.163	14.394	45.742	132.7	-55.3	-86.8	5.2	-56.3	-86	5.2
2880	155.0369	16.5029	2013.7935	16.967	17.574	4.062	30.8	-116.6	-83.4	5.2	-117.2	-88.5	5.3
2881	155.0615	81.2003	2013.8165	11.863	14.897	17.31	72.8	-79.8	-13.5	5.9	-79.1	-18.7	5.9
2882	155.0913	15.7132	2013.9235	14.467	15.457	10.144	350.2	-101.7	-33.2	5	-101.4	-33	5
2883	155.1232	27.9984	2013.8915	11.989	12.096	12.043	180.8	-10.8	-100.5	5.1	-10.2	-101.3	5.1
2884	155.1807	3.2555	2014.246	16.625	17.694	33.883	217	-80.6	-5.2	6.1	-77.1	-5.4	7.5
2885	155.2188	7.2447	2014.1855	14.132	14.251	24.631	311.8	154.4	-37.6	5.7	151.1	-37.1	5.7
2886	155.2888	24.4370	2013.904	11.152	13.573	8.17	286.6	-71.4	-102.8	5.1	-73.2	-104.4	5.1
2887	155.3338	12.2702	2013.94	13.891	14.826	5.397	273.2	44.2	-66.2	5	46.3	-65.2	5
2888	155.4387	46.8485	2013.7315	10.091	11.582	13.113	187.7	-92.6	-72.8	5.5	-93.1	-76.2	5.5
2889	155.4588	35.9171	2013.88	15.238	16.513	13.213	144.6	-156.5	-93.3	5.2	-160	-91.6	5.2
2890	155.5373	34.4505	2013.846	16.185	16.367	28.153	70.5	-69.2	-33.9	5.2	-69.1	-32	5.2
2891	155.5397	51.4234	2013.764	8.694	15.543	12.605	31	-70	-53	5.4	-75.3	-51.3	5.6
2892	155.5693	-6.1931	2014.1275	13.667	15.838	14.05	255.3	-68.4	30.5	5.4	-70.3	33.1	5.6
2893	155.5719	75.7681	2013.7	9.975	14.497	34.541	80.8	110.6	28.5	5.6	106.5	22.8	5.6
2894	155.5973	6.9382	2014.194	14.579	14.704	22.686	18	-161.9	10	5.8	-165	13.8	5.8
2895	155.6952	57.0444	2013.8075	14.277	17.226	12.876	266.4	-108.5	-93.3	5.9	-110	-94.8	5.9
2896	155.8150	54.6685	2013.82	14.359	16.019	8.873	258.3	-68.3	-74.7	5.9	-71.6	-72.3	5.9
2897	155.8245	60.7209	2013.8765	8.755	15.837	12.824	236.9	-201	-51	5.8	-204.6	-49	5.9

2898	155.8789	27.9644	2013.902	11.988	18.215	12.982	147.5	-67.4	-27.7	5.1	-70.4	-20.2	5.3
2899	155.8981	26.8510	2013.873	14.533	15.084	19.542	175.3	-74.6	-29.9	5.1	-76.3	-28.3	5.1
2900	155.9301	27.2002	2013.828	10.305	14.274	10.921	44.4	78.6	-65.2	5.1	76.9	-65.2	5.2
2901	155.9601	31.8793	2013.8655	14.711	14.973	39.328	38.7	-82.1	12.9	5.5	-79.9	15.8	5.5
2902	156.0095	30.2040	2013.9045	8.94	10.63	59.55	122.4	-106.1	46.8	5.5	-110.2	47.5	5.5
2903	156.0208	40.5630	2013.918	8.694	10.858	6.648	143.2	23.6	-82.8	5.2	22	-83.5	5.1
2904	156.0431	55.2136	2013.8925	17.697	18.462	32.357	156.3	0.7	-66.9	6	9.4	-69.6	6.3
2905	156.1237	44.8630	2013.58	16.686	17.852	45.982	307.2	-20.4	-74	6	-32.4	-70.3	7.4
2906	156.1287	49.4176	2013.8145	14.124	16.215	4.927	318.3	7.5	-79	5.4	10.3	-79.4	5.5
2907	156.1654	33.2575	2013.907	11.29	14.032	5.401	322.2	-0.3	-62.9	5.5	3.7	-63.7	5.6
2908	156.1944	51.3268	2013.532	16.55	17.195	3.975	50.3	2.8	-73.4	5.4	-5	-74.3	7.1
2909	156.2041	22.9793	2013.819	15.143	15.955	6.675	339.1	-73.4	23.1	5.1	-71.5	22.8	5.1
2910	156.2491	17.8528	2013.8755	14.021	16.418	5.035	73.3	-93.1	-7.6	5	-91.6	-9.8	5.9
2911	156.2901	8.9227	2014.086	11.284	13.045	47.7	107.6	96.8	-93	5.8	101.9	-94.3	5.8
2912	156.2924	-13.2133	2014.215	12.317	14.503	9.263	101.9	-62.7	-18.6	5.5	-64.4	-22.7	6
2913	156.2979	-8.4993	2014.084	12.479	14.498	9.652	189.9	-66.4	0.9	5.4	-67.6	0.8	5.4
2914	156.3254	11.6563	2013.8655	15.466	17.722	58.829	167.1	-122.9	-8.2	5.8	-118.2	-13.4	6.1
2915	156.3338	30.9615	2013.8605	16.562	17.061	35.079	112.7	-89.1	-70.4	5.6	-86	-68.8	5.6
2916	156.4524	51.3586	2013.842	9.897	15.076	20.044	9.9	-85	-24.3	5.4	-85.7	-26.3	5.4
2917	156.4586	4.7760	2014.194	15.676	15.772	5.539	87.9	-69.2	6.9	5.8	-69.4	5.5	5.9
2918	156.5419	39.1725	2013.8035	13.149	14.495	7.204	247.9	-60.8	20.4	5.2	-61	20.7	5.2
2919	156.5514	67.1631	2013.738	10.5	17.981	29.058	352.1	-55.8	-109.5	5.5	-59	-110.2	5.8
2920	156.5799	4.8497	2014.198	13.688	14.714	18.394	226.9	0.4	-62.5	5.7	-1.8	-63.2	5.7
2921	156.6095	8.3958	2014.0865	16.449	16.914	4.681	194.1	4	-70.2	5.8	4.1	-71.2	6.5
2922	156.6536	10.3554	2014.016	16.135	16.219	26.31	93.2	-8.1	-73.1	5.9	-8.9	-73.6	5.8
2923	156.6745	16.7649	2014.0175	9.41	14.091	7.881	51.2	2.1	-66.1	5	-3.4	-67	5.2
2924	156.7200	45.1839	2013.7055	16.725	17.441	8.536	324	-67.8	-69.5	6	-64.9	-68.8	6.1
2925	156.7213	42.7282	2013.64	16.348	17.575	7.599	326.6	-109.5	-58.5	6.1	-111.2	-54.4	6.1
2926	156.7414	67.0501	2013.7075	8.993	14.237	8.104	267.3	-48.3	-81.8	5.4	-53.7	-77.6	6.3
2927	156.7640	49.3664	2013.7075	16.251	17.324	22.403	192.9	-92.9	-21.8	5.5	-93.4	-20.4	5.5
2928	156.7845	18.0620	2013.9935	8.989	9.014	6.447	281.8	-118.9	-104.6	5.1	-115.7	-98.6	5.1
2929	156.7934	22.7660	2014.0185	8.381	12.084	7.691	158.9	-116.3	-75.5	5.1	-115.1	-83.6	5.1
2930	156.8205	36.2276	2013.9015	15.263	16.911	8.473	289.6	32.5	-61.7	5.1	30.5	-61.5	5.2
2931	156.9577	26.5934	2013.775	15.401	17.233	6.17	59.5	-60.5	-32.1	5.2	-60.3	-29.9	5.2
2932	156.9624	49.1988	2013.77	12.014	14.079	31.273	247.4	-68.3	-87.9	5.4	-66.6	-84	5.5
2933	156.9719	53.8516	2013.8395	11.538	12.3	30.696	43.5	169.2	-7.1	5.4	172.8	-6.5	5.4
2934	157.0047	28.0000	2013.861	10.69	17.884	27.825	128.3	14.9	-107.7	5.1	14.8	-105.6	5.3
2935	157.0062	30.4832	2013.853	11.356	15.001	14.677	159.8	32.3	-69	5.8	28.5	-67.3	5.8
2936	157.0437	19.9484	2013.9035	12.578	15.29	9.423	337.7	-53.9	-65.6	5.1	-54	-62.9	5.1
2937	157.0775	15.3883	2013.9025	10.81	18.802	27.848	248.8	-38	-83.6	5	-35.9	-86.8	5.5
2938	157.0918	9.3167	2014.101	17.524	17.781	5.485	86.6	-65	-67	6	-61.5	-70.5	6.2
2939	157.1058	38.2738	2013.844	11.748	12.828	7.558	211	-83.4	-21.3	5.2	-83.6	-18.4	5.2
2940	157.1264	12.9514	2013.9005	15.198	17.214	59.24	249.1	-70.2	-45.7	5	-71.9	-47.2	5.1
2941	157.1658	28.7269	2013.9045	16.294	16.495	4.614	90.2	-124.4	-27.6	5.2	-128.8	-29.5	5.3
2942	157.1725	29.7376	2013.8445	10.972	11.183	6.873	211.7	-66.8	-43.9	5.1	-66.3	-43.4	5.1
2943	157.2282	33.8003	2013.851	13.335	15.471	20.311	285.8	-55.5	-77	5.8	-56.4	-75.8	5.8

2944	157.2762	28.7055	2013.917	17.472	17.613	4.254	348.7	-99.7	-76.9	5.3	-99.6	-77.2	5.4
2945	157.3418	29.0369	2013.9395	11.697	16.048	9.978	317.9	-65.1	44.8	5.1	-66.5	41.2	5.1
2946	157.3494	79.5448	2013.819	12.613	13.837	10.673	38.8	-102.2	-119.6	5.9	-101.6	-115.8	5.9
2947	157.3809	62.9234	2013.658	11.8	13.513	5.662	99.5	-37	-62.3	5.5	-39.1	-62.2	5.6
2948	157.3974	37.9581	2013.8005	10.066	10.268	13.027	69.7	63.5	-52.9	5.2	61.7	-54.1	5.2
2949	157.4321	-13.4553	2014.269	11.814	13.39	24.02	107.7	-128.4	29.7	5.1	-128.9	31.5	5.2
2950	157.4355	25.7601	2013.8575	12.954	15.461	32.493	31.9	87.4	-75.7	5.1	85.8	-77.1	5.1
2951	157.4548	-3.9154	2014.1765	7.629	8.986	15.19	345.8	-94.2	-13.5	5.5	-93.9	-14.1	5.4
2952	157.4761	24.0423	2013.797	15.129	18.26	6.936	137.2	-16.1	-66.1	5.2	-27	-64.3	6
2953	157.4843	54.3855	2013.819	13.284	13.395	21.679	243.2	-50.2	-69.1	5.8	-51.8	-65	5.8
2954	157.4849	-4.5795	2014.193	11.88	15.705	13.641	64.3	-60.3	16.1	5.3	-63.5	16	7.4
2955	157.5815	9.6147	2014.078	17.441	18.284	43.351	9	58	-62	6.9	48.1	-70.2	6.4
2956	157.6357	31.4970	2013.825	16.206	16.747	17.29	254.7	-64.3	-16.1	5.8	-67.2	-14.2	5.9
2957	157.6945	42.6956	2013.7985	12.206	16.67	13.559	104.9	-85.7	-5.3	5.2	-89.9	-2.1	5.2
2958	157.7251	4.7339	2014.1885	13.445	13.617	5.266	206.9	-77.9	44.8	5.8	-80.1	40.7	5.8
2959	157.7276	52.6538	2013.711	10.84	16.65	11.756	116.5	-88.1	-88.2	5.5	-86.9	-86.2	5.6
2960	157.7626	5.9940	2014.096	16.913	16.986	28.282	32.1	-51.5	-98.9	5.8	-48.4	-96.5	5.9
2961	157.8363	30.5149	2013.892	12.615	17.776	8.747	294.4	-21.6	-125.5	5.8	-20.6	-125	5.9
2962	157.8426	22.3686	2013.8265	14.15	17.633	9.304	275	-67.6	-33.9	5.8	-67.6	-39.7	6
2963	157.8669	45.9279	2013.831	14.019	17.768	37.263	196.9	-69.8	0.8	5.2	-68.6	4.8	5.3
2964	157.9138	56.2422	2013.8385	10.923	15.722	7.133	185.1	-91.3	3.3	5.9	-91.9	0	6.3
2965	157.9265	-1.5936	2014.177	12.415	12.443	6.871	52.9	-85.7	-38.8	5.3	-84.7	-35.8	5.4
2966	157.9357	0.9373	2014.154	11.015	12.299	42.671	255.9	60.4	-67.3	5.7	63	-63.7	5.7
2967	157.9558	19.0691	2013.895	14.304	15.812	6.313	287.2	-72.6	-6.7	5.1	-73.9	-7.2	5.1
2968	158.0149	34.0610	2013.868	14.081	14.541	5.694	199.2	-55	-73.1	5.4	-57.1	-70	5.4
2969	158.0332	53.5006	2013.8035	11.687	15.033	32.098	147.6	-61.9	-24.5	5.5	-60.7	-19.8	5.5
2970	158.0478	32.2373	2013.873	10.308	13.968	31.2	273.5	-36.1	-119.3	5.4	-38.2	-115.6	5.4
2971	158.0784	27.9906	2013.862	12.515	14.884	30.409	265.6	-73.3	17.5	5.1	-75	17.4	5.1
2972	158.0980	50.2325	2013.791	16.484	17.989	13.524	281.6	-72.9	-4.2	5.5	-75.5	-7 7	5.8
2973	158.1393	23.9593	2013.7425	16.868	17.083	4.983	15.6	-61.1	-19	5.2	-61.9	-22.4	5.2
2974	158.1527	-4.6547	2014.1785	11.182	12.82	10.039	355.1	50.8	-84	5.3	47.2	-86.4	5.6
2975	158.1994	21.8815	2013.5865	16.145	16.684	4.027	57.7	-73.5	-16.4	5.1	-75.7	-22.9	5.5
2976	158.2066	27.7159	2013.7835	17.014	17.592	4.701	313.3	-67.6	-17.3	5.2	-69.4	-23.1	5.3
2977	158.2418	8.5133	2014.1055	9.841	13.473	48.859	38.8	62.7	-25.4	5.8	61.5	-29.5	5.8
2978	158.3151	36.6373	2013.871	13.854	17.613	17.398	208.8	-63	2.3	5.5	-64.1	1.6	5.6
2979	158.3212	37.6733	2013.7205	9.668	9.675	5.047	160.2	21.4	-77.7	6.3	20.4	-76.2	5.9
2980	158.3617	78.7739	2013.908	18.25	18.627	11.117	119.8	-73.1	-72.5	6.1	-74.6	-66.4	6.2
2981	158.3737	0.8476	2014.0905	12.856	16.972	9.118	48.4	-145.9	-36.7	5.8	-143.4	-37.2	5.9
2982	158.4195	48.9852	2013.5755	17.197	17.976	11.601	181.2	57.7	-99.3	5.6	58.9	-95	5.6
2983	158.4360	6.1191	2014.141	13.324	16.22	59.801	235	-11.2	-60.6	5.7	-2.1	-62.2	5.8
2984	158.4765	58.6440	2013.8385	10.112	16.092	46.215	284.6	-84.3	64.9	5.8	-81	64.6	5.8
2985	158.4873	57.5173	2013.898	15.285	16.406	4.536	342.3	-60.3	-71.6	5.8	-62.4	-74.3	5.9
2986	158.4909	-4.3867	2014.157	11.351	15.401	27.016	215.9	-100	5.8	5.3	-102.9	3.3	5.9
2987	158.4916	67.0562	2013.7515	14.781	15.374	4.26	52.7	71.9	9.7	5.5	71.2	6	5.7
2988	158.5132	-10.8839	2014.2015	12.707	13.402	5.328	188.4	-90	30	5.4	-86.7	36.4	5.4
2989	158.5859	-13.8571	2014.269	9.458	15.607	11.76	1	-91	-2.2	5.1	-94.8	7.2	6.6

2990	158.6693	22.0323	2013.8855	13.049	16.304	6.474	302.1	108.6	-85.6	5.1	109.2	-86.5	5.3
2991	158.7047	1.9780	2014.1095	13.457	13.526	9	125.1	125.7	-151.9	5.9	124.6	-152.4	5.8
2992	158.7325	54.0306	2013.8365	11.938	12.101	33.383	206.5	-108.1	-57.1	5.8	-109.2	-54.9	5.8
2993	158.7479	10.2023	2014.0845	12.105	18.264	42.249	251.1	-91.7	19.9	5.8	-92.8	22.3	6.2
2994	158.7812	24.3597	2013.7145	16.654	16.833	6.743	223.2	-144.3	82.9	5.3	-142.2	80.7	5.3
2995	158.8443	2.3354	2014.112	11.818	14.467	9.428	85.1	-75.3	-4.1	5.7	-73.4	-2	5.8
2996	158.8461	43.3617	2013.763	15.617	17.013	28.717	196.1	-180.3	-45.7	5.5	-180.4	-45.8	5.5
2997	158.8800	31.8997	2013.903	12.252	14.989	42.392	104.6	26.5	-72.5	5.5	27.2	-72.6	5.5
2998	158.9018	74.1232	2013.5805	16.69	16.755	18.207	5	-89.8	-100.4	6.2	-84.7	-98.7	6.1
2999	158.9288	12.4497	2014.0985	12.156	13.897	5.283	34.1	-78.1	30.1	5.9	-78.8	30.2	5.9
3000	158.9366	77.0102	2013.7665	16.842	18.096	12.338	296.5	-61.9	-14.9	6.1	-60.8	-15.5	6.1
3001	159.0032	45.2932	2013.75	14.897	16.106	5.835	326.8	-88.7	9.8	5.4	-85.5	5.6	5.5
3002	159.0111	43.3272	2013.786	11.003	15.575	39.956	269.4	8.5	107.1	5.4	7	105.3	5.4
3003	159.0219	29.1048	2013.89	14.657	14.868	16.329	271.2	-78.1	-43.1	5.1	-78.5	-41.7	5.1
3004	159.1525	-12.7711	2014.1655	12.706	14.554	14.492	222.1	49.3	-66.1	5.3	52.1	-65.1	5.4
3005	159.1618	54.6804	2013.837	14.049	17.487	9.174	311.3	-86.5	-74.5	5.8	-83.1	-71.2	5.8
3006	159.1918	34.1740	2013.963	13.491	14.716	4.419	104.7	-88.4	74.6	5.2	-93.5	72.5	5.3
3007	159.1953	15.0762	2013.828	14.805	16.213	8.496	66.4	-73.1	-37.3	5.5	-77.2	-35.3	5.6
3008	159.2061	1.3144	2013.7355	13.71	15.606	16.613	35.1	-78.8	31.5	5.9	-74.5	34.8	6.3
3009	159.2272	50.6978	2013.779	9.585	16.006	10.468	112.2	-77	-42.9	5.4	-82.1	-41	5.6
3010	159.2798	51.3373	2013.775	13.087	15.029	41.29	10.1	-32.1	-157.4	5.4	-32.6	-160.8	5.5
3011	159.3073	29.9424	2013.892	13.94	17.77	14.656	333.6	-93.4	-32.2	5.1	-95.7	-31	5.3
3012	159.3676	29.5256	2013.9235	12.425	15.465	43.94	324.8	-84.8	-61.8	5.1	-84.8	-61.5	5.1
3013	159.4560	28.4094	2013.9175	11.646	13.808	8.674	16.6	-61.9	53.2	5.1	-63	54	5.1
3014	159.4676	29.5326	2013.826	11.296	17.097	28.12	288	42.6	-86.1	5.1	37.6	-83.8	5.2
3015	159.5132	-2.7650	2014.1345	11.037	17.043	18.028	174.7	-71.9	-49.9	5.4	-72.4	-49.7	5.8
3016	159.5295	-2.9252	2014.063	11.492	15.585	9.757	67.4	-89.7	16.6	5.4	-92.2	17	5.5
3017	159.6250	25.0057	2013.8575	14.509	14.641	26.816	113.5	66.7	-89.3	5.1	68.4	-90.2	5.2
3018	159.6469	2.5931	2014.102	9.778	11.282	25.705	82.7	-68.6	-37.3	5.7	-70.3	-36.4	5.8
3019	159.6507	3.6367	2014.0685	12.927	17.751	15.497	221.9	-3.8	-64.9	5.8	0.6	-65	6
3020	159.6699	11.5388	2013.968	12.885	15.779	8.006	122.6	-11.9	-170.4	5.8	-6.3	-173.6	5.9
3021	159.6712	2.1107	2014.108	11.664	16.226	24.719	302.8	-80.5	-3.1	5.8	-84.4	-4.6	5.8
3022	159.7259	22.0992	2013.8495	15.221	15.517	3.954	80.9	-36.6	-165.3	5.2	-39.4	-162.5	5.5
3023	159.7339	4.0022	2014.108	16.907	18.171	53.063	129.3	-73.7	-20.1	6.7	-70.5	-32	6.5
3024	159.7366	71.2143	2013.691	10.316	15.603	47.821	184.9	-73.2	-30.7	5.5	-72.1	-34	5.5
3025	159.7434	14.1739	2013.9565	17.571	18.319	9.132	157.6	-88	18.6	5.7	-89.3	11	6.5
3026	159.7557	51.0735	2013.7965	9.681	12.872	17.678	295.2	-115.5	-30.3	5.4	-116.3	-23.9	5.4
3027	159.8037	6.1639	2014.1225	14.445	15.658	5.295	265.5	-71	-15.1	5.8	-70.2	-17.8	6
3028	159.8073	14.4793	2013.904	17.641	17.95	21.609	298.9	-60.1	-42.5	5.8	-60.4	-41	5.7
3029	159.8547	26.2340	2013.9095	14.173	19.043	33.251	315.8	-64.7	6.4	5.1	-62	15	6.9
3030	159.8622	31.7008	2013.9005	9.057	9.077	18.1	36.6	-110.6	-34.5	5.1	-111.8	-34.6	5.1
3031	159.8971	30.8816	2013.8955	13.01	13.778	18.223	298.7	118.9	-37	5.1	118.3	-37.9	5.1
3032	159.9079	-8.9659	2014.232	13.548	14.948	15.615	285.5	63.7	-102.7	5.4	65.8	-104.6	5.3
3033	159.9184	32.4513	2013.926	12.027	13.205	48.733	316.6	110.4	-177	5.1	109	-178.4	5.1
3034	159.9658	42.8317	2013.7835	12.719	17.255	49.263	188.3	-68.6	-73.4	5.5	-70.5	-69.3	5.6
3035	160.0009	14.1157	2013.976	10.996	14.914	19.556	308.3	22.4	-91.5	5.7	15.8	-91.9	5.8

3036	160.0635	36.0919	2013.9125	12.383	15.284	19.922	177.2	-28.2	-82.3	5.1	-27.7	-82.9	5.1
3037	160.1065	61.8901	2013.7955	11.278	16.074	40.867	127.8	-82.3	-104.6	5.5	-83.6	-101.2	5.5
3038	160.1293	41.3019	2013.8575	13.971	17.23	59.31	70.1	-76.3	11.1	5.8	-79.5	14.3	5.9
3039	160.2341	37.6363	2013.8405	16.402	18.322	29.817	150.8	-37.6	-62.6	5.8	-36.9	-64.7	6.1
3040	160.2369	6.0309	2014.0205	12.553	15.779	22.259	131.3	-129.4	71.2	5.8	-131.5	71.9	5.9
3041	160.2479	57.3402	2013.8825	14.99	15.113	5.716	138.8	-99.8	10.6	5.8	-101.3	14.1	5.8
3042	160.2589	3.5958	2014.088	9.705	10.928	7.441	120.2	122.5	-137.3	5.8	122.1	-139	5.8
3043	160.3161	63.0543	2013.645	11.531	17.737	39.547	230.6	-71.7	15	5.5	-71.6	4.8	5.7
3044	160.3188	74.0419	2013.831	12.951	19.121	17.528	297.2	-61.7	-5.1	5.5	-63.5	1.1	6.5
3045	160.3246	21.9050	2013.903	15.23	15.716	24.144	8.3	-81.3	-51.7	5.1	-82.5	-55.6	5.1
3046	160.4240	26.5858	2013.7845	15.826	16.465	6.804	149	-9.4	-66.9	5.2	-6.1	-66.1	5.2
3047	160.4948	2.1450	2013.8455	13.729	16.725	22.903	84.4	-73.4	-7.8	5.9	-73.1	-9.9	6
3048	160.5349	0.4792	2014.1145	14.125	15.323	5.942	283.3	1.8	-84.1	5.8	4.4	-82.5	5.9
3049	160.5352	51.4826	2013.7215	11.967	12.503	11.494	182.3	-79	-52.4	5.4	-82.4	-52.7	5.4
3050	160.5456	-3.9424	2014.029	15.902	16.559	56.435	184	-76.3	-19.8	5.4	-76.9	-15.1	5.7
3051	160.5751	11.6755	2013.959	13.711	15.614	8.124	289.8	-80.1	-36.4	5.8	-79.5	-37	5.9
3052	160.5814	-10.7108	2014.2505	14.01	15.811	43.381	235	-76.1	10.4	5.3	-74.9	6.9	5.4
3053	160.5942	-1.2300	2013.9775	14.101	16.66	15.252	302.1	76.9	-81.1	5.4	72.9	-80.1	5.6
3054	160.6155	32.0906	2013.9155	10.854	12.28	14.794	327.5	-76.9	-21.5	5.2	-76.5	-23.3	5.2
3055	160.6278	29.1912	2013.9445	15.544	15.864	5.597	230.9	-94.2	13.6	5.2	-94.8	11.9	5.2
3056	160.6337	65.1092	2013.737	14.16	15.905	19.705	131.8	-61	-21.8	5.5	-61.2	-23	5.5
3057	160.6575	60.6155	2014.023	14.948	17.85	5.776	69.2	3.8	-141.9	5.5	13.2	-140	5.9
3058	160.7500	54.7044	2013.7645	14.122	14.214	21.406	186.4	-79.2	-18.4	5.8	-80.6	-16.4	5.8
3059	160.7712	29.4613	2013.904	14.175	15.071	7.679	107.6	-78.2	-15.6	5.4	-76.6	-15.2	5.4
3060	160.8211	49.0170	2013.6695	12.223	12.588	20.726	334.8	-64.3	13.3	5.5	-61.5	16.1	5.5
3061	160.8566	43.7988	2013.7775	12.955	14.533	12.573	290.4	-162.1	-24.4	5.5	-162.5	-23	5.5
3062	160.8675	8.2475	2014.1	11.571	11.853	36.453	293.9	-40.6	-80.7	5.8	-37.1	-80.8	5.8
3063	160.8725	4.1609	2013.949	14.656	15.586	6.15	261.5	-141	-158.6	5.8	-140.1	-155.5	6
3064	160.9186	40.6006	2013.8245	15.57	18.593	46.991	179.7	-64.9	-68	5.2	-61.7	-68.8	6
3065	160.9301	16.2304	2013.9115	15.697	16.173	15.747	286.8	94.2	-46.2	5.4	93.6	-44.7	5.4
3066	160.9511	1.3326	2013.904	13.225	13.531	48.912	352.7	-72.6	52.1	5.8	-70.6	54.7	5.9
3067	160.9972	21.8741	2013.9105	9.624	12.234	17.734	193.5	-119	-94.2	5.1	-122	-91.9	5.1
3068	161.0085	8.4135	2013.9995	13.222	16.35	6.601	163.3	-112.3	-9.9	5.8	-109.8	-7.2	6
3069	161.0165	28.0770	2013.8385	15.193	15.635	4.646	154.2	-46.1	-107.7	5.4	-45.6	-107.7	5.5
3070	161.0396	10.9426	2013.9985	12.28	14.715	10.684	116	13.4	-80.8	5.8	15	-80.1	5.8
3071	161.1185	7.6519	2014.0055	15.717	16.202	5.166	3.3	36.6	-77.9	5.9	37	-76.9	6
3072	161.1206	21.8076	2013.823	9.759	17.239	36.42	168.1	97.6	-136.5	5.1	99.1	-127.2	5.4
3073	161.1686	48.5499	2013.682	13.75	14.697	16.018	31.9	34.5	-75.7	5.5	34.3	-73.4	5.5
3074	161.2325	3.6301	2014.1095	14.859	14.915	34.507	187.8	-47	-83.5	5.8	-49.6	-81.1	5.8
3075	161.2545	25.6209	2013.8405	14.157	16.697	9.845	298.7	-91.9	-26.6	5.4	-95.7	-20.8	5.5
3076	161.2821	-7.5818	2014.2365	11.275	13.935	12.336	358.6	11.2	-100.1	5.3	11.9	-100.4	5.3
3077	161.4493	37.7893	2013.8815	15.059	18.145	10.385	189.6	-65.4	-138.8	5.2	-70.2	-142.7	5.3
3078	161.4914	19.4771	2013.7345	10.954	16.517	9.14	308	-83.3	-30.5	5.1	-80.5	-31.1	5.4
3079	161.6033	60.3244	2013.82	12.848	16.574	8.983	159.9	-77.3	-66.6	5.8	-78.9	-66.1	5.9
3080	161.6152	9.3244	2013.9625	14.865	16.965	22.823	217.5	-95	-39.6	5.9	-95.1	-42.2	5.9
3081	161.6386	19.1100	2013.8625	12.186	16.55	36.322	115.4	-151.5	-92.6	5.1	-151.1	-92.7	5.3

3082	161.7017	45.4165	2013.732	16.535	17.716	11.589	94.7	-76.9	-13.8	5.5	-75.3	-15.2	5.6
3083	161.7113	68.8006	2013.814	16.085	18.194	14.607	5.4	50.7	-82.9	5.8	56.4	-80.9	5.9
3084	161.7364	36.8492	2013.8955	13.301	16.736	8.684	4.4	-81.2	-12.5	5.1	-80.6	-9.2	5.2
3085	161.7437	24.6145	2013.9275	12.276	13.578	9.739	49.6	-168.4	-21.9	5.5	-175.2	-17.7	5.1
3086	161.8244	18.2942	2014.079	9.503	13.26	5.868	286.6	-85.2	-53.3	5.1	-85.1	-44.1	5.5
3087	161.8388	33.8265	2013.9475	15.379	17.358	24.534	334.4	44.8	-72.9	5.8	41.1	-75.3	5.8
3088	161.8473	21.2946	2013.77	13.484	18.405	21.457	359.2	-163.6	-39	5.1	-162.6	-36.9	5.4
3089	161.9120	2.8315	2013.952	10.566	14.192	8.444	289.2	-77.3	-11.6	5.8	-73.9	-12.3	5.9
3090	161.9313	65.4601	2013.8545	7.522	10.021	10.876	324.2	-99.9	-6	5.5	-98.2	-4.1	5.5
3091	161.9654	27.4082	2013.801	14.569	15.565	4.416	93	-98.4	-14.6	5.1	-94.4	-18.4	7
3092	161.9806	12.7273	2013.9895	12.704	14.054	12.357	206.5	-97.8	-29.8	5	-97.4	-27.8	5
3093	162.0154	49.2861	2013.688	11.669	17.024	55.292	115.2	-79	-16.9	5.5	-78.3	-9.6	5.5
3094	162.0436	36.6079	2013.7995	16.334	16.724	5.3	315.3	-112.1	-44.8	5.2	-113.1	-43	5.2
3095	162.0462	61.2351	2013.9145	10.197	12.384	5.164	225.1	-66.4	-17.8	5.4	-63.9	-17.1	5.4
3096	162.0908	4.6268	2014.105	14.12	14.449	3.581	337.2	-95	-21.3	5.8	-92.2	-22	6.1
3097	162.1172	8.7192	2013.8895	10.779	16.513	26.652	52.5	-133	40.4	5.9	-136.5	37	5.9
3098	162.1229	14.5609	2013.694	16.339	16.489	3.75	307.2	-72.5	-18.2	5.4	-70.3	-24.1	5.2
3099	162.1326	42.6899	2013.7315	13.462	15.299	16.193	170	-67.1	13.3	5.6	-69.2	13.8	5.6
3100	162.1972	1.4849	2013.833	15.396	17.02	23.382	52.5	-120.6	14.7	6	-120.6	9.4	6
3101	162.2111	26.0011	2013.8465	15.002	18.085	12.291	248.2	9.4	-66.2	5.1	12.1	-64.1	5.3
3102	162.2170	-1.0378	2013.892	16.156	16.484	11.277	61.1	-73.5	16.3	5.5	-73.6	17	5.5
3103	162.2557	16.8994	2013.7925	14.04	16.144	5.02	83.9	-129.3	-7.1	5.1	-131.4	-13.4	7.7
3104	162.3196	-4.4997	2014.256	13.292	16.473	13.137	210.2	-103.8	85.7	5.3	-103.4	84.6	5.4
3105	162.3380	56.6466	2013.7815	16.551	16.946	5.913	243.9	-37.2	-74.7	5.8	-40.8	-73.3	5.9
3106	162.3726	24.7501	2013.883	10.691	14.751	10.594	110.2	-38.4	-134.7	5.1	-39.6	-133.6	5.1
3107	162.4243	3.7105	2014.1375	12.126	12.585	5.703	158.9	-62.4	-25.9	5.8	-62.3	-25.9	5.7
3108	162.4414	-7.5648	2014.2565	11.883	17.833	10.565	49.2	-129.4	9.3	5.3	-130.7	6.4	5.9
3109	162.4529	44.6312	2013.721	11.201	12.941	22.023	163	-109.7	-4	6	-108.8	-1	6
3110	162.4566	39.1530	2013.913	14.404	15.254	14.409	214.8	-79.1	-2	5.1	-79.5	-0.6	5.1
3111	162.4761	4.4789	2014.052	11.216	15.794	9.045	151.1	-85.8	18.8	5.8	-82.7	19.6	6
3112	162.4944	-9.8482	2014.265	11.82	12.757	20.32	52.8	-60.1	-24.7	5.3	-60.5	-22.2	5.3
3113	162.5055	9.5564	2013.9405	16.719	17.195	7.374	169.4	-99.9	59.4	5.9	-97	60.2	6
3114	162.5215	43.2634	2013.7095	12.302	17.217	7.305	271.3	-116.4	-40.4	6	-123.6	-36.3	6.3
3115	162.5231	54.0760	2013.745	13.08	13.439	7.403	142.6	-70.5	-36.4	5.8	-71.1	-36.3	5.8
3116	162.5500	32.6862	2013.913	14.307	16.692	8.919	287.9	-79.8	11.7	5.1	-77.8	15.8	5.2
3117	162.6155	-6.6753	2014.2215	8.475	10.796	49.923	6	-17.2	-64.1	5.3	-20.7	-61.4	5.3
3118	162.6317	20.6122	2013.905	12.095	14.874	8.892	353.4	-94.9	-6	5.1	-95.7	-7.5	5.1
3119	162.6372	73.2863	2013.9095	12.459	14.481	5.166	356.9	-69.3	-20	5.5	-71.4	-23.5	5.4
3120	162.6617	-2.8294	2014.098	15.75	17.529	19.824	219.6	102	-77.8	5.4	99.4	-71.8	5.6
3121	162.6961	62.0801	2013.7365	16.667	16.759	19.258	48.9	-73.2	-2.4	5.6	-75	2.7	5.6
3122	162.7043	58.8853	2013.7675	16.552	17.174	4.826	202.9	-97.2	48.4	5.9	-93.6	47.4	5.9
3123	162.7580	38.9603	2013.9345	16.429	17.547	8.699	2.8	-154.4	-32.7	5.2	-159.7	-32.8	5.2
3124	162.7652	14.9715	2013.875	14.419	15.963	13.938	73.8	37	-149.7	5.1	31.3	-149.4	5.1
3125	162.7688	26.3102	2013.7865	16.258	16.394	36.715	281	-67	-22.8	5.8	-64.7	-21	5.8
3126	162.7943	57.3608	2013.8425	14.883	16.286	4.627	16.6	-128.7	-146.4	5.9	-131.7	-154	6.3
3127	162.8690	0.0611	2013.896	16.344	17.009	6.497	295.9	-74.5	24.3	5.9	-74	23.2	6.2

3128	162.8995	11.1589	2014.0245	17.209	17.509	9.53	227.7	-84.6	-12.4	5.9	-81.4	-12.4	5.9
3129	162.9293	34.7235	2013.889	14.817	15.751	7.067	12.1	122.9	-46.2	5.1	124.3	-45.2	5.1
3130	162.9572	29.0429	2013.8015	13.986	13.995	15.692	264.3	44.1	-66.4	5.1	45.2	-65	5.2
3131	162.9615	34.5974	2013.9135	16.143	17.09	14.905	98	-71.1	6.2	5.1	-69.1	6	5.1
3132	162.9735	29.7311	2013.8455	15.257	15.678	3.836	184.3	-75	-20.1	5.2	-74.3	-17.7	6.1
3133	163.0095	15.3546	2013.9135	15.279	16.893	11.645	219.1	122.9	-149.1	5.1	123.3	-145	5.2
3134	163.0246	27.9845	2013.687	9.695	16.646	18.538	98.8	121.9	-112.5	5.2	122	-111	5.3
3135	163.0678	22.5033	2013.8305	14.295	15.784	6.988	181.4	-58.2	-97.6	5.1	-60	-98.7	5.1
3136	163.0796	11.5422	2013.926	13.437	14.757	16.93	117.8	-67.7	-7	5.9	-66.3	-4.6	5.8
3137	163.1062	71.2743	2013.693	13.26	14.607	7.102	354.6	-69.3	-52.9	5.5	-71.8	-54	5.6
3138	163.1274	-9.6772	2014.259	13.734	13.947	3.59	273.7	21.5	-79.7	5.5	23.2	-81	5.5
3139	163.1591	16.1871	2013.9715	15.099	17.314	8.439	311.1	-69.5	-6	5	-68.9	-8	5.1
3140	163.2007	-5.2742	2014.2295	11.276	14.599	15.099	229.2	-60.7	-20.9	5.3	-62.3	-21.1	5.3
3141	163.2017	-8.9520	2014.2505	11.644	14.52	7.831	206	-93.5	0.9	5.3	-96.4	0.9	5.3
3142	163.2199	44.3338	2013.6965	13.19	15.912	7.413	167.4	-66.3	-113.2	5.2	-66.7	-110.3	5.3
3143	163.2242	23.0183	2013.7545	16.106	16.57	28.747	0.3	-82	-27.1	5.1	-83.1	-29.2	5.2
3144	163.2610	10.0755	2014	12.628	13.361	5.244	309.2	-150.4	-23.3	6	-151.6	-26.4	5.8
3145	163.2872	7.4765	2013.8785	14.535	14.596	9.144	327	-80.9	5.3	5.9	-82.9	6	5.9
3146	163.2944	82.4213	2013.944	11.4	13.143	40.3	153.1	72.6	-9.2	5.9	71.9	-9.7	5.9
3147	163.3919	22.5650	2013.807	14.632	17.594	46.41	62.2	-126.4	-10.3	5.1	-123.4	-10.2	5.3
3148	163.4314	19.8012	2013.876	13.993	17.226	32.336	106.1	-85.1	-77.1	5.1	-84.1	-76.3	5.2
3149	163.5033	8.9046	2013.771	13.721	16.902	8.503	145.4	-84.9	60.5	6	-84.6	55.9	6
3150	163.5311	1.3992	2013.886	13.099	14.146	22.935	275.3	9.1	-210.7	5.9	8.7	-212.1	5.9
3151	163.5394	48.3416	2013.654	14.829	16.498	31.917	190.9	-61.3	-19.8	5.5	-60.6	-21	5.5
3152	163.5398	22.1485	2013.811	16.203	16.629	8.85	324.2	-64.4	-38	5.1	-67.8	-37.2	5.2
3153	163.5677	0.0495	2014.0455	11.868	15.199	42.08	117.4	98.5	-101.2	5.8	99.8	-97.3	5.8
3154	163.5816	71.9664	2013.7445	15.431	16.835	43.625	22.6	-87.9	-38.1	5.6	-87	-41.8	5.6
3155	163.6237	0.9586	2013.5845	15.953	16.144	4.791	162.2	31.8	-100	6.2	35.2	-102.5	6.4
3156	163.6378	10.5741	2013.557	10.63	15.523	10.106	88.8	-82.6	-10.3	5.9	-86.2	-7.3	6.2
3157	163.6771	65.1714	2013.867	16.956	17.514	27.29	280.7	65.8	-46.6	5.5	64	-49	5.6
3158	163.6996	17.3270	2013.9185	14.061	16.398	6.307	47.8	-68.3	1.8	5.5	-71.6	-2.7	5.6
3159	163.7098	26.9123	2013.8735	14.428	16.157	7.759	109.7	121.8	-67.7	5.1	119	-68.3	5.2
3160	163.7141	77.6818	2013.7505	8.824	13.814	15.005	255.2	-73.6	-38.9	5.5	-74.9	-43.5	5.5
3161	163.7196	-2.6071	2014.02	11.546	16.638	18.874	117.5	-67.1	-11.5	5.3	-65.2	-12.3	5.5
3162	163.7224	33.9664	2013.9335	15.914	16.455	6.454	348.8	-71.7	11.6	5.1	-70.2	12.6	5.1
3163	163.7496	9.2297	2013.9255	15.166	17.31	6.762	261.7	-67.8	-8.7	5.9	-65.7	-2.3	6.1
3164	163.7873	6.3301	2013.991	10.488	17.948	32.817	331.9	-94.4	-75.7	5.8	-100.5	-63.9	6.3
3165	163.8490	34.3277	2013.8575	10.801	11.733	14.167	187	-36	-62.7	5.1	-37	-63.2	5.1
3166	163.8705	-8.2211	2014.2545	13.776	16.498	16.56	202.2	-112.4	94.1	5.6	-113.3	95.5	5.5
3167	163.9403	47.2384	2013.5955	15.017	16.623	5.89	102.6	-81.4	-14.7	5.3	-78.5	-15.3	5.3
3168	163.9573	-8.6828	2014.2625	15.207	15.797	6.124	4.5	-66.1	-65.2	5.4	-65.1	-66.2	5.4
3169	164.0211	-1.6180	2014.137	7.536	16.235	31.432	234	-123.9	11.7	5.3	-122.6	14.4	5.4
3170	164.0609	23.7320	2013.7555	12.208	16.978	11.814	195	-80.5	-0.6	6	-77.8	4.2	6.2
3171	164.1449	58.9918	2013.8155	15.66	16.704	17.033	262.5	-42.7	-112.9	5.8	-40.9	-115	5.8
3172	164.1608	14.7683	2013.955	14.936	15.294	5.869	65.2	37.6	-85.9	5.5	34	-85.8	5.5
3173	164.1612	24.3688	2013.855	13.291	16.15	7.503	311.1	-116.1	-1.1	5.1	-115.9	-2.3	5.2

3174	164.1788	-5.7074	2014.242	13.756	14.789	9.205	85.4	-57.4	-169.2	5.3	-62.2	-170.5	5.3
3175	164.1803	28.6816	2013.7975	15.588	16.728	34.261	89.8	-84.8	-42.5	5.2	-92.6	-36	5.2
3176	164.2139	37.7176	2013.9505	9.814	16.63	13.316	228.2	42	-126.1	5.9	39.2	-124.3	6
3177	164.2329	19.0662	2013.7445	16.19	16.191	5.362	91.9	-44.2	-63.5	5.1	-42.5	-64.6	5.2
3178	164.2404	-0.8517	2014.119	9.795	10.861	26.315	114.3	74.2	-133.2	5.4	75.6	-140.4	5.4
3179	164.2413	12.1391	2013.8015	15.189	16.091	7.764	3.8	-74.3	-25.3	5.6	-75.7	-26.1	5.5
3180	164.2700	36.6166	2013.8655	13.354	16.38	5.82	342.6	-63.2	-29.5	5.9	-64.3	-29.9	6.1
3181	164.2795	-2.2310	2013.999	14.387	16.565	59.877	84.5	-66.2	-16.7	5.4	-66.4	-8.8	5.4
3182	164.2882	33.0692	2013.799	11.345	14.699	45.87	193.3	66.3	-38	5.2	66.5	-35.5	5.1
3183	164.2981	9.9193	2013.9725	12.749	13.768	5.777	273.6	95.3	59.9	5.9	93.9	57.1	5.9
3184	164.3477	12.6054	2013.8845	12.83	13.008	9.115	301	-68.5	-113.4	5.5	-72.8	-113.7	5.5
3185	164.3504	-3.9798	2014.2475	14.247	17.551	47.226	28.8	-72	-18.9	5.3	-68.2	-20.6	5.8
3186	164.3756	25.9536	2013.8735	11.335	17.72	36.734	14.3	-106	-18.6	5.1	-106	-17.1	5.4
3187	164.3821	57.5503	2013.8175	10.87	12.975	23.482	24.6	-76.1	-29.9	5.9	-76.1	-26.6	5.9
3188	164.4163	4.4705	2014.1355	17.572	17.951	10.173	315	-100.6	56.8	5.8	-93.2	62.1	6.1
3189	164.4196	-3.8812	2014.2435	11.517	14.052	5.156	142.7	-22.3	-77.7	5.3	-26.2	-76.3	5.4
3190	164.4491	-1.1188	2014.0305	14.918	16.327	50.656	185.7	-74.4	12.4	5.4	-71.1	11.3	5.5
3191	164.4498	23.7187	2013.797	14.368	15.99	9.374	149.3	-42.8	-101.5	5.1	-47.5	-99.6	5.2
3192	164.4521	-5.9179	2014.254	16.401	16.721	7.575	170.1	-51.6	-73.9	5.4	-50.3	-74.6	5.6
3193	164.5144	33.1338	2013.9005	14.414	17.304	11.539	187.7	-62.5	-29.1	5.1	-65.8	-23.3	5.2
3194	164.6057	37.0284	2013.9155	8.424	9.859	10.666	330.7	-136.4	-74.5	5.9	-139	-80.9	5.9
3195	164.7014	5.0891	2013.9445	10.026	12.229	20.466	7.6	-144.1	-32.5	5.8	-136.2	-39.4	5.8
3196	164.7305	8.1717	2013.9375	14.877	18.729	19.616	201.6	55.5	-108.2	5.9	62.9	-104.3	7.5
3197	164.8150	-2.2305	2013.9765	13.88	14.39	8.029	96.9	-70.1	-11.7	5.4	-71.2	-13.2	5.4
3198	164.8187	71.9914	2013.739	13.41	15.471	12.466	226.5	-107.4	-32.6	5.5	-104.5	-31.5	5.5
3199	164.8247	50.3308	2013.6	10.861	16.183	10.594	291	-61.1	-9.4	5.4	-60.6	-8.7	5.6
3200	164.8702	17.0414	2013.8615	17.866	18.348	8.68	70.2	-101.2	-41.7	5.3	-99.1	-34.5	6
3201	164.9083	-0.7267	2013.8715	13.209	16.31	9.179	218.2	-74.1	7.3	5.4	-72.3	6.3	5.6
3202	164.9448	22.7128	2013.7755	16.244	16.506	25.129	352.4	39.9	-100.6	5.2	39	-104.7	5.2
3203	164.9515	-7.8005	2014.2015	14.116	14.336	5.139	348.8	-102.3	5.8	5.4	-102.5	-2.3	5.8
3204	164.9646	4.8218	2013.9145	16.182	17.289	8.156	253.6	-67.8	-26.5	5.9	-69.3	-27.9	5.9
3205	164.9985	13.7463	2014.003	14.632	17.581	31.258	13.4	80.9	-77.9	5	84.9	-78.6	5.1
3206	165.0219	42.7215	2013.8	14.98	16.973	12.036	322.7	-30.6	-80.3	5.4	-32.5	-81.7	5.4
3207	165.0617	34.3303	2013.921	14.455	14.854	4.583	264	69.5	-29.7	5.1	70.2	-27.4	5.3
3208	165.0801	10.1588	2013.7335	16.454	16.767	5.666	188	-75.1	-67.9	6	-75.5	-66.2	6.1
3209	165.0856	42.9112	2013.807	6.061	11.615	37.19	151.3	-104.4	-118.3	5.4	-106.4	-112.1	5.4
3210	165.1057	14.7565	2013.995	17.216	17.62	7.57	40.1	-82.7	-49.7	5.1	-80.8	-52.1	5.1
3211	165.1301	64.2893	2013.8515	13.109	15.245	28.371	67	-101	-47	5.5	-97.7	-46.1	5.5
3212	165.1779	2.3110	2013.5515	17.481	17.553	8.938	329.7	-62.6	-6.7	6.2	-64	-9.9	6.7
3213	165.2532	15.5560	2013.8555	12.315	12.562	4.937	53.6	-66.8	-33.6	6	-67.9	-33.6	6.1
3214	165.2592	12.5775	2013.8525	14.704	16.185	7.201	153.5	46.2	-71.6	6	41.7	-70	6.1
3215	165.2843	12.7211	2013.898	14.322	14.831	4.603	232.6	-61	-3.1	6	-60.3	-3.6	6
3216	165.3052	-10.2913	2014.264	16.893	17.216	11.069	258.1	46.4	-156.7	6.4	42.1	-155.4	6.1
3217	165.3353	36.3226	2013.8845	10.354	14.976	17.965	131.2	-3.3	-68.9	5.1	-4.2	-67.4	5.2
3218	165.3560	29.9771	2013.8795	14.308	16.167	7.733	227.4	46.9	-79.2	5.1	47	-81.1	5.2
3219	165.3620	2.5316	2013.948	13.744	15.116	4.904	311.9	34	-62.2	5.9	34.3	-62.8	5.8

3220	165.3719	38.2643	2013.892	10.932	13.436	19.656	1.4	-18.7	-96.2	5.2	-16.7	-95.9	5.1
3221	165.4397	7.0897	2013.88	11.73	16.252	10.362	269.1	-23.5	-110.7	5.9	-19.1	-109.5	6
3222	165.4492	-3.8307	2014.214	11.391	12.863	9.775	209.8	-76.4	-19.2	5.3	-76.5	-19.4	5.3
3223	165.4826	34.5398	2013.944	13.434	17.797	16.955	182.8	-66.3	6.1	5.1	-67.9	4.1	5.2
3224	165.5554	46.0723	2013.692	14.7	15.278	10.482	66.2	-71.3	-21.7	5.4	-68.2	-22.4	5.5
3225	165.5804	16.5076	2014.0015	10.256	10.355	18.689	281.4	-22.9	-170.2	6.2	-23.6	-174.9	6.2
3226	165.7047	1.6053	2013.7715	15.578	17.168	9.615	178.4	-85.4	-14	5.9	-85.9	-12.1	6.3
3227	165.7313	27.0714	2013.901	13.849	14.777	5.318	96.5	-8.4	-65.4	5.1	-8.8	-63.3	5.1
3228	165.7327	18.9405	2013.806	14.1	17.658	53.959	93.7	-70.6	10	5.1	-70.1	7.2	5.2
3229	165.7434	-3.3820	2014.217	8.473	11.601	14.827	175.1	137.9	-152.7	5.3	135.9	-160.4	5.3
3230	165.7489	-10.2794	2014.264	12.5	12.809	10.587	199.1	2.1	-61.2	5.5	2	-62.7	5.5
3231	165.8085	-1.8832	2013.82	9.155	16.473	55.834	195	-86.9	79.1	5.4	-87.8	80.7	5.8
3232	165.8665	-0.1747	2014.07	11.317	13.694	5.453	306.9	102.5	-84	5.4	101.2	-85.3	6.1
3233	165.8882	-4.7058	2014.2185	14.374	15.958	16.417	315.2	-74.4	-27.3	5.3	-73.5	-26.6	5.4
3234	165.9546	48.7824	2013.6595	11.897	15.332	5.937	295.9	-62.7	3.7	5.5	-63.5	3.8	5.6
3235	165.9664	37.3579	2013.8915	11.543	16.067	8.644	141.3	-18.7	-123	5.7	-25.6	-123.1	5.8
3236	165.9865	43.3586	2013.6435	14.576	17.204	19.061	123.8	-164.3	27.3	5.5	-165.9	29.2	5.5
3237	165.9867	8.9176	2013.685	12.824	15.958	7.157	128.4	-112.8	11.8	5.9	-116.5	8.5	6.2
3238	165.9999	58.7195	2013.819	17.004	17.689	20.596	276.6	-68.9	-29.8	5.5	-67.4	-27.5	5.5
3239	166.1011	6.6812	2013.8475	10.125	17.179	33.765	330.1	-114.8	-16.6	5.9	-115.7	-14.3	6
3240	166.1075	26.0551	2013.853	12.963	14.298	29.13	167	-100.5	-47.8	5.1	-101.7	-47.8	5.1
3241	166.1245	42.7790	2013.782	14.504	15.639	6.39	254.7	-67.6	-17.7	5.4	-66.3	-18.3	5.4
3242	166.1411	63.0611	2013.782	15.467	17.501	25.732	95.3	26	-72.7	5.5	26.5	-71.2	5.6
3243	166.1591	28.6316	2013.715	10.841	12.219	19.355	197.1	-126.8	36.6	6	-124.2	33.8	6
3244	166.2523	61.7956	2013.6775	14.799	16.262	7.086	351	-88.7	1.7	5.5	-89.4	-1.4	5.6
3245	166.2568	70.7335	2013.922	13.979	15.497	4.647	232.2	-116.9	-74	5.5	-117	-65.8	7.2
3246	166.3030	-6.3617	2014.2215	11.756	14.268	14.194	43.1	-140	-136.7	6.1	-139.5	-135.7	6.1
3247	166.3736	-3.9545	2014.248	16.246	16.421	7.658	38.8	-77.2	10.2	5.5	-78.7	9.5	5.4
3248	166.3785	14.3937	2013.903	12.557	15.016	11.187	50.6	-85.9	-42.3	5	-82.5	-43.6	5.1
3249	166.4847	74.9067	2013.626	17.217	17.826	7.372	82.5	-73.6	-8.8	5.7	-73.2	-14.1	5.7
3250	166.4910	78.6172	2013.75	12.285	13.066	18.295	15.8	-169	-71.9	5.6	-167.9	-72.1	5.6
3251	166.4922	30.2496	2013.802	10.75	13.88	8.26	345.9	-64.1	-47.8	5.2	-65.3	-45.7	5.2
3252	166.5309	7.0332	2013.936	8.59	9.992	8.336	165.3	-23.9	-62.4	5.8	-22.3	-60.5	5.9
3253	166.5570	47.4350	2013.701	14.169	15.095	12.888	146.5	67.1	-37.9	5.5	68.2	-37.8	5.5
3254	166.5668	57.8104	2013.754	14.048	14.389	5.39	276.4	-95.1	-44.6	5.6	-94.4	-43.9	5.6
3255	166.5706	7.8381	2013.752	12.593	12.621	25.144	340.8	-179.5	33.8	6	-178.7	31.4	6
3256	166.5747	-8.4382	2014.2295	13.208	14.186	5.86	159.5	11.5	-65.9	5.3	9.5	-65.9	5.3
3257	166.5939	69.4132	2013.73	13.954	15.586	8.308	201.8	-61.8	14.6	5.5	-63.5	13	5.5
3258	166.6019	23.0885	2013.679	15.328	16.616	5.036	324.2	42.2	-62.2	6	49.5	-61	6.2
3259	166.7136	68.7358	2013.84	16.117	17.444	5.934	191	-73.7	-34.7	5.5	-74.6	-34.1	5.5
3260	166.7952	16.6361	2013.9185	12.162	14.522	7.961	268.7	-91.9	-5	5	-90.7	-5.2	5
3261	166.8163	31.7625	2013.6985	15.464	16.459	15.04	161.1	41.3	-79.8	5.2	41.3	-75	5.2
3262	166.8450	10.1954	2013.7935	9.466	14.622	15.296	230.1	-69.1	-2.1	5.9	-72.1	-3.8	5.9
3263	166.8764	66.8270	2013.8365	14.697	15.024	35.102	338.1	-62.8	-34.4	5.5	-64	-31.5	5.5
3264	166.8934	23.7505	2014.0045	10.761	10.96	4.804	340.8	-87.5	58.6	5.3	-87	57.8	5.2
3265	166.8941	-2.3484	2014.204	12.295	16.646	6.502	207	-65.9	-28.5	5.3	-69.2	-24.6	5.6

3266	166.9096	68.9827	2013.8575	16.568	17.183	5.721	14.8	-75.5	-38.8	5.5	-76.8	-32.1	5.5
3267	166.9199	72.0379	2013.733	15.381	17.402	9.491	54.6	122.7	-40.5	5.5	123.4	-36.7	5.6
3268	166.9240	57.5334	2013.8385	13.505	15.706	6.992	218.7	-83.7	-18.6	5.9	-83.3	-17.8	5.9
3269	166.9953	36.0618	2013.8895	12.977	13.425	8.163	176.7	-76.8	-35.2	5.2	-77.6	-36.1	5.1
3270	167.0177	-5.8916	2014.2465	15.761	16.763	17.781	248.6	-89.9	-55.2	5.3	-90.3	-53.2	5.4
3271	167.0412	10.3762	2013.853	15.347	18.215	23.127	302.4	20.2	-84.1	5.9	27.5	-82.7	6.5
3272	167.0526	2.5519	2013.873	16.388	16.901	9.916	131.2	-101.3	-11.9	6	-99.7	-11.6	5.9
3273	167.0603	36.4546	2013.8845	7.985	14.779	54.084	267.3	-151.4	-21	5.2	-157.6	-18.9	5.2
3274	167.0663	67.4755	2013.7935	11.147	16.326	9.87	197.4	-91.6	-8.6	5.5	-93.6	-10.4	5.7
3275	167.0883	47.8600	2013.62	12.773	14.595	23.013	222.9	-121.7	-177.6	5.7	-119.7	-180.7	5.7
3276	167.1215	-4.5034	2014.2325	12.769	17.049	6.743	174.6	-93.1	-42.7	5.3	-90.6	-44.2	6
3277	167.1373	-9.4640	2014.2455	13.067	13.953	4.254	291.4	-84.8	-107.4	5.3	-89.9	-111.1	6.3
3278	167.1498	6.7258	2013.8295	12.406	14.656	5.21	107.6	-17.9	-76.2	5.9	-28.1	-75.3	6.2
3279	167.1550	6.4572	2013.856	16.803	17.065	14.892	31.1	-21.2	-104.5	6	-18.7	-107.2	6.1
3280	167.1583	9.7535	2013.8455	11.144	15.717	13.352	345.6	-2.5	-88.5	5.9	-2.6	-89	5.9
3281	167.1947	26.1589	2013.8505	13.997	14.574	18.271	355.9	-189	-43.6	5.2	-186.3	-44.1	5.2
3282	167.3618	15.8737	2013.8975	12.168	14.497	5.738	282.1	-53.3	-123.6	5.1	-52.4	-124.7	5.1
3283	167.3668	-2.7002	2014.205	10.13	17.91	19.798	306.9	-53.3	91	5.4	-57.9	89.7	6.1
3284	167.4314	16.5045	2013.94	14.484	15.921	5.421	156.3	-79.2	2.6	5.1	-77.6	3.7	5.1
3285	167.4367	47.7704	2013.5625	16.607	16.972	44.968	315.1	-64.3	2.9	5.6	-61.3	-5.3	5.6
3286	167.4502	10.1572	2013.8605	8.13	13.496	54.018	183.7	-101.3	-26.5	5.9	-102.2	-25.9	5.9
3287	167.4570	49.7038	2013.6545	13.964	15.256	15.335	168.3	27.8	-69.3	5.9	27	-70	5.9
3288	167.4623	47.6190	2013.6235	13.247	16.799	10.524	301.4	-115.2	-9.5	5.6	-119.3	-6.9	5.7
3289	167.4662	8.8454	2013.5125	13.788	15.914	6.332	187.9	82.1	-15	5.9	84.3	-11.2	6.2
3290	167.4895	34.7583	2013.9605	11.189	13.141	9.284	127.3	-75.1	-129.4	5.1	-76.2	-129.2	5.1
3291	167.4961	-2.8095	2014.1545	14.14	15.104	49.409	165.4	-81.7	-57.1	5.4	-83.1	-60.3	5.3
3292	167.5078	-8.8561	2014.244	15.141	16.883	7.586	156.3	26.5	-70.7	5.3	29.8	-72.8	5.6
3293	167.5571	37.0646	2013.933	12.541	13.07	5.723	98.9	-38.6	-83.6	5.2	-43.1	-81.8	5.1
3294	167.5738	56.9412	2013.7255	16.882	17.152	12.069	79.1	-65.8	-20.4	5.8	-68.7	-19.8	5.8
3295	167.5804	38.6666	2013.918	9.994	12.281	14.769	223.4	-87.8	40.6	5.1	-91.8	43.2	5.1
3296	167.6610	36.8775	2013.88	14.389	16.687	8.513	53.6	-85.1	-59.8	5.2	-83.9	-54	5.2
3297	167.6979	16.5137	2013.9085	17.643	17.991	12.469	170.7	-72	-93.7	5.1	-74.2	-94.8	5.2
3298	167.7873	-5.4656	2014.2055	11.26	17.155	22.199	257.7	-77.4	-37	5.3	-75.8	-36.8	5.5
3299	167.7904	2.3519	2013.8045	14.423	18.142	8.011	284.4	-150.6	-143.3	5.9	-151.5	-143.4	7
3300	167.8324	50.6649	2013.595	13.483	17.04	9.326	190.9	-56.5	-73.8	5.9	-58.5	-74.6	6
3301	167.8850	77.0506	2013.7605	12.139	14.716	33.368	315.1	-35.8	-105	5.5	-36.5	-101.6	5.5
3302	167.8945	-13.4359	2014.236	10.646	14.923	28.027	186.1	-96.7	56.5	5.2	-98.1	53.4	5.1
3303	167.9274	8.8732	2013.868	13.981	14.33	13.045	109	-80.7	-72.1	5.9	-78.9	-74.7	5.9
3304	167.9289	6.5458	2013.9915	13.613	17.259	13.217	166.1	41.1	-83.6	5.8	34.4	-86.7	6
3305	167.9874	4.9727	2013.9315	14.165	16.397	24.555	289	-86.6	-55.8	5.9	-84.4	-58	6
3306	168.0237	41.6446	2013.7705	12.551	17.329	18.812	294.3	69.8	-12.5	5.2	71	-8.3	5.3
3307	168.0797	-1.2093	2013.944	14.996	15.451	6.762	347.2	-70.7	-26.7	5.5	-73.2	-23.1	5.5
3308	168.0817	37.2777	2014.0035	14.784	15.63	4.911	52.6	-109	-19.1	5.2	-113.8	-20.6	5.1
3309	168.1202	1.5873	2013.79	15.974	16.393	4.368	263.1	-78.2	-28.1	6	-78.8	-29.8	6.1
3310	168.1415	25.7152	2013.923	12.762	13.019	15.919	68	-101.6	-32.8	5.1	-103.1	-31.5	5.1
3311	168.1798	19.2809	2013.887	11.821	13.046	6.09	305.5	51.6	-111.4	5.8	50.5	-112.7	5.9

3312	168.1901	46.6745	2013.8285	8.93	12.419	16.149	102.2	-144	-25.3	5.5	-142.2	-28.3	5.6
3313	168.1957	7.9599	2013.863	13.036	16.065	33.146	346.5	-63.5	-70.7	5.9	-66.4	-71.4	5.9
3314	168.1995	-7.3365	2014.211	12.69	13.081	13.565	108.4	-97.8	25	5.3	-95.9	27	5.3
3315	168.2802	23.7886	2013.7955	16.72	17.834	11.445	256.1	-74.8	29.9	5.6	-72.8	28.1	5.8
3316	168.2852	26.4778	2013.955	13.027	16.785	9.268	120.1	-80.9	-38.6	5.2	-83.1	-40.7	5.2
3317	168.3436	15.5472	2013.844	14.165	14.612	8.422	173.2	-64.6	38.6	5.1	-63.3	38.1	5.1
3318	168.3474	16.6516	2013.875	13.213	14.706	5.862	262.2	-66.3	-2.7	5	-64.2	-3	5.1
3319	168.3561	-2.7709	2014.1165	13.316	14.496	4.403	221.7	-147.4	12.4	5.4	-141.3	7.7	5.6
3320	168.4420	8.4137	2013.7885	13.634	17.787	59.845	300.6	-63.2	-17.1	5.9	-64.6	-21.3	6.2
3321	168.4760	21.9759	2013.921	10.892	12.018	15.546	189.8	-79	-11.9	5.5	-79.9	-12	5.5
3322	168.4970	59.2815	2013.779	15.444	17.06	14.408	110.3	-70.4	-1.2	5.8	-69.8	1.2	5.8
3323	168.5006	36.5769	2013.9185	11.293	15.105	23.058	60.8	-82.9	-11.4	5.1	-82.6	-8.7	5.1
3324	168.5209	22.7377	2013.9205	11.769	13.183	17.192	244.5	-64.9	8.7	5.5	-64.7	8.5	5.5
3325	168.5364	53.7770	2013.7265	15.794	16.592	9.87	78.1	-141	-25.3	6	-138.5	-26	6.1
3326	168.5688	4.4132	2013.755	10.839	13.88	6.383	324.1	54.2	-141	5.9	54.9	-146.7	6.1
3327	168.5782	5.9640	2014.0135	7.798	11.775	8.052	100.2	-19.8	-117.9	5.8	-27	-121.5	5.9
3328	168.5864	58.8975	2013.8085	13.763	14.684	4.382	314.9	-61.7	-29.2	5.8	-61.2	-29.8	5.8
3329	168.5991	51.3568	2013.76	11.693	16.48	23.477	216.3	-66.6	5.3	6	-69.1	8.4	6
3330	168.6094	45.3444	2013.5665	11.072	17.876	20.9	154.7	-76.9	-8.8	5.5	-74.2	-5.9	5.9
3331	168.6331	42.8525	2013.7245	10.819	16.053	8.507	312.4	-10.6	-64.7	5.5	-8.4	-65.7	5.8
3332	168.6497	9.4624	2013.8605	11.169	14.492	9.579	324.5	-82.5	-14.2	5.9	-83.4	-18.4	5.9
3333	168.7177	51.1411	2013.7185	13.192	15.044	7.527	271.3	192.2	24.1	6.1	193.4	24.9	6
3334	168.7417	61.6141	2013.7785	16.472	17.372	39.642	218.5	9.8	-60.6	5.5	6.1	-61.7	5.5
3335	168.7685	23.7043	2013.9055	14.507	14.622	42.316	103.9	-72.1	-104.7	5.5	-71.6	-105.5	5.5
3336	168.7949	21.3420	2013.8865	13.163	15.26	11.52	275.3	-55.4	-174.4	5.5	-55.4	-174.1	5.6
3337	168.7956	-5.3695	2014.135	17.047	17.657	17.996	51.5	-71.1	1.3	5.6	-68.7	-3.7	5.5
3338	168.7997	19.9419	2014.013	14.634	14.674	4.515	178.8	-62.1	-0.1	5.5	-62.8	-2.3	5.5
3339	168.8275	18.8524	2013.7835	15.533	16.303	12.379	131.6	-78.9	-8.4	5.6	-78.2	-8.2	5.6
3340	168.8965	-8.1836	2014.234	13.511	16.813	41.316	174.5	-61.6	12.3	5.7	-64.6	5.2	5.8
3341	168.8971	26.6520	2013.8885	12.956	15.601	8.908	38.1	-78	-36.1	5.1	-77.7	-38.3	5.2
3342	168.9015	54.1478	2013.8155	14.209	14.796	4.418	96.4	-52.7	77.8	5.8	-50.6	74.6	5.9
3343	168.9191	59.3822	2013.688	15.844	15.871	8.813	317	-81.9	-16.3	5.8	-81.2	-19.2	5.9
3344	168.9283	16.9366	2014.118	10.326	12.223	5.039	26.3	-119.3	44	5.4	-119.4	36.8	5.7
3345	168.9289	0.7804	2013.901	14.184	15.278	19.529	155.6	-90	-15.4	5.8	-88.6	-16.1	5.9
3346	168.9541	32.0371	2013.878	12.71	17.198	38.828	157.1	8.1	-80.1	5.1	8	-79.7	5.2
3347	168.9612	-7.5777	2014.2275	14.074	14.14	11.779	280.8	-10	-74.7	5.7	-10.2	-75.6	5.7
3348	169.0043	19.6441	2013.8425	12.411	18.08	14.827	339.3	39.2	-83.9	5.8	28.6	-84.7	6.2
3349	169.0329	54.2395	2013.879	9.578	17.012	18.069	320.2	82.9	-77.1	5.8	76.7	-76.2	5.8
3350	169.0525	5.6551	2013.928	15.009	17.511	28.254	151	-123.5	-6.7	5.8	-128	-5.9	6.1
3351	169.0607	6.1710	2013.8175	11.167	14.528	11.833	203.8	-63.7	-27.1	5.9	-64.5	-28.3	5.9
3352	169.2459	44.1802	2013.7645	14.656	17.219	47.982	250.9	-96.4	20.2	5.2	-96.5	17.8	5.3
3353	169.3036	8.9112	2013.872	17.618	17.812	5.931	55.4	13.9	-64.9	6.2	19.3	-65.1	6.2
3354	169.3249	-2.7968	2013.853	11.011	16.456	14.719	237	11.3	-70.3	5.4	14.5	-70.4	5.5
3355	169.3410	64.5647	2013.771	15.443	15.703	6.825	327.4	-99.4	-10.2	5.5	-97.6	-9.3	5.5
3356	169.3580	68.0345	2013.7135	13.314	15.214	25.694	221	-103.4	-22	5.5	-102.6	-23.1	5.5
3357	169.3708	-0.6635	2014.028	17.697	18.346	51.324	71.9	-8.3	-89.2	5.7	0.1	-85.4	6.2

3358	169.3825	18.0805	2013.8005	16.156	16.915	12.201	141.1	-76.2	-50.3	5.2	-79	-49.2	5.2
3359	169.5237	8.1692	2013.891	11.055	11.643	38.106	147	68.7	-100.5	5.9	68.4	-105.2	5.9
3360	169.5874	29.2188	2013.848	16.436	17.173	16.495	237.8	-106.5	-30.7	5.8	-104.1	-35.5	5.9
3361	169.6184	-4.7417	2014.1065	11.419	14.669	7.772	261.1	-101	-12.7	5.4	-103	-12.4	5.3
3362	169.6243	32.4391	2013.9505	11.351	12.11	5.139	134.7	-73.6	-0.6	5.1	-75.9	-0.9	5.1
3363	169.6568	15.5698	2013.9105	14.136	14.89	6.756	153	53.3	-67.3	5.4	51.7	-69.6	5.4
3364	169.6622	2.8212	2013.9895	9.879	9.994	30.232	0.4	-72.1	-56	5.8	-75	-57.2	5.8
3365	169.7273	-2.5985	2014.0605	14.833	15.214	3.367	354.8	73.2	-60.6	5.9	66.3	-65	5.5
3366	169.8112	-10.0228	2014.2425	12.73	15.655	6.362	300.1	-131.2	57.1	5.7	-130.5	56.5	5.8
3367	169.8376	23.6847	2013.767	15.028	15.571	13.831	353.5	-80	-28.7	5.1	-81.5	-27.2	5.2
3368	169.8419	-6.1383	2014.191	13.773	15.408	10.173	116	22.9	-81.6	5.8	22.2	-85.4	5.8
3369	169.8482	33.5167	2013.8465	14.636	17.895	17.286	46.1	-10	-74.1	5.2	-10.7	-76.3	5.2
3370	169.8721	52.2903	2013.7145	10.531	16.003	10.259	76.2	94.8	-21.4	5.9	98.4	-16.8	6
3371	169.9111	28.2276	2013.791	12.827	13.088	26.44	176.9	-135.3	-6.7	5.6	-136.5	-10.6	5.6
3372	169.9326	-6.8985	2014.177	8.403	8.76	7.598	313.4	-110.7	3.6	5.9	-110.1	-0.1	5.8
3373	169.9487	-1.4056	2013.712	16.33	17.301	5.98	7	-118.2	-197.2	5.8	-114.8	-195.9	5.9
3374	169.9537	-12.7802	2014.2285	11.664	12.408	9.104	214.4	-66.5	26.7	5.1	-63.8	28.4	5.1
3375	169.9726	1.3795	2013.775	10.552	15.616	20.286	262.1	-83.8	-13.2	5.9	-84.7	-12.7	6
3376	170.0266	15.1865	2013.75	15.125	15.497	3.874	48	33.7	-96	6	26.2	-100.1	6.2
3377	170.0507	25.4724	2013.801	14.404	14.997	21.599	116.3	-20.8	-100.6	5.6	-23.6	-95.3	5.6
3378	170.0589	28.1473	2013.749	15.396	16.56	5.322	285.3	-86.5	-42.5	5.6	-89.4	-43.8	5.7
3379	170.0673	11.1550	2013.925	10.81	14.719	6.224	287.7	45.5	-85.6	5.9	47.4	-81	6.7
3380	170.0732	7.1149	2013.7205	15.836	16.307	29.604	20.3	-69.2	-37.4	6	-71	-32.5	6
3381	170.0766	19.8433	2013.8605	14.565	16.734	17.101	220	61.7	-35.2	5.1	63.3	-33.2	5.2
3382	170.1044	33.6856	2013.939	17.45	18.155	23.786	72.7	-66.9	-35.8	5.2	-62.5	-37.6	5.2
3383	170.1141	21.6198	2013.913	14.126	18.255	19.886	101.8	-72.3	-0.6	5.1	-70.5	3.7	5.4
3384	170.1205	33.4664	2013.9265	15.597	18.19	10.107	67.9	-7.3	-108.8	5.1	-9.4	-110.6	5.4
3385	170.1207	5.5056	2013.926	9.832	12.353	16.404	152.8	-155.9	204.6	6	-163	200.8	5.9
3386	170.1747	20.7476	2013.9205	12.704	14.489	7.374	30.7	-63	14	5.1	-62.8	14.5	5.1
3387	170.1865	25.1592	2013.7725	14.92	17.943	33.251	308.6	-69.9	12.3	5.6	-70.7	16.9	5.9
3388	170.2237	-5.2971	2014.0865	13.365	14.881	12.682	357.4	-114.5	11.5	5.4	-116	12.8	5.4
3389	170.2249	2.4804	2013.975	12.787	16.213	8.172	237.2	-75.4	11.5	5.8	-73.1	11.9	5.9
3390	170.2788	59.4697	2013.712	16.131	17.734	7.571	209.4	-78.3	14.7	5.8	-80.9	17.3	5.9
3391	170.3218	5.7416	2014.032	8.744	14.745	34.04	128.6	-73.2	-99.8	5.8	-71.9	-103.8	5.8
3392	170.4379	2.2824	2013.896	13.681	17.763	10.515	99.1	29.9	-120.4	5.8	25.1	-122.2	6.4
3393	170.4547	20.6748	2013.8695	12.226	17.037	48.99	233	-65.6	1	5.1	-68.6	-1	5.1
3394	170.4566	30.3034	2014.0345	12.552	13.815	5.177	292.4	79.3	-36.8	5.1	84.4	-35.7	5.1
3395	170.4940	54.7467	2013.7415	13.41	14.056	30.965	163.5	-72.9	-39.2	6	-73.7	-41.3	6
3396	170.5242	1.4679	2013.7385	15.04	15.519	9.613	34.8	-94.3	84.4	6	-95.9	85.7	5.9
3397	170.5581	18.7027	2013.72	15.631	16.742	10.168	38	-120.4	89.7	5.2	-116.8	90.7	5.3
3398	170.5598	44.8848	2013.828	12.158	18.126	12.863	56.3	-4.8	-101.1	6	0.6	-98.6	6.2
3399	170.6116	52.7800	2013.7475	13.935	14.094	15.477	98.5	-142.3	90.4	6	-147.1	88.6	6
3400	170.6330	-6.5560	2014.2115	13.755	14.679	6.298	170.3	-28.3	-62.4	6.1	-28.1	-63.4	6.1
3401	170.6736	40.1755	2013.8235	6.669	10.857	43.818	33.5	-72	-5.3	5.2	-75.3	-0.6	5.2
3402	170.6876	-7.4395	2014.225	10.517	11.801	11.824	38.5	-2.8	-69.8	6.1	-6	-68	6.1
3403	170.7603	4.1383	2014.0745	7.935	11.296	6.325	60.6	-103.7	31.6	5.8	-102.1	33.5	5.8

3404	170.7767	40.0683	2013.6985	15.085	15.183	3.309	152.7	-63.2	-59.3	5.2	-69.8	-52.8	5.6
3405	170.8380	4.6127	2013.8675	15.671	17.245	6.045	203.7	-66.1	-35.5	5.9	-71.2	-31.6	6.1
3406	170.9036	-10.8345	2014.19	10.295	16.368	57.313	248	-109.2	-6.7	6.2	-112.1	-7.7	6.6
3407	170.9269	-7.4616	2014.1995	13.305	15.016	15.609	305.6	-17.9	-128.1	6.1	-23.6	-128.2	6.1
3408	170.9447	53.6369	2013.859	17.96	18.023	5.898	2.1	-73.8	-14.8	6	-76.4	-15.5	6
3409	171.0043	30.8697	2013.933	13.768	14.278	22.873	80.5	-74.4	0.3	5.1	-75.3	1	5.1
3410	171.0070	21.5214	2013.705	15.826	16.425	4.648	321.2	3.2	-124.9	5.2	3.1	-128.8	5.5
3411	171.0188	38.1363	2013.874	12.631	17.273	8.21	130.5	116	-10.9	5.2	119.8	-5.1	5.3
3412	171.0252	4.3804	2013.937	11.997	14.386	9.16	193.6	-58.4	-123.1	5.8	-61.2	-124.8	5.9
3413	171.1501	37.8776	2013.881	10.426	10.485	16.281	209.5	-145.8	-2.7	5.2	-144.1	-4.7	5.2
3414	171.1508	34.9756	2013.9135	13.141	13.811	29.792	353.8	-126.4	-13.4	5.1	-125.5	-13.1	5.1
3415	171.1756	23.9120	2013.7995	15.968	17.512	13.249	314.2	-70.3	28.2	5.1	-69.7	30.3	5.2
3416	171.2413	34.8479	2013.917	15.427	17.816	7.589	284.6	-86.1	-35.3	5.2	-87.5	-34.1	5.2
3417	171.2490	2.8977	2013.8845	10.391	15.23	11.712	227.7	-69.9	2.5	5.8	-72.8	3.8	5.9
3418	171.2596	4.6376	2013.85	16.312	16.537	4.88	314.7	-8.4	-70.8	6	-8	-69.4	5.9
3419	171.2751	-4.4122	2013.8475	15.471	15.97	34.056	343.8	128	-12.7	5.6	128.8	-14.9	5.7
3420	171.3311	74.5470	2013.637	14.53	15.597	11.844	130.1	56.4	-92.3	5.6	56.8	-93.6	5.6
3421	171.3323	44.9375	2013.842	11.756	12.16	4.042	292.1	43.9	-64.8	5.5	50	-63.9	5.8
3422	171.3574	25.2006	2013.793	14.505	16.788	33.027	25.7	-38.3	-67.5	5.2	-36.7	-67.3	5.3
3423	171.3821	10.0746	2013.9405	10.337	10.516	7.243	194.4	-88.8	24	5.8	-90.8	19.9	5.8
3424	171.4254	-1.4575	2013.804	14.511	17.913	19.464	225.2	-23.8	-62.9	5.5	-18.3	-67.8	5.9
3425	171.5022	-4.8175	2014.178	15.341	16.296	22.762	204.5	-73.1	6.1	5.4	-76.3	-0.7	5.6
3426	171.6166	41.7737	2013.823	10.701	14.046	7.88	158.9	46.7	-130.1	5.2	43.3	-128.1	5.2
3427	171.6347	-2.6203	2013.712	11.929	15.386	8.234	240.5	-167.2	-19.9	5.5	-174.5	-20	5.7
3428	171.6634	39.7754	2013.8595	11.367	11.529	24.424	293.9	-37.2	-99.1	5.2	-35.5	-100.3	5.2
3429	171.6863	25.0544	2013.76	13.825	15.938	12.343	355.3	-65.5	-53.3	5.9	-63.4	-53.8	6
3430	171.7083	37.4922	2013.8825	9.994	14.258	9.088	35.9	97.4	-180.9	5.2	89.3	-180.8	5.2
3431	171.7325	3.7416	2013.9485	11.318	13.471	8.454	70.1	-55	112.6	5.9	-57.4	110.5	5.8
3432	171.7353	-2.7322	2013.825	14.833	17.558	7.433	319.7	51.3	-73.5	5.5	53.8	-70.4	5.8
3433	171.7757	23.2205	2013.938	13.077	14.805	6.004	211.5	41.5	-155	5.1	45.9	-154.4	5.1
3434	171.8107	3.4074	2014.082	10.171	11.155	3.829	308.1	-109.3	15.9	5.8	-106.6	12.9	7.7
3435	171.8817	24.0410	2013.8745	8.951	16.102	19.428	184.6	-189.7	-102.2	5.8	-195.8	-94.9	6
3436	171.9145	46.2923	2013.8395	7.785	9.771	7.797	27.4	14.1	-86.1	5.6	9.9	-87.5	5.5
3437	171.9634	50.2239	2013.755	16.133	16.53	23.589	196.3	-75.5	30	5.9	-76.5	31.6	5.9
3438	171.9755	44.4251	2013.782	10.889	13.042	9.322	171.9	-102	-19.4	5.5	-104.3	-18.6	5.5
3439	171.9987	24.1253	2013.776	10.216	16.805	19.967	263.1	-68.7	31.8	5.8	-70.3	31.6	6
3440	172.0142	39.6727	2013.8515	10.424	12.525	26.048	92.5	-72.6	-8	5.2	-73.8	-4.6	5.2
3441	172.0552	17.0274	2013.8835	10.969	13.794	7.686	157.4	-5.2	-73.9	5.1	-4.4	-74.6	5.1
3442	172.0600	40.8452	2013.657	16.637	17.148	7.483	195.1	-65.4	-11.1	5.3	-67.3	-10.1	5.3
3443	172.0719	45.9762	2013.6655	13.548	16.143	16.493	323.8	-64.4	-8.3	5.5	-64.7	-6.8	5.6
3444	172.1388	61.9513	2013.859	15.768	18.319	43.35	153.1	-70.6	25.4	5.5	-67.5	32.2	5.9
3445	172.1566	45.4385	2013.7055	13.24	14.247	37.99	236.2	-65.7	-6.1	5.5	-65.5	-10.9	5.6
3446	172.1620	0.1240	2013.6135	14.211	15.91	19.777	79.8	-88.5	4.7	6	-90.8	5.3	6
3447	172.1781	32.3931	2013.876	9.9	12.273	45.424	83.4	47.4	-65.5	5.1	44.6	-67.7	5.1
3448	172.1809	-0.3048	2013.788	10.477	13.76	16.834	342	-72	-16.3	5.5	-75.1	-14.9	5.5
3449	172.2206	-3.1855	2013.7775	11.481	13.685	5.935	241.6	-67.8	20.4	5.5	-67.9	21.1	5.6

3450	172.2231	20.1676	2013.9215	13.55	13.957	8.752	31.3	-59.4	-76.4	5.1	-59.5	-74.3	5.1
3451	172.2256	17.8725	2013.957	12.794	14.024	11.4	314.1	-0.3	-68.9	5	-1.7	-66.6	5
3452	172.2283	2.7496	2013.749	13.991	16.446	6.497	164	-83.2	-19.9	5.9	-80.7	-25.5	6.1
3453	172.2365	11.9357	2013.962	13.201	14.119	5.369	124.8	-83	-10.4	5.1	-87.1	-11.9	5.9
3454	172.2486	8.3992	2013.802	15.322	17.58	8.163	200.1	64.7	-93.4	5.9	59.2	-93.1	6.1
3455	172.2884	54.8521	2013.6765	12.246	14.944	23.281	40.7	-109	-10.5	6	-110.5	-13.8	6
3456	172.3156	39.5273	2013.7405	15.528	17.375	14.746	275.8	87.2	-146.6	5.3	88.4	-152.7	5.3
3457	172.3195	54.0114	2013.9335	10.188	11.668	4.203	185.7	66	-20.6	5.9	64.3	-20.9	5.8
3458	172.3539	9.0445	2013.7585	10.616	16.811	30.83	59.3	-64.2	-37.4	5.9	-68.7	-35.3	6
3459	172.3755	-3.9933	2013.7945	13.486	14.06	5.588	99.7	-83.2	-84.9	5.5	-81.6	-85.2	5.6
3460	172.3969	61.5474	2013.8025	16.612	17.033	3.981	333.5	-3.1	-98.7	5.5	-9.3	-95.8	5.8
3461	172.4075	20.1176	2013.806	15.991	18.092	8.034	325.4	-60.5	-33.5	5.8	-62.8	-34.3	6.1
3462	172.4196	37.2768	2013.994	17.066	18.719	57.403	174.8	-65.1	-23.2	5.2	-65.8	-21.7	5.4
3463	172.4225	41.9800	2013.661	12.784	16.573	19.259	256.8	-29.8	-73.1	5.2	-32.4	-75.5	5.3
3464	172.4649	25.9056	2013.834	11.513	14.45	8.95	51.6	-88.6	-58.6	5.2	-90.1	-58.4	5.2
3465	172.4670	-4.9590	2014.1	9.948	10.103	7.896	18.5	-66.4	-21.4	5.4	-62.7	-28.1	5.4
3466	172.5710	53.0104	2013.7265	11.043	14.359	28.579	350.1	89.7	70	5.5	93.5	67.8	5.5
3467	172.5738	87.3288	2014.167	11.382	14.902	10.199	333.6	109.5	-32.5	5.4	110	-29.6	5.4
3468	172.5833	84.8215	2014.004	15.125	16.224	35.898	289	-67.9	4.1	5.4	-65.2	3.2	5.5
3469	172.6401	35.0097	2013.903	13.462	16.181	8.909	141.2	35.1	-63.5	5.1	39	-63	5.1
3470	172.6446	16.6880	2013.998	10.239	10.586	5.261	196	-82.2	-19.5	5.5	-78.7	-20	5.4
3471	172.6801	-11.2145	2014.217	14.554	16.994	58.227	251.9	96.2	-66.5	6.2	95.8	-71.9	6.4
3472	172.6824	54.9580	2013.6815	10.622	16.441	33.626	197	39.5	101.6	6	37.2	101.6	6
3473	172.6868	-10.7828	2014.2105	13.735	16.039	27.256	156.7	-141.7	10.4	6.1	-148.5	4.9	6.4
3474	172.7020	-4.4530	2014.0185	12.66	15.406	12.558	313.5	-41.1	-85.9	5.4	-38.3	-84.6	5.5
3475	172.7021	45.1292	2013.8055	13.385	15.755	51.481	192.6	23.9	-165.3	5.2	24.1	-164.9	5.2
3476	172.8511	-3.2812	2013.8865	11.529	12.329	9.586	308.4	-73.9	-48.4	5.5	-76.8	-48.9	5.5
3477	172.8766	-10.0294	2014.213	14.01	15.989	24.81	21.8	-100.5	-1.1	5.4	-97.6	-0.3	5.4
3478	172.8983	22.3452	2013.894	12.349	16.427	7.895	164.6	-86	-26.6	5.7	-85.2	-25.8	5.9
3479	172.9254	-8.7476	2014.221	15.265	16.922	25.57	184	-104.2	8.2	5.4	-102.5	7	5.6
3480	172.9552	35.2185	2013.948	14.755	14.812	7.386	143.9	-161.5	-11.5	5.1	-159.8	-12.1	5.1
3481	172.9724	15.5336	2013.859	10.649	15.219	13.866	169	-60.7	14.9	5.5	-61.7	15.6	5.6
3482	173.0184	2.7033	2013.8755	13.517	14.751	36.145	259.3	-43.5	-67.5	5.9	-41.6	-66.4	5.9
3483	173.0713	17.8929	2013.908	12.327	15.208	6.096	151.1	91.1	-130.1	5.5	91.5	-127	5.8
3484	173.0745	-4.1834	2013.792	12.205	16.469	41.192	118.2	-63.4	14.3	5.4	-67	11.9	5.7
3485	173.1312	25.6883	2013.768	13.997	17.186	8.241	69.6	-73.8	-36	6	-75.4	-41.3	6.1
3486	173.2150	-12.0357	2014.153	8.624	14.668	11.562	246.1	33.1	-142.1	5.4	37.4	-145.6	5.9
3487	173.2212	37.4977	2013.9435	11.065	14.987	6.694	304.2	-60.1	-72.8	5.8	-60.6	-74.8	5.8
3488	173.2222	-4.0034	2014.0635	15.747	16.817	5.451	141	-101.5	-31.1	5.5	-105.3	-21.8	5.7
3489	173.2275	8.5753	2013.747	13.233	17.5	46.232	266.9	-99.3	-1.8	5.9	-100.9	-2.8	6.3
3490	173.2337	6.8878	2013.8575	12.182	12.802	30.045	70.7	-140.2	12.3	5.9	-141.3	11.3	5.9
3491	173.2808	30.0564	2013.68	12.934	16.487	9.511	98.3	-73.6	-28.8	5.2	-75.2	-33.6	5.3
3492	173.3199	-7.3564	2014.181	11.996	16.741	19.463	168.4	-47.5	-69.7	5.4	-46.9	-68	5.5
3493	173.3356	32.3630	2014.0225	11.874	18.302	9.786	295.5	-55.7	-75.2	5.1	-58	-71.9	6.1
3494	173.3365	20.7578	2013.8185	17.961	18.324	10.99	177.5	-72.9	8.7	6.1	-74.8	8.2	6.5
3495	173.3396	59.7866	2013.8105	16.448	16.717	5.762	109.4	-173	-45.1	5.9	-173.3	-44	5.9

3496	173.3761	33.1415	2013.801	11.913	15.009	56.853	211.3	-143.1	-29.6	5.2	-141.7	-35.8	5.2
3497	173.4046	14.2290	2013.9265	11.275	11.452	47.674	356.7	-46.4	-175.2	5.1	-47.2	-176	5.1
3498	173.4193	30.0163	2013.871	15.507	15.594	5.228	282.9	-2.3	-90.3	5.2	-5.9	-88.7	5.2
3499	173.5144	46.9214	2013.711	11.608	12.389	55.071	54.6	95.5	-103.5	5.3	93.2	-104.9	5.3
3500	173.5682	-0.0191	2013.6565	15.184	15.611	41.006	255.3	-87.2	2.7	5.6	-87.2	2.3	5.5
3501	173.5710	42.0903	2013.66	15.201	15.767	5.744	213.2	-120.5	-74.4	6.1	-119.1	-78.1	6.1
3502	173.5719	7.9273	2013.7175	10.136	13.942	7.306	317.9	-133.2	-12.9	5.9	-132.9	-12.1	6
3503	173.5848	4.2252	2013.7395	16.206	16.588	5.563	331.9	-62.5	-57.2	6	-60.9	-55.5	6
3504	173.5995	1.8600	2013.687	10.082	14.082	10.837	140.6	-97.1	20.4	6	-97.6	18.9	6
3505	173.6221	19.0727	2013.8035	15.089	16.14	6.372	76.7	-66.8	11	5.8	-66.6	12.4	5.9
3506	173.6395	40.8564	2013.6965	11.402	11.835	6.394	326.7	83.5	20.3	5.2	82.1	18.6	5.2
3507	173.6789	-0.3578	2013.653	15	15.381	21.653	355.5	-69.1	-41.5	5.5	-70.7	-39.3	5.5
3508	173.7274	19.3529	2013.921	10.599	11.323	10.825	267.1	67.1	-145.3	5.5	68.5	-144.4	5.5
3509	173.7654	20.4415	2013.966	6.254	12.989	49.326	81.1	-62.3	-2.4	5.7	-62.4	-4.6	5.5
3510	173.9381	43.1265	2013.7535	11.139	14.693	29.522	252.4	-75.3	-17.4	5.9	-76.5	-17.1	6
3511	173.9411	12.4770	2013.875	15.113	15.513	3.847	324.7	-103.3	-8.1	5.8	-102.2	-7.4	6
3512	173.9694	-4.0173	2013.945	9.988	11.948	12.673	343.8	-109.7	71.2	5.4	-106.5	71.2	5.5
3513	174.0198	23.3925	2013.811	14.632	17.137	5.783	123	-62.6	-44.5	5.5	-64.4	-46.4	5.6
3514	174.1067	7.7530	2013.7175	12.743	15.775	18.872	250.4	-69	-24.7	5.9	-67.7	-26.5	6
3515	174.1161	41.7364	2013.5795	16.124	16.856	8.379	326.2	-88.7	-22.4	5.3	-89.9	-21.8	5.3
3516	174.1475	56.1350	2013.849	7.428	7.75	6.053	165.3	-182	-97.1	5.5	-180.6	-93.9	5.9
3517	174.1531	20.2679	2013.7605	16.254	16.801	6.093	287	-80.7	-40.9	5.6	-80.2	-42.1	5.6
3518	174.1690	56.3421	2013.97	18.61	18.977	36.756	348.8	-66.3	-35.6	6.7	-67.1	-36.1	7.7
3519	174.2002	40.0132	2013.715	15.84	16.578	15.168	47	-71.3	12.7	5.2	-73.4	14.6	5.3
3520	174.2364	-6.1037	2014.0595	10.495	13.924	6.737	171.5	87.7	-69.7	5.4	87.4	-67.6	5.5
3521	174.2630	29.4651	2013.6745	15.177	17.273	13.62	210.3	39.3	-79.9	5.9	33.1	-78.7	6.1
3522	174.2691	39.7942	2013.7015	15.178	15.883	23.22	147.4	-80.2	-12.3	5.2	-79	-13.1	5.2
3523	174.3061	21.1121	2013.7595	13.812	14.825	9.056	208.2	-70.3	-0.1	5.6	-71.4	-1.5	5.6
3524	174.3154	31.9478	2013.6915	14.06	15.396	26.287	261.5	-148.7	-18.4	5.3	-146.4	-19.6	5.3
3525	174.3519	28.8870	2013.7795	11.022	11.834	27.004	296.2	-99.2	-47.7	5.9	-99.7	-47.7	5.9
3526	174.3780	17.7940	2013.887	13.525	15.355	39.748	355.9	64.7	-150.3	5.8	61.7	-149.7	5.9
3527	174.4052	27.3773	2013.649	16.955	17.839	6.497	140.6	-75.2	-45.4	5.3	-74.7	-43.8	5.3
3528	174.4195	21.1087	2013.7925	14.29	15.732	10.319	4.2	-74.6	-16	5.2	-75.8	-18	5.2
3529	174.4448	7.4383	2013.7355	13.399	15.631	5.991	100.5	11.4	-66.7	5.9	12.4	-66.7	6.1
3530	174.5345	17.0145	2013.9045	10.7	15.845	11.823	41.4	-119.5	-13.3	5.8	-122	-16.5	5.8
3531	174.5778	61.2103	2013.7215	14.998	15.16	19.882	216	-89.6	-84.5	5.5	-88.4	-84.2	5.5
3532	174.5962	-4.7239	2013.9505	15.24	15.969	7.236	217.4	-76.4	1.1	5.5	-76.4	1.7	5.4
3533	174.6073	39.8563	2013.9875	12.563	14.9	4.579	164.4	-61.9	-26.6	5.1	-62.5	-28.9	5.7
3534	174.6392	28.4905	2013.797	9.507	13.049	10.154	118.6	-66.5	5.9	5.1	-67	9.4	5.2
3535	174.6666	-13.6187	2014.233	14.346	15.314	10.268	336.1	60.3	-40.6	5.7	61.1	-39.2	5.9
3536	174.7517	4.1060	2013.6605	14.535	14.659	10.842	0.4	-80.4	39.2	6	-78.7	40.4	6
3537	174.7855	15.2108	2013.7	16.588	17.134	4.978	70.7	-83.8	40.4	6	-87.9	38.5	6
3538	174.8307	-2.8167	2013.808	13.742	15.478	31.735	166	-61.1	-22.1	5.4	-63.1	-25.5	5.5
3539	174.8576	24.8793	2013.692	17.456	17.578	8.719	210.8	-17.8	-66	5.3	-20.5	-63.8	5.3
3540	174.8639	7.7947	2013.7735	14.297	14.782	34.365	12.5	-69.6	-10.5	5.9	-71	-6.9	5.9
3541	174.8786	-2.1377	2013.7435	13.999	16.3	13.273	128.9	-92.9	59.8	5.5	-94.5	58.8	5.6

3542	174.9086	21.0790	2013.716	16.536	16.631	13.644	93.5	72	-79.2	6	76.3	-77.7	6
3543	174.9521	-2.4846	2013.897	11.831	12.947	20.95	67.7	51.4	-103.3	5.5	51.4	-102.3	5.4
3544	174.9918	34.9059	2013.753	13.636	15.49	8.25	100.3	175	-139.1	6.1	176.1	-139.8	6.2
3545	175.0087	15.4416	2013.7755	16.887	17.403	6.119	280.6	-121.1	-2.3	5.2	-121.6	-7.2	5.2
3546	175.0290	65.7048	2013.881	11.745	14.721	15.807	32.3	-94.8	13.8	5.5	-93.7	12.7	5.5
3547	175.0311	37.9620	2013.7735	11.949	15.114	18.562	196.4	-96.2	-1	5.2	-97.1	-0.8	5.2
3548	175.0419	27.6906	2013.7545	10.523	12.819	6.081	213.3	79.5	-81.4	5.1	80	-84.9	5.4
3549	175.0718	14.1505	2013.6885	15.206	16.664	4.85	3.5	89.2	-100.6	5.1	90.7	-103.3	5.4
3550	175.0908	13.6840	2013.936	11.001	12.978	55.37	119.2	39.7	-78.6	5.1	40.7	-78.4	5.1
3551	175.1021	70.0039	2013.762	12.963	16.278	41.126	135.2	-62.2	-22.9	5.5	-64.9	-19.6	5.5
3552	175.1113	-0.1438	2013.693	13.662	15.957	54.806	266	-75	-47.9	5.5	-80.1	-47.2	5.6
3553	175.1291	3.6538	2013.856	8.194	12.436	10.034	307.8	151.4	-78.1	6.1	156	-84.1	5.9
3554	175.1442	36.1073	2013.857	12.856	14.921	7.154	81.7	25.9	-67.8	5.2	28.9	-65.4	5.2
3555	175.1528	-3.7114	2014.089	13.038	17.542	8.801	158.6	-156.4	-20.7	5.4	-154.3	-9.3	7.8
3556	175.1691	13.0571	2013.6705	16.165	16.658	4.392	47.2	-64	-7.7	5.1	-65.1	-13.6	5.3
3557	175.2112	57.3188	2013.656	11.993	13.686	4.631	34.9	76	-59.1	5.5	73.7	-58.7	5.9
3558	175.2135	-1.0220	2013.7605	13.044	13.099	7.676	212.2	24.3	-151.6	5.5	26.5	-151.3	5.5
3559	175.2594	57.3152	2014.0025	13.648	14.091	4.048	156.1	-70.3	-39.6	5.7	-72	-38.7	5.6
3560	175.3145	26.6094	2013.8285	10.854	13.437	9.987	7.6	56.3	-116.6	5.2	57.8	-118	5.2
3561	175.3560	58.9237	2013.8195	8.6	9.063	30.458	352.4	-196.5	-42.5	5.5	-195.4	-32.4	5.5
3562	175.3783	19.5745	2013.701	14.888	15.616	23.705	297.1	16.3	-80.9	5.2	17.1	-79.8	5.3
3563	175.3907	51.0089	2013.669	14.373	14.64	11.628	95	-25.2	-72.4	5.9	-27.1	-73.8	5.9
3564	175.3962	-0.9931	2013.8415	15.836	18.547	13.672	265.2	-65.4	-1.2	5.6	-68.4	-3.5	6.6
3565	175.4714	20.8752	2013.92	9.784	14.369	11.946	159.1	37.6	-110.3	5.2	37.2	-112	5.2
3566	175.5237	-0.6011	2013.7325	16.908	17.031	5.234	335.4	-61.6	-51.7	5.7	-60.7	-53.1	5.8
3567	175.5322	23.8458	2013.6345	11.972	12.846	5.393	227.7	-77.2	-65.2	5.3	-75.9	-64.7	5.4
3568	175.5417	55.0996	2013.94	18.335	18.961	44.654	181.9	-65.9	-10.2	5.7	-65.2	0.9	7.1
3569	175.5939	11.8927	2013.905	14.24	14.461	8.207	42.5	81.5	-60	5.9	81.6	-60	5.9
3570	175.5942	21.8783	2013.7785	13.462	16.004	5.52	219.3	-113.8	-53	5.3	-113.4	-50.2	5.3
3571	175.6499	5.6215	2013.7705	11.788	12.127	5.331	246.4	-118.8	15.2	5.9	-119.1	19.7	5.9
3572	175.6658	47.9877	2013.5715	14.385	15.113	5.304	96.6	-98.3	-57	5.9	-96.9	-58.5	6
3573	175.6798	-9.1674	2014.228	14.359	14.705	37.641	296.8	-78.6	-53.7	5.3	-80.3	-52.9	5.4
3574	175.7113	30.6474	2013.7925	10.685	12.907	13.7	269.5	-118.9	8.6	5.2	-120.3	8.8	5.2
3575	175.7318	-4.9945	2013.9055	13.567	16.308	11.129	167	-63	-1.7	5.4	-60.8	-3.9	5.5
3576	175.7693	39.8663	2013.8375	13.829	16.257	8.41	234.6	-110	-33.8	5.2	-109.4	-34.8	5.2
3577	175.7852	71.0931	2013.8495	17.436	17.58	4.664	298.2	-84.4	6.9	5.6	-87.9	-0.3	5.6
3578	175.8051	28.9761	2013.8435	10.459	15.867	23.793	3.2	-64.1	-11.5	5.1	-63	-13.1	5.2
3579	175.8268	22.4492	2013.8005	11.553	17.686	13.433	176	-80.7	-3.1	5.8	-80.9	1.5	6
3580	175.8539	-0.6737	2013.844	10.229	15.578	25.037	136.1	-108.1	-8.5	5.9	-110.6	-10.8	6
3581	175.8588	9.6777	2013.7265	14.85	16.94	9.626	331.9	-86.6	-31.9	5.9	-85.2	-27.9	6.1
3582	175.8664	16.6825	2013.948	10.907	12.82	26.708	305	-61.7	-0.2	5	-62.1	-1.2	5
3583	175.9146	11.4258	2013.9065	10.865	17.294	52.124	198.4	-79.8	21.4	5.8	-75.3	23.8	6
3584	175.9250	54.9624	2013.7115	14.675	16.177	29.446	175.9	-92.2	26.5	5.5	-94.9	28.7	5.6
3585	175.9732	16.4813	2013.91	12.715	17.185	7.685	265.5	-65.7	0.3	5	-63.6	-3.1	5.3
3586	176.0320	3.7086	2013.929	16.909	17.274	51.824	257.1	-89.6	-16	6.2	-90.5	-5.8	6.7
3587	176.0365	19.2390	2013.808	12.65	15.004	15.29	212.9	-106	-18.8	5.6	-108.5	-18.8	5.6

3588	176.0472	13.9493	2013.693	16.027	17.541	10.532	205	-84	-41.8	5.1	-79.6	-45.2	5.3
3589	176.0819	-7.4383	2014.1585	11.653	14.389	28.571	47.7	-101.8	-11.2	5.4	-102.9	-12.8	5.4
3590	176.0849	19.9988	2013.6765	14.278	15.218	5.887	42.5	-2.6	-75.9	5.6	-2.7	-76	5.6
3591	176.1040	24.9160	2013.667	15.112	18.659	24.659	258.2	59.4	-63.5	5.1	51.8	-71.7	5.9
3592	176.1396	86.5192	2014.052	10.366	11.927	10.626	135.8	-79.1	-20.4	5.4	-80.6	-19.7	5.4
3593	176.1745	-6.6271	2014.2165	15.44	18.546	8.513	219.4	-72.6	10.9	5.4	-71.4	17.6	7.2
3594	176.1890	60.9829	2013.792	16.096	16.647	4.984	47.3	63.2	-26.2	5.5	60.6	-24.9	5.5
3595	176.2118	11.3849	2013.95	8.773	11.172	8.976	286.1	37.9	-156	5.9	47.4	-152.9	5.9
3596	176.2349	5.9442	2013.7235	12.083	16.565	15.427	10.8	-22.6	-71	5.9	-25.7	-67.4	6.1
3597	176.2590	51.1506	2013.7385	10.322	17.384	23.915	69.9	-113.6	-28.9	5.5	-119.4	-25.3	5.6
3598	176.2921	31.8338	2013.725	12.861	15.982	29.12	138.9	-70.3	46.2	5.2	-67.6	43.4	5.2
3599	176.2938	9.5074	2013.8605	10.734	16.535	33.539	338.1	-72	39.8	5.9	-75.7	39	6
3600	176.3547	38.9367	2013.729	12.491	14.087	33.099	12.7	17	67.2	5.2	15.3	66.4	5.2
3601	176.3633	18.9591	2013.515	17.167	18.241	4.694	291.2	-85.6	8.3	5.7	-87.9	7.2	6.3
3602	176.3801	32.2038	2013.7	9.902	17.424	13.107	261.4	67.2	-139.6	5.2	72.6	-142.1	5.8
3603	176.3812	18.1643	2013.7815	11.346	15.282	10.76	192.4	-62.1	2.3	5.6	-62.5	5.8	5.6
3604	176.3867	25.5253	2013.7665	14.715	14.821	54.312	32.2	7.2	-83.2	6	6.4	-80.8	6
3605	176.4247	-2.8294	2013.8445	11.871	12.809	24.378	20.1	-40.7	-93.5	5.5	-40.2	-94.4	5.5
3606	176.4507	23.6833	2013.764	11.871	15.087	19.187	86.7	73.9	-44.6	5.6	76.7	-44	5.6
3607	176.4929	12.3392	2013.913	10.838	14.418	8.024	250.5	-142.5	-2.3	5.1	-141.3	-2.3	5.1
3608	176.4971	30.2339	2013.783	10.809	12.81	34.198	308.9	-180.6	98	5.3	-179.3	97.4	5.3
3609	176.5299	8.8895	2013.676	17.063	17.737	30.378	44.6	-108.2	-27.7	6.1	-109.6	-35.5	6.3
3610	176.5683	21.0211	2013.679	12.809	15.719	13.592	62.1	-86.4	-18.4	5.6	-87.8	-15.6	5.7
3611	176.5694	28.2377	2013.9835	10.85	15.371	7.047	1.1	-84.5	22.4	5.9	-85.2	17.6	5.9
3612	176.5901	30.9616	2013.7085	14.392	14.577	21.722	2.1	-148.2	85.1	6	-148.5	86.5	6
3613	176.6319	0.9569	2013.6805	12.378	13.876	6.364	13.4	-15.3	-119.1	5.9	-12.3	-120.3	6
3614	176.6437	56.0554	2013.764	8.77	16.061	24.141	40.2	-70.8	-52.8	5.5	-71.8	-48.1	5.5
3615	176.6573	51.0953	2013.7705	16.865	17.419	5.246	7.3	-73.8	-106.9	5.9	-69.9	-114.2	6.1
3616	176.6736	12.1974	2013.803	14.247	16.735	6.874	119.7	-73.6	-27.2	5.1	-69.8	-30.4	5.2
3617	176.6905	9.5299	2013.8095	12.806	16.096	9.711	154.5	9	-83.2	5.9	11.3	-82.9	5.9
3618	176.7062	25.0309	2013.4135	16.561	17.898	31.456	238.2	63.1	-20.3	6.1	67.1	-16.6	6.6
3619	176.7289	35.7146	2013.987	13.104	17.958	12.32	229.8	-74.2	-50.8	5.2	-74.9	-53.7	5.3
3620	176.7321	-11.8505	2014.2835	14.87	16.04	55.886	155.9	12.6	-154.6	5.6	15.7	-157.6	5.7
3621	176.7379	20.9385	2013.7475	14.084	14.12	25.308	261.6	-94	-7.3	5.5	-93.6	-8.2	5.5
3622	176.7453	6.6195	2013.623	16.786	17.809	7.129	180.5	-75.7	-39.2	6.1	-75.1	-37.8	7
3623	176.7593	35.0757	2013.8125	13.324	15.947	12.169	303.5	-69.7	23.1	5.2	-69.5	25.4	5.2
3624	176.8163	52.8220	2013.7275	10.161	16.399	12.501	291.4	-105.6	10.3	5.8	-106.9	8.6	5.9
3625	176.8633	8.4088	2013.8145	13.92	14.129	10.188	50	-92.9	6.9	5.9	-91.6	7.8	5.9
3626	176.8795	35.9739	2013.867	12.504	16.537	17.745	41.9	27.7	62.2	5.2	23.9	63.6	5.2
3627	176.8902	64.7620	2013.717	14.801	14.901	7.247	63.4	-63.1	26.2	5.5	-60.7	26.9	5.5
3628	176.9090	16.6749	2013.836	12.274	13.386	4.262	250.7	-75.2	-55.2	5.1	-74.9	-53.5	5.2
3629	176.9695	5.0659	2013.601	15.826	15.99	6.651	292.4	-87.4	4.1	6	-87.4	-0.4	6
3630	176.9894	-9.1775	2014.238	14.072	15.732	4.62	154.3	-6.5	-88.4	5.4	-6.2	-88	6.7
3631	177.0174	2.2230	2013.7065	13.992	14.806	6.301	144.2	-75.8	-6.7	5.9	-75.5	-6.3	6
3632	177.0261	3.1050	2013.794	13.649	15.28	22.668	237.8	-61	-161.6	6	-57.4	-161.4	6
3633	177.0794	78.2565	2013.6205	15.139	15.901	26.414	253.9	-105.7	-2.4	6	-108.8	-1.1	6

3634	177.0938	42.2821	2014.01	11.272	16.624	7.805	277.4	-107.3	-50.3	5.4	-104.2	-52.4	5.6
3635	177.1077	78.5961	2013.634	12.579	15.797	44.859	216.7	-66.8	-133.2	5.9	-71.5	-131.3	6
3636	177.1512	0.9842	2013.648	12.707	17.139	10.286	206.5	-197.3	-30.9	6	-197	-35.2	6.3
3637	177.1668	28.5162	2013.8215	11.379	14.687	8.376	179.3	-132.7	6	5.9	-132.6	15.7	6
3638	177.2010	29.8881	2013.6585	12.641	16.781	39.371	176.4	-62.6	-7	5.9	-63.3	-9	6.1
3639	177.2142	0.3010	2013.482	14.839	15.861	4.523	153.8	-170.4	-21.1	6	-170.3	-11.6	6.6
3640	177.2224	21.9648	2013.6175	16.23	16.761	5.56	207.5	-69.3	6	5.2	-69.9	8.3	5.2
3641	177.2383	-9.8190	2014.24	13.079	16.003	40.376	163.9	-63.8	45.1	5.4	-61.1	45.8	5.5
3642	177.2545	-0.3186	2013.7885	6.107	12.375	27.46	127.9	-198.9	6.7	5.6	-195.8	0.1	5.5
3643	177.2627	38.3569	2013.809	12.834	15.697	9.388	129	-35.7	-88	5.2	-37.1	-89.9	5.2
3644	177.2949	48.5638	2013.635	15.481	16.798	7.072	38.8	-42.2	-69.6	6	-40.9	-73.6	6
3645	177.3059	27.9689	2014.056	12.014	12.922	4.226	152.4	-65.8	-20	5.9	-70.8	-13	5.9
3646	177.3282	17.9927	2013.431	18.005	18.114	6.046	42.3	-84.4	-31.4	5.4	-83.7	-41	5.6
3647	177.3683	6.9436	2013.643	15.024	15.689	7.151	41.3	-108.8	19	6	-108.1	19.8	6.1
3648	177.4183	18.0025	2013.1525	17.409	19.113	15.152	117.4	19.6	-91.4	5.1	28	-85.1	6
3649	177.4293	1.4762	2013.737	13.107	16.196	11.49	150.1	47.9	-82.5	5.9	47.2	-85.5	6
3650	177.4669	37.9050	2013.844	9.531	16.748	54.572	270.2	-63.1	-56.9	5.1	-68.7	-52.7	5.3
3651	177.4728	5.9392	2013.745	11.825	12.798	7.995	154.4	-126	-202.7	6.1	-127.2	-205.9	6.1
3652	177.5185	50.3849	2013.743	13.098	16.64	8.1	265.6	47.2	-176.7	5.9	45.3	-175.5	6
3653	177.5785	37.6807	2013.9015	10.563	11.116	25.194	252.6	-146.4	-65.6	5.9	-144.9	-65.8	5.9
3654	177.5852	51.5543	2013.7565	12.011	12.516	5.891	60.3	-131.5	-78.1	5.9	-131.7	-78.5	5.9
3655	177.6035	15.0945	2013.735	14.092	17.082	10.92	316.9	-81.2	26.4	5.1	-82.6	27	5.2
3656	177.6338	22.0646	2013.5505	17.925	17.926	3.891	14.4	40.4	-84.9	6.4	46.8	-81.8	6.3
3657	177.6711	29.0410	2013.866	16.335	16.371	4.889	225.9	-76.8	-14.8	5.2	-73.1	-17.5	5.2
3658	177.7178	33.6002	2013.8955	16.783	18.822	44.681	121.6	-64.1	12	5.2	-64.2	1.3	5.8
3659	177.7179	13.8376	2013.5795	16.822	18.424	31.299	90.1	-98.4	34	5.2	-95.8	28.7	5.9
3660	177.7331	65.8502	2013.8615	14.8	16.553	15.969	303.5	-70.7	1.6	5.5	-73.8	3.3	5.5
3661	177.7679	-8.4370	2014.2115	14.886	16.701	31.753	346.1	-63.9	-8.1	5.4	-63.8	-6.1	5.4
3662	177.7763	-4.7361	2013.9645	13.648	15.851	13.066	87.4	-66.2	9	5.4	-68.9	10	5.4
3663	177.8003	37.1304	2014.078	11.408	13.593	4.646	8.9	-175.4	-64.3	5.2	-178.1	-67.7	7.7
3664	177.8027	46.3034	2013.8265	15.841	18.812	10.839	196.5	-78.6	-36.6	5.5	-72.4	-41.6	6.2
3665	177.8338	61.0439	2013.717	13.331	15.641	26.378	91.7	-65.5	-51	5.5	-65.3	-53.4	5.5
3666	177.8886	26.5058	2013.7485	11.375	16.358	9.691	36.2	-41.2	-138.7	5.2	-44.3	-138.2	5.4
3667	177.9020	30.7647	2013.7305	12.649	14.657	23	336	-34.6	-80.6	5.2	-35	-80.3	5.2
3668	177.9024	3.2600	2013.787	8.786	13.941	21.438	175.8	-150.5	-23.8	5.9	-155.6	-30.4	6
3669	177.9035	41.8569	2013.757	14.708	16.397	6.656	44.5	-65	-14.8	6	-62.7	-13.3	6
3670	177.9041	35.1903	2013.886	9.3	12.961	12.931	79.1	-137.4	-32	5.1	-140.6	-26.8	5.2
3671	177.9468	17.0033	2013.6595	13.779	16.02	12.184	157.1	-68.8	10.6	5.1	-66.4	10.1	5.2
3672	177.9638	40.6509	2014.025	12.626	16.234	5.946	123.8	-98.5	-4.3	5.9	-103.2	6.1	6.2
3673	178.0550	33.6998	2013.7675	16.05	17.177	14.231	182.2	-45.3	-84.2	5.2	-49.2	-80.8	5.3
3674	178.0806	38.4354	2013.741	15.837	16.198	24.838	78.2	-110.9	-15.7	6	-111	-14.1	6
3675	178.1111	18.2045	2013.7955	12.131	13.932	19.877	306.9	-134.7	74	5.7	-135.2	72.9	5.6
3676	178.1495	20.2203	2013.6405	15.143	15.167	59.728	205.9	-156	39	5.7	-159.4	39.6	5.7
3677	178.1946	21.2835	2013.901	11.756	11.956	5.743	22.6	-76.8	9.1	5.6	-76	7.5	5.5
3678	178.1972	7.6037	2013.62	8.669	12.091	6.55	305.5	96.8	-74.8	6	89.1	-83.1	6.2
3679	178.2054	3.0471	2013.609	13.366	13.72	4.592	164.6	-98.3	-12	6	-100.2	-2.1	6.2

3680	178.2091	18.4693	2013.695	11.466	17.394	26.945	139.4	-87.6	-15.8	5.6	-88.3	-14.7	5.8
3681	178.2098	-8.2705	2014.1815	10.269	12.216	7.282	8.6	-50.2	-88.9	5.4	-55.5	-87.4	5.3
3682	178.2189	30.8274	2014.121	8.868	9.517	5.306	241.5	-62.1	16.5	5.1	-61	18	5.1
3683	178.2570	45.2141	2013.672	14.246	15.365	30.441	94.6	-81.6	-55.5	5.5	-82.5	-54.1	5.5
3684	178.2765	61.3977	2013.773	14.61	15.315	26.575	359.4	-51.8	-62	5.5	-50.6	-63.8	5.5
3685	178.3152	7.4713	2013.5155	16.211	16.289	3.659	357.4	-114.3	-49.1	6.1	-109	-53	6.4
3686	178.3763	58.7622	2014.043	9.147	14.134	6.738	320.4	-129.6	-22.8	5.9	-130.3	-20	5.9
3687	178.3822	-1.8487	2013.8285	13.566	14.123	16.922	12.7	-86.7	29.4	6	-86.1	29.5	6
3688	178.4545	25.9680	2013.949	11.349	15.234	7.446	351.3	-92.2	11.4	5.5	-96.2	9.9	5.5
3689	178.4752	-12.8780	2014.2575	15.7	15.833	14.94	160.6	-146.8	-21.6	5.3	-148.5	-21.4	5.2
3690	178.5052	44.6302	2013.7755	10.157	14.38	18.296	51.5	-77.2	-65.9	5.4	-79.3	-59.6	5.5
3691	178.5257	54.6391	2013.6765	14.678	16.765	5.993	108.9	-67.5	23.4	5.9	-70.1	22.5	6.1
3692	178.5284	6.4720	2013.7595	11.484	14.561	10.325	8.2	-163.7	-96.9	6	-165	-96.5	6
3693	178.5443	9.9769	2013.849	14.265	15.034	5.106	122.1	92.8	-37.6	6	92	-33.6	6
3694	178.5594	20.3901	2013.698	13.915	13.92	14.702	254.7	-72.8	3.9	5.6	-71.7	3.7	5.6
3695	178.5823	21.4731	2013.7225	12.203	12.793	6.142	107.8	-66.2	-0.9	5.5	-65.7	-2.3	5.5
3696	178.6906	18.2090	2013.7335	12.602	16.38	8.673	245.8	89.3	-45.8	5.5	90.2	-42.3	5.7
3697	178.7317	11.6949	2013.6965	8.673	13.338	7.594	236.2	-113	24.5	6	-108.7	34.8	6.1
3698	178.7673	39.3177	2014.0365	12.4	18.237	9.27	229.6	-67.2	-24.9	5.5	-70.6	-17.7	6.2
3699	178.7743	37.7302	2013.77	13.804	16.267	12.717	248.6	-61	-15.4	5.5	-62.2	-15.3	5.5
3700	178.9092	64.7106	2013.9255	15.051	15.6	4.885	114.7	-75.6	-9.7	5.5	-77.5	-6.7	5.5
3701	178.9529	-11.7808	2014.25	14.554	17.106	26.958	351.7	-135	-43.3	5.4	-134.8	-42.4	5.7
3702	178.9596	-1.8755	2013.6385	12.682	13.084	19.199	356.9	-125.1	-55	5.6	-125.3	-55.7	5.6
3703	178.9691	43.1398	2013.8085	14.177	14.595	27.726	272.3	-126	-36.8	5.5	-125.3	-34.3	5.5
3704	179.0012	45.4388	2013.7405	11.811	13.226	11.102	131.7	-115.1	-40.8	5.5	-112.7	-38.9	5.5
3705	179.0187	22.9805	2013.4725	13.037	13.946	3.722	229.3	-21.1	-81.2	5.5	-16.6	-79.3	5.8
3706	179.0214	72.0543	2013.6955	15.766	15.901	4.718	185.2	83.1	-32.2	6	84.6	-28.2	6
3707	179.0531	20.4510	2013.5175	15.233	17.377	6.696	58.7	10.9	-91.5	5.6	11	-90	5.8
3708	179.0607	23.1099	2013.5495	15.821	16.428	10.719	259.9	51.5	-60.3	5.6	52.6	-62.2	5.7
3709	179.0790	3.9484	2013.9495	14.895	18.182	50.313	216.7	43.5	-60.2	6	36.6	-68.6	7.2
3710	179.0840	11.0289	2013.735	10.119	14.838	16.277	101.5	-162.2	-12.5	6	-159.8	-14.4	6
3711	179.1532	44.9195	2013.814	13.783	13.85	5.552	168.7	-65.1	10.3	5.5	-66.5	11.6	5.5
3712	179.1538	-4.6089	2013.972	8.508	9.975	13.027	188.2	-109.5	-45.2	5.4	-110	-44.4	5.4
3713	179.1913	14.6952	2013.653	13.31	17.845	18.933	349.1	-2.9	-79.5	5.1	-0.5	-79.7	5.3
3714	179.2714	35.4902	2013.7425	13.984	14.749	9.19	34.5	-18.4	-63.2	5.5	-17.8	-64.6	5.5
3715	179.2824	36.8524	2013.8475	11.215	15.535	27.458	9.5	-92.4	19.1	5.9	-94.7	19.3	5.9
3716	179.3795	18.1435	2013.614	13.68	16.596	9.732	184.2	-82.1	-1.9	5.6	-81.7	-2.3	5.7
3717	179.3863	61.4047	2013.733	12.734	13.888	38.91	242.2	-88.9	13.4	5.5	-89.7	13.2	5.5
3718	179.4236	43.3369	2013.785	14.232	14.786	45.964	27.5	-67.2	-20.9	5.5	-65.7	-18.1	5.5
3719	179.4265	14.4065	2013.3565	15.708	16.222	4.341	31.4	-70.7	-35.7	5.2	-71	-39.7	5.5
3720	179.4957	32.5080	2013.7775	12.689	13.252	7.389	19.4	38.8	-69.5	5.5	35.2	-71	5.5
3721	179.5610	-11.3376	2014.281	14.028	15.873	8.454	332.9	-129.5	-142.8	5.6	-127.9	-139.3	5.6
3722	179.5866	41.8309	2013.8645	9.504	15.914	19.207	42	-71	-23.6	5.8	-70.9	-26.4	5.9
3723	179.6549	7.3330	2013.6265	14.51	15.103	9.245	301	-92.6	-30.1	6	-92.4	-28.6	6
3724	179.6839	-1.4609	2013.7835	10.864	17.077	20.332	66.4	61.3	-73.9	5.5	60.9	-74.3	5.6
3725	179.6856	47.3127	2013.706	12.024	16.184	8.515	264.1	-108.2	36.7	5.5	-106.2	39.4	5.6

3726	179.7157	20.2325	2013.5825	15.456	18.597	51.683	345.4	-64.8	-4.5	5.6	-66.4	-8.7	6.1
3727	179.7174	15.9610	2013.836	13.859	14.68	3.966	333.4	-123.1	-22	5.1	-120.7	-22.9	5.6
3728	179.7196	2.7238	2013.715	10.625	14.786	39.35	132.8	-84.2	-75.1	6.3	-86	-70.3	6.5
3729	179.7862	-13.1311	2014.287	9.836	13.354	8.883	236.6	64.2	-12.9	5.9	60.4	-23	5.6
3730	179.8399	-0.1642	2013.683	9.818	12.018	4.805	249.4	-74.3	-31.9	5.4	-79.2	-26.1	5.7
3731	179.8459	6.3284	2013.709	12.938	15.772	6.295	40.6	-62	-3	5.9	-63.4	-3.4	6.2
3732	179.8628	20.8709	2013.9415	8.995	11.555	10.319	138.3	106.8	32.2	5.5	106.2	34.1	5.5
3733	179.8718	-4.4859	2013.8785	12.392	15.839	7.566	162.5	-97.7	7.1	5.4	-98.6	7.9	5.5
3734	179.8819	3.6296	2013.718	12.807	16.037	14.304	197.5	-19.3	-80.4	5.9	-22.1	-81.1	6
3735	179.9092	15.7630	2013.812	13.754	14.382	11.63	252.4	-178.2	-83.9	5.2	-176.9	-83.8	5.2
3736	179.9504	-3.2547	2013.746	14.01	16.093	24.738	353.6	-67.4	14.5	5.5	-70	12.3	5.5
3737	179.9605	51.6171	2013.681	9.073	10.926	16.77	288.4	-92.7	70	5.6	-91.4	68.5	5.6
3738	179.9676	48.1543	2013.594	15.368	17.109	9.246	242.8	-69.7	-9.9	5.6	-69.8	-9.7	5.7
3739	180.0465	11.2553	2013.63	16.837	16.87	54.604	293	-71.8	5.1	6	-70.6	-3.5	6.1
3740	180.0688	6.2961	2013.565	15.353	16.229	19.091	121.3	86.7	-53.4	6.1	93.9	-50.1	6.1
3741	180.0708	16.5559	2013.595	16.819	17.985	21.451	343.5	-64.4	23.1	5.3	-68.5	21.7	5.4
3742	180.0883	9.1589	2013.671	14.633	17.759	9.91	157.6	-106	-12.2	5.9	-106.6	-16.3	6.2
3743	180.0957	15.5289	2013.7055	13.965	14.477	22.53	350.9	-138	1.4	5.2	-139.6	0.5	5.2
3744	180.1453	17.3681	2013.7075	14.519	18.611	9.124	308.4	-162.7	30.9	5.2	-159.1	34.2	6
3745	180.1507	-4.8225	2013.8435	15.385	16.372	42.517	333.5	-72.9	16.6	6	-71	19.2	6.1
3746	180.1618	-0.3833	2013.4765	15.211	15.451	3.948	218.2	-121.7	35.4	6.1	-114.8	40.9	6.5
3747	180.1919	26.1935	2013.82	13.384	13.389	21.804	70.8	-76.5	-25.8	5.1	-78.9	-22.5	5.1
3748	180.2363	26.4526	2013.7185	13.098	17.472	58.369	313.5	-86.3	-55.1	5.2	-85.6	-55	5.3
3749	180.2841	22.6045	2013.759	15.069	15.417	3.707	69	-94.1	-44	5.6	-91.2	-55	7.7
3750	180.2985	48.3484	2013.8285	11.232	18.781	19.453	68	-21.8	-100	5.6	-27.7	-94.5	5.9
3751	180.3405	-0.6246	2013.883	13.393	18.081	43.092	20.7	-85.7	26.3	5.5	-81.8	27.5	6
3752	180.3647	21.4575	2013.6485	11.644	17.31	21.962	295	24.3	-114.8	5.6	29.2	-118.3	5.9
3753	180.3936	64.8252	2013.85	15.411	17.557	8.051	290.6	-98.7	11.4	5.6	-98.7	19.3	5.6
3754	180.4197	-12.7706	2014.274	12.704	15.8	25.027	150.2	-88.1	13.3	5.4	-92.6	8.5	5.6
3755	180.4682	26.7280	2013.823	11.87	12.553	26.402	215.3	-91.2	-22.1	5.1	-95.8	-17.5	5.2
3756	180.4796	40.7580	2013.8435	10.607	14.301	31.206	276.5	-80.1	-11.9	5.2	-81.1	-13.3	5.2
3757	180.5406	7.7145	2013.663	13.214	14.192	7.697	217.4	115.2	-119.3	6.1	113.9	-115.9	6.1
3758	180.5431	28.1921	2013.677	16.8	17.792	12.519	19.9	-65.8	-0.1	5.2	-62.6	-5.5	5.3
3759	180.5529	24.0431	2013.6125	15.659	16.775	10.337	227.7	-101.3	-47.2	5.2	-101.5	-51.9	5.4
3760	180.5572	7.8019	2013.758	17.081	17.991	12.208	295.6	-76.8	-5.8	6.1	-77.5	-6.3	6.4
3761	180.5664	27.0706	2013.651	12.668	17.835	41.381	247.9	-77.2	15.1	5.1	-80.2	17.8	5.4
3762	180.6144	27.0506	2013.4945	17.124	17.819	10.232	325.9	-7.9	-60.5	5.3	-11.7	-61.1	5.3
3763	180.6554	44.2991	2013.738	15.252	17.308	12.871	239.3	-170.4	-68.9	5.5	-168.4	-67.6	5.6
3764	180.7109	35.6412	2013.6415	9.322	16.686	12.996	125	-90.4	-7.5	5.1	-93.8	-2.4	7.4
3765	180.7326	0.0683	2013.7725	13.462	15.815	12.065	99.2	-80.8	51.7	5.9	-80.2	51.7	5.9
3766	180.7566	11.0863	2013.68	11.987	14.864	17.528	144.7	-56.5	-109.6	6	-60.5	-111.6	6
3767	180.7769	17.2660	2013.7	12.488	15.532	8.374	272.1	-65	-10.7	5.1	-65.6	-12.8	5.2
3768	180.7891	30.1690	2013.8505	12.547	16.5	8.584	260	-66.7	38.4	5.2	-66.2	36.9	5.2
3769	180.7951	-13.4345	2014.367	13.798	14.98	10.869	133.9	-44.2	-63.1	5.5	-47.2	-65.2	5.9
3770	180.9690	18.1827	2013.6585	12.155	16.516	21.442	195.5	55.5	-114.2	5.5	54.5	-112.2	5.8
3771	180.9945	42.9379	2013.795	12.482	14.806	16.949	226.4	-23.9	-194.7	5.5	-24.9	-193	5.5

3772	181.0707	13.1575	2013.7685	13.947	17.343	18.676	304.5	-69	12.8	5.1	-71.3	13.3	5.2
3773	181.0938	3.6555	2013.8125	15.599	16.834	5.181	279.5	-77	-59.5	6	-77.1	-57.7	6.4
3774	181.1088	81.3801	2013.8215	15.333	15.748	39.666	339.9	-59.2	-68.1	5.5	-57.2	-69.9	5.5
3775	181.2505	62.9718	2013.931	9.626	10.626	5.931	35.7	-120.3	12.9	5.5	-123.7	4.1	5.5
3776	181.2598	13.2245	2014.0125	12.161	12.425	4.727	346.6	45.5	-70.7	5.1	48.7	-66.5	6.2
3777	181.3376	-9.7998	2014.2905	10.489	11.943	20.254	58.1	-13.8	-78.4	5.3	-13	-76.8	5.3
3778	181.4136	4.5195	2013.8155	16.591	17.372	6.584	77.6	15.4	-116.3	6.1	18.5	-118	6.2
3779	181.4476	53.9153	2013.7955	8.719	10.513	54.652	350.1	-104.6	-80	5.4	-108	-73.3	5.5
3780	181.4586	22.4456	2013.82	14.928	15.048	6.263	239.8	-60.5	-30.5	5.5	-60.2	-29.1	5.5
3781	181.4715	53.2632	2013.7665	14.843	16.939	7.844	220.6	-138.9	-66.4	5.5	-139.9	-67	5.6
3782	181.4751	43.2060	2013.7345	15.731	16.297	3.904	244.2	-42.7	-115.5	5.6	-37.1	-111.3	6
3783	181.5088	12.3045	2013.6405	13.005	15.682	38.099	269.5	-2	-173.5	5.2	-3.7	-173	5.2
3784	181.5122	13.1307	2013.774	10.118	17.235	23.848	73.3	-87.5	-21.3	5.1	-88.7	-16.9	5.2
3785	181.5405	70.3798	2013.7185	9.92	10.49	22.329	221.9	-116	32.1	5.6	-120.9	33.3	5.6
3786	181.5524	27.4398	2013.7385	13.129	14.778	4.968	14.9	-110.2	-7.5	5.2	-112.8	-10	5.2
3787	181.5656	77.2685	2013.6545	12.994	15.736	12.124	348	-76.4	-10	5.6	-79.1	-6.7	5.6
3788	181.5928	44.5701	2013.858	15.671	16.244	3.662	10.3	-74.3	-27.7	5.6	-78.3	-27.6	6.7
3789	181.6473	-2.4569	2013.687	13.366	16.136	12.616	321.2	-63.6	-3.7	5.5	-64	-2.3	5.6
3790	181.6531	38.1103	2013.7745	14.212	16.807	9.872	330.7	-81.9	-13.8	5.2	-85.2	-13.8	5.2
3791	181.6729	66.3493	2013.7595	12.037	16.216	48.376	56.4	-65.7	-16.3	5.5	-65.4	-18.4	5.5
3792	181.6853	19.5744	2013.9975	10.49	11.196	5.689	145.8	36.9	-69.7	5.9	37	-72	5.8
3793	181.6938	41.4226	2013.784	15.276	17.689	10.301	296.4	-62.6	-9.1	5.2	-63.5	-8.7	5.2
3794	181.6980	15.7985	2013.7085	11.132	15.218	25.384	319.1	-4.7	64.9	5.1	-2.4	65.4	5.2
3795	181.6991	-0.3562	2013.6205	14.192	17.677	35.295	24.2	-74.5	-14.1	5.5	-73.9	-18.8	5.7
3796	181.7040	42.1942	2013.8335	14.676	14.83	5.989	42.6	-70	-27.7	5.2	-70.3	-26	5.2
3797	181.7650	0.2142	2013.6585	13.976	17.601	7.347	193.2	-173.9	21.4	6	-173.2	17.6	6.1
3798	181.7970	83.9091	2014.0345	10.613	13.488	6.533	339.9	84.5	6.8	5.4	84.4	7.6	5.4
3799	181.7991	63.0295	2013.9065	13.69	15.147	5.037	10.6	-142.2	-41.9	5.9	-145.6	-38.9	5.9
3800	181.8314	40.8972	2013.806	14.631	18.409	13.968	301.2	-114.4	-11.5	5.2	-114.5	9.5	5.4
3801	181.8766	50.6012	2013.804	13.705	17.877	13.918	20.6	-38.9	-90.2	5.5	-37.1	-89.7	5.5
3802	181.8890	13.4251	2013.638	15.822	16.764	4.321	278.3	-98.7	-18	5.2	-96.9	-11.8	5.4
3803	181.8956	25.5758	2013.642	15.469	16.465	5.165	39.3	-78.3	-50.1	5.2	-78.5	-47.9	5.3
3804	181.9028	15.5507	2013.738	14.275	14.731	14.54	104.5	-100.1	-44.6	5.1	-101.4	-42.7	5.1
3805	181.9162	1.6016	2013.813	9.149	14.222	9.084	265.3	-217	-161.9	6	-219.4	-159.6	6.2
3806	181.9473	68.5824	2013.693	12.655	17.101	37.484	125.2	-64.2	-26.9	5.5	-64	-27.4	5.6
3807	182.0265	-6.6197	2013.963	16.822	17.681	20.515	307.2	-71	-46	5.6	-66.4	-46.9	5.8
3808	182.0712	41.8898	2013.8065	11.691	13.824	11.999	65.8	-83.4	-7	5.2	-80.8	-8	5.2
3809	182.0726	-0.6193	2013.7515	10.568	11.49	49.08	245	29.2	-81.2	5.5	31.2	-82.6	5.5
3810	182.0810	30.9260	2013.833	9.844	12	14.118	328.6	-54.3	-75.1	5.1	-52	-77.2	5.2
3811	182.1770	21.7776	2013.7285	15.57	15.658	7.375	95.7	-128.6	-29.3	5.5	-127	-32.1	5.5
3812	182.2403	0.1222	2013.802	17.382	18.179	53.447	183.2	-77.3	-3.6	6.1	-73.6	3.2	5.7
3813	182.2479	3.9483	2013.916	16.218	17.648	18.548	194	4.8	-65.6	6	0.5	-67.8	5.9
3814	182.3077	31.0257	2013.8865	9.06	12.524	11.169	358.2	93.7	-101.7	5.1	93	-99.2	5.2
3815	182.3177	1.2812	2013.6825	17.03	18.013	13.094	72.2	-66.1	15.9	6	-65.7	18.4	6.1
3816	182.3355	14.4095	2013.5875	13.878	15.74	13.558	205	-21	-65	5.1	-22.3	-65.2	5.2
3817	182.3381	6.6545	2013.6805	14.104	16.313	8.512	74.3	-70.7	35.2	6	-72.2	33.6	6

3818	182.3522	30.7066	2013.66	18.7	18.772	56.467	272.6	-35	-72.1	6.7	-39	-69.1	5.8
3819	182.3549	20.5269	2013.5875	14.717	17.402	23.303	267.1	91.6	-64.7	5.9	92.2	-63.1	6.1
3820	182.3621	31.4103	2013.9105	12.541	13.718	5.492	130.8	30	-118.3	5.2	30.9	-117.3	5.2
3821	182.3673	62.9783	2013.8565	8.995	15.702	25.233	9.2	104.8	-43.1	5.5	101.5	-49.6	5.5
3822	182.4094	24.8034	2013.6615	14.207	16.584	7.193	125.1	-74.5	-170.6	5.3	-73.6	-170.4	5.3
3823	182.4191	54.1960	2013.963	10.551	14.413	6.503	311.5	-91	-66.1	5.4	-85.8	-66.6	5.5
3824	182.4234	28.3022	2013.723	17.291	17.792	13.473	193.9	-172.9	-48.8	5.5	-172.5	-50.5	5.3
3825	182.4325	35.5954	2013.838	12.333	15.088	24.735	308.4	-62.1	-39.7	5.1	-65.2	-38.9	5.2
3826	182.4344	79.1188	2013.746	10.951	18.185	19.15	77.9	-106	-73.1	5.9	-108	-72.7	6
3827	182.4949	52.3015	2013.8735	14.838	15.67	5.194	151.9	-66.3	-10	5.5	-69.1	-8.6	5.4
3828	182.5367	20.0208	2013.7565	14.473	14.907	4.739	27.5	57.5	-70.6	5.9	57.1	-75.3	6
3829	182.5448	42.9039	2013.8535	10.064	13.943	19.802	277.2	-103.9	-36.7	5.9	-103.7	-36.1	5.9
3830	182.5909	10.4371	2013.5025	12.116	15.557	32.031	195.8	-99.3	50.3	6	-91	59.1	6.2
3831	182.5913	-6.4668	2014.14	10.462	10.551	23.347	282.7	-46.8	-125.1	5.4	-42.7	-127.7	5.4
3832	182.5965	0.6165	2013.793	13.485	14.169	25.653	26.6	-73.5	-25.7	5.9	-72.5	-29.1	5.9
3833	182.6360	33.2703	2013.8305	13.74	15.252	13.77	303.9	-90.2	-15.3	5.9	-88.4	-13.7	5.9
3834	182.6468	30.9332	2013.8475	11.182	15.473	50.467	11.4	-120	1.8	5.9	-119	6.5	6
3835	182.6830	38.7903	2013.8755	11.773	15.099	7.153	207.2	85.6	-45	5.2	85.6	-45.5	5.2
3836	182.7389	-7.3229	2013.849	16.512	16.783	7.564	350.6	-71.3	18.7	5.6	-70.7	14	5.7
3837	182.7792	49.2096	2013.677	13.71	15.348	11.45	115.9	-62.4	-2.1	5.5	-62.1	-3.6	5.5
3838	182.7802	-5.8277	2013.9765	13.473	16.376	11.415	104.9	41.8	-138.7	5.5	45.9	-141.6	5.6
3839	182.7972	33.2609	2013.82	10.976	11.084	11.774	342.2	-166.6	30.2	6	-167.8	28.4	6
3840	182.8305	49.3546	2013.6885	12.809	13.732	7.765	211	-89.2	-149.6	5.5	-83.6	-145.8	5.5
3841	182.8559	58.6115	2013.717	14.311	17.077	10.723	220.3	-119.4	-19.5	5.9	-117.4	-23.1	5.9
3842	182.8573	57.9636	2013.798	11.205	18.191	22.558	303.3	43.9	-108.7	5.6	32.5	-112.4	6
3843	182.8800	-8.5705	2014.306	10.686	15.928	35.41	329.5	-75	3	5.4	-75.2	0.1	5.4
3844	182.8851	40.1272	2013.719	15.057	16.734	14.85	288.8	-102.3	8.3	5.2	-105.4	9.3	5.2
3845	182.9037	11.0637	2013.364	17.969	18.576	5.392	40.8	-63.7	-32.7	6.3	-63.4	-33.4	7.2
3846	182.9346	-1.4228	2013.8925	12.558	12.57	6.238	138.2	16.6	-65.4	5.7	15.6	-65.9	5.5
3847	182.9470	63.9273	2013.6775	12.353	14.56	40.662	222.8	-115.2	-39.6	5.6	-119.5	-39	5.6
3848	182.9551	64.2511	2013.903	13.636	13.691	5.069	190.8	44.8	-82.1	5.6	44.6	-79.3	5.5
3849	182.9605	3.0647	2013.7035	11.422	12.775	6.461	335.7	-67.4	-0.7	6	-68	1.6	6
3850	182.9713	38.7676	2013.8955	8.309	10.426	29.493	330.9	75	-45.2	5.1	73.5	-45.1	5.2
3851	182.9750	34.3964	2013.7375	18.219	18.931	37.519	136.5	-74	-34.8	6.2	-67.1	-39.5	7
3852	182.9936	81.1010	2013.8005	15.143	16.734	19.532	257.2	-71	44.7	5.9	-74.9	43.4	5.9
3853	183.0088	38.8193	2013.741	13.788	17.039	18.542	9.4	-139.7	-29.9	5.2	-143.9	-31.6	5.3
3854	183.1512	23.7889	2013.779	13.657	14.24	27.829	173.1	-88.2	-58.1	5.9	-90.1	-57.3	5.9
3855	183.1624	13.0519	2013.62	10.834	12.478	5.056	241.4	-112.1	8.7	5.9	-109.3	13	6.4
3856	183.1752	30.5553	2013.91	14.342	16.192	6.827	230	59.3	-92.5	5.9	58.8	-93.2	5.9
3857	183.1785	26.2995	2013.7585	12.297	13.562	16.853	343.2	-73.5	-6.9	5.2	-74.1	-8	5.2
3858	183.2345	49.0719	2013.812	11.488	12.431	5.144	301	-80.4	-188.3	5.5	-77.3	-188	5.6
3859	183.2781	13.3297	2013.573	15.492	16.006	4.502	31.7	-110.6	-29.1	6.1	-109.8	-28.9	6.2
3860	183.2820	-12.7262	2014.2745	13.347	14.857	13.804	166.8	7.6	-181.8	5.2	7.6	-181.3	5.4
3861	183.2952	-3.5828	2014.032	10.39	14.669	8.988	169.4	-102.9	14	5.4	-107.8	15.1	5.4
3862	183.2998	18.9784	2013.867	11.052	11.543	9.992	43.2	-68.3	-29.4	5.9	-66.6	-29.4	5.9
3863	183.3215	50.1567	2013.705	13.547	15.774	9.238	94	93.4	-101.1	5.5	91.6	-102.9	5.5

3864	183.3236	34.7937	2013.7615	11.512	16.913	18.334	293.7	-116.8	-48.2	6	-123.9	-46.9	6
3865	183.3287	29.3050	2013.8025	12.008	14.681	11.767	323.4	-83.7	-17.1	5.1	-83	-14.8	5.1
3866	183.3664	30.9511	2013.859	12.422	17.294	13.209	291.6	-71.7	9.2	5.9	-74.8	13.4	6
3867	183.4100	3.9717	2013.9685	8.757	17.965	21.692	173.1	-75.8	10.3	5.9	-76.1	13.1	6.4
3868	183.4133	13.1291	2013.513	14.829	15.678	14.128	142.5	45.2	-79.7	6.1	44.2	-75.8	6.1
3869	183.4563	10.3779	2013.5855	12.235	12.3	18.093	80.2	-185.5	20.2	6.1	-183	21.8	6.1
3870	183.4654	69.4119	2013.7025	11.83	13.842	13.771	188	-120.8	-13.7	5.6	-121.5	-12.8	5.6
3871	183.5267	32.7842	2013.9545	6.862	8.627	26.834	87.6	-105.9	-9.5	5.9	-103.3	-4.2	5.9
3872	183.5550	38.0322	2013.795	10.063	13.623	18.505	7.8	57.4	-69	5.9	52.2	-72.5	5.9
3873	183.5727	27.8673	2013.835	10.698	14.465	10.043	346.5	-74.2	-11.6	5.8	-73.6	-10.6	5.8
3874	183.6070	27.4001	2013.7095	14.586	15.425	28.255	333	-111	5.9	5.9	-113.2	7.1	5.9
3875	183.6141	8.7819	2013.8135	7.331	9.345	23.286	295.8	50.3	-130.9	6	54.8	-135.7	6
3876	183.6721	6.3668	2013.7575	10.914	14.008	7.45	167.5	-101.7	31.5	5.9	-106.7	33	6.1
3877	183.7054	-8.1093	2014.183	11.302	13.546	18.084	138.2	-68.9	0.1	5.4	-71.9	-3.3	5.4
3878	183.7338	58.4950	2013.7815	14.287	17.27	13.144	155.7	89.2	-94.2	5.6	84.6	-96.5	5.6
3879	183.7626	39.2400	2013.7155	13.182	16.891	32.466	48.8	-181	-136.7	5.3	-181.5	-137.5	5.5
3880	183.7650	48.9630	2013.6605	16.708	17.093	10.71	20.8	-68.4	25.1	5.5	-65.4	30.5	5.6
3881	183.8249	4.5839	2013.9185	9.362	12.175	7.479	138.6	-79.5	1.5	5.9	-76.8	4.9	5.9
3882	183.8691	19.6530	2013.785	11.691	13.59	5.552	293	-72.9	-76.8	5.2	-66.5	-78.6	5.3
3883	183.8774	-5.4006	2013.893	15.349	16.795	9.412	73.6	-79.7	7.7	6	-83.8	5.1	6.1
3884	183.9240	-3.6379	2013.9185	11.93	16.861	32.788	188	-75.9	57	5.9	-79	53.3	6.3
3885	183.9711	22.1559	2013.685	15.512	16.533	30.486	316.6	-67.8	-6.6	5.3	-70.2	1.3	5.3
3886	184.0141	56.3999	2013.786	13.482	15.317	5.553	163.1	-75.7	-3.3	5.6	-74.9	2.6	5.6
3887	184.0176	7.7963	2013.5775	13.254	16.112	32.94	300.8	-134.4	-14	6	-132.9	-9.8	6.1
3888	184.0395	18.5718	2013.7305	12.761	14.072	4.728	64	72.6	-42.5	5.2	69.1	-42.9	5.4
3889	184.0445	-1.4026	2013.6115	15.386	16.185	5.089	274.8	-70.1	-24	6	-69.3	-23.1	6.1
3890	184.0456	4.7107	2013.8265	8.717	10.111	45.972	181.4	-132.6	-12.3	5.9	-127.9	-9.9	6
3891	184.1011	35.6959	2013.8345	10.967	13.611	24.909	29.2	-142.1	0.6	5.2	-145.2	1.9	5.2
3892	184.1353	13.9881	2013.8125	11.918	14.551	43.39	187.7	86	-45.3	5.9	81.6	47	6
3893	184.2267	69.0258	2013.7215	11.171	13.69	12.574	303.6	80.3	30.2	5.9	79.4	26.2	5.9
3894	184.2532	-1.6209	2013.73	17.469	18.261	8.868	6.1	-69.8	-20.4	5.6	-66.1	-22.3	5.6
3895	184.2731	29.9803	2013.7285	16.691	18.08	5.612	334.9	18.2	-89.6	5.2	15.1	-86.5	5.3
3896	184.2732	55.8581	2013.786	11.515	15.524	10.06	89.3	-101.3	-0.3	5.6	-99.9	-0.1	5.6
3897	184.3183	47.0685	2013.676	15.469	16.38	5.158	253.8	-85.2	37	5.9	-85.8	36	6
3898	184.3465	6.2536	2013.6895	15.288	16.179	5.42	238.6	-27.2	-74.5	6	-26.7	-74.7	6
3899	184.3497	22.0451	2013.688	13.535	17.475	16.622	17.6	-160.5	-54.1	5.3	-162.8	-52.2	5.5
3900	184.3659	2.8875	2013.787	9.776	13.228	29.664	293.5	-50.1	-91.3	6.4	-58.7	-89.6	6.4
3901	184.4921	27.8577	2013.761	11.539	15.74	36.993	330.6	-77.7	15.6	5.5	-80.4	15.3	5.5
3902	184.5526	25.8550	2013.8115	13.993	16.312	4.954	139	-64.4	-4.9	5.5	-66.4	5.3	5.9
3903	184.5819	1.2666	2013.676	14.691	17.135	41.153	64	-62.5	3	6.4	-64.7	10.2	6.6
3904	184.6003	-5.0322	2013.978	8.511	10.97	53.042	281.4	25.1	-146	5.4	27.4	-153.3	5.5
3905	184.6093	79.6470	2013.684	15.82	17.493	10.446	12.6	-98.7	-21.3	5.6	-94.1	-21	5.6
3906	184.6266	-6.5220	2013.8475	16.274	16.978	20.977	116.8	-89.5	-17.9	5.7	-93.6	-9.7	6.2
3907	184.6681	14.8532	2013.7125	15.682	16.58	57.687	194.8	-62.1	3	5.2	-63.8	0.3	5.3
3908	184.6761	12.7288	2013.51	16.165	16.784	40.041	267.7	-65.2	-4.6	5.3	-66.8	-12	5.4
3909	184.7108	34.9348	2013.666	16.248	16.713	22.262	233.7	-64.3	-3.7	5.2	-62.4	-5.2	5.2

3910	184.7360	13.9150	2013.75	10.625	15.667	14.122	214.2	-93.7	-32.9	5.2	-96.5	-34.4	5.3
3911	184.7867	36.7826	2013.8175	16.154	16.548	4.109	275.2	27	-76.8	5.2	26.2	-76.2	5.4
3912	184.8701	1.7555	2013.662	14.071	17.855	39.495	290.1	-62.6	-30.3	5.5	-65.3	-27.6	5.7
3913	184.9094	43.3083	2013.6685	14.265	15.206	7.682	272.9	-93	-54.7	6	-92.3	-55.9	6
3914	184.9223	44.3383	2013.792	10.587	14.81	14.943	249.8	56.1	-88.9	5.9	55.2	-92.8	5.9
3915	184.9718	13.6544	2013.8815	10.082	13.925	7.644	279.6	-114.9	21.2	5.2	-117.1	18.2	5.2
3916	184.9834	62.0314	2013.6725	14.187	14.734	7.948	58.1	-35.7	-100.9	5.6	-34.7	-100.2	5.6
3917	184.9956	4.4567	2013.977	9.335	13.245	6.53	33.8	-132.4	-70.5	5.9	-130.1	-70.9	6.2
3918	184.9981	52.6851	2013.946	11.15	15.134	6.378	71	-86.4	-31	5.5	-91.7	-25.7	5.6
3919	185.0559	37.9021	2014.05	6.656	9.989	10.048	192.4	-65.1	-2.4	5.1	-63.4	1.5	5.1
3920	185.1699	27.0531	2013.93	7.284	7.698	8.926	65.4	13.2	-114.6	5.9	9.4	-114.1	5.8
3921	185.2056	51.7317	2013.6815	12.296	16.489	27.065	138.8	-61.1	24	5.6	-60.9	26.2	5.6
3922	185.2703	44.5853	2013.741	13.759	15.401	17.542	301.9	28.8	-69.3	5.9	26.4	-69.9	6
3923	185.3909	0.0258	2013.712	11.912	15.842	7.983	287.2	-71.5	-29.8	5.5	-72.2	-32.1	5.6
3924	185.4503	46.2897	2013.582	16.887	17.042	33.739	83.1	-73	11.8	6.1	-71.2	8.3	6
3925	185.5316	6.5614	2013.7315	11.668	14.216	6.334	120.3	-108.2	17.1	6	-111.4	16.6	6
3926	185.6205	10.9254	2013.4005	17.238	17.415	6.794	313.9	-74	36.9	6.2	-72.4	34.7	6.3
3927	185.6356	-5.4069	2013.9955	10.76	12.39	10.37	175.6	-122.2	73.8	5.5	-123.5	76.2	5.5
3928	185.6847	3.4113	2013.8405	12.978	15.979	39.716	338.9	-127.2	51.2	5.9	-122.8	55.2	6
3929	185.6953	2.7716	2013.837	12.087	12.249	40.818	53.3	-71.6	-19	5.9	-75.4	-18.6	5.9
3930	185.6989	23.1755	2013.715	12.251	16.185	10.654	227.4	-121.5	-1.4	5.2	-119.2	-0.2	5.3
3931	185.7116	25.7235	2013.6545	15.226	16.928	5.226	5.4	-96	-37.4	5.9	-93.7	-40.2	6.1
3932	185.7633	-8.8715	2014.3205	11.405	12.171	4.931	248.3	-90	-5.6	5.3	-92.1	-4.3	5.3
3933	185.7884	23.3912	2013.8105	10.457	11.019	17.403	342.5	-61.6	-19.1	5.2	-63.4	-20.1	5.2
3934	185.7917	4.7043	2013.701	12.944	14.668	4.89	233	-166.5	47.6	6.1	-173	42.1	6.5
3935	185.8181	66.7606	2013.6705	14.204	16.031	33.451	313.1	-84	-44.4	5.6	-85.3	-45.9	5.6
3936	185.8557	-10.7118	2014.347	14.757	15.769	13.698	82.8	6.5	-61.3	5.3	4	-62.7	5.6
3937	185.8920	9.1757	2013.6785	10.942	14.695	11.662	32	90	-18.3	6	89.7	-20.2	6
3938	185.8989	-4.8047	2013.759	11.949	17.117	10.198	155.3	-121.8	11.9	5.5	-126.1	11.7	5.8
3939	185.9170	-11.3442	2014.3035	12.144	16.158	24.198	321.1	-76.7	-1.8	5.4	-77.2	-6.3	5.8
3940	185.9305	6.4194	2013.593	13.216	16.885	23.399	159.5	-184.3	-35.7	6	-189.2	-34.6	6.3
3941	185.9507	-1.9375	2013.5935	15.396	16.777	39.808	169.6	-105	-14.3	5.6	-100.4	-12	5.7
3942	186.0040	36.1961	2013.8475	17.578	18.053	23.191	23.7	-69.4	-11.6	5.2	-67.1	-18.4	5.3
3943	186.0371	24.5651	2013.801	11.414	11.8	12.21	100.7	-93.8	-41.2	5.8	-93.2	-38.5	5.8
3944	186.0498	3.9302	2013.8035	9.781	14.368	37.167	292.6	-11.7	-130.4	5.9	-14.7	-130.6	6
3945	186.0630	66.9104	2013.6515	14.813	16.57	11.069	207	-62.1	19.3	5.6	-62	19.3	5.6
3946	186.0836	53.3706	2013.793	16.779	17.037	27.112	33.3	-67.7	-51.6	5.5	-68.7	-48.9	5.5
3947	186.1096	-8.8087	2013.9735	16.673	16.829	10.128	5.8	-20.6	-91.9	5.7	-23.4	-87.7	5.8
3948	186.1475	49.6483	2013.6955	15.734	17.224	5.764	13.9	-69.5	-103.3	5.6	-71.9	-100.9	5.6
3949	186.1626	39.7154	2013.751	14.911	15.245	4.544	250.3	-77.6	0	5.2	-79.3	3.9	5.3
3950	186.2216	-2.8554	2013.849	11.035	15.357	41.689	85	-61.4	-9	5.4	-62.5	-8.5	5.5
3951	186.2581	15.4120	2013.6415	14.969	17.673	20.397	92.2	-112	16.8	5.2	-108.5	15.7	5.3
3952	186.2738	60.4184	2013.6695	11.734	11.959	19.996	65.5	-188.8	-11.1	5.9	-190.8	-11.5	5.9
3953	186.3566	31.4855	2013.7975	14.272	17.562	6.857	158.2	-52.7	-66.1	5.8	-53.6	-64.4	5.9
3954	186.3903	-5.2193	2013.9175	11.579	13.188	46.312	268	47.9	-78.7	5.4	51.6	-81.6	5.4
3955	186.4786	47.6652	2013.645	15.441	15.678	6.741	123.6	-63.4	3.8	5.9	-63.2	6.2	5.9

3956	186.4939	45.8668	2013.7925	14.032	16.16	14.512	308.9	-155.5	-50.9	5.8	-157.4	-48.2	5.9
3957	186.5393	27.4489	2013.661	15.263	15.762	47.707	231.7	-92	-55.6	6	-93.2	-56	6.1
3958	186.5464	53.0716	2013.8095	9.994	14.553	7.448	234.9	-87.2	-35.9	5.5	-82.7	-36.1	5.5
3959	186.5551	3.9088	2013.8395	10.239	12.862	7.314	53.7	-64.4	51.1	5.9	-66.6	53.4	6
3960	186.5714	10.9688	2013.531	16.562	17.089	8.895	100.1	-86.3	-65.1	6.1	-88.9	-60.3	6.2
3961	186.6002	2.5680	2013.732	15.116	15.478	28.319	74.3	3.3	64.1	5.9	-1.3	64	5.9
3962	186.6007	12.6664	2013.625	13.922	17.328	16.59	113.8	37.3	-77.3	5.8	28.6	-76.5	6.1
3963	186.6387	-2.1687	2013.7735	12.986	13.212	12.717	336.2	51.1	-117.1	5.5	54.1	-116.8	5.5
3964	186.7341	28.2681	2013.9435	3.902	12.009	16.109	206.4	-67.8	-101.8	5.9	-74.7	-102.1	5.9
3965	186.7537	-3.5353	2014.0305	8.153	9.9	24.742	146.4	128	-106.5	5.4	130.4	-100	5.5
3966	186.7889	15.7824	2013.6115	14.709	15.189	6.815	74.3	-70.4	31.1	5.5	-70.1	33.2	5.5
3967	186.8083	31.3253	2013.81	10.495	15.829	10.665	219.3	-143.6	-37.9	5.2	-140.9	-35.9	5.3
3968	186.8200	79.3071	2013.7605	14.042	16.359	8.632	273.5	-80.5	31.9	5.6	-82.4	34.4	5.6
3969	186.8477	21.8006	2013.792	13.072	14.323	12.128	186.1	-80.2	22.3	5.2	-79.3	20.8	5.2
3970	186.8642	-2.7475	2013.826	14.56	15.032	6.995	203.9	125.1	-79.1	5.4	128.8	-78.2	5.5
3971	186.8996	31.3918	2013.7125	15.625	15.926	11.601	359.3	-78.5	0.4	6	-80.8	0.7	6
3972	186.9370	75.8989	2013.6855	11.734	17.044	35.266	338.1	-141.7	-128.6	5.6	-140.3	-134	5.7
3973	186.9500	9.5889	2013.6855	13.73	13.913	11.003	303.3	-91.1	99.7	6	-95.7	99.5	6
3974	186.9927	31.1531	2013.749	12.423	12.525	45.43	273.9	23.9	-178.6	6	21.4	-177.2	6
3975	187.0171	23.1436	2013.8405	9.702	15.519	54.378	232.5	-185.6	22.7	5.8	-184.3	12.4	5.9
3976	187.0176	44.7943	2013.8355	7.432	7.804	9.634	157.4	-170.9	-1.9	5.8	-176.5	1.4	5.6
3977	187.0596	21.2714	2013.771	12.733	16.102	20.152	317	-185.3	-166	5.9	-181	-173.7	6.1
3978	187.0711	31.7074	2013.822	11.964	12.017	17.853	218.8	-71.7	1.8	5.9	-72.8	1.7	5.9
3979	187.0755	12.3763	2013.6335	12.337	18.102	15.957	25.4	-170.3	-117.3	5.5	-167.8	-106.6	6.4
3980	187.0880	65.3570	2013.866	11.38	15.253	6.425	171.2	-73.4	22.1	5.6	-71.3	25	5.6
3981	187.1477	47.4639	2013.6825	13.702	15.731	8.188	5.5	-52.5	-81.6	5.6	-51.8	-78.8	5.6
3982	187.1524	11.2890	2013.792	11.041	12.069	22.04	359.7	-86.7	-35.5	5.9	-89.1	-33.8	5.9
3983	187.1643	-0.5807	2013.729	11.167	11.64	24.853	277.6	23.8	-82.9	5.5	24.4	-85.7	5.5
3984	187.1900	50.9134	2013.7605	12.959	14.716	17.965	168.7	-70.5	-136.3	5.5	-68.4	-134.2	5.5
3985	187.2348	-4.5988	2013.7105	15.699	15.702	4.348	108.2	-79.4	-41.4	5.7	-82.2	-38.7	5.6
3986	187.2429	4.4191	2013.624	14.171	15.374	12.977	248.7	-64.7	-158.4	6.1	-68.1	-164.4	6.1
3987	187.2923	41.1393	2013.689	15.217	17.474	16.435	98.5	57	-104	6	60	-96.3	6.1
3988	187.3064	37.3077	2013.7575	17.908	18.311	5.921	37.1	-62.3	13.1	6.2	-61	13.7	6.2
3989	187.3831	56.9190	2013.7375	11.039	15.289	36.723	162.4	134	-61	5.6	136.1	-58.4	5.6
3990	187.4264	27.9885	2013.744	12.542	17.139	8.661	191.8	-68.6	-27.9	5.9	-69.6	-27.4	6
3991	187.4367	63.2015	2013.677	14.796	15.938	5.215	70.3	-57.1	88.5	5.9	-56.7	85.7	6
3992	187.4625	46.6206	2013.9495	15.801	15.802	3.564	205.7	-76.3	8	5.9	-72.6	15.8	5.7
3993	187.4739	4.1638	2013.7915	16.548	16.779	4.028	4.4	26	-121.1	6.1	23.3	-124.8	6.3
3994	187.5039	22.9553	2013.8185	12.242	12.45	58.286	152	-69	21.3	5.8	-71.4	24.1	5.8
3995	187.5073	85.3026	2014.006	14.314	14.379	14.035	309.8	-79.6	52	5.4	-78.8	51.7	5.4
3996	187.5076	-6.0237	2014.0685	7.396	13.575	18.876	292	-112.4	-59.9	5.4	-108.7	-65.1	5.5
3997	187.5332	-9.3177	2014.1595	14.878	15.119	3.381	309.2	-160.4	-52.3	5.8	-161.2	-51.9	6.7
3998	187.5372	-6.8847	2014.0335	9.25	14.499	49.336	333.3	-73.1	-21.5	5.4	-75.5	-20.8	5.5
3999	187.5515	39.2467	2014.049	7.72	11.779	6.203	18	-115.1	-109.2	6	-118.2	-112.9	5.9
4000	187.6466	47.4326	2013.605	14.369	16.634	8.226	253.5	-24	-69.8	5.9	-25.5	-66.8	5.9
4001	187.6587	-10.6940	2014.2295	10.703	14.656	20.042	256	-30.6	-67.2	5.9	-35.1	-68.5	5.8

4002	187.6927	36.6597	2013.9235	10.988	11.238	5.165	172.7	-111.7	16.7	5.9	-111.9	17.3	5.9
4003	187.7097	54.1334	2013.7425	15.199	16.169	8.82	139.7	-96.4	7.1	5.6	-94.9	5.1	5.6
4004	187.7282	8.1122	2013.5495	15.155	16.269	25.253	302.2	-63.1	11.8	6	-65.6	13.2	6.1
4005	187.7349	-10.4360	2014.305	14.925	15.854	13.405	72.6	-69.2	-1.3	5.9	-68.6	-2.3	5.9
4006	187.7407	2.2281	2013.7845	17.151	17.696	8.412	292.6	-8.6	-61.8	6	-11.3	-64.3	6.1
4007	187.7465	41.2609	2013.683	13.804	15.197	19.363	169.7	-60.6	18.1	6	-61.3	16.8	6
4008	187.7720	61.2732	2013.6055	10.109	13.698	22.77	218.8	-102.1	187.5	6	-105.6	194.9	6
4009	187.8063	-3.5521	2013.519	11.408	18.172	38.951	252.1	25	-84.6	5.5	26.7	-80.5	7.1
4010	187.8137	25.9332	2013.6935	12.086	14.832	11.935	222.2	-118.7	10.7	6	-119.2	9	6
4011	187.8338	16.3185	2013.7705	10.961	17.176	17.619	54	-93	20.9	5.8	-96.7	20.8	6
4012	187.8771	17.1059	2013.7085	14.451	18.026	7.561	47.3	-93.2	1.4	5.8	-93.6	-4.9	6.3
4013	187.8930	25.6137	2013.622	15.612	15.773	10.899	298.6	-79.9	54.5	6.1	-79.8	53	6
4014	187.9304	63.9086	2013.744	11.986	15.382	17.866	122.8	-98.8	-7.9	5.9	-95.8	-10.3	5.9
4015	187.9522	-2.3256	2013.513	14.986	16.144	11.483	121.4	-65.1	-37.2	5.6	-67.8	-34.6	6
4016	187.9531	28.2439	2013.815	13.167	14.253	4.927	177.9	-65.5	13.1	6	-64.6	10.9	5.9
4017	187.9708	17.1460	2013.8495	11.392	16.071	9.929	90.6	-139.3	68.6	5.1	-139.5	66.2	5.1
4018	187.9790	35.9679	2013.7785	15.084	18.163	6.648	234.6	-68.9	-64.8	6	-74.5	-61.9	6.3
4019	187.9883	37.3787	2013.8135	11.204	13.123	7.886	92.7	-93.9	38.6	6	-97.1	40.4	6
4020	188.0044	64.6981	2013.994	13.489	18.355	24.231	15.3	-124.7	37.1	5.5	-126.9	41.5	6
4021	188.0149	14.3356	2013.9065	10.389	11.573	30.443	207.3	-90	1.6	5.1	-92.6	2.6	5.1
4022	188.0152	37.5098	2013.717	16.617	18.072	6.44	62.8	-137.5	-47.3	6.1	-137	-39.5	6.3
4023	188.0549	31.7889	2013.745	13.28	14.567	10.426	310	-76.1	22.3	5.2	-76.9	21.1	5.2
4024	188.0838	1.9986	2013.6685	16.356	17.599	7.708	222.3	-114.3	0.7	6.1	-111.7	-0.1	6.1
4025	188.1281	30.8414	2013.7195	13.585	13.776	17.992	120	-110.4	-27.3	5.2	-108.8	-22.5	5.2
4026	188.1302	11.0528	2013.6625	17.632	18.339	39.008	0.2	16.4	-66.3	6.1	19.7	-65.9	6.6
4027	188.1822	38.9566	2013.93	13.121	14.635	4.235	196	-65.6	55	6	-69.3	55.9	6.7
4028	188.1844	28.0615	2013.8225	14.541	18.221	55.442	336.4	-72.4	-13.8	6	-75.7	-13.7	6.2
4029	188.2241	-3.5144	2013.6855	15.028	15.131	7.284	76.3	-77	-2	5.6	-78	-0.6	5.5
4030	188.2816	36.2393	2013.746	14.333	14.617	13.714	247.8	70.8	-66.6	6	69.9	-66.5	6
4031	188.3020	70.0523	2013.7185	12.544	13.589	26.294	26.9	-82.6	11.4	5.6	-83	13.1	5.5
4032	188.3275	68.7730	2013.76	12.818	17.885	25.987	162.7	-113.8	-2.9	5.6	-115.7	-1.8	5.7
4033	188.3399	44.0153	2013.6255	14.003	16.467	7.386	231.6	8.5	-99.2	5.9	6.2	-97.2	5.9
4034	188.3499	-1.5185	2013.657	12.639	15.324	5.436	21	-150.3	-19.2	5.6	-150.2	-23.6	6.1
4035	188.3519	14.2630	2013.8765	12.958	14.878	35.098	262.8	-100.2	-113.2	5.1	-99.8	-113.5	5.1
4036	188.3577	52.4495	2013.7735	13.713	14.649	17.864	356.6	-129.7	-52.5	5.6	-133.9	-52.1	5.6
4037	188.3878	51.2729	2013.6865	15.232	16.812	8.758	57.4	-75.7	-34.9	5.6	-73.2	-33.1	5.7
4038	188.4138	1.6145	2013.629	12.969	18.307	46.542	228	13.6	-139.7	5.9	14.1	-137.7	7.9
4039	188.4889	6.0050	2013.7755	15.013	16.342	17.065	347.9	60.7	-57.1	6	66.2	-55.4	6.6
4040	188.4935	14.6886	2013.939	10.196	10.842	12.039	3.8	-168.7	-18.6	5.1	-164.7	-16.7	5.1
4041	188.5468	-3.5305	2013.6075	13.564	15.426	9.127	116.9	-61.8	-30.1	5.5	-64.6	-29.9	5.6
4042	188.5667	0.8581	2013.549	15.274	15.875	12.012	167.1	25	-60.2	6	22.2	-60.2	6.1
4043	188.6515	47.4382	2013.6005	15.445	16.428	7.685	184.7	-66.9	-41.4	5.3	-66.2	-42.2	5.3
4044	188.6790	25.8244	2013.772	12.702	14.419	7.64	164.8	-92.6	54.8	6	-89.8	56.4	6
4045	188.7488	20.4407	2013.7695	15.03	16.967	32.185	191.5	-71.1	19.6	5.6	-69.6	21	5.6
4046	188.7768	25.7855	2013.748	13.757	13.793	9.293	271	45.4	-65.2	6	45.4	-66	6
4047	188.7783	7.5297	2013.621	13.48	18.282	54.101	61.6	-116	-38.9	6	-116.8	-33	6.4

4048	188.8242	16.8153	2013.8355	15.034	15.542	4.392	124.6	-109.4	31.3	5.1	-109.9	31	5.2
4049	188.8474	44.5886	2013.645	17.795	17.932	7.593	313.3	-101.8	-1.4	5.4	-97.7	-10.5	5.4
4050	188.8863	17.8712	2013.7205	14.945	16.573	22.926	325.9	70.1	-57.8	5.2	62.2	-64.4	5.7
4051	188.9036	80.7306	2013.8465	15.328	18.061	9.441	52.6	-26.1	-83.2	5.6	-30.2	-81	5.7
4052	188.9155	29.2757	2013.664	13.206	16.479	12.849	105.6	70.8	-25.2	6	70	-25.3	6.1
4053	188.9182	7.5176	2013.585	12.89	15.467	22.909	283.3	-111.3	41.6	6	-112.8	38.3	6.1
4054	188.9248	37.4549	2013.887	11.646	13.244	5.095	200.7	-75.3	-6.2	6	-72.2	-0.1	6.6
4055	188.9314	-12.0249	2014.2965	7.864	8.186	27.527	351.2	-150.9	32.2	5.2	-146.9	30.1	5.1
4056	188.9492	0.5611	2013.7465	13.997	17	10.353	74.9	-31.5	-137.4	5.9	-28.3	-134.8	6.1
4057	188.9680	-9.0550	2014.22	8.784	12.591	30.765	133.1	-68.7	41.4	5.3	-67.4	44.6	5.3
4058	188.9843	-10.6066	2014.322	12.423	13.016	4.246	78.2	-42.3	-78	5.3	-45.1	-81.1	7.8
4059	188.9860	37.9673	2013.5545	14.248	14.926	7.839	178.2	-107.3	-4.4	6.1	-107.6	-4.8	6.1
4060	188.9919	54.6203	2013.7695	12.05	15.695	6.721	160.9	-75.9	2.6	5.9	-75.7	2	5.9
4061	189.0135	62.4954	2013.6155	14.706	16.873	16.457	302.9	5.4	-91.5	5.6	4.6	-92	5.6
4062	189.0663	21.2211	2013.7865	9.835	17.038	37.058	311.4	-173.1	-114.7	5.6	-173.6	-122	5.9
4063	189.0836	11.5045	2013.5975	16.856	17.353	8.635	35.4	-192.1	-39.1	6.3	-197.6	-42.3	6.4
4064	189.1048	31.7515	2013.799	13.704	14.766	14.179	182.7	-119.7	12.5	5.2	-120.5	14.1	5.2
4065	189.1566	47.1074	2013.7285	13.009	15.028	6.055	120.5	50.9	-155.7	5.6	47.2	-151.2	5.6
4066	189.1799	1.8361	2013.807	15.245	15.395	14.361	297.6	-26.7	-82.5	5.9	-32.1	-83.8	5.9
4067	189.2105	-6.8496	2014.071	13.517	15.473	18.58	313.1	77.5	-53.1	5.4	74	-57.6	5.5
4068	189.2501	34.2630	2013.8625	14.87	16.877	5.422	52.8	-65.6	-13.6	5.1	-63.6	-15.4	5.4
4069	189.2763	15.7596	2013.5945	16.131	16.831	21.977	259.4	-132.4	-17.4	5.3	-137.5	-15.7	5.3
4070	189.2978	20.1572	2013.9055	9.627	12.114	15.809	284	-87.4	-13.9	5.5	-87.4	-14.4	5.5
4071	189.3300	-0.2516	2013.3515	16.894	17.938	15.237	129.8	-86.8	-63.2	5.9	-79.6	-74.4	7.3
4072	189.3578	57.2679	2013.9115	18.41	18.48	40.132	106.8	-64.6	-5.2	6.4	-62.4	-11.8	6.4
4073	189.3630	29.8784	2013.6775	12.234	16.226	9.524	263.6	-165.1	-92.7	5.2	-170.2	-96.7	5.3
4074	189.3665	-7.4022	2014.1795	11.343	13.93	6.975	215.2	-63	18.8	5.4	-60.6	18.1	5.4
4075	189.4656	-5.8798	2014.054	14.443	15.371	4.911	136.5	-81.9	1.3	5.5	-80.6	2.7	5.8
4076	189.4773	17.9690	2013.816	14.329	17.605	42.57	154.2	-75.1	-31.3	5.5	-77.1	-35.2	5.6
4077	189.5264	74.9502	2013.682	13.077	15.686	11.734	13.2	-90.7	-6.8	5.6	-89.2	-10.5	5.6
4078	189.5431	42.5620	2013.664	11.759	15.76	14.607	97.9	-29.2	-69.5	5.3	-29.9	-70.8	5.3
4079	189.5549	23.3758	2013.6685	16.382	17.747	18.401	307.3	-4.8	-62.4	5.6	-3.1	-64.8	5.7
4080	189.5592	37.3230	2013.821	14.156	17.574	8.229	329.5	-66.5	-26.4	5.6	-66.5	-28.6	5.5
4081	189.6176	13.7926	2013.852	12.522	13.167	7.181	197.9	-80	-23.9	5.1	-80	-23.6	5.1
4082	189.6519	14.4064	2013.754	12.22	16.064	23.401	344.6	8.7	-63.8	5.1	8.8	-63.9	5.2
4083	189.7850	-1.5519	2013.865	10.219	10.994	6.92	52.3	-159.3	-11.3	5.5	-161.9	-10.2	5.5
4084	189.8193	1.4846	2013.724	11.061	13.245	6.377	297	65	-65.5	5.9	67	-65.3	6
4085	189.8534	19.0682	2013.9455	15.684	18.353	24.566	305.3	-77.3	23.5	4.9	-78.6	27.8	5
4086	189.8724	62.7356	2013.628	12.113	14.058	8.867	130.6	-66.9	-40.6	5.6	-67.3	-38.5	5.6
4087	189.8940	18.2880	2013.841	14.986	16.602	44.586	282.1	-69.8	-76.5	4.9	-68	-77.1	5
4088	189.9090	58.6494	2013.739	13.857	17.413	14.518	137.1	-3	-64.9	5.5	-5.4	-67.2	5.6
4089	189.9304	49.3945	2013.767	11.594	13.761	5.64	312.5	-60.3	-1.3	6	-61.7	-1	6
4090	189.9462	37.1888	2013.783	11.141	11.769	6.15	272.3	-75.9	-22.4	6	-72.6	-20.5	6
4091	189.9813	75.4107	2013.5345	18.658	19.004	39.862	127.2	0.9	-63.5	6.9	4.7	-66.2	6.2
4092	190.0052	-4.5729	2013.6485	14.277	14.537	22.067	205.3	-66.1	23.6	5.5	-67.2	25.2	5.5
4093	190.0348	37.3623	2013.742	12.715	15.202	33	76.6	118.6	-194.2	6	117.3	-193.7	6.1

4094	190.0412	1.1538	2013.602	15.064	18.338	9.448	243.8	-85.9	-81.1	6	-79.5	-84.2	7
4095	190.0414	-1.1335	2013.571	11.586	16.127	12.805	32	33.8	-82.2	5.5	29.4	-81.2	5.7
4096	190.0422	3.4329	2013.672	14.347	17.636	32.122	297.2	-63.7	8.4	5.9	-65.9	-0.1	6.1
4097	190.0455	55.0571	2013.7495	9.785	11.876	13.547	246	-70.5	-15	5.5	-69.5	-7.4	5.5
4098	190.0793	21.5555	2013.7655	11.618	15.297	14.863	97.6	-69.3	36.7	5.9	-71.5	27.5	6
4099	190.0935	30.6961	2013.6635	15.634	16.08	4.988	22.7	-69.5	-11.7	5.6	-71.2	-10.1	5.7
4100	190.1248	-0.5336	2013.671	15.211	15.395	4.458	100.2	-79.7	61.1	5.7	-77.6	59.5	5.6
4101	190.1493	34.7392	2013.775	14.333	14.479	11.145	104.4	-20.2	-65.1	5.6	-19.8	-68	5.6
4102	190.1558	40.2899	2013.9225	9.557	12.537	38.086	341	-24.3	61.1	6	-25.1	64.5	6
4103	190.1668	-2.1606	2013.321	17.092	18.101	10.382	4.7	-76.2	-25.3	5.9	-78.4	-29.9	7.4
4104	190.1760	-3.0408	2013.9385	11.28	11.346	6.917	41.4	37	-63.3	5.5	39.9	-63.5	5.4
4105	190.1796	46.4485	2013.628	15.519	17.245	11.514	257.2	-28.3	-73.2	5.3	-30.1	-76	5.3
4106	190.1946	8.0860	2013.5995	15.255	16.418	5.326	244.1	62.2	-34.7	6	66.5	-33.3	6.3
4107	190.2078	3.8696	2013.8755	7.691	13.164	49.29	60.3	-72.2	-24.7	5.8	-69.4	-35.7	5.9
4108	190.2647	43.3800	2013.7125	10.412	14.897	15.773	175.6	61.1	-143.7	5.3	66.7	-142.9	5.3
4109	190.2710	43.2585	2013.826	11.483	11.903	5.071	359.1	-69.5	20.6	5.3	-71.5	16	5.2
4110	190.2750	50.5050	2013.9675	10.141	12.508	4.929	254.6	-125	-15.1	6	-123.7	-16.6	5.9
4111	190.3012	50.6386	2013.6495	14.744	16.238	13.047	138.1	-73	30.1	6	-71	26.3	6.1
4112	190.3470	-5.8420	2013.6905	13.663	15.331	42.665	68.8	-72.3	-96.1	5.6	-72.4	-96.8	5.6
4113	190.3924	-7.2195	2014.044	12.352	13.518	7.326	77.5	-85.3	37	5.6	-83.2	36.1	5.6
4114	190.4393	21.5287	2013.698	14.882	16.683	25.719	183.4	19.7	-64.5	4.9	17.5	-63.1	5
4115	190.4755	25.3162	2013.751	16.094	16.433	4.611	212.2	-81.5	-51.1	5.6	-78.2	-51.4	5.6
4116	190.5002	-9.6842	2013.596	16.559	18.027	40.322	270.9	-15.3	-69.8	5.6	-8.7	-70.9	6
4117	190.5164	52.4631	2013.699	15.307	16.312	13.813	8.2	-149.2	-23.5	6.1	-148.7	-22.7	6
4118	190.5599	15.9727	2013.8695	11.966	16.3	25.421	301.8	-138.7	-15.4	5.1	-143	-13.7	5.1
4119	190.5772	81.0822	2013.861	12.687	13.168	7.109	20.7	71.1	-46.6	5.9	71.2	-50.5	5.9
4120	190.5898	-6.5166	2013.5265	12.94	16.625	10.374	206.9	-96	-7.7	5.5	-93.5	-1.6	5.8
4121	190.5919	29.2482	2013.928	13.975	14.916	4.231	310.1	-70.5	12.6	5.5	-69.1	8.6	5.5
4122	190.5976	24.2141	2013.6855	15.982	16.95	24.171	141.3	-80.4	4	5.5	-81.7	8.5	5.6
4123	190.5990	6.6947	2013.8455	14.351	14.614	4.42	351.2	-97.5	-24.1	6	-99.3	-25.8	5.9
4124	190.6533	55.7552	2013.765	11.68	15.009	6.568	247.8	-75	-85.8	5.5	-75	-85.8	5.5
4125	190.6956	2.7443	2013.735	10.591	15.776	14.772	338.2	-203	-66.9	6	-203.4	-65.4	6
4126	190.6984	31.5561	2013.821	14.568	14.665	6.109	2.7	-92.6	-0.7	5.9	-90.4	-1.6	5.9
4127	190.7559	21.2900	2013.868	10.33	16.284	12.804	136.9	-169.3	-135.4	5.9	-172.6	-136.2	6.1
4128	190.7701	10.1672	2013.7685	11.459	14.753	35.066	27.4	-116.5	-57.1	6	-122	-56.8	6
4129	190.7763	50.1260	2013.8995	17.28	19.117	17.565	118.2	-65.3	-8.6	5.6	-66.2	-12.8	6.3
4130	190.7913	-12.0146	2014.2945	9.718	10.752	5.606	77.4	-86.9	30	5.6	-91.7	23.9	5.6
4131	190.8881	15.3217	2013.6155	15.717	17.417	7.312	64.6	-89.3	-73.7	5.2	-90	-71.9	5.3
4132	190.9724	14.2386	2013.857	14.541	15.873	5.444	226	-66	22	5.1	-64.2	24.6	5.3
4133	191.0020	85.4492	2013.965	8.568	11.258	28.073	147.8	-129.7	43.9	5.5	-126.1	42.4	5.5
4134	191.1320	-8.2437	2013.815	14.401	16.875	9.78	191.2	-130.2	-52.2	5.6	-132.6	-50.3	5.8
4135	191.1439	40.3630	2013.749	11.319	17.716	11.428	209.3	-45.6	-133.3	6	-45.7	-131.8	6.2
4136	191.1816	-3.4129	2013.736	13.054	15.451	5.656	341.4	-50.1	-139.7	6.1	-49.1	-132.7	6.3
4137	191.2108	19.4368	2013.863	12.442	15.17	8.166	241.8	-77.2	12.4	5.8	-78.9	17.6	5.8
4138	191.2120	77.6718	2013.777	11.042	14.364	51.379	295.8	-21.7	-164.3	5.5	-22.6	-162.2	5.5
4139	191.2253	-6.7294	2013.778	16.356	16.595	4.924	272.6	-70.7	6.2	5.5	-69.6	3.9	6.2

4140	191.2450	42.8035	2013.551	14.081	16.562	22.618	70.4	-103.1	-47.3	5.6	-101.5	-44	5.7
4141	191.2754	11.2196	2013.576	14.563	15.697	4.653	301.1	34	-96.2	6	42.8	-98.1	6.4
4142	191.2867	12.9021	2014.039	9.973	12.899	5.029	297.9	82.8	-37.5	5.1	81.2	-39.4	5.4
4143	191.2955	47.2908	2013.6785	12.265	16.801	18.377	124.8	70.5	-93.8	5.6	71.8	-92.5	5.7
4144	191.3457	1.0179	2013.578	12.456	14.835	5.663	331.4	-190.8	-19.6	6.4	-189.6	-16.9	6.6
4145	191.3470	60.3816	2013.629	11.581	16.047	11.709	15.7	-72.8	41.7	5.6	-71.4	38.5	5.6
4146	191.3637	26.7294	2013.704	11.948	13.554	4.683	6.6	-68.5	21.9	5.6	-69.7	19.4	5.6
4147	191.3842	-1.9417	2013.6245	11.333	15.747	7.608	120.9	15.3	-61.3	6	3.2	-65.4	6.6
4148	191.3865	71.5648	2013.6245	15.811	15.989	7.546	188.6	-120.3	3.8	5.6	-115.4	6.5	5.6
4149	191.4091	19.6886	2013.68	14.911	15.84	4.438	1.7	-44.5	-71.7	4.9	-46.2	-71.8	5.5
4150	191.4148	43.2649	2013.69	11.536	14.033	9.785	133.6	-117.5	-40.4	6	-115.2	-39.6	6
4151	191.4603	63.3327	2013.6505	15.61	15.847	15.045	144	-160.1	81.6	5.6	-160.2	82.5	5.6
4152	191.4941	2.1399	2013.866	15.599	18.437	10.187	288.3	-94.7	-3.3	6	-97.1	-4	6.7
4153	191.5132	30.0863	2013.6795	14.366	17.179	8.083	42.2	-71.9	12.2	5.2	-69.8	15.3	5.2
4154	191.5223	-0.9815	2013.6465	12.27	16.7	12.085	192.8	-114.2	-47.3	6	-116.9	-47.5	6.2
4155	191.5257	-7.4245	2013.8885	10.481	14.582	10.435	153.4	-61.3	-7.3	5.9	-60.1	-5.7	6
4156	191.5461	3.2630	2013.7775	14.274	16.104	6.395	82.8	67.1	-70.1	5.9	60.9	-71.6	6
4157	191.6393	41.1198	2013.635	12.773	15.803	57.145	152.6	105.8	-43.1	6	107.2	-43.6	6.1
4158	191.6640	47.6212	2013.7215	12.582	18.226	8.949	323.4	-16.1	-63.8	6	-18.9	-66.5	6.4
4159	191.6642	3.5412	2013.4525	15.368	16.892	17.457	170	-61.3	-3.8	6	-60.5	-7.7	6.1
4160	191.6921	31.8161	2013.6835	12.263	16.343	23.779	149.7	-61.5	18.5	5.2	-63.1	20	5.2
4161	191.7057	29.9922	2013.7675	12.198	14.445	8.347	230.6	-68.6	44.2	5.2	-71.1	42.2	5.2
4162	191.7064	55.1019	2013.678	15.317	17.492	7.741	202.4	5.1	-70.8	5.6	11.4	-66.7	5.6
4163	191.7522	32.8409	2013.861	10.757	12.353	54.069	20.1	38.4	-62	5.1	36.3	-60.7	5.1
4164	191.8296	2.4942	2013.786	13.639	15.456	18.885	83.9	-85.1	-17.4	5.9	-87.8	-19.7	5.9
4165	191.8405	17.5681	2013.666	16.611	17.999	12.979	58.7	-65.2	4.7	5.2	-65.8	10	5.5
4166	191.8718	19.6591	2013.7645	16.206	17.625	10.583	11.9	-60.7	-19.5	6	-63.8	-16	6.1
4167	191.8945	7.3786	2013.499	15.497	15.902	6.331	265.4	63.4	-84	6.1	64.1	-81.1	6.1
4168	192.0166	25.2834	2013.702	16.562	17.017	5.217	12.2	-86.1	-18.1	5.7	-83.9	-20 7	5.7
4169	192.0729	-12.9212	2013.2755	16.276	17.486	17.974	314.4	-60.1	27.6	5.5	-64.7	25.1	7.5
4170	192.1202	-7.7138	2013.8215	14.729	16.564	56.47	65.5	-80.7	18.4	6	-81	19.6	6.1
4171	192.1875	40.5027	2013.6695	12.81	13.974	6.725	315.8	-94.6	-11	6	-92.7	-6.9	6
4172	192.2044	35.8733	2013.8115	12.445	15.542	7.339	130.3	-68.5	-29.1	5.2	-68.6	-27.5	5.1
4173	192.2523	24.0222	2013.729	15.025	17.329	27.961	155.4	-63.7	34.9	5.6	-63	39.2	5.6
4174	192.2895	-5.8965	2013.9545	15.169	15.695	7.906	282.6	-124.4	-12	5.5	-125.3	-5.5	5.5
4175	192.3063	74.1294	2013.795	12.471	17.23	7.497	22.4	-135.1	5.7	5.6	-136.3	-3.4	5.7
4176	192.3134	72.5355	2013.6295	15.598	18.735	14.216	211.3	-68.4	31.7	5.6	-70	27.7	6.1
4177	192.3439	40.3743	2013.617	13.397	14.226	10.965	306.5	-63.1	5.7	6	-61.7	6.7	6
4178	192.3483	0.0126	2013.852	14.741	15.321	4.136	45.3	-35	-73.6	5.5	-38.4	-72.8	5.6
4179	192.3583	4.2596	2013.666	13.783	14.773	5.888	287.6	-67.7	0.1	6	-70.2	3.2	6
4180	192.4048	3.9587	2013.6775	12.395	13.81	14.056	289.8	-170.6	8.3	6	-173.6	5.7	6
4181	192.4148	20.8978	2013.8545	12.338	13.355	11.09	158.8	-186.5	103.8	5.6	-186.9	103.3	5.6
4182	192.4542	7.0832	2013.7245	14.041	14.237	18.91	200.3	-31	60.3	6	-31.8	60.4	5.9
4183	192.4693	26.7180	2013.669	15.703	17	15.08	281.6	-80.2	24	5.9	-79.9	22.7	6
4184	192.4830	0.2505	2013.7485	11.444	14.102	9.369	32	-63.2	24.4	5.9	-64.5	25	5.9
4185	192.5148	36.2948	2013.767	13.572	15.91	9.481	255.9	-110.1	45.9	6	-112.6	44.5	6

4186	192.5704	67.2662	2013.668	11.373	14.874	7	223.7	-111.7	4.9	5.6	-109.2	2.3	5.6
4187	192.6408	71.7099	2013.6615	12.64	14.974	18.974	123.8	-62.3	26.1	6	-62.5	23.3	6
4188	192.7035	7.9651	2013.6185	10.505	16.691	27.992	293.7	49.5	-168	6	46.9	-170.6	6.3
4189	192.7301	-4.4808	2013.8635	11.946	13.889	7.946	269.2	66.6	-131.4	5.5	67.6	-131.2	5.5
4190	192.7461	49.4066	2013.617	16.973	18.212	5.291	101.6	-63.9	-18.4	5.5	-66.9	-8.5	6
4191	192.7711	-3.8567	2013.9805	17.217	17.505	22.846	147.5	-14.8	-82	5.8	-4.8	-83.9	6.8
4192	192.7716	1.8682	2013.946	9.583	11.896	8.207	52	-87.9	-2.2	5.8	-83.8	-8.7	5.9
4193	192.7909	19.1493	2013.83	11.069	14.226	10.662	296.9	-66.1	39.7	5.6	-66.6	39.4	5.6
4194	192.8011	15.6914	2013.932	13.58	14.381	5.594	227.6	-77.5	0.9	5.1	-77	-0.4	5
4195	192.8152	-9.2038	2013.883	9.658	14.89	18.001	211.6	33.6	-65.1	5.5	35.4	-62.7	5.5
4196	192.8485	23.8742	2013.8725	13.575	14.623	5.961	128.5	-89.3	-10.6	5.6	-89.9	-7	5.5
4197	192.9606	24.1972	2013.699	13.713	15.257	45.268	117.4	-76.9	-43.6	5.5	-73.8	-48.9	5.6
4198	192.9807	-4.3286	2014.013	11.191	14.843	14.626	336.3	-23.2	-86.3	5.4	-26.9	-88.5	5.4
4199	193.0103	51.0803	2013.6725	12.594	16.082	8.11	338.6	44.6	-70.8	5.5	44.2	-71.1	5.5
4200	193.0339	37.9245	2013.689	14.117	15.793	16.644	206.9	-134.3	17.8	5.9	-133.2	17.4	5.9
4201	193.0683	38.5921	2013.72	13.021	13.061	10.631	339.1	-84	115.8	5.9	-75.2	121	5.9
4202	193.1309	40.6618	2013.7435	15.253	16.007	6.551	251.4	-1.9	-68.6	5.9	-4.7	-68.8	5.9
4203	193.1594	-5.1572	2013.744	14.835	17.678	11.039	112.2	-62.9	-38.7	5.5	-66.2	-38	5.7
4204	193.2330	-12.6623	2013.46	12.692	16.016	41.952	36.1	-106	6.5	5.4	-107.1	8	5.5
4205	193.2453	3.8821	2013.6455	11.332	17.413	12.958	45.5	-75.7	-2.9	6	-75	-7.8	6.1
4206	193.2544	48.2564	2013.701	14.127	14.598	7.506	120.8	27.3	-124.5	5.9	28.4	-123	5.9
4207	193.2651	40.4474	2013.6805	13.155	17.521	11.429	3.4	36	-70.6	5.2	35.8	-67.1	5.3
4208	193.3636	10.3892	2013.774	11.31	14.42	8.462	111	30	-71.8	5.9	20.7	-71.4	6
4209	193.3681	36.2085	2013.7575	13.157	14.208	15.054	278	-70.1	3.4	5.2	-69.9	3.6	5.2
4210	193.3702	29.1378	2013.792	12.455	14.042	5.669	262.4	-75.9	4.8	5.8	-76.6	2.9	5.8
4211	193.4291	4.7464	2013.9485	10.196	14.195	6.281	152.8	-72.1	22	6	-72.2	22.9	5.9
4212	193.4554	7.6679	2013.2075	17.518	18.112	6.313	128.1	-162.1	-2.2	6.6	-156.8	-1.7	6.8
4213	193.5008	25.6018	2013.6645	11.197	16.665	15.224	105.4	-12.3	-63	5.8	-10.7	-60.5	5.9
4214	193.5277	55.0085	2013.673	16.035	16.057	10.777	278.4	-108.5	-3.5	5.6	-106.9	-5.7	5.6
4215	193.5735	78.4340	2013.849	16.908	17.479	3.989	194.2	-76.2	11.8	5.6	-73.8	15.9	6
4216	193.5837	25.8616	2013.6775	14.172	16.843	7.644	88.3	-110	68.9	5.9	-106.1	64.7	6
4217	193.5858	-10.9849	2013.775	14.744	15.018	54.156	31.5	-84.9	27.6	5.5	-84.5	25	5.5
4218	193.7399	13.7949	2013.683	15.172	15.534	39.529	331.2	60.1	-29.9	5.1	60.8	-29.2	5.2
4219	193.7873	-9.4168	2013.431	15.233	17.098	6.902	152.7	-69.7	-0.5	5.6	-66.8	-7	6
4220	193.7980	26.3013	2013.754	14.746	15.102	5.326	354	-152.2	-68.2	6	-150.5	-67.4	6
4221	193.8195	15.2933	2013.6145	13.611	15.114	4.127	339.8	-69.2	-2.1	5.1	-68.6	1.9	5.3
4222	193.8435	18.7740	2013.906	18.107	18.22	13.527	340.5	-33.6	-71	5.7	-35.3	-74	5.9
4223	193.9049	6.3478	2013.6365	14.805	15.193	37.77	286	-145.8	32.7	6.1	-140.9	34.9	6.1
4224	193.9183	56.2081	2013.7305	14.684	16.64	5.831	207.2	-92.1	43.7	5.6	-90.9	47.3	5.6
4225	193.9451	5.0063	2013.535	14.743	17.169	9.973	26.7	-83.1	-6.1	6	-85.7	-8.9	6.1
4226	193.9570	-3.4949	2013.882	13.936	15.992	28.623	202.7	21.9	-65.9	5.4	23.9	-68.4	5.5
4227	193.9681	16.8750	2013.8545	9.45	12.495	51.33	89	-110	-24.8	5.1	-101.8	-34.5	5.1
4228	193.9748	49.7074	2013.7165	16.866	16.894	4.362	341.3	-88.8	28.3	6	-90.7	25.1	5.9
4229	194.0026	55.6544	2013.7205	15.75	17.325	14.157	349.1	-185.1	43.9	5.6	-186	38.6	5.7
4230	194.0199	7.8933	2013.6595	11.388	13.865	16.742	235.1	-106.1	3.8	6	-110.7	4.2	6
4231	194.0296	6.1793	2013.7195	11.531	16.195	8.863	84.5	10.1	-92.7	6	3.6	-92.5	6.1

4232	194.0533	0.8351	2013.991	9.955	11.579	5.52	97.5	-96.9	-12.4	5.8	-94.2	-19.7	5.9
4233	194.0875	6.8927	2013.8585	8.49	14.265	16.019	64.3	-162.6	-114	5.9	-160.5	-106.5	6.1
4234	194.1048	36.3068	2013.6715	15.291	17.209	31.481	349.4	25.6	-66.3	5.2	29.1	-67.6	5.3
4235	194.1388	21.0114	2013.8015	9.081	17.099	35.384	244.5	43.5	-164.9	5.5	49.2	-158.9	5.7
4236	194.1438	3.7768	2013.583	14.137	17.612	26.618	316.2	-77.4	8.5	5.9	-80.6	10.5	6.3
4237	194.1980	11.7723	2013.31	17.455	17.465	4.681	25.4	-73.4	-16.8	6.3	-74.8	-22	6.4
4238	194.2304	31.2224	2013.8215	12.037	15.897	7.92	73.1	-105.6	-39.9	5.9	-104.8	-40	6
4239	194.2971	1.6818	2013.7595	13.402	15.465	26.096	348.8	17.3	71.6	5.9	21.9	67.5	5.9
4240	194.3358	-1.7991	2013.867	14.595	16.389	7.498	138	-91.3	7.8	5.4	-87.7	9.7	5.5
4241	194.3779	13.2839	2013.519	15.514	17.441	40.875	276.3	-81.7	-8.8	5.2	-81	-10.1	5.3
4242	194.3874	65.5275	2013.5535	14.211	14.659	7.191	50.5	-200.4	17.9	5.6	-197.3	18.7	5.6
4243	194.4254	-2.7288	2013.5825	16.638	16.841	4.281	178.1	-92.4	-21.1	5.8	-93.6	-18.4	5.9
4244	194.4540	30.6215	2013.6495	11.639	13.737	5.542	281.7	76.9	-177.2	5.2	76.8	-172.2	5.3
4245	194.4876	86.6976	2014.0085	14.659	17.546	32.232	80.6	-78.8	-16.3	5.5	-78	-15.2	5.5
4246	194.4917	0.4520	2013.7045	16.861	17.853	9.094	293.5	-78.7	-51.8	6.1	-85	-48.2	6.1
4247	194.4975	4.8291	2013.705	17.606	18.146	38.979	350.5	-65.3	-16.4	5.9	-61.8	-20.7	6.2
4248	194.5427	38.2385	2013.801	18.021	18.99	59.543	17.3	3.8	-67.9	6	-3.2	-67.2	7.6
4249	194.5610	5.6109	2013.6545	11.905	15.676	22.421	143	112.5	-62	6	110.4	-62.7	6
4250	194.6326	38.2786	2013.7805	8.34	8.788	36.035	82.1	-126.5	-51.2	5.2	-126.4	-48.1	5.2
4251	194.6541	31.4159	2013.8155	12.221	15.304	8.752	261.5	-101.7	-46.9	5.2	-102.3	-49.6	5.1
4252	194.7611	9.4062	2013.898	13.79	14.582	3.961	209	-136.5	-45	6	-133	-49.7	6.2
4253	194.7927	54.6818	2013.669	11.12	13.902	54.335	168.9	-171.2	14.8	5.7	-171	13.2	5.7
4254	194.8721	39.1656	2013.7145	11.514	15.979	43.781	129.5	-62.5	39.2	5.2	-60.9	38.8	5.2
4255	194.9170	51.3481	2013.796	10.764	13.04	5.775	206.2	-67.9	-15	5.9	-68.1	-12.3	5.8
4256	194.9702	31.5138	2013.725	11.387	15.861	8.135	218.9	10	-106.2	5.2	12.8	-103.6	5.2
4257	194.9770	36.3348	2013.752	14.171	16.415	13.209	168.6	-17.1	-112.5	5.2	-21.2	-111.1	5.2
4258	195.0140	35.1713	2013.8275	13.7	15.969	5.198	235.8	-148.2	26.6	5.2	-151.4	21.4	5.7
4259	195.0177	50.4540	2013.615	14.946	16.029	4.481	62.3	-196.1	-39.3	6	-200.7	-37.9	6.3
4260	195.0312	3.7833	2013.6405	12.914	17.444	10.1	165	-73.9	-43.8	5.9	-76.7	-38 2	6.2
4261	195.0363	-10.6876	2013.333	17.237	18.068	49.784	87.1	-93.7	-15.8	5.9	-90.8	-17.5	7
4262	195.0743	40.0492	2013.754	10.564	13.9	6.36	138.6	-62.1	-13.8	5.2	-62	-13.3	5.2
4263	195.0808	10.6763	2013.8005	11.572	13.128	8.645	256.4	-64.2	-18.2	5.9	-64.7	-17.3	5.9
4264	195.1225	3.0565	2013.713	10.879	11.096	5.404	215.9	-117.9	21.3	5.9	-116.3	19.1	6.3
4265	195.1253	28.9943	2013.787	14.207	14.797	35.124	310.2	-68.2	-22.5	5.9	-68	-18.9	5.9
4266	195.1307	46.1348	2013.634	14.477	16.529	23.238	20.9	-112.9	49.6	5.5	-112.4	49.1	5.6
4267	195.1320	-10.0799	2013.752	15.085	15.384	35.009	190.1	-60.7	14.5	5.5	-63.2	15.4	5.6
4268	195.1487	3.4616	2013.631	16.792	17.248	23.146	68.3	-64	32.9	6.1	-62	30.6	6.1
4269	195.1538	5.9163	2013.771	11.258	16.662	9.582	122.5	-90.6	-81.2	6	-95.8	-82.7	6.3
4270	195.1779	73.6854	2014.134	9.474	9.699	4.542	90.6	-38.5	-108.5	5.5	-45.2	-111.7	5.7
4271	195.1793	12.8159	2013.8305	9.599	12.596	15.791	188.5	-76.3	-7.8	5.1	-76.6	-5.9	5.1
4272	195.1936	14.3770	2013.9155	8.752	9.679	26.807	187.7	5.5	-69.7	5.1	6.7	-71.3	5.1
4273	195.2252	9.6607	2013.714	13.816	14.552	24.417	171	46.7	-116.8	6	48.8	-122.6	6
4274	195.2260	26.2219	2013.745	12.265	16.485	9.784	194.6	-87.4	-32.1	5.9	-88.1	-30.9	6
4275	195.2739	9.2243	2013.7335	11.372	14.134	30.92	42.4	-140.3	115.6	6	-145.9	117.9	6.1
4276	195.3417	9.3180	2013.8175	13.782	15.366	4.817	228.2	-5.5	-62.8	6	-6.9	-61.2	6
4277	195.4180	36.9653	2013.7545	13.469	14.468	9.919	65.8	-72.6	0	5.2	-72.3	0.3	5.2

4278	195.5323	7.2798	2013.82	8.988	10.014	11.476	294.1	-121.2	19.3	6	-119.9	20.2	5.9
4279	195.5554	23.9300	2013.596	13.457	17.898	53.075	308.8	8.8	-63	5.5	4.7	-64.6	5.9
4280	195.5862	-5.7553	2014.037	14.724	15.934	6.73	195.8	-59.6	-82.3	5.4	-61.8	-80.3	5.5
4281	195.5950	41.8696	2013.339	13.696	16.981	6.147	17.9	-92.9	-6.6	5.2	-95.9	-1.4	5.8
4282	195.6239	7.7862	2013.467	16.135	18.087	12.835	156.1	-74.3	-12.6	6.1	-76.8	-10.8	6.3
4283	195.6825	-11.3604	2013.442	16.74	16.927	58.837	10.6	-64.4	20.2	6	-65.7	17.5	6
4284	195.6936	-5.9544	2013.9785	11.205	14.47	24.98	269.1	-85.8	-142.4	6	-86.1	-142.7	6
4285	195.6978	-1.9849	2013.9225	11.104	11.687	14.843	111.2	-70.1	-20	5.9	-67.7	-22.6	5.9
4286	195.7155	32.6321	2013.768	11.245	16.577	56.331	96.2	-77	20.3	5.2	-74.2	18.7	5.2
4287	195.7556	30.7103	2013.8035	12.993	13.501	22.976	7.9	-129.8	-102.8	5.2	-131.3	-99.4	5.2
4288	195.8350	67.7819	2013.633	17.334	18.221	33.236	112.6	-26.4	-67.8	5.7	-14.3	-69.1	7
4289	195.8658	42.8363	2013.6115	14.675	17.082	43.145	170	-55.9	106.7	5.5	-55.8	109.5	5.6
4290	195.8769	37.6597	2013.7495	12.972	14.112	5.626	252.3	-71.5	-81.2	5.2	-69.7	-79	5.2
4291	195.8864	11.8440	2013.6375	12.418	18.083	29.207	334.1	-120.5	53.5	5.9	-125.3	52	6.7
4292	195.8914	47.5943	2013.702	11.586	15.069	36.668	250.3	-31.6	-132.1	5.5	-33.3	-128.9	5.5
4293	195.9346	13.5937	2013.687	12.559	16.339	22.253	284	-76.6	-2	5.1	-77.5	-1.1	5.2
4294	195.9372	21.5728	2013.552	16.804	17.859	9.336	104.7	-165.9	-98.4	5.8	-169	-96.3	6.1
4295	195.9378	40.8878	2013.725	10.649	10.686	12.096	190.8	-149.1	26	5.2	-148.2	25.1	5.2
4296	195.9583	32.1667	2013.7665	14.751	15.007	9.814	232.8	-10.7	-67.7	5.2	-11.1	-66.9	5.2
4297	195.9741	49.0124	2013.7065	11.158	12.079	10.326	343	-29.4	-87.8	5.6	-30.1	-84.3	5.6
4298	195.9757	67.8546	2013.613	15.737	17.292	13.473	25.5	-68	-17.5	5.6	-67.6	-12.5	5.7
4299	195.9957	22.8214	2013.6065	16.43	16.56	53.951	196.1	-87.2	19.8	5.7	-88.2	18.6	5.6
4300	196.0135	-3.5918	2013.9905	12.047	16.291	47.874	63	-98.4	5.7	5.4	-102.2	9.1	5.5
4301	196.1135	-8.7475	2013.9465	12.324	14.409	41.008	164.8	-125.4	-27.1	5.5	-127.2	-26.5	5.5
4302	196.1274	-6.2189	2013.9165	11.67	12.857	15.593	31.7	-166.9	43.6	5.5	-165.2	45.7	5.5
4303	196.1558	13.3537	2013.879	10.299	11.861	7.658	142.2	123.5	-82.8	5.1	124.8	-83.6	5.1
4304	196.2394	-11.0951	2013.8795	12.482	15.802	18.489	38.5	-108.2	-22.1	5.5	-104.6	-23.5	5.5
4305	196.2612	54.1842	2013.672	12.081	16.334	11.429	250.1	-89.8	-9.2	5.5	-90.3	-9.9	5.6
4306	196.3449	54.8734	2013.741	12.626	13.524	43.283	183.8	-86.9	19.6	5.5	-82.4	22.8	5.5
4307	196.3573	61.8952	2013.548	11.368	15.915	17.436	220.9	-75.7	22.5	5.6	-76.1	20.4	5.6
4308	196.3825	-14.4685	2013.273	13.219	16.756	14.721	32.4	-65.2	-10.9	5.4	-65.2	-10.4	5.8
4309	196.3874	-4.0938	2013.934	12.781	15.357	17.841	45.5	-19.9	-65.9	5.4	-16.2	-66.8	5.4
4310	196.4835	3.7264	2013.733	13.022	15.166	11	64.3	-85.1	8.3	5.9	-81.8	6.8	5.9
4311	196.4868	15.8199	2013.8355	11.246	11.772	27.406	205.6	-85.3	23.6	5.1	-84.3	20	5.1
4312	196.5424	29.0292	2013.817	6.964	10.169	6.149	222.5	-72.8	-19.9	5.9	-77.6	-15.9	6
4313	196.5424	9.3332	2013.624	15.827	17.537	15.308	246.7	-86.7	-24.9	6	-83.6	-25.2	6.2
4314	196.5479	70.4272	2013.889	12.67	13.439	3.878	102.3	62.9	12.6	5.7	60.8	8.2	5.8
4315	196.5553	-13.1697	2013.6715	15.609	15.856	10.227	268.9	-68.8	-17.8	5.3	-68	-17.1	5.4
4316	196.6191	28.3233	2013.763	12.38	13.359	6.702	151.3	74.4	-34.6	5.9	73.1	-32.8	5.9
4317	196.6317	22.4450	2013.581	14.919	16.679	7.322	244.1	-102.6	-10.2	5	-103.4	-6.3	5
4318	196.7131	35.9887	2013.809	14.155	15.782	5.139	148.4	12.1	-67	5.2	12.9	-64.3	5.3
4319	196.7264	56.2017	2013.718	13.413	13.816	9.922	344	-63.6	-60.7	6	-64.8	-59.1	6
4320	196.7848	-8.2472	2014.085	15.052	15.403	3.593	352.6	-66.8	-2.2	5.5	-67.7	-3.8	5.7
4321	196.8010	44.5018	2013.616	14.167	14.998	54.774	222.4	-136	-2.8	5.6	-132.1	-7.4	5.6
4322	196.8771	31.0005	2013.7525	12.92	15.085	11.999	222.5	-79.5	43.7	5.2	-78.3	45.6	5.2
4323	196.8953	-2.7524	2013.884	14.28	17.026	44.464	278.6	17.8	-85.1	5.5	15.5	-86.4	5.5

4324	197.0658	8.2127	2013.5045	15.825	16.613	11.82	329.7	-16.8	-84.5	6.1	-16.9	-82.1	6.1
4325	197.1008	35.8953	2013.756	15.17	16.923	16.54	353.8	-80.7	-33.7	5.5	-83.2	-30.7	5.6
4326	197.1209	-12.5902	2013.8155	10.172	15.119	10.818	65.6	34.2	-86.9	5.6	36.4	-85.2	5.7
4327	197.1251	25.8344	2013.6095	12.625	15.89	21.774	249	-122.1	53.7	5.7	-122.3	55.1	5.7
4328	197.1404	14.4085	2013.623	16.024	16.234	49.856	228.8	-88.1	-3.1	6	-92.4	-4.8	6
4329	197.2126	-1.8017	2013.9085	10.6	16.09	27.522	197.2	-75.5	2.8	5.4	-72.2	2.8	5.5
4330	197.2833	26.7744	2013.7765	13.306	14.461	8.032	160.7	-15.1	-102.2	5.6	-9	-104.4	5.6
4331	197.3858	44.9125	2013.578	16.186	16.653	4.797	43.8	-90.8	28.2	5.6	-89.2	27.2	5.6
4332	197.4054	45.3665	2013.7985	18.244	18.758	44.421	230	-73	14.1	5.6	-74	10.3	6
4333	197.4557	51.8370	2013.6825	11.74	14.965	15.538	86.8	-125.6	-10.3	5.6	-123.5	-13.8	5.7
4334	197.4619	3.3404	2013.916	16.209	16.215	5.342	267.9	-61.3	-123.6	6	-59	-123.6	6
4335	197.4624	-6.4022	2013.9425	14.472	15.859	10.591	194.3	-65.3	-12	5.4	-64.1	-12.9	5.5
4336	197.4702	31.8617	2013.757	10.347	16.975	52.614	235.9	-80.2	-38.2	5.2	-79.7	-36.6	5.2
4337	197.5269	9.7276	2013.7615	11.698	15.87	9.137	199.4	-62.4	2.8	5.9	-60.8	4.2	6
4338	197.5572	37.4116	2013.7195	11.932	15.711	23.886	13.3	22.6	-97.5	5.2	22.4	-99.3	5.2
4339	197.6019	17.4373	2013.713	11.61	17.047	13.693	221.4	-14.1	-77.6	5.1	-8.6	-76.1	5.2
4340	197.6063	-4.1769	2014.113	17.587	18.087	27.368	200.7	-76.9	-49.8	5.9	-77.4	-45.7	6
4341	197.6218	11.0443	2013.775	9.797	15.12	19.639	101.8	-121	-27.1	5.9	-119.8	-31.8	6
4342	197.6253	-6.6156	2013.9895	11.331	13.368	7.566	309.3	-82.9	-12.2	5.4	-81.1	-13.2	5.4
4343	197.6467	30.5494	2013.7585	13.517	17.132	8.263	295.9	-138.3	25.3	5.2	-137.2	28.2	5.3
4344	197.6641	7.2390	2013.7515	13.935	14.368	23.331	301.5	-78.8	-72.6	6	-80.8	-74	6
4345	197.6835	-3.2921	2013.95	8.621	12.372	13.371	6.7	95.3	-52.2	5.4	95.7	-56.2	5.4
4346	197.7224	54.6571	2013.5985	14.034	16.75	8.525	344.4	-76.3	-8.9	5.6	-78.5	-16.6	5.7
4347	197.7350	12.0094	2013.4155	17.188	17.408	3.944	155.3	-128.3	-34.4	6.2	-127	-28.2	6.6
4348	197.7376	85.1955	2013.8915	14.283	16.196	7.05	45.7	30.3	-61.2	5.5	29.7	-62.1	5.5
4349	197.7743	42.2868	2013.82	15.035	17.141	5.049	234	-60.8	-20.8	5.9	-60.4	-19.7	6.1
4350	197.8375	58.9370	2013.5975	12.706	15.01	34.072	234.9	-83.4	0.4	5.6	-80.7	0.8	5.6
4351	197.8392	-8.2049	2013.931	14.346	15.876	24.5	45.9	-117.5	6.7	5.4	-118.2	6	5.5
4352	197.9172	44.9703	2013.481	16.114	16.437	3.971	19.3	-114.5	-61.8	5.7	-113.5	-61.9	5.9
4353	198.0294	32.2213	2013.746	11.991	15.716	25.843	306.7	121.1	-88.4	5.2	126.7	-78.6	5.2
4354	198.1173	-4.8160	2013.8495	14.551	16.863	20.8	86.7	-75.2	-31.7	5.9	-76.8	-30.3	6.1
4355	198.1432	35.1293	2013.791	16.221	17.295	7.106	84.6	-61.2	2.8	5.2	-62.2	6.6	5.2
4356	198.1583	53.8546	2013.844	16.93	18.737	17.078	255.4	-64.5	-6.9	5.6	-67.7	-6.4	6.3
4357	198.1685	2.5418	2013.9355	12.065	13.97	5.394	15.9	-14.7	-179.4	6	-17.3	-182.5	5.8
4358	198.2406	3.9430	2013.6945	12.622	17.965	21.973	97.5	-77.3	7.5	6	-75.9	14.1	6.1
4359	198.2922	5.4255	2013.568	16.007	16.278	13.422	143.3	-78.5	-5.5	6	-81	-6.7	6.1
4360	198.2970	59.6361	2013.784	16.487	17.491	4.035	264.9	-113.1	-11.7	5.7	-106.8	-20.7	6.3
4361	198.3049	59.9630	2013.5455	8.977	15.751	20.08	134.4	-84	-42.2	5.6	-86.9	-44.9	5.6
4362	198.3068	-2.5559	2014.091	7.632	9.566	9.464	302.6	-121.6	-33.6	5.9	-119.4	-23.8	5.9
4363	198.3429	46.6895	2013.675	11.61	15.077	13.557	248.5	-78.4	36.6	6	-76.6	33.1	6
4364	198.3702	-12.2246	2014.146	10.05	14.157	8.385	242.6	-77.3	16.5	5.1	-77	16.9	5.1
4365	198.3783	16.9411	2013.682	15.578	15.854	55.944	76.5	-76	-1.3	6	-76.6	-1.6	6
4366	198.3893	47.7974	2013.632	9.519	16.711	55.086	206.7	53.6	-83.9	6	53.5	-83.2	6.1
4367	198.4141	59.2403	2013.602	12.399	17.166	16.214	290.1	-71.4	-26.1	5.6	-68.9	-24.9	5.7
4368	198.4307	-4.4907	2013.8945	12.115	13.968	20.13	211.7	-107.5	-62.3	6	-105.7	-67.8	6
4369	198.4379	30.1607	2013.7025	12.492	17.024	8.109	326.2	-112.8	-15.8	6	-110.2	-19.7	6.2

4370	198.4481	-2.5865	2013.935	14.628	15.154	4.994	332.1	121.9	-61.2	6	120	-61.5	6
4371	198.4546	78.3545	2013.7895	15.339	17.965	31.789	294.5	42.7	-64.2	5.5	46.5	-62.1	5.6
4372	198.4955	-5.9773	2013.8885	17.194	18.133	8.094	338.9	-6.7	-117.4	6.3	-4.9	-118.8	6.7
4373	198.5640	12.1680	2013.5845	9.971	18.487	44.656	325.3	-62.7	-12.6	5.8	-60.8	-5.2	6.4
4374	198.5789	17.7174	2013.6555	14.488	15.735	50.053	326	-80.6	-89.4	5.9	-80.7	-94.8	6
4375	198.6071	13.0945	2013.629	15.201	18.309	51.204	302.9	-74.2	-32.3	6.1	-77.8	-26	6.2
4376	198.6145	23.9674	2013.7875	12.719	13.9	13.536	55.3	3.2	-80.3	6	4.5	-80.1	6
4377	198.6607	14.2788	2013.798	14.536	14.83	5.096	181	-95.8	-70.5	5.9	-97.6	-69.6	5.9
4378	198.6754	37.2400	2013.587	14.364	18.265	39.117	267.1	-127.9	53.5	5.2	-123.4	51.3	5.7
4379	198.6952	25.7407	2013.7105	14.869	16.134	8.556	109.9	-104.5	6.4	6	-104.4	5.7	6
4380	198.7485	58.7288	2013.7205	14.068	18.304	35.365	327.3	-130.1	12	5.6	-131	12.8	6
4381	198.7923	84.7522	2013.934	7.396	12.636	17.229	339.2	-136.8	15.6	5.5	-134.1	8.7	5.5
4382	198.8501	28.7205	2013.73	11.349	14.057	11.363	83.9	-4.4	77.4	5.9	-5.1	80.6	5.9
4383	198.8566	-4.4209	2014.058	13.898	17.342	11.334	347	-90.1	28.9	6.3	-91.3	26.5	6.6
4384	198.8668	85.1386	2013.8905	14.918	15.739	7.15	116.1	-70.5	10	5.5	-71.8	6.7	5.5
4385	198.9085	30.7037	2013.7525	12.435	12.963	18.214	217.6	97.6	-67.6	6	97.5	-66.5	6
4386	198.9382	36.9318	2013.731	14.348	17.465	20.661	325	0.6	-68.4	5.2	2.5	-68.6	5.2
4387	198.9439	55.3569	2013.864	16.592	18.488	30.695	53.1	-106.8	15.5	5.6	-111.4	10	6
4388	198.9564	36.1889	2013.7225	12.639	13.698	39.71	279.2	-118	-120.5	5.3	-120.6	-114.2	5.2
4389	199.0595	15.9412	2013.7855	11.987	14.391	19.372	273.4	-72.4	-20.9	5.8	-74.8	-15.3	5.8
4390	199.0789	36.0520	2013.7505	12.483	14.109	26.104	267.1	39.6	-69.8	5.2	37.9	-74.2	5.2
4391	199.1187	48.0004	2013.6745	11.772	17.61	14.855	153.7	-10.4	-60.4	5.9	-12.9	-61.4	5.9
4392	199.1273	48.1594	2013.514	14.1	14.483	3.517	193.1	-103.2	36	6	-97.2	38.3	6.6
4393	199.1557	38.3047	2013.7495	11.376	12.27	6.702	12.7	-72.5	25.5	5.2	-72.9	23.8	5.2
4394	199.1978	8.1737	2013.7005	10.682	17.25	32.76	259.7	198.3	-139.2	6	196.9	-139.5	6.5
4395	199.2049	-8.6368	2013.911	13.775	15.191	22.469	326.8	-60.9	-8	5.4	-61.7	-10.5	5.5
4396	199.2223	-5.9281	2013.9495	12.394	15.782	7.591	28.6	-104.1	-45.8	5.4	-105.8	-51.1	5.6
4397	199.2982	22.3705	2013.8585	16.318	16.375	3.727	308	-17.3	-66.6	5.9	-14.1	-69.5	6
4398	199.3094	36.5486	2013.786	14.231	17.621	14.279	165	-74	20.4	5.2	-77.2	20.2	5.3
4399	199.3242	21.2880	2013.65	15.034	16.372	8.988	179.8	-87.5	20.2	5.8	-87.8	20.6	5.9
4400	199.3635	-10.5462	2014.1795	7.162	11.19	9.224	132.9	-84.8	19.8	5.4	-82.5	12.4	5.4
4401	199.3642	21.3295	2013.6995	12.74	13.172	5.269	271.6	-124.4	-40.2	5.9	-128.5	-42.7	5.9
4402	199.3785	62.4179	2013.603	13.8	16.856	6.389	329.9	-64.7	-10.4	5.6	-61.4	-13.6	5.6
4403	199.3972	-11.9502	2014.2555	10.87	11.519	24.296	344.8	-87.5	42.2	5.3	-86.1	42.9	5.4
4404	199.4553	32.3244	2013.541	13.524	15.342	4.796	213.6	57	-107	5.5	63	-105.5	6
4405	199.4840	-5.4532	2013.9955	10.272	13.157	11.14	303.4	96.1	-96.8	5.4	98.2	-95.7	5.5
4406	199.4902	44.4108	2013.6515	16.379	18.261	12.421	212.8	-1.6	-63	6	-1.9	-63.3	6.2
4407	199.5100	34.4955	2013.6855	16.946	17.143	4.108	117.8	-66.7	24.5	5.6	-68.7	28.6	5.6
4408	199.5596	0.2988	2013.773	11.157	12.617	19.959	98.5	-97.2	-18.2	5.9	-97.8	-19.1	5.9
4409	199.6105	-5.2848	2014.012	12.351	16.121	57.586	147.9	-162.1	31.5	5.4	-165	33.4	5.6
4410	199.6178	-6.1644	2014.082	14.358	17.619	36.12	205.8	-120.1	38	5.4	-115.4	37.1	5.9
4411	199.6531	41.9743	2013.62	15.046	17.219	6.567	153.7	-70.2	-59.9	5.2	-75	-55.9	5.4
4412	199.6914	28.1263	2013.756	9.213	11.484	5.869	70.9	-81	27.8	4.9	-78.6	25.1	5
4413	199.6930	41.9057	2013.736	17.937	18.548	11.595	23.7	-26.5	-78	6.1	-26.3	-81.8	6.3
4414	199.7265	0.5055	2013.797	8.291	12.67	13.215	232.7	-80.9	-38.4	5.9	-73.1	-45	5.9
4415	199.8113	17.9808	2013.785	13.153	14.005	39.918	268	-177	-47.9	5.9	-179.1	-47.7	5.9

4416	199.9131	28.3725	2013.6845	13.947	14.134	27.396	193.3	-116.5	-72.7	5	-116.8	-72.9	5
4417	199.9437	53.9642	2013.666	13.194	14.543	16.006	171	-21.9	-79.8	5.6	-23.9	-78	5.6
4418	199.9729	66.6402	2013.695	10.643	15.532	17.83	36.2	-84	-17.7	5.6	-84.4	-23	5.6
4419	200.0326	61.0573	2013.6315	13.813	16.24	11.717	153	-43.5	-134.4	5.6	-44.9	-134.1	5.6
4420	200.1297	-7.1331	2013.871	11.167	17.335	15.555	340.2	-92.3	0.7	5.4	-93.5	-1	5.9
4421	200.1526	32.7415	2013.8645	9.968	15.453	9.56	103	-97.8	-54	5.5	-102.2	-53.8	5.5
4422	200.1875	32.9813	2013.8835	14.446	15.912	4.759	46.9	-50.3	-83.6	5.2	-46.8	-85.5	6.6
4423	200.2066	5.6497	2013.733	13.1	15.469	9.793	187.2	76.7	-17.2	5.9	72.8	-17.5	6
4424	200.2072	-13.0844	2014.221	11.908	14.158	33.281	308.5	-66.2	-32.5	5.1	-66.6	-35.5	5.1
4425	200.2553	9.6894	2013.6475	12.328	18.344	42.307	351.4	-109.5	29.1	5.9	-110.6	27.7	6.4
4426	200.2836	34.8496	2013.8025	9.33	15.508	13.736	70.3	-150.2	37.5	5.2	-145.5	37.8	5.2
4427	200.2951	15.0562	2013.763	12.071	13.455	10.414	226.6	-73	-33.5	5.1	-73.6	-34.7	5.1
4428	200.3186	53.0852	2013.9615	13.082	13.651	3.974	58.8	-120.8	11	5.6	-126.2	8.5	6.2
4429	200.3241	53.3546	2013.7885	9.723	10.761	6.176	211.3	-40.9	-61.7	5.6	-43.7	-62.6	5.6
4430	200.3488	-0.8361	2013.878	11.964	18.099	19.282	172.2	-131.7	28.3	5.5	-136.4	20.4	6
4431	200.3589	45.0113	2013.6465	14.826	16.803	36.1	295.5	-67.9	-22.5	6	-69.2	-21.9	6.1
4432	200.4095	39.6390	2013.6795	12.796	16.042	9.481	165.5	-76.8	-5.3	5.2	-74.2	-2.9	5.3
4433	200.4170	15.6935	2013.897	9.881	9.934	17.068	86.7	-74.5	6.3	5.1	-72.6	4.6	5.1
4434	200.4648	55.9012	2013.6965	9.101	11.495	25.067	76.6	-73.1	6.7	5.6	-71.2	-2.2	5.6
4435	200.4761	17.9169	2013.806	14.185	16.589	40.607	110.5	-113.9	88.6	5.2	-109.3	87.6	5.7
4436	200.4833	23.0788	2013.7115	13.461	15.067	30.861	304.5	-66.4	0.5	5.6	-66.2	-2.7	5.6
4437	200.4918	-3.4592	2013.8575	15.09	16.379	10.043	269	-97.1	-98.4	5.5	-97.5	-100.6	5.6
4438	200.5314	67.8120	2013.6525	14.916	15.29	28.233	8.1	-59.7	85.4	5.6	-60.8	84.9	5.6
4439	200.5658	5.5211	2013.474	14.13	16.508	46.899	236.6	39.9	-144.5	6	41.1	-140.7	6.3
4440	200.5688	-11.6242	2014.031	9.757	10.143	15.147	278.1	102.7	-69.3	5.5	105.1	-70	5.5
4441	200.5872	70.1717	2013.606	12.34	13.273	15.193	308.2	92.4	-179.7	5.7	94.2	-179.1	5.7
4442	200.6137	60.0537	2013.6715	12.632	13.844	6.893	168	-77.7	-1.1	6	-79	-3.4	6
4443	200.6293	32.5138	2013.7835	12.178	13.047	40.895	346.2	1.7	-142.2	5.9	2	-140	5.9
4444	200.6416	26.1158	2013.8975	7.547	11.227	9.926	244.5	-168.9	25.2	5.8	-169.1	24.1	5.9
4445	200.6924	37.8743	2013.693	16.263	18.463	21.19	127.8	29	-65.8	5.2	30.4	-67.3	5.5
4446	200.7050	-7.8952	2013.973	14.927	15.195	4.502	40.2	-67.4	-2.8	5.5	-70.7	-2.7	5.5
4447	200.7055	-10.7927	2013.945	14.019	14.235	15.213	6.9	-95.1	-50.8	5.5	-93.6	-50.5	5.5
4448	200.7445	-4.4887	2013.4625	13.291	14.898	4.903	35.8	-118.1	3.5	5.5	-117.2	1.1	6
4449	200.7551	26.9924	2013.6715	14.064	16.373	11.984	138.1	-90.7	74.3	5.9	-88.3	76.2	6
4450	200.7784	50.5509	2013.627	16.341	16.607	13.131	231	-76.8	-5.5	5.6	-75.7	-2.4	5.7
4451	200.8034	36.5609	2013.7245	12.31	15.906	28.017	94.1	-167	12.4	5.2	-166.1	12	5.3
4452	200.8069	34.3963	2013.866	10.943	16.401	8.214	239.5	-85.3	40.2	6	-91.4	37.1	6.1
4453	200.8359	-0.2704	2013.735	15.638	16.741	16.833	51.1	-66.1	0.8	5.5	-66.1	-2.2	5.6
4454	200.8959	-1.9258	2013.888	12.15	16.865	29.419	172.2	-123	-16.1	5.5	-124.8	-19.7	5.5
4455	200.9069	26.2006	2013.7735	14.301	14.748	7.027	35.4	-63.7	-21.6	5.9	-64	-22.5	5.8
4456	200.9132	2.7241	2013.776	6.978	7.136	26.192	74.4	-1	212.7	6.1	-10	209.2	6.1
4457	200.9363	32.5305	2013.7325	10.71	18.885	32.513	263.3	-204	-79	6	-204	-78.4	7.8
4458	200.9447	42.4146	2013.667	12.285	15.494	10.601	32.3	-128.5	52.2	5.6	-130.3	54.9	5.6
4459	200.9632	10.6566	2013.8225	13.627	14.287	5.356	237.1	-69	41.6	5.9	-67.3	43.1	5.9
4460	201.0270	-3.8735	2013.7955	11.196	11.475	5.142	358.9	-80.9	17.8	5.5	-80.2	19.8	5.5
4461	201.0493	-12.2801	2013.8725	13.464	14.454	14.861	31.6	-86	-28.4	5.2	-86.3	-29.6	5.2

4462	201.0607	46.0639	2013.3025	15.66	15.747	3.866	189.6	-67.5	-16.9	5.7	-70.8	-11.9	5.9
4463	201.0853	43.2171	2013.624	14.275	15.249	8.281	76.1	-42.7	-83.9	5.6	-46.5	-83.8	5.6
4464	201.0898	-2.3926	2013.738	12.966	15.701	24.801	47.8	-164.5	40.6	5.5	-167.9	39.5	5.6
4465	201.0998	43.9802	2013.6945	12.033	12.48	7.256	207.2	-90	-20.8	5.6	-89.3	-20	5.6
4466	201.1585	-1.8931	2013.638	13.599	15.109	34.786	230.3	-168.7	-147.9	5.6	-165.8	-148.8	5.7
4467	201.1670	17.9147	2013.7775	11.043	14.162	23.704	234.1	-140.4	26.8	5.2	-135.6	36.6	5.9
4468	201.1862	13.0463	2013.6795	13.574	18.422	11.708	5	-73.6	-4.9	5.1	-74.8	-12.5	5.7
4469	201.2002	52.7444	2013.7535	14.081	18.047	51.855	42.1	36.9	-85.6	5.6	33	-83.1	5.7
4470	201.2630	82.8266	2013.778	15.722	16.636	6.11	63.7	-36.1	-94.1	5.9	-43.5	-89.1	5.9
4471	201.2741	51.8825	2013.6595	15.78	16.426	36.516	88.3	-8.5	-139	5.6	-12	-143	5.7
4472	201.2779	6.5601	2013.601	14.314	16.279	8.336	115.1	11.5	-81.9	6	12	-81.4	6.2
4473	201.2941	24.9566	2013.5455	17.483	17.903	11.088	255.3	-23.1	-61.9	6	-22.7	-61.4	6.2
4474	201.3194	-7.7542	2013.6265	14.039	15.841	10.928	303.4	-59.8	-127.4	5.6	-56.8	-126.7	5.7
4475	201.3451	-6.0138	2013.8025	12.316	15.346	11.987	239.5	-76.5	-13.9	5.5	-77.2	-17.4	5.6
4476	201.3890	22.8942	2013.7205	14.731	15.085	21.975	324.5	51.9	-91.3	5.6	54	-91	5.6
4477	201.4040	43.7435	2013.613	15.034	15.344	9.682	165.7	-17.2	-88.7	5.6	-16.7	-87.6	5.6
4478	201.4778	57.7321	2013.6245	16.108	16.569	26.048	189.7	-74	0.9	6	-75.7	2.2	6
4479	201.5485	30.4407	2013.9315	18.094	18.676	5.318	117.5	-111.9	39.7	5.5	-115.6	38.8	6.7
4480	201.5656	57.1771	2013.6395	15.52	17.41	11.25	118	-83.2	9.4	6	-84.4	10.2	6.1
4481	201.5824	18.2911	2013.7425	11.061	15.753	16.59	149.2	-94	-124.4	5.6	-97.6	-124.6	5.6
4482	201.6156	80.3972	2013.704	15.364	16.102	13.228	324.7	-81	24.1	5.9	-79.1	22.3	5.9
4483	201.6257	20.6539	2013.8325	11.929	16.508	8.167	115.1	-91.4	0.5	5.5	-91.7	2.2	5.6
4484	201.6495	51.9077	2013.6145	14.865	15.214	5.412	77.6	-73.4	-173.9	6	-73.9	-174.5	6.1
4485	201.6546	26.6122	2013.624	15.849	17.863	10.174	289.5	71.8	-57.3	5	72.2	-54.9	5.2
4486	201.7250	14.0530	2013.798	13.105	13.814	5.224	311.8	-185.1	-67.4	5.2	-182.4	-72.8	5.2
4487	201.7355	10.7498	2013.779	14.214	14.414	11.132	358.6	-67.5	26.7	5.9	-67.4	24.4	5.9
4488	201.7423	23.0971	2013.736	13.68	16.555	7.071	22.4	-72.7	-8.6	5.6	-71.8	-7.1	5.6
4489	201.7468	2.4947	2013.534	14.844	17.206	15.891	176.6	-158.2	-47.9	6.1	-157.1	-49.8	6.1
4490	201.8249	46.8203	2013.6845	8.865	15.455	48.151	88.2	-69	40.5	5.9	-68	37.2	6
4491	201.8682	9.2754	2013.535	12.633	18.517	36.049	223.3	-146.9	-67.4	6	-143.4	-70.1	6.7
4492	201.8934	71.7790	2013.711	11.985	12.167	6.73	322.7	-103.8	12.6	5.6	-105.3	8.6	5.5
4493	201.9139	3.0242	2013.579	9.362	16.318	53.697	258.5	-134	6.4	6	-133.5	4.5	6.1
4494	201.9772	-6.4444	2013.6595	14.793	15.468	13.892	164.6	-58.1	-67.9	5.6	-58.6	-69	5.6
4495	201.9957	-2.3815	2013.868	14.796	15.062	3.812	339.4	-106.2	-50.6	5.5	-109.5	-53.6	5.8
4496	202.0173	28.1320	2013.651	13.772	15.915	28.19	70.6	-150	41.6	5	-151.5	41.3	5
4497	202.0198	13.5554	2013.786	10.905	15.106	26.482	349.1	18.4	-67.5	5.1	19.5	-68.9	5.1
4498	202.0565	12.7449	2013.62	16.228	17.256	34.198	259	-83.3	-31.7	5.2	-74.3	-42	5.3
4499	202.0718	45.3970	2013.61	15.731	17.356	25.347	42.2	-72.9	-50.2	5.3	-78.4	-45.2	5.4
4500	202.0732	30.0454	2013.712	10.13	17.15	52.501	51.3	-169.6	-181.1	5.6	-173.9	-181.5	5.9
4501	202.1193	21.8652	2013.658	14.587	16.048	6.626	319.9	-133.2	32.2	5.6	-134.5	30	5.6
4502	202.1365	39.9604	2013.6575	10.894	14.296	5.147	276.1	-36.4	-90	5.2	-36.7	-87.5	5.3
4503	202.1745	9.7389	2013.8165	11.038	14.395	10.25	359.8	-73.9	-28.3	5.9	-73.7	-28.6	5.9
4504	202.1816	41.8887	2013.659	15.585	18.566	58.047	341.8	-57.2	72.6	5.2	-51	79.8	5.8
4505	202.2121	21.5160	2013.7735	12.938	13.313	26.626	323.8	-165.5	-199.2	5.7	-165.8	-202.3	5.7
4506	202.2132	-11.5926	2013.951	12.85	14.739	14.294	204.1	-146.8	-68.2	5.5	-144.9	-68.1	5.5
4507	202.2513	-4.6101	2013.7295	12.048	13.797	18.137	166.9	-85.8	37.4	5.5	-86.1	33.8	5.5

4508	202.2714	62.0219	2013.5945	13.835	18.004	21.811	38.2	-101.4	-118.5	5.7	-96.9	-115.7	5.7
4509	202.2982	24.0746	2013.715	11.323	13.002	7.438	209.1	-95.9	-16.1	4.9	-97.7	-15.2	5
4510	202.3557	33.6831	2013.7375	12.55	17.201	9.531	43.7	-84.1	-7.5	5.5	-83.6	-8.9	5.5
4511	202.4007	29.7186	2013.76	11.453	14.151	7.246	238.5	63.8	-74	4.9	62.4	-72.5	4.9
4512	202.4290	45.4795	2013.7335	10.255	13.165	47.559	235.4	-100.1	21.9	5.2	-95.4	25.3	5.3
4513	202.4716	26.7919	2013.6475	17.11	17.464	23.695	127.4	43.2	-60.2	5	45.8	-61.8	5.1
4514	202.5915	38.7491	2013.697	14.095	15.948	38.277	181.5	-30.8	-61.4	5.2	-32.1	-63.4	5.2
4515	202.6883	-4.7649	2013.7705	13.153	13.625	8.697	115.5	-180.6	-144.2	5.6	-183.2	-142.4	5.6
4516	202.7028	17.0795	2013.6175	16.637	16.818	7.767	233	7.6	-106.5	5.2	7.7	-106.7	5.2
4517	202.7168	-7.7209	2013.5235	14.577	17.364	32.523	46	-75.7	-13.3	5.5	-70.4	-23	6.1
4518	202.7443	58.8500	2013.763	13.666	13.901	3.793	121	-81	-5.4	5.6	-84.5	-1.1	5.6
4519	202.7509	21.5639	2013.752	13.566	14.634	16.266	200	-78.8	23.9	5.6	-78.1	25.3	5.6
4520	202.7512	36.4410	2013.681	9.514	9.944	6.052	161	-135.6	-19.1	5.2	-134.7	-11.6	5.3
4521	202.7743	12.2604	2013.8345	10.299	10.795	10.479	291.8	-19.2	-80.4	5.1	-15.1	-79.9	5.1
4522	202.8152	42.1063	2013.776	5.86	13.457	32.411	117.7	-98.5	24.7	5.2	-101.4	18.9	5.2
4523	202.8956	44.5373	2013.652	12.036	14.677	30.939	201.5	-71.6	64.2	5.3	-71.7	69.1	5.3
4524	202.8970	54.4642	2013.7865	17.66	17.665	7.84	1.8	-63.4	-36.3	5.6	-65.1	-26.4	5.6
4525	202.9016	32.9232	2013.832	9.854	15.847	25.486	108.8	-75.6	51	5.9	-75.4	48.2	6
4526	202.9078	38.3611	2013.725	14.612	17.529	10.457	186.5	132.5	-98.6	5.3	128	-100.3	5.3
4527	202.9279	-6.9813	2013.6215	13.946	15.877	18.544	235	-68	-8.8	5.6	-69.5	-6	5.7
4528	202.9421	65.7003	2013.5605	13.026	16.821	6.539	4.1	-68.6	-83.7	5.6	-67.2	-90.4	5.8
4529	202.9810	4.7237	2013.7635	11.979	14.508	6.049	185.6	-72.4	-14.7	5.9	-70.9	-13.8	5.9
4530	202.9880	53.7078	2013.9	9.668	13.668	7.239	159.9	-87.6	-28.1	5.9	-90.6	-21.6	5.9
4531	203.0198	-12.2405	2013.991	14.871	15.157	9.187	107.8	-97.9	-63.2	5.2	-98.1	-60.2	5.3
4532	203.0407	7.7376	2013.5115	14.774	16.824	18.898	281.9	-71.5	34	6	-71.4	33.1	6.2
4533	203.0705	13.5854	2013.7215	10.812	17.977	37.789	151.4	-67.4	18	5.1	-68.9	14.1	5.3
4534	203.0747	-7.2416	2013.9165	10.028	14.407	20.055	164.8	93.1	-66.9	5.4	95.5	-72.7	5.5
4535	203.0981	15.0722	2013.816	10.252	14.825	10.581	35.6	-79	6.1	5.1	-82	3.7	5.1
4536	203.1434	15.0004	2013.811	11.659	14.308	19.313	229.9	28.9	-81.9	5.1	25.3	-79.3	5.1
4537	203.1718	73.4329	2013.544	12.15	13.683	13.591	287.9	-107.1	3.8	6	-106.9	3.1	6
4538	203.1920	46.7035	2013.6225	12.124	18.016	18.233	16.4	-66.9	4.5	5.3	-67.5	4.7	5.4
4539	203.2009	-2.4388	2013.4815	11.884	14.294	6.433	264.4	-68	-31	5.6	-66.1	-32.6	5.7
4540	203.2672	38.2186	2013.662	14.514	16.741	9.094	233.2	-49	-63.1	5.2	-48.4	-66.1	5.3
4541	203.3096	0.8662	2013.4595	13.804	15.305	20.621	198.2	117	-70.2	6.1	117.9	-74.4	6.1
4542	203.3225	-4.4085	2013.5975	11.651	13.579	18.927	4.1	-11.9	-86.7	5.6	-11.6	-87.8	5.6
4543	203.3289	23.0141	2013.7985	9.459	10.061	14.441	177	-106.4	-0.4	5.8	-110.1	-7.4	5.8
4544	203.3727	-3.4714	2013.678	12.317	14.664	45.031	249.6	-81.9	0.5	5.5	-80.7	-0.6	5.5
4545	203.3984	59.1721	2013.6895	11.476	12.468	8.854	342.3	-132.3	-59.2	5.9	-138.3	-61.1	5.9
4546	203.4445	19.2797	2013.7285	15.431	16.718	54.928	108.1	-98	52.9	5.5	-94.2	50.1	5.6
4547	203.5519	3.5895	2013.6555	15.535	17.899	57.028	41.4	-66.6	-6.9	6.1	-62.8	-13.8	6.4
4548	203.6140	1.1814	2013.539	14.856	15.454	19.736	129	-91.3	-14.5	6.1	-89.8	-14	6
4549	203.6266	37.3423	2013.6265	14.193	15.891	8.287	324	-46.4	-66.8	5.2	-44.1	-67.3	5.3
4550	203.6339	68.5057	2013.6515	11.476	13.744	5.537	154.7	-162.3	13.5	6	-158.5	17	6
4551	203.6558	11.6306	2013.6845	16.221	18.559	53.613	195	-66.3	18.4	6	-68.1	16.3	7.5
4552	203.6584	19.5734	2013.755	13.735	15.711	10.6	261.4	45.3	-96.5	5.6	41.4	-96	5.6
4553	203.6625	-7.6059	2013.811	10.078	14.624	12.825	46.5	-40.3	-64.2	5.5	-39	-66.5	5.5

4554	203.6845	-0.8086	2013.5635	11.41	16.767	32.376	126.4	-63.6	20.3	5.6	-63.7	15.6	5.7
4555	203.7220	1.6878	2013.5495	12.974	15.684	7.786	41.2	-55.4	-64.5	6	-54.5	-62.1	6.1
4556	203.7236	37.7465	2013.6995	11.992	16.605	28.547	146.4	-64.5	-27.1	5.2	-65.2	-26.9	5.2
4557	203.7706	42.6089	2013.7295	12.813	15.943	22.818	219.1	-106.2	-7.9	5.3	-106.1	-10.9	5.3
4558	203.8428	-9.0554	2013.604	16.434	16.741	56.373	147	-95.8	-91.7	5.7	-94.5	-93.6	6.1
4559	203.8583	43.7657	2013.7345	10.17	13.713	20.593	326.1	-126.8	-6.6	5.2	-131	-0.9	5.3
4560	203.8659	74.6406	2013.499	16.499	16.672	4.568	296.3	-70.6	8.5	6.1	-73.1	5.3	6.1
4561	203.8980	1.4866	2013.537	15.593	15.784	5.361	253.4	-15.3	-79	6.1	-15	-81.1	6.1
4562	203.9206	76.5860	2013.592	11.774	12.998	12.358	202.7	-103.6	46.9	6	-100	49	6
4563	203.9747	-9.7852	2013.945	8.27	13.48	14.439	3.1	-80.9	-5.7	5.5	-79.8	-7.3	5.5
4564	203.9971	23.5670	2013.4345	16.219	18.558	20.25	94.9	-85	4.4	5.7	-89.1	7.3	6.1
4565	204.0100	18.7644	2013.791	12.886	16.219	32.612	54.7	-134.4	27.3	5.6	-131.5	25.9	5.6
4566	204.0398	11.0141	2013.8265	11.708	13.36	19.192	230.5	49.8	-82	5.9	53	-84.2	5.9
4567	204.0676	33.7083	2013.737	12.688	15.191	42.277	321	-158.6	-46.6	5.5	-162.2	-49.6	5.5
4568	204.1060	21.9280	2013.8725	9.283	14.98	20.995	66.4	-177.3	-100.6	5.5	-172.1	-104.7	5.6
4569	204.1222	9.4068	2013.555	12.64	17.283	10.055	157.9	-95.7	-3.9	5.9	-98.8	2.3	6.2
4570	204.1259	35.7096	2013.8315	12.146	13.69	6.042	287.8	-75.7	9.2	5.9	-73.9	9.6	5.8
4571	204.1861	-8.2064	2013.772	15.135	15.426	3.362	234.3	7.5	-84.1	5.7	15.2	-80.4	6.2
4572	204.2825	25.9697	2013.716	12.327	15.272	9.369	199.3	-63.4	20	5.9	-62.3	22.4	5.9
4573	204.3008	30.0853	2013.8655	9.021	10.076	20.231	67.8	-159.1	49.5	5.4	-157.6	47.2	5.5
4574	204.3146	33.9677	2013.777	12.857	17.767	12.511	261.9	-77.8	-3.3	5.5	-78	-5.1	5.5
4575	204.3586	40.2245	2013.693	12.7	13.297	11.405	218.9	-4.9	-82.2	5.2	-4.4	-80.8	5.2
4576	204.4165	26.3795	2013.648	13.71	16.797	18.388	222.8	-94.9	-142.1	6	-97.9	-141.5	6.2
4577	204.4398	-11.1925	2013.7995	12.762	14.187	6.075	78.6	-79.2	0.5	5.6	-82.6	-0.1	5.6
4578	204.4735	6.7186	2013.7355	9.861	14.441	16.193	115	18.4	-79.9	5.9	12.9	-78.2	6
4579	204.4864	50.0781	2013.7405	12.045	17.822	56.901	210.9	-120.7	-29.6	5.9	-123.8	-32.5	6.1
4580	204.5207	26.8383	2013.8695	9.402	13.487	8.546	277	-71.7	-60.8	5.8	-68.7	-62.3	5.9
4581	204.5748	-9.5236	2013.7365	11.315	16.147	28.75	122	-83.9	19.8	5.5	-80.9	13.6	5.6
4582	204.5766	71.9855	2013.626	10.572	15.242	51.786	170.3	-81.2	48.8	5.6	-85	51.6	5.6
4583	204.6531	43.2656	2013.699	12.672	14.168	5.75	186.4	-9.9	-90.1	5.9	-10.3	-87.3	5.9
4584	204.6795	49.4964	2013.684	12.987	16.69	8.794	203.7	20	-94.9	5.9	19.2	-99.3	6
4585	204.7261	37.5666	2013.6915	15.149	16.226	8.982	63.6	-188.3	-97.9	5.3	-186.7	-98.7	5.3
4586	204.7999	60.6311	2013.468	11.328	15.463	24.142	154.4	-107.4	75.3	6.1	-108.2	72.8	6.1
4587	204.8228	55.5107	2013.6095	16.304	16.563	4.344	331	-82.1	-35.3	6	-81.4	-36	6
4588	204.8473	-5.2865	2013.7835	10.457	14.686	7.785	299.6	-76.4	-10.4	5.6	-78.2	-8.9	5.7
4589	204.8724	0.5198	2013.5225	12.752	12.998	17.432	324.4	-120.9	22.4	6.1	-120.8	25.3	6.1
4590	204.9241	23.3308	2013.7505	9.46	16.68	49.183	299.5	-97	89.7	5.6	-96.9	88.8	5.7
4591	205.0162	45.9194	2013.7105	11.017	13.358	8.89	195.7	72.4	-205.6	6	69.9	-201.8	6
4592	205.0240	23.6334	2013.656	13.345	15.381	6.048	239.6	-77	42.2	5.6	-79.6	41.9	5.7
4593	205.0360	68.7409	2013.6425	8.857	8.917	34.357	111	65.7	-73.7	5.6	63.4	-78.2	5.6
4594	205.0414	16.2266	2013.577	16.729	17.632	5.87	42.5	-84.2	8.8	5.6	-84.6	11.3	5.7
4595	205.0621	35.8169	2013.699	13.533	14.492	18.01	125.3	-113.8	-1.6	5.5	-115	-2	5.5
4596	205.0841	-9.2975	2013.9915	9.367	10.95	6.277	214.7	-116.4	-6	5.5	-121.2	-11.3	5.5
4597	205.2064	30.1905	2013.951	13.623	16.174	5.668	198.2	-27.2	-65.7	5.5	-24.8	-65.1	5.5
4598	205.2335	9.9871	2013.96	13.664	15.456	5.478	42.9	1	-61.2	5.9	-3.2	-63.2	5.9
4599	205.2600	-2.3668	2013.758	14.021	15.138	12.784	17.9	-113.2	3.4	5.5	-116.7	4.1	5.5

4600	205.3304	45.3561	2013.641	13.04	16.275	12.875	129.6	-120.9	-49.1	5.3	-116.3	-47.9	5.3
4601	205.3540	17.6869	2013.8745	11.052	14.227	12.417	217.1	-76.8	-84.3	5.4	-77.8	-86.3	5.4
4602	205.3752	48.0281	2013.7	12.165	15.803	25.945	37.6	-71.5	1	5.9	-70.8	-3.2	6
4603	205.3834	20.3545	2013.6955	16.189	16.844	28.666	90.4	-161.5	66.4	5.6	-159.9	64.8	5.6
4604	205.3986	42.4828	2013.59	16.871	18.013	21.257	334.4	-78.2	0.3	6	-77.4	1.6	6.1
4605	205.4068	37.2720	2013.7125	16.295	16.543	5.767	333.7	-67.2	-6.8	5.2	-64.5	-6.2	5.2
4606	205.5040	72.3398	2013.6385	14.688	16.128	6.037	58.9	-74.2	-36.5	5.6	-71.8	-38.8	5.6
4607	205.5261	13.9809	2013.669	12.732	17.409	11.897	99.1	-88	2.6	5.4	-87.1	1.9	5.7
4608	205.5321	67.3891	2013.62	15.212	17.096	11.223	228.8	-107.5	73.9	5.6	-109.9	71.8	5.7
4609	205.5933	-12.2069	2014.04	14.943	16.494	8.907	315.5	-65.8	-11.6	5.2	-63.1	-8.4	5.4
4610	205.6439	62.2952	2013.587	11.468	11.708	49.675	249.2	-77.1	30.1	6	-77.9	28.5	6
4611	205.6549	-10.7653	2013.9135	13.565	17.114	59.23	102.8	6.7	-111	5.5	6.1	-111.7	5.6
4612	205.6689	41.3195	2013.597	15.214	16.775	30.289	332.5	66.5	-27.6	5.2	67	-29.4	5.3
4613	205.6861	-3.9985	2013.6295	10.682	14.064	27.005	126.4	-108.8	17.8	5.5	-109.6	15.6	5.6
4614	205.7722	10.8314	2013.76	13.013	14.962	6.684	103.7	-204.4	0.3	6	-205.3	-2.2	6.1
4615	205.8045	67.6335	2013.6735	9.556	11.733	6.396	100.4	-61.7	10.2	5.6	-65.4	4.4	5.6
4616	205.8101	3.7059	2013.9575	17.377	18.453	25.191	297.9	-60.7	-29.1	6	-61	-22.1	7.6
4617	205.8428	-5.2091	2013.69	12.204	13.947	8.18	128	-75.3	-68.8	5.5	-75.8	-69.9	5.5
4618	205.8557	30.7788	2013.8235	13.188	13.229	9.392	288	-103	-38.2	5.2	-102.7	-39.2	5.2
4619	205.8592	-5.9492	2013.804	10.794	15.358	45.113	355	124.9	-162.9	6.5	120.2	-160.8	6.5
4620	205.8670	39.1924	2013.7075	10.96	14.63	7.636	7.7	-122.7	-14.8	5.2	-123.6	-17.7	5.2
4621	205.9924	-9.4031	2013.961	17.897	17.957	55.899	125	-61.6	-10.2	7.1	-60.3	1.1	7.3
4622	206.0261	65.0358	2013.7085	7.804	13.172	10.322	222.6	-94.9	46.5	5.9	-91.4	55.8	6
4623	206.0494	36.3235	2013.703	15.509	16.801	10.127	71.1	-73.5	21.4	5.2	-72.9	19.7	5.3
4624	206.0744	6.0596	2013.6595	9.545	16.743	51.522	231.7	-61.3	-96.4	6.3	-60.3	-92.4	6.7
4625	206.0748	12.0004	2013.64	11.581	11.754	4.887	304.6	-82.7	11.1	5.9	-82.2	12.7	5.3
4626	206.0772	86.0563	2013.907	10.223	14.749	26.716	302.2	-61.6	24.6	5.5	-63.9	24.8	5.5
4627	206.1220	10.4212	2013.712	12.677	14.413	14.433	62.5	-44	-91.7	6	-47.1	-89.9	6
4628	206.1466	24.5995	2013.7435	11.679	13.063	49.401	17.6	-61.8	-51.9	5.6	-60.7	-50.6	5.6
4629	206.1477	26.3601	2013.7505	12.519	17.772	9.392	199.4	54.4	-66.5	5.6	55.5	-61.7	6
4630	206.1693	-7.9533	2013.8995	14.615	16.399	5.409	327.5	-53.4	-160.4	6.4	-49.4	-162.5	6.9
4631	206.2134	21.7143	2013.817	10.039	16.724	16.178	228.6	-90.8	-64.2	5.9	-95.8	-64.2	6.1
4632	206.3280	60.5550	2013.697	13.791	15.484	5.908	80.4	-114.6	50.9	6	-118.5	55.2	5.9
4633	206.3467	24.0697	2013.7655	13.992	14.322	4.679	269.8	-77.8	-7.1	5.6	-79.5	-5.7	5.6
4634	206.3520	10.8589	2013.5465	17.679	18.575	38.725	211	-67.7	-28.3	6.1	-60.4	-37.8	6.4
4635	206.3579	28.9567	2013.74	13.735	15.224	20.321	89.3	-84	35.6	5.6	-81.6	45.7	5.6
4636	206.3597	28.7913	2013.6725	15.618	17.732	28.019	97.2	-65.6	-3.1	5.6	-66.1	-0.6	5.7
4637	206.3630	36.7251	2013.715	14.619	14.781	8.093	39.7	-101.4	34	5.2	-102.4	33.5	5.2
4638	206.3694	81.4718	2013.7985	15.957	16.322	6.467	189.2	-69.3	-38.7	5.5	-66.1	-39.5	5.5
4639	206.3748	19.9268	2013.812	12.165	14.657	37.285	86	-120.9	1.2	6	-120.6	4.7	6
4640	206.4053	34.7363	2013.746	9.363	14.852	15.787	64.2	-80.3	69.6	5.2	-83.5	71.3	5.2
4641	206.4742	45.0046	2013.5465	12.037	16.664	23.116	323.9	-83.3	22.8	5.6	-83.6	27.5	5.7
4642	206.4823	-2.4125	2013.7215	11.047	14.406	15.458	174.1	38.8	-134.8	6.1	38.2	-137.2	6.1
4643	206.4898	75.8950	2013.544	16.272	16.3	29.755	195.4	-69.6	-108.7	5.7	-71.1	-109.2	5.7
4644	206.5096	31.5378	2013.702	12.252	18.178	17.112	68.1	-5	-180	5.2	-6	-179.4	5.5
4645	206.5150	20.2155	2013.778	15.466	17.822	11.363	280.9	-69.7	-20	6	-66.9	-20.2	6.1

4646	206.5217	5.8917	2013.776	17.535	18.229	38.546	9.9	9.6	86.9	7.3	16.4	83.2	6.9
4647	206.5524	56.9421	2013.7665	17.718	18.183	7.229	122.5	-131.3	36.8	5.6	-126.6	38.6	6
4648	206.5621	36.9923	2013.8615	13.234	13.28	5.34	286.6	42.3	-78.3	5.2	40.3	-78.3	5.2
4649	206.6085	2.5104	2013.364	13.212	13.493	5.77	332.6	63.8	-41.1	6.1	63	-42.2	6.2
4650	206.6122	38.7294	2013.7105	16.981	17.818	41.623	1.8	-102.1	-91.8	5.3	-100.4	-86.3	5.3
4651	206.6269	15.3727	2013.563	14.607	17.126	11.966	234.8	-70	-40.8	5.2	-71.5	-39.1	5.5
4652	206.6287	14.9385	2013.6965	14.952	17.271	10.884	48.4	-85.7	4.5	5.2	-83.8	3.7	5.3
4653	206.6364	34.7221	2013.7795	13.257	15.521	5.979	178	-138.7	39.8	5.2	-144.9	36.7	5.2
4654	206.6646	10.0331	2013.6265	13.279	18.088	19.831	207.1	-81.7	9.5	6	-79.9	7.5	6.3
4655	206.6652	57.0939	2013.683	11.123	11.47	12.3	287.6	-14.9	93.1	5.5	-16.5	93.9	5.5
4656	206.6837	37.3878	2013.7835	11.239	13.36	6.186	113.9	-73.3	-28.4	5.2	-71.2	-30.3	5.2
4657	206.7054	70.1294	2013.6005	15.003	15.294	35.653	348.6	-34.2	-75.7	5.6	-34.4	-78.3	5.6
4658	206.7169	53.3032	2013.9005	17.457	17.594	5.574	22.6	-71.9	-8.1	5.6	-69.9	-0.9	5.6
4659	206.7726	5.9152	2013.6195	14.983	15.111	37.228	358.9	-5.9	-65.8	6	-2.7	-65.2	6
4660	206.7939	28.3015	2013.6205	15.738	18.53	16.918	202.2	-63.5	5.9	5.5	-65.1	10.9	5.8
4661	206.8130	-13.0889	2013.7535	15.255	15.916	12.53	316.8	-208.4	-157.6	6.7	-212.3	-156.9	6.7
4662	206.8232	3.4127	2013.857	17.344	18.011	45.719	18.5	-18.7	61.2	6	-6.4	66.6	7.3
4663	206.8278	7.7701	2013.515	16.888	17.922	11.442	171.6	-157.1	27.2	6.2	-152.7	31.3	6.4
4664	206.8462	41.8920	2013.7215	10.418	14.88	10.181	70.8	-131.7	2.2	5.2	-133	2.7	5.2
4665	206.8748	1.5554	2013.661	10.062	13.752	6.748	321.4	27.8	-110.5	6	35.2	-113.2	6.4
4666	206.8771	56.8637	2013.817	11.319	15.233	6.661	153.2	-111.1	-11.4	5.5	-112.1	-9.1	5.7
4667	206.9366	9.3723	2013.445	16.238	17.394	5.403	28.6	-74.5	42.8	6.1	-75.8	32.3	6.2
4668	206.9481	1.2055	2013.833	8.597	11.343	6.471	86.9	-29.8	-84.1	5.9	-35.5	-80.6	6
4669	207.0078	-8.0254	2013.8895	7.751	14.789	21.675	142.9	-112.2	-111.8	5.5	-112.7	-115.5	5.6
4670	207.1116	-6.8505	2013.788	14.635	14.911	22.094	3	-97.3	-29.8	5.5	-98.3	-29.1	5.5
4671	207.1220	38.2857	2013.744	13.002	13.086	9.737	204.7	-94.2	7.6	5.2	-97	5.9	5.2
4672	207.1327	60.0864	2013.7055	14.287	15.723	5.628	91.1	-157	88.7	5.7	-160.7	91.4	5.7
4673	207.1639	29.0225	2013.828	13.886	15.545	14.495	56.9	-149.3	34.9	6	-151.1	34.9	6
4674	207.1775	14.5496	2013.663	11.815	15.382	6.058	187	14.6	-148.8	5.2	19.3	-145.6	5.4
4675	207.2420	-1.5931	2013.6225	8.178	11.123	15.752	20.8	-137.1	-44.6	5.7	-138.6	-39.1	5.8
4676	207.2477	-6.5857	2013.5925	11.316	16.877	9.334	337.7	-20.2	-116.5	5.5	-24.1	-120.2	6.4
4677	207.2499	17.9868	2013.74	12.987	15.874	31.606	347.1	22.9	-68.2	4.9	24	-68.1	5
4678	207.2730	69.1120	2013.8015	12	14.359	4.92	323.3	-42.5	76.1	5.6	-40.7	74.4	5.8
4679	207.3290	30.7457	2013.7875	12.568	13.8	6.801	1.9	-90.7	-10.4	5.2	-90.5	-7.4	5.1
4680	207.3474	2.7866	2013.7385	16.586	16.91	4.665	47.6	-1	62.1	6	5.1	63.5	6.6
4681	207.3570	14.9140	2013.7805	12.312	15.709	18.878	150.8	-78.2	2.4	5.2	-79.7	2.9	5.2
4682	207.4065	44.2988	2013.7	10.021	14.553	20.819	340.3	-81.6	74.2	6	-89.5	72.2	6.1
4683	207.4477	-11.0134	2013.9005	9.201	11.676	30.9	111.8	-164.5	-118.2	5.6	-160.2	-125.4	5.6
4684	207.4497	48.8543	2013.7065	13.925	14.775	4.261	87.1	-98.5	17.9	5.6	-102.8	20.1	5.6
4685	207.4521	88.1277	2013.9265	10.443	14.2	47.818	14.4	-70.4	-9.6	5.5	-72	-3.6	5.5
4686	207.4723	28.9156	2013.859	13.39	13.596	53.995	358.1	5.7	-98.4	5.9	6	-101.9	5.9
4687	207.4724	-3.5806	2013.665	11.646	15.756	24.875	126.2	-75	-10.7	5.7	-75.7	-13.4	5.7
4688	207.5192	18.6662	2013.436	18.486	19.334	55.214	94.1	-69.7	3.1	6.2	-72.9	-5.7	7.5
4689	207.5513	-2.9197	2013.8245	17.691	17.969	11.622	301	-70.1	-26.7	6.2	-68.9	-23.2	5.9
4690	207.6166	65.0898	2013.56	14.36	17.308	15.636	103.3	-67.3	-9.6	5.6	-67.2	-0.5	5.7
4691	207.6392	43.3066	2013.734	12.591	16.051	10.332	61.6	-61.7	-14.3	6	-61.6	-13.3	6

4692	207.6703	-1.0117	2013.655	16.212	16.97	23.67	99.1	-23.5	-72.7	6.4	-21.2	-71.5	6.8
4693	207.6887	18.5051	2013.8245	10.608	14.869	47.462	339.3	-46.1	65	4.9	-47.7	65.8	4.9
4694	207.7055	13.6163	2013.5935	16.404	17.158	24.216	177.3	-67.6	4.2	5.3	-71	4.9	5.3
4695	207.8177	21.0284	2013.8165	12.989	15.356	12.514	86.3	49.5	-65	5.8	50.1	-66.4	5.8
4696	207.8505	43.6196	2013.757	14.824	17.553	6.38	45.5	-173	23.6	6	-168.1	24.2	6.2
4697	207.9066	8.0870	2013.6255	16.256	17.344	5.189	20.6	75.2	-52.2	6.1	77.1	-53.9	6.4
4698	207.9172	1.3548	2013.669	13.917	14.026	42.486	233.8	-98.3	-31.6	6	-97.8	-32.1	6
4699	208.1097	-3.7113	2013.741	13.806	16.856	25.425	233.1	-185.2	-52.5	6.4	-187.4	-56.5	6.7
4700	208.1634	61.1013	2013.529	14.65	15.44	8.569	202.3	-121.2	98.5	6	-119.2	100.4	6
4701	208.1650	-7.2534	2013.746	9.646	15.471	12.32	288.4	-119.3	49.1	5.6	-111.4	54.7	5.6
4702	208.2993	25.5494	2013.7495	12.231	16.753	28.295	290.3	9.3	-65.9	5.6	8.1	-67.5	5.7
4703	208.3411	51.3896	2013.7755	12.681	14.644	5.277	280.3	-149.9	35.9	5.6	-148.5	31.4	5.9
4704	208.3452	48.6185	2013.647	11.931	17.598	42.454	318.3	8.2	-61.1	5.6	10.4	-60.4	5.7
4705	208.3641	46.3782	2013.5845	11.493	17.179	37.94	26.7	-6.9	-63.1	6	-8.2	-62.6	6.1
4706	208.3710	11.5541	2013.843	9.56	13.611	7.541	137.8	-172.7	34.5	5.9	-167.9	29.9	6.1
4707	208.4528	22.8097	2013.8095	12.841	15.331	29.858	173.4	-71.8	-21.2	6	-70.8	-23.2	6
4708	208.4696	46.7566	2013.601	16.315	16.657	23.453	277.8	30.8	-64.9	6.1	31.7	-61.3	6.1
4709	208.4864	24.3728	2013.7665	13.268	14.231	6.114	186	-76.7	-27.6	5.6	-78	-30.8	5.6
4710	208.4890	40.2612	2013.7605	12.856	13.204	8.521	238.5	-65.3	2.5	5.2	-66.1	2.7	5.2
4711	208.4952	61.6217	2013.522	11.475	15.531	55.691	356	-60.9	83.3	6	-65.8	84.2	6
4712	208.5160	16.5196	2013.705	14.753	16.049	36.246	8.8	-73.8	-21.5	5.2	-71	-23.5	5.3
4713	208.5221	32.8266	2013.8525	8.368	8.717	6.657	327.2	113.7	26.2	5.1	111.3	28.9	5.2
4714	208.5223	25.6712	2013.618	14.59	17.23	11.084	145.2	-75.5	15.1	5.6	-78.7	16.7	5.8
4715	208.5934	66.8581	2013.6615	13.904	14.24	3.901	34.8	95.9	-43.2	5.7	90.7	-45.1	5.7
4716	208.7133	15.7585	2013.755	13.477	15.23	13.4	187.1	114.5	-47	5.2	111.5	-47	5.3
4717	208.7152	51.4709	2013.682	13.746	16.192	10.322	333.5	-72.1	-44.3	5.6	-74.3	-41.6	5.6
4718	208.7346	13.7481	2013.95	17.85	19.093	30.74	234.3	-80.7	23.4	5.5	-75.8	30.6	5.9
4719	208.7353	61.3679	2013.68	11.803	13.1	6.37	254.5	58.5	-111.1	6	57.5	-118.1	5.9
4720	208.7583	8.4991	2014.0565	13.371	15.551	5.395	271.8	-97.8	23	5.9	-98.7	26.2	6.4
4721	208.7864	18.6035	2013.8345	10.53	15.264	17.686	141.8	90	-92.6	5.8	90.7	-92.1	5.9
4722	208.8144	7.3156	2014.117	17.453	18.573	15.949	115.7	-79.1	-10.3	6.1	-79.4	-0.1	6.6
4723	208.8217	38.5764	2013.653	11.848	17.344	24.828	182.7	-65.3	1.8	5.2	-66.2	1.9	5.3
4724	208.9855	59.8212	2013.667	10.845	11.486	27.75	311.1	-143.6	67.3	6	-142.4	66.2	6
4725	208.9874	-13.0948	2013.93	12.145	15.354	11.66	95.6	-72.9	-29.3	5.3	-72.2	-30.6	5.1
4726	208.9927	42.8432	2013.55	11.43	17.269	32.34	57.5	43.2	-63.6	5.6	41.5	-62.6	5.7
4727	209.1189	9.6730	2013.8215	9.807	14.276	9.232	100.1	61	-28.5	5.9	60.7	-35	6
4728	209.1314	22.8342	2013.619	17.246	19.007	20.025	172.3	-68.3	24.7	6.1	-69	13.2	7.2
4729	209.1499	13.3411	2013.7555	14.493	17.371	18.174	342.6	22.8	66.3	5.8	21.7	64	6
4730	209.1784	64.6651	2013.676	9.421	12.067	7.937	79.6	76.2	-160.8	6	67.8	-156.9	6
4731	209.1848	-10.9041	2013.776	14.071	17.451	19.027	168.4	-77.2	-64.1	5.5	-78.7	-62.6	5.8
4732	209.2060	59.7361	2013.876	9.817	12.196	4.702	352.6	-66.4	-73	5.5	-68.7	-73.4	5.4
4733	209.2135	58.4133	2013.6505	13.864	15.906	8.847	348.1	-176.9	35.7	5.6	-176.4	35.9	5.6
4734	209.2563	14.7938	2014.034	15.042	15.452	4.128	166	64.7	-50.3	5.8	67.6	-48.3	5.8
4735	209.2766	10.6521	2013.667	15.274	18.151	36.479	343.5	-1.5	-67.5	6	6.4	-67.5	6.3
4736	209.3021	56.5260	2013.7055	15.436	15.818	4.024	107.1	-53.2	62.9	5.5	-59.2	62.1	5.6
4737	209.3120	19.8818	2013.7375	14.428	15.368	13.868	211.6	-87.2	-37.3	6	-86.7	-39.6	6

4738	209.3316	60.9431	2013.548	12.641	15.522	8.629	20.4	-91.1	12.5	5.6	-90.2	14.5	5.7
4739	209.3634	17.2757	2013.6005	17.703	17.916	8.715	68.6	19.3	-82.6	6.1	13.3	-84.6	6.1
4740	209.3707	17.6913	2013.8855	10.841	12.222	20.498	266.3	46.1	-128.5	5.8	47.9	-127.2	5.8
4741	209.4066	-9.2906	2013.8375	12.166	12.572	46.342	15.5	-102.8	-6.4	5.5	-101.4	-6.7	5.5
4742	209.4286	65.7397	2013.5695	12.45	13.057	16.173	98.8	-72.7	18.5	5.6	-72.9	20.4	5.7
4743	209.4514	24.7935	2013.6585	15.584	15.866	36.45	147.5	-66.6	33.3	6	-67	34.9	6
4744	209.5020	57.0642	2013.7495	13.13	14.349	4.424	166.5	-124	-32.3	5.5	-125.3	-25	5.7
4745	209.5230	51.0345	2013.693	14.869	16.68	49.285	114.4	52.8	-87.4	6	43.5	-95.3	6
4746	209.5334	42.9037	2013.687	11.126	13.291	18.441	102.8	-113.3	63.3	5.6	-113.6	67.5	5.6
4747	209.5367	7.5002	2013.9005	13.917	15.311	6.89	211.7	-75.3	24.3	5.9	-76.8	25.7	5.9
4748	209.5443	48.8787	2013.6855	9.526	11.053	41.618	356.6	-128	-76.2	6	-126.3	-79.7	6
4749	209.5748	11.4364	2013.814	11.497	17.002	17.806	271.3	37.6	-71.6	5.9	36.5	-73.5	6.1
4750	209.6084	26.5773	2013.599	17.203	17.728	5.68	72.9	-65.8	1.4	5.3	-68.4	4.3	5.3
4751	209.6268	53.1149	2013.8505	12.638	14.163	5.054	217.2	-66.5	-158.8	6	-72.7	-161.2	6
4752	209.6435	28.2011	2013.708	14.159	17.638	15.709	236.2	-95.5	10.8	5.2	-92.3	12.4	5.3
4753	209.6478	-8.1694	2013.749	13.324	13.375	13.522	323.8	-79.1	52.9	5.5	-80.6	55.2	5.5
4754	209.6534	30.4628	2013.82	10.799	14.339	12.739	288.1	-12.8	-139.9	5.2	-14.2	-139.1	5.2
4755	209.6863	38.7090	2013.6365	13.209	16.773	39.75	348	-175.8	-32.9	5.2	-181.2	-36.8	5.4
4756	209.7243	4.2210	2013.898	11.701	12.38	4.736	220	-97.2	-122.9	5.9	-93.7	-120	6.8
4757	209.7434	5.5147	2013.9325	15.074	15.3	49.398	281	-103.4	-59.6	5.9	-101.6	-58	5.9
4758	209.7498	21.0923	2013.8985	12.882	15.73	6.408	24.9	-90.7	-5.9	5.2	-91.9	-7.9	5.3
4759	209.7613	59.2955	2013.8525	10.966	12.243	4.751	51.8	-6.4	-60.1	5.5	-15.3	-60.4	5.5
4760	209.8026	2.4581	2013.8015	13.378	13.698	7.963	90.7	-78.5	-106.9	5.9	-80.2	-106	5.9
4761	209.8039	19.3459	2013.9645	11.815	16.198	6.072	283.3	17.9	-79.1	5.2	23.4	-76.4	5.2
4762	209.8164	13.5675	2013.8325	12.625	14.139	22.424	206.6	-68.3	-8.9	5.1	-68.1	-9.9	5.1
4763	209.8357	-6.0427	2013.6795	14.111	16.057	39.517	335.5	-94.5	31.6	5.5	-91.1	31	5.6
4764	209.9269	16.6370	2013.858	13.628	14.916	7.342	199.6	-44.5	-83.9	5.1	-42.8	-83.9	5.1
4765	209.9948	68.2755	2013.7815	17.984	18.597	35.724	96.3	-138.8	1.6	5.7	-138.6	10.8	5.8
4766	209.9989	8.7683	2013.862	12.553	14.553	22.12	5.9	26	-61.4	5.9	25.3	-61.5	5.9
4767	210.0349	3.9368	2013.932	10.612	11.391	21.556	138.2	65.1	-62	5.9	70.1	-58.7	5.9
4768	210.0896	3.5472	2013.7095	15.383	16.838	6.277	179.4	-106.5	-38.6	6	-105.1	-40.1	6.2
4769	210.0933	39.4076	2013.835	12.608	13.018	32.531	213	-71.4	-5	5.2	-72.3	-2.7	5.2
4770	210.0991	66.6241	2013.6325	10.954	16.026	51.254	76.7	77.9	-134.7	5.6	75.3	-135.2	5.7
4771	210.1327	71.7089	2013.5665	12.261	13.055	47.638	284.4	-49.3	81.5	5.6	-47.4	81	5.6
4772	210.1754	17.8988	2013.8895	10.321	12.341	8.152	169.5	-88.3	14.8	5.2	-86.1	16.5	5.2
4773	210.1990	23.4423	2013.7505	14.632	15.246	7.009	92.1	-85.8	-8.5	5.2	-82.2	-4.7	5.2
4774	210.2630	43.9380	2013.6565	9.969	14.281	47.747	34	-94.6	54.4	5.6	-94.3	52.1	5.6
4775	210.2746	9.0374	2013.8305	10.024	12.663	18.726	336.3	87	-53.5	5.9	84.9	-57.9	6
4776	210.3179	14.5964	2013.7735	14.721	14.841	6.131	243.4	90.3	-69.7	5.1	85.8	-67.6	5.1
4777	210.3720	53.7782	2013.8745	13.045	14.243	4.845	313.4	-95.8	129.8	5.6	-98.2	127	5.5
4778	210.3764	-12.0454	2014.023	13.455	14.119	17.707	347.8	45.2	-97.2	5.5	44.8	-99.3	5.5
4779	210.4100	0.5619	2013.7925	14.342	14.409	6.557	18.5	-65.2	-16.8	5.9	-65.7	-19.1	5.9
4780	210.4992	31.0141	2014.039	16.024	16.275	3.764	268.8	-103.9	76.9	5.5	-103	74.5	5.7
4781	210.5237	56.2221	2013.7045	15.615	16.09	4.297	135.3	-62.7	37.4	5.5	-71.2	28.5	5.6
4782	210.5478	0.7641	2013.793	10.785	15.629	45.832	248.9	-105.5	16.2	6	-109.8	12.3	6
4783	210.5802	8.7706	2013.885	12.84	13.955	5.523	320.8	-115.5	-7.7	5.9	-115.2	-8.3	5.9

4784	210.5890	3.3436	2013.952	15.118	15.269	3.549	359.5	-141.8	21.2	6	-142.6	22.8	6.2
4785	210.6121	15.3427	2013.82	13.242	13.375	6.827	312.4	-182.8	-0.3	5.2	-182.3	2.5	5.3
4786	210.6181	-9.7064	2013.91	14.649	16.17	7.799	97.8	-43.9	-74.8	5.5	-45.8	-73.3	5.7
4787	210.6187	4.4722	2013.7905	10.573	17.866	58.203	143.9	-77.9	-96.3	5.9	-74.3	-96.3	6.3
4788	210.6190	39.2160	2013.7865	10.581	17.148	12.23	339.7	-94.6	-79.1	5.2	-93.8	-83.8	5.3
4789	210.6641	9.1071	2013.954	14.144	15.435	4.319	171.8	-64.5	49.7	6	-62.6	50.2	6.1
4790	210.7376	-4.4999	2013.5175	18.052	18.513	41.85	7	76.9	-86.1	5.9	66.7	-91	7.1
4791	210.7506	-7.8098	2013.759	12.31	15.723	14.469	90.1	-110.4	-65.3	5.5	-112.6	-69.6	5.6
4792	210.7677	11.8969	2013.8525	8.39	12.071	35.182	9.5	110.2	-143	5.2	105.5	-140.3	5.3
4793	210.7792	24.3856	2013.727	15.542	16.781	10.208	37.6	-77.4	-77.5	5.2	-78	-76.8	5.3
4794	210.7867	-6.4744	2013.8425	11.053	16.953	13.589	255.5	-28.5	-64.2	5.4	-28.7	-66.1	5.9
4795	210.8431	85.3737	2013.8355	16.274	16.783	16.698	43.3	-73	-29.7	5.5	-77	-29.4	5.5
4796	210.8736	7.0612	2013.7185	10.437	12.063	4.702	125.1	63.8	-126.6	5.9	61.6	-124.2	6.5
4797	210.8875	6.3001	2013.8705	16.227	18.395	7.097	315.5	-82.5	2.8	5.9	-78.6	2.2	7.2
4798	210.9043	-12.0550	2013.996	13.892	14.336	5.332	47.4	-126.3	2.3	5.7	-129.4	0.9	5.6
4799	210.9213	14.6823	2013.7395	14.357	16.763	54.716	358.7	-62.5	37.2	5.4	-64.3	35.1	5.5
4800	210.9275	54.8438	2013.7285	10.322	14.625	8.817	326.3	-1.8	120.1	5.5	0.3	121.9	5.5
4801	210.9401	-11.1850	2013.816	12.36	12.489	16.845	259.9	-71.6	-13.3	5.5	-74.1	-14	5.5
4802	210.9544	47.4458	2013.613	10.415	15.516	11.819	280.6	-108.7	-40.8	5.6	-111.1	-36.6	5.7
4803	210.9558	50.1891	2013.7405	12.444	14.141	29.449	188.4	-15.1	-116.1	5.6	-18.7	-114.4	5.6
4804	210.9840	24.8608	2013.647	16.456	18.086	6.222	139.4	-74.8	-19.3	5.2	-70.6	-24.9	5.4
4805	210.9978	43.2295	2013.745	9.853	14.057	23.414	217.1	-110.1	30.3	5.6	-111.5	34.6	5.6
4806	211.0005	31.1319	2013.6725	12.382	18.073	19.47	84.8	-74.6	44.4	5.6	-71.1	51.3	5.9
4807	211.0114	60.0199	2013.604	11.707	15.762	16.006	170.4	-7.7	-90.4	5.6	-5.3	-90.3	5.6
4808	211.0528	30.8634	2013.7115	14.009	16.711	11.162	115.3	-63.6	19.5	5.6	-63	21	5.6
4809	211.0733	87.7277	2014.394	14.215	14.398	3.632	264.2	24.5	-66.8	6.4	26.3	-68.9	6.7
4810	211.0747	72.2402	2013.608	14.501	17.182	5.667	358.1	-84.2	-93.5	5.6	-88.4	-90	6.2
4811	211.0869	43.9901	2013.737	12.098	15.079	7.857	22.3	-72.3	-47.9	5.6	-73.6	-44.6	5.6
4812	211.1011	53.3277	2013.738	12.043	14.408	18.532	78.4	-73	-15	5.6	-75.2	-15.9	5.6
4813	211.1074	-9.8153	2013.908	13.824	17.326	23.025	55.7	-49	-70	5.5	-53.2	-69.1	5.8
4814	211.1672	-0.6723	2013.699	15.688	17.512	7.673	158.6	-139.9	-88.2	5.6	-148.4	-85.6	6.3
4815	211.1916	29.4434	2013.745	10.872	15.352	11.123	287.6	-66.7	16.9	5.2	-67.2	15.1	5.2
4816	211.2437	1.9497	2013.8875	8.453	12.186	49.038	299.4	-215.3	8.3	5.9	-208.1	7.9	6
4817	211.2447	-8.1413	2013.7375	16.29	16.738	8.432	8.6	-100.6	-4.8	6	-99.1	-7.6	6.1
4818	211.2958	8.5371	2013.9775	14.132	17.431	18.79	137.5	-141	-33.3	5.9	-135	-39.5	6
4819	211.3310	21.9955	2013.7885	11.02	16.016	48.465	146.2	-78.6	28	5.2	-78.8	23.6	5.2
4820	211.3698	12.9351	2013.8865	18.52	18.805	59.122	127.2	-63.8	-24.7	6	-62.3	-20.6	7.8
4821	211.3822	25.5537	2013.9225	11.148	16.16	7.558	348	-89.4	7.9	5.2	-91.6	7	5.9
4822	211.4223	-2.2629	2013.735	14.553	16.899	14.566	165.5	-74.7	9.1	5.5	-71.9	7.8	5.6
4823	211.4264	15.6190	2013.8285	14.089	14.395	48.867	339.8	-54.7	-74.1	5.4	-53.4	-74.6	5.5
4824	211.4781	19.4946	2013.6795	16.457	17.596	7.471	17.2	-62.5	37	5.3	-62.9	38	5.3
4825	211.5074	31.0913	2013.5875	15.777	17.576	28.218	132.7	-61.3	-17.9	5.5	-64.8	-14.6	5.7
4826	211.5269	4.6474	2013.8855	11.015	13.83	6.5	246.8	-137.2	-58.7	5.9	-137.3	-59	6
4827	211.5941	47.9543	2013.5735	14.673	15.163	5.585	97.4	-3.5	-116.1	5.7	-4.9	-116.8	5.7
4828	211.6032	11.6078	2013.805	15.642	16.855	14.832	94.7	-153.9	50	6	-154.4	48.8	6
4829	211.6250	62.0997	2013.554	11.203	12.732	8.744	162.3	-65.6	60.7	5.6	-65.3	58.6	5.6

4830	211.6285	8.0876	2013.959	12.837	13.803	14.786	208.7	11.5	62	5.9	13.1	62.1	5.9
4831	211.6706	28.8726	2013.751	10.367	13.998	11.433	17.4	-77	37.3	5.2	-75.1	39.9	5.2
4832	211.6940	62.9499	2013.7655	13.927	14.869	4.527	216.6	-161.1	34.3	5.7	-158.9	39.3	5.6
4833	211.7061	46.0114	2013.82	12.071	16.258	7.134	78.4	-82.1	38.8	5.6	-83.2	41.1	5.8
4834	211.7177	-3.6083	2013.844	12.736	13.486	33.803	129.4	-114.6	-35.1	5.5	-115.3	-34.5	5.5
4835	211.7591	83.8096	2013.866	11.603	17.413	11.966	176.7	-71.7	33.6	5.8	-71	35.9	6
4836	211.7816	-0.4131	2013.905	12.777	14.585	5.283	323.2	69.7	-31.9	5.5	70.2	-34.7	5.6
4837	211.7863	42.3426	2013.5905	14.393	15.858	24.175	293	-109.7	-14.9	6.1	-110.5	-14.3	6.1
4838	211.8248	24.1864	2013.7455	11.748	13.625	23.112	149.5	-105	-3.2	5.2	-106.1	-4.1	5.2
4839	211.8602	18.5519	2013.6945	15.473	16.44	6.308	145.2	-3.8	-78.2	5.3	-4.1	-79.1	5.3
4840	211.8666	69.4174	2013.5625	11.259	12.011	4.575	263.8	-61.6	13.4	5.6	-60.4	13.2	5.6
4841	211.8984	53.3698	2013.6795	14.674	16.457	40.331	248.4	-153.5	56	5.3	-149.3	51.3	5.7
4842	211.9162	26.0449	2013.6415	17.489	18.697	24.845	116.2	4.1	-74.7	5.3	4.3	-75.5	5.7
4843	211.9516	19.3784	2013.7275	15.472	16.724	5.961	128.7	-121.8	-38.9	5.5	-121.9	-37.6	5.6
4844	211.9522	18.8809	2013.4535	17.857	18.706	40.302	293.6	-20.3	-62.7	5.6	-11.9	-66.1	6.3
4845	211.9888	2.7178	2013.796	16.658	17.525	4.689	310.7	72.7	-117	6.1	65.6	-117.5	6.3
4846	212.0051	-13.2730	2014.2905	11.899	12.001	13.25	0.6	-64.9	-10.3	5.5	-63.4	-9.8	5.5
4847	212.0252	0.0632	2013.884	15.989	16.445	3.366	202.1	-90.7	28.3	6.2	-84.1	37.6	7
4848	212.0340	61.9328	2013.538	10.511	16.453	14.816	199.6	-60.2	50.4	5.6	-61.6	51.8	5.7
4849	212.0842	28.4729	2013.6835	15.951	16.578	4.319	179.8	15.7	-76.1	5.9	8.9	-74.7	6.1
4850	212.0974	13.4731	2013.738	16.22	17.317	8.049	199.5	-78.1	-31.8	5.5	-76	-31.6	5.7
4851	212.1013	66.8591	2013.5665	10.894	14.012	17.668	339.7	-68.7	-2	5.6	-70.8	-3.1	5.6
4852	212.1256	36.4299	2013.878	14.027	15.793	5.012	328	-81.7	-28.9	5.2	-80.2	-29.8	5.3
4853	212.1275	13.8123	2013.691	15.574	17.451	13.058	324.9	-28.4	-81.6	5.6	-34.9	-74.6	5.7
4854	212.1573	5.4507	2013.6805	15.077	17.776	52.285	133.9	33.4	-65.8	5.9	28.3	-65.3	6.6
4855	212.1918	-9.8025	2013.9575	9.102	16.39	15.5	191.3	-83	36.5	5.4	-89.1	32.1	5.8
4856	212.3161	14.9188	2013.9195	10.887	14.183	7.31	352.2	-74	-120.8	5.5	-71.6	-124.5	5.5
4857	212.3540	86.3353	2013.9315	15.669	15.766	6.168	296.8	-70.4	26.7	5.5	-72.7	27.4	5.5
4858	212.4074	13.8209	2013.8675	14.01	16.271	19.167	235.4	-78.3	-2.2	5.5	-78.5	-1.9	5.5
4859	212.4425	44.8949	2013.5925	12.56	13.275	4.725	245.4	-144.3	71.9	6.1	-144	69	6.2
4860	212.4666	62.0253	2013.559	14.077	15.138	38.859	256.5	-70	26.6	5.6	-67.8	26.3	5.6
4861	212.4860	43.1960	2013.308	16.752	17.666	5.584	200.1	-74.5	50	6.2	-74.8	44.7	6.4
4862	212.5337	39.4552	2013.7	12.243	16.259	7.728	20.5	-101.7	-67	6	-101.8	-62.8	6.2
4863	212.5526	-4.3275	2013.789	16.049	17.532	36.021	115.1	-64.6	-9.2	5.6	-62.3	-10.4	5.7
4864	212.6215	72.5898	2013.7355	8.375	10.588	11.462	155.5	-68.5	-80.9	5.6	-63.6	-78.4	5.6
4865	212.6594	15.0474	2013.7735	17.252	17.96	6.105	187.1	-62.1	-54.1	5.6	-64.7	-54.7	5.7
4866	212.7031	-3.0893	2013.797	12.269	12.76	18.013	314.3	-93.6	-58.6	5.6	-92.3	-58.7	5.6
4867	212.7375	-5.9739	2013.754	7.299	17.248	48.266	147.2	-60.2	-6.5	5.4	-62.3	-8.7	6.5
4868	212.7424	10.1962	2013.8935	14.719	15.943	20.005	18.9	-87.5	28.3	5.9	-87.8	23.3	5.9
4869	212.7530	50.2529	2013.7815	8.983	9.551	11.217	256.4	-174.4	135.1	6	-169.2	138.3	6
4870	212.8085	63.1927	2013.517	12.936	16.406	9.844	16.4	-60.1	35.7	5.6	-64	34.6	5.7
4871	212.8230	10.5151	2013.7765	16.546	18.206	32.929	288.8	-68.1	-1.5	6	-71.2	7	6.1
4872	212.8547	18.0593	2013.699	13.597	16.282	15.852	44.3	-158.7	-37.2	5.9	-152.8	-41.1	6.1
4873	212.8618	8.2519	2013.957	15.608	18.278	12.958	124.7	61.9	-40.8	5.9	60.1	-39.2	6.6
4874	212.8670	34.6600	2013.6425	11.201	17.933	32.027	240.6	23.7	-78.4	5.5	25.3	-77.4	5.8
4875	212.9139	-4.4227	2013.7565	13.34	16.63	11.578	104.8	-61.7	-31.5	5.5	-61.8	-37.8	5.8

4876	212.9668	0.2931	2013.895	12.734	13.534	7.442	269.1	-74.1	-11.4	5.9	-76.4	-12.7	5.9
4877	213.0166	4.1950	2013.952	10.853	14.742	14.735	281.7	-68.8	8.6	5.8	-70.2	13.7	5.9
4878	213.0225	61.6217	2013.5535	12.141	16.101	21.257	46.9	-122.4	182.5	5.7	-119.2	176.7	5.7
4879	213.1108	13.5772	2013.9245	11.107	11.988	8.094	336.1	-89.4	29.2	5.8	-92.1	29.2	5.8
4880	213.1192	16.3291	2013.8	14.937	16.479	7.763	45.3	-103.5	-40.4	5.9	-103.7	-39.1	5.9
4881	213.1195	5.2906	2013.9035	12.977	18.028	15.672	317.5	-65.5	-178.4	5.9	-66.3	-179.7	6.5
4882	213.1342	8.9612	2013.875	15.983	16.534	4.366	259.4	-68.5	17.5	5.9	-69	13	6.1
4883	213.1354	7.2958	2013.921	13.094	15.714	11.826	355.3	63.3	-100.3	5.9	58.6	-100	5.9
4884	213.1463	45.5206	2013.6485	15.419	15.781	6.289	274.6	-66.4	-38.5	5.3	-67.6	-39.4	5.3
4885	213.1566	-1.8835	2013.716	14.865	17.208	15.018	306.2	-65.3	-16.6	5.5	-61.6	-18.8	5.6
4886	213.1739	62.2874	2013.6705	13.617	15.486	5.408	285.8	-76.7	55.1	5.6	-75.1	54	5.6
4887	213.1987	51.3277	2013.691	16.475	17.85	44.575	326	-78.9	31.6	6	-78	33.2	6
4888	213.2334	30.5050	2013.89	10.277	14.774	27.94	206.2	-47.1	-87.5	5.5	-48.7	-91.6	5.5
4889	213.2825	61.0117	2013.657	9.316	13.672	7.622	40.7	-105.6	58.1	5.6	-107.7	59.1	5.6
4890	213.3204	9.0132	2014.0105	14.07	17.844	13.455	337.5	-110.9	-31.8	5.9	-110.2	-34.9	6.1
4891	213.3285	49.4013	2013.731	11.782	15.624	44.278	336.3	-16.3	-115.9	5.9	-25.4	-113	6
4892	213.3365	46.8272	2013.547	16.105	18.061	48.053	147.6	-108.9	-4.2	5.3	-103.2	-13.7	5.4
4893	213.3662	51.7878	2013.9005	6.648	6.731	13.611	55.6	60.5	-18.7	5.8	64.7	-11.1	5.9
4894	213.3694	23.5201	2013.7805	13.353	15.868	11.378	299.2	-73.8	-48.9	5.8	-73.3	-48.4	5.9
4895	213.3876	23.5690	2013.8045	11.36	13.342	55.701	226.5	-97.6	62.2	5.8	-102.4	57.3	5.9
4896	213.4021	51.0649	2013.65	12.071	17.031	25.094	239.5	-123.6	18.9	6	-123	19.7	6.1
4897	213.4088	20.9663	2013.897	14.514	15.7	4.723	336.1	12.1	-85.6	5.9	11.4	-85.1	5.8
4898	213.4116	79.4228	2013.686	15.045	17.117	14.312	55.1	-74.9	71.7	5.6	-73.6	76.1	5.7
4899	213.4427	71.0535	2013.5515	13.607	14.585	4.617	149.1	-17	110.5	5.7	-15.3	109.3	5.7
4900	213.4954	-8.2595	2013.65	15.777	16.929	8.223	82.9	-67.3	9.5	5.6	-64.1	16.1	5.8
4901	213.5330	0.5281	2014.1145	14.268	14.648	3.869	145.3	-64.8	-22.8	6.2	-66.8	-17.7	6
4902	213.5675	33.9284	2013.885	8.681	9.697	5.378	330	-106.1	28.4	5.3	-104.2	26.7	5.5
4903	213.5734	59.3385	2013.369	11.152	16.968	9.406	51.1	19.9	-167.3	5.5	10.3	-173.9	6.3
4904	213.6512	1.6305	2013.743	9.016	17.33	21.922	33.3	31.8	-65.7	5.8	35.6	-63.4	6.2
4905	213.6534	43.0168	2013.535	15.47	16.159	15.211	116.2	-62.5	-15.7	5.3	-61.4	-14.3	5.3
4906	213.7431	45.7701	2013.593	12.039	15.105	13.184	122.5	-70.6	92.8	5.3	-73.7	92.7	5.3
4907	213.7494	2.3745	2013.842	13.389	15.889	26.351	299.1	-94.8	30.7	5.9	-98.1	33.8	6
4908	213.7907	22.2667	2013.8115	11.194	11.51	23.728	197.9	-161.1	112.8	5.5	-162.9	112.6	5.6
4909	213.8196	65.0890	2013.5265	13.127	13.731	22.018	8.1	-89.3	46	5.6	-88.8	45.5	5.7
4910	213.8628	25.1773	2013.727	15.09	16.722	26.197	312	3.1	-61.5	5.9	1.7	-64	5.9
4911	213.8825	39.8807	2013.571	13.384	18.105	26.048	154	-62.4	37.5	5.2	-65.2	34.5	5.4
4912	213.9099	12.4251	2013.9605	8.868	13.971	35.981	21.7	-177.9	-18.2	5.8	-176	-19.6	5.9
4913	213.9253	12.5705	2013.8265	13.24	15.052	17.212	258.7	-174.5	-11.8	5.9	-173.8	-14.7	5.9
4914	213.9328	12.1796	2013.86	14.718	15.656	6.367	301.1	-61.9	-12.9	5.8	-60.5	-12.5	5.9
4915	213.9454	17.4973	2013.823	10.95	13.482	5.943	240.6	-170.5	19.6	5.9	-174.5	22.3	6
4916	213.9839	5.2033	2013.8985	14.212	15.502	6.977	220.4	-76.8	-17	5.9	-78.2	-21.2	5.9
4917	213.9988	5.5835	2013.8615	15.673	15.826	12.433	70.7	-61.2	-32.5	5.9	-61.6	-32.7	5.9
4918	214.0280	58.5691	2013.8355	15.383	16.093	3.779	7.5	-115.3	46.7	5.9	-118.1	41.8	6
4919	214.0405	51.3676	2013.845	6.059	7.883	38.926	32.6	-144.9	99.8	6	-153.4	102.2	5.9
4920	214.1376	13.0264	2013.5565	17.14	17.415	26.973	155.6	-3.2	-67.8	6.1	-1.6	-67.9	6.1
4921	214.1512	-4.7495	2013.706	14.607	16.449	42.265	262.5	-89	-25.6	5.9	-91.1	-25.4	6.2

4922	214.1713	32.5414	2013.7515	12.719	15.69	10.432	291.1	-36.7	94.7	5.2	-37.7	95.2	5.2
4923	214.1784	37.6910	2013.6855	18.054	18.179	54.002	352.1	29.2	-68.5	6	28.6	-69.6	6.2
4924	214.1845	14.0853	2013.7905	13.293	17.722	9.481	57.2	-70.8	-94.3	5.9	-70.3	-94	6.2
4925	214.1872	49.8072	2013.7065	11.029	13.945	45.849	300.1	30.6	-65.8	5.9	28.9	-65.9	5.9
4926	214.2118	-4.0575	2013.7185	12.046	16.457	55.163	12.8	-60.9	3.7	5.9	-61.9	1.4	6.2
4927	214.2392	-4.3695	2013.507	15.961	17.821	8.019	87.2	50.8	-190.1	6.1	43.9	-195.5	6.6
4928	214.2445	40.9277	2013.5635	12.938	17.704	13.011	184.6	-58.7	-64.2	6	-62.1	-65.8	6.2
4929	214.2677	8.5723	2013.9615	13.458	15.259	8.083	290.2	-14.5	-82.9	5.9	-17.1	-78.6	5.9
4930	214.2780	13.5163	2013.749	16.459	18.12	9.49	85.1	-62	-48.7	6	-60.6	-48.4	6.1
4931	214.2802	1.0533	2013.97	9.186	9.901	18.746	4.3	45.7	-98.2	5.9	47.6	-101.6	5.9
4932	214.2934	16.7650	2013.7895	15.737	16.859	21.347	159.2	-102.8	28	5.9	-105.7	30.9	6
4933	214.3149	-0.1396	2013.9135	13.913	14.753	5.223	327.8	-150.2	-69.3	6	-152.2	-70.2	6
4934	214.3296	14.4071	2013.7435	12.473	13.716	5.627	98.8	-76.6	-4.6	5.9	-74.6	-6.6	5.9
4935	214.3426	13.5011	2013.731	15.129	15.366	10.525	55.2	-79.6	25.1	5.9	-78	28.4	5.9
4936	214.3855	-7.3155	2013.789	14.578	16.39	8.728	256.2	-79.6	1.5	5.5	-80.6	-1	5.7
4937	214.3894	51.3078	2013.7195	16.155	16.848	7.117	303.3	-81.4	-83.3	5.9	-85	-85.1	6
4938	214.4037	-4.8747	2013.855	12.73	15.899	38.464	105.4	-122.2	-41.8	5.9	-119	-42.9	6
4939	214.4459	-8.5182	2013.805	12.776	16.257	32.985	319.2	-1.4	-94.8	5.5	-0.1	-96.9	5.8
4940	214.4887	7.4266	2013.855	14.766	16.526	18.207	148.3	-80.7	47.4	5.9	-79.3	47.4	6
4941	214.5039	35.7349	2013.6845	13.07	16.586	9.44	98.5	-116.6	-16	5.2	-120.1	-13.8	5.3
4942	214.5137	5.1116	2013.7995	11.785	17.759	13.985	245	-57.3	-70.1	5.9	-58.2	-69.5	6.2
4943	214.5146	41.0178	2013.2865	18.018	19.044	48.977	182.3	-66.6	-9.4	5.4	-68.4	-6.6	6.7
4944	214.5552	20.1591	2013.6995	14.716	16.578	14.749	299.2	42.5	-128.9	5.2	40.7	-133.8	5.3
4945	214.5857	63.9784	2013.5465	14.242	14.859	11.489	158.6	33.4	-151.5	5.7	34.4	-154.5	5.7
4946	214.6194	7.3850	2013.967	13.093	14.478	16.96	353.3	-80.5	-46.3	5.9	-80.2	-44.6	5.9
4947	214.6218	32.3236	2014.016	10.159	13.77	6.244	135.3	112.2	-150.8	5.3	108.5	-151.2	5.1
4948	214.6427	17.3334	2013.7015	13.251	16.747	56.607	296	122.5	-9	5.5	121.4	-10.3	5.6
4949	214.6563	69.2429	2013.526	12.565	12.955	52.443	124.2	-78.4	154.7	5.7	-68.7	152.9	5.7
4950	214.6897	-3.3991	2013.513	15.681	17.345	34.354	116.5	-10.8	-75.9	6.1	-4.5	-78.8	6.5
4951	214.7310	-4.6209	2013.7155	15.826	16.139	14.362	137.9	-61.8	-3.7	6	-62.3	-5.2	6.1
4952	214.7615	49.2983	2013.679	12.647	16.74	22.511	125.3	-44.2	-62.5	6	-46.7	-62.4	6
4953	214.8061	42.3887	2013.688	10.02	10.46	7.59	130.3	-105.5	22.9	5.3	-102.1	24.1	5.3
4954	214.8570	20.6870	2013.688	10.681	17.783	57.842	325.7	-73.3	34.7	5.5	-74.7	38.9	5.9
4955	215.0283	16.1369	2013.784	13.901	15.501	6	31	-137.9	-27.6	5.5	-136.9	-26.5	5.5
4956	215.0335	-11.0586	2014.1085	9.509	15.804	14.156	240.6	-123.3	-42	5.4	-122.5	-50.9	5.8
4957	215.0530	33.9207	2013.627	12.375	17.863	18.18	232.8	-81.5	-39.4	5.6	-80.7	-32.2	5.8
4958	215.0621	9.9412	2013.7845	15.789	16.872	4.775	275.6	-96.7	-21.5	5.9	-94.5	-25.2	6.1
4959	215.1619	-10.5894	2014.2265	11.732	14.982	6.71	179.1	-111.9	-51.4	5.5	-110.4	-53.5	6
4960	215.1736	12.9495	2013.8615	12.025	14.363	15.859	108.3	5.2	-150	5.4	-3.1	-150.6	5.4
4961	215.1818	46.9458	2013.634	15.478	15.917	4.185	203.3	-97.5	108.5	5.3	-93.1	113.5	5.4
4962	215.2153	-4.9683	2014.157	9.669	10.247	4.674	354.1	-129.4	-57.4	5.9	-124	-60.3	5.8
4963	215.2345	39.2188	2013.531	17.167	17.323	22.889	109.8	-80.1	-3.7	5.3	-83.5	-2.9	5.3
4964	215.2371	34.9976	2013.789	9.608	11.036	28.027	278.1	-45.7	-79.2	5.6	-49.5	-75.1	5.6
4965	215.2442	37.1286	2013.874	14.379	14.608	3.6	250.2	-152.8	93.2	5.5	-144.5	99.3	5.3
4966	215.2608	30.9170	2013.812	10.779	18.715	58.616	49.7	-97	-42.2	5.6	-100.8	-41.7	7.1
4967	215.2727	49.7052	2013.644	11.571	12.407	9.638	316.8	-28.8	-68.4	5.6	-27.9	-69.5	5.6

4968	215.3208	47.0645	2013.644	11.281	13.583	11.822	126.3	-62.9	-18.7	5.3	-63.2	-18	5.3
4969	215.3219	67.1111	2013.516	12.433	16.455	56.003	289	-65.1	-16.2	5.7	-67.4	-17.5	5.7
4970	215.3506	58.9170	2013.6365	15.178	17.099	4.968	36.4	-14.3	-85	6	-11.9	-84.6	6.1
4971	215.5096	16.0576	2013.634	12.563	18.25	40.324	75.5	73.3	-81.5	5.9	72.9	-79	6.5
4972	215.5219	50.4521	2013.717	12.517	14.783	7.143	67.8	-21.4	-65	5.6	-23	-65.4	5.6
4973	215.5223	38.3806	2013.681	14.679	15.317	15.697	38.2	-81.6	11.8	5.2	-84.1	10.3	5.2
4974	215.5757	51.1993	2013.691	14.793	15.198	5.198	190	-164.1	69.8	5.6	-161.6	65.8	5.7
4975	215.6050	7.0063	2013.8665	13.185	16.424	12.629	29.1	-62.1	31.5	5.9	-61.8	28.8	6
4976	215.6204	60.9654	2013.6175	7.32	9.75	6.049	282.1	-14	-77.6	5.6	-21.6	-78.9	5.7
4977	215.7624	17.9151	2013.7195	14.786	15.724	6.436	196.2	62.6	-36.4	6	64.2	-38.9	6
4978	215.7885	22.7207	2013.815	14.994	16.127	32.579	295.8	-102.4	-164	6	-104.7	-165.1	6.1
4979	215.8209	23.8166	2013.8195	13.068	15.023	9.37	133.8	-71.1	14.9	5.9	-70.2	12	6
4980	215.8719	20.3674	2013.7725	15.321	17.216	5.869	147.7	-77	6	5.8	-74.9	7.1	5.9
4981	215.8939	8.4947	2013.894	14.233	16.178	8.329	189	-63.1	-30.7	5.9	-61.2	-30.1	6
4982	215.9133	4.1246	2013.952	12.586	14.834	16.223	300.8	40.2	-109.4	5.9	40.9	-110.7	5.9
4983	215.9278	35.4066	2013.7045	16.126	17.447	5.408	129.9	-63.3	-0.5	6	-64.5	0.8	6
4984	215.9418	47.0197	2013.403	12.51	13.084	5.182	92.9	-109.2	-2.1	5.3	-112.9	-2	5.5
4985	216.0016	26.6221	2013.691	16.155	18.654	7.915	312	-110.3	26.6	5.9	-113.1	28.7	6.7
4986	216.0243	11.2470	2014.1465	7.302	7.443	9.337	334.7	68.3	-0.3	5.8	67.8	4.2	5.8
4987	216.0537	8.1399	2013.9745	10.397	17.537	14.956	39.6	-95.4	34.4	5.9	-99.5	34.5	6.1
4988	216.1470	49.7521	2013.6975	10.56	15.354	31.017	56	-81.8	7.1	5.6	-83.1	9.2	5.6
4989	216.1495	18.0703	2013.7095	15.015	16.937	33.295	224.8	-157	-79.4	5.9	-156.6	-79.8	6.1
4990	216.1976	10.4857	2013.8395	11.84	18.845	42.241	359.5	-61.8	7	5.9	-63.1	4	7
4991	216.2182	39.0518	2013.6925	11.453	16.864	10.564	215.7	-76.9	58.1	5.2	-73.8	58.7	5.4
4992	216.2231	2.7497	2013.903	14.038	14.094	29.509	163.6	-25.3	-97.1	5.9	-21.3	-96.9	5.9
4993	216.2371	57.8990	2013.625	16.207	18.054	29.537	325.8	61.2	-69.8	6	57.1	-71.6	6.1
4994	216.2485	35.9505	2013.608	10.351	14.452	7.232	351.3	-16.6	-116.2	5.2	-16.2	-120.6	5.5
4995	216.2980	79.1459	2013.766	11.862	12.998	10.027	323	-170.2	-89	6	-170	-89.6	6
4996	216.3133	39.2447	2013.907	14.046	17.365	5.368	2.9	-112.8	50.3	5.2	-107.9	52.4	5.9
4997	216.4230	10.1715	2013.9925	12.474	13.264	12.027	273.7	-60.8	54	5.9	-63.3	54.2	5.9
4998	216.5019	7.2310	2014.0065	7.949	11.082	22.889	216.1	-61.5	-47.7	5.9	-64.5	-47.7	5.9
4999	216.5336	49.6418	2013.616	14.572	16.063	6.662	234.3	32.8	-65.6	5.6	32.4	-67	5.7
5000	216.5390	-11.6029	2014.409	15.228	15.375	4.042	305.3	-80	12.2	5.9	-77.2	11.5	6.1
5001	216.5578	23.6631	2013.7395	16.332	16.782	3.921	209.4	-153.6	1.9	6.1	-152.1	3.1	6.3
5002	216.6344	-3.1470	2013.841	12.457	14.768	19.738	282.1	-81.9	-139.1	6	-86.7	-139.4	6.1
5003	216.6475	2.1301	2014.007	10.384	10.48	6.194	243	-137.8	92.9	6	-134.2	89.4	5.9
5004	216.6837	50.1072	2013.694	12.543	14.166	7.889	348.3	-16	-101.1	5.6	-21	-95.3	5.6
5005	216.7006	-5.1783	2013.917	8.497	14.938	41.853	299.4	-149.2	-147.3	6	-147.1	-149.7	6.1
5006	216.7387	55.6540	2013.654	15.942	15.952	14.536	339.6	-78	24.8	5.9	-75	24.1	5.9
5007	216.7441	49.3942	2013.66	13.76	16.985	13.057	46	-80.5	2.4	5.6	-82.5	3.7	5.7
5008	216.8406	33.9569	2013.6455	16.572	17.919	55.108	3.6	23.4	-102.7	6.1	11.5	-105.1	6
5009	216.8439	3.2304	2013.94	14.469	15.069	17.141	126.8	-83.4	-12.3	5.9	-83.9	-9.4	5.9
5010	216.8460	39.3545	2013.6245	11.171	18.424	39.094	178.2	159.3	-143.7	5.3	160.1	-139.1	6.2
5011	216.8758	19.1436	2013.841	12.338	16.302	6.718	295.4	16.2	-79.3	6	16.8	-82.1	6.1
5012	216.8968	-7.4690	2013.5955	11.414	16.835	11.024	278.1	19.8	-157.5	5.5	18.6	-161.8	6
5013	216.8997	52.0689	2013.6905	10.723	12.907	13.448	289.2	-63.1	7.1	6	-66	7.3	6

5014	216.9197	26.2266	2013.7145	15.594	15.891	8.778	244.8	-109.9	103.8	6	-109.1	106.5	6
5015	216.9258	55.0159	2013.566	14.199	15.432	4.307	342.8	-134.1	16.8	6	-134.5	16.7	6.9
5016	216.9669	-13.7334	2014.395	12.307	12.461	9.303	27.2	61.5	-76.7	5.1	59.6	-75.8	5.1
5017	217.0480	61.3639	2013.5705	12.676	14.253	34.708	269.8	-98.6	53.4	5.6	-97.8	51.7	5.6
5018	217.0550	10.2224	2013.932	15.073	16.291	19.578	205.2	-88.5	-26.6	5.9	-90.7	-23.8	5.9
5019	217.0754	63.8177	2013.5865	12.439	12.718	19.658	306	-137.4	190.1	5.7	-145.6	183.6	5.7
5020	217.0781	54.7093	2013.6605	14.413	16.51	49.419	89.7	-106.4	44.5	5.9	-108.3	46	6
5021	217.0956	12.6380	2013.924	12.603	13.479	11.051	153.9	-73.9	52.1	5.4	-74.8	51.8	5.5
5022	217.2110	58.3645	2013.809	9.106	13.046	8.665	299.4	-36	-61.8	5.9	-39.6	-60.5	5.9
5023	217.3404	-4.5952	2013.6765	14.356	14.627	14.727	65.1	2.9	-83.2	5.6	6.3	-80.4	5.6
5024	217.3647	16.9172	2013.383	15.931	18.077	5.91	170.2	-72.2	11.6	5.9	-73.8	11.7	6.8
5025	217.3658	30.6799	2013.8325	7.78	12.295	56.683	218.3	-78.4	51.8	5.9	-79.9	40.6	5.9
5026	217.3659	12.4216	2013.852	13.22	14.335	26.75	154.2	31.2	-60.5	5.8	31.8	-61.2	5.8
5027	217.3674	74.1468	2013.4655	14.042	15.608	4.924	49.3	-71.6	74.2	5.6	-72.2	71	5.8
5028	217.4652	73.4020	2013.507	15.503	15.893	4.47	247.2	67.2	-29.7	5.7	70.6	-27.5	5.7
5029	217.4692	6.2831	2013.8875	12.775	16.99	8.156	357	-100.5	-39.5	5.9	-97.7	-37.1	6.3
5030	217.4806	50.0570	2013.6985	10.388	13.842	6.256	103.5	-39.8	-149.3	6	-44.2	-146.5	6.1
5031	217.4821	31.7273	2013.7265	16.072	17.118	5.133	300.9	-78.9	-3	6	-77.4	-7.7	6
5032	217.4979	59.1342	2013.746	11.815	16.267	7.032	326.9	-130.2	-11.3	6	-133.3	-13.4	6
5033	217.5002	-4.2473	2013.9265	7.166	10.53	24.994	267	-88.1	-11.2	5.5	-91.9	-10.9	5.9
5034	217.5109	25.4678	2013.6915	15.979	16.444	21.834	29.4	-60.9	-14.9	5.6	-62.4	-10.3	5.6
5035	217.6716	62.7219	2013.6795	14.537	17.167	5.661	181	-145.9	12.3	5.7	-142.3	16.6	5.7
5036	217.7433	25.1318	2013.785	14.197	15.683	5.789	121.2	-90	-43	5.6	-93.4	-40.8	5.6
5037	217.7574	32.8763	2013.8925	11.983	17.051	7.556	247.7	-76.5	-59	5.9	-80.1	-59.7	6.3
5038	217.7853	15.0821	2013.9485	13.744	14.328	4.698	225.1	-83.1	-1.9	5.2	-82.4	-0.7	5.1
5039	217.8209	-6.7120	2013.6435	10.648	12.485	6.148	31.1	-94.7	-14.4	5.5	-94.3	-20.9	5.8
5040	217.8719	-10.5371	2014.1705	11.337	11.674	4.448	339.9	28.1	-103.6	5.5	33.3	-106.3	5.9
5041	217.9175	24.7879	2013.9505	13.56	14.271	3.734	306.5	-164.5	-96.6	5.6	-161.9	-97.2	5.8
5042	217.9381	73.7054	2013.4905	13.86	16.251	23.01	196.8	-76.9	19	5.6	-75.1	16.8	5.7
5043	217.9449	1.0061	2013.573	14.868	18.442	11.411	210.1	-143.9	-64.3	6	-138.8	-71.9	7.3
5044	218.0648	52.4337	2013.7385	14.957	16.115	5.234	216.3	-8.3	-79.1	5.9	-6.4	-79	5.9
5045	218.1146	4.7343	2013.8955	14.272	15.169	5.569	308.4	-75.8	21.3	6	-76.1	21.1	6
5046	218.1207	58.9455	2013.69	14.121	18.497	16.556	10.6	6.2	-69.4	6	3.4	-70.5	6.1
5047	218.1256	-2.3665	2013.7765	10.124	15.407	38.793	100.9	76.2	-77.7	5.5	72.9	-82.5	5.6
5048	218.1462	-1.3871	2013.6095	16.116	16.325	35.759	73.5	-76.3	7.5	5.7	-75.5	3.5	5.6
5049	218.1727	55.8908	2013.6845	11.22	11.264	15.117	192.6	77.4	-90.4	6	76.4	-93.3	6
5050	218.1909	-6.8714	2013.7195	13.475	13.927	21.014	72.6	114.4	-22.2	5.6	112.9	-22.1	5.6
5051	218.1933	36.6557	2013.649	16.145	16.745	8.172	95.4	-6.6	-64	5.2	-7	-62.7	5.3
5052	218.2711	55.4588	2013.7635	9.361	10.936	29.042	333.7	-114.9	-134.2	6	-119.4	-133.2	6
5053	218.2776	28.3144	2013.7585	12.437	17.064	11.816	310	-125.4	20.6	5.5	-123.8	15.9	5.6
5054	218.2881	68.5938	2013.514	17.86	18.162	38.051	332.1	-61.5	-25.2	5.8	-66.1	-15.3	5.7
5055	218.2965	16.7997	2013.631	18.284	19.005	35.174	209.7	-66.7	-16.1	5.3	-65.5	-10.5	6.1
5056	218.3023	23.2316	2013.689	15.885	16.385	4.815	100.9	-64.6	-29.7	4.9	-65.1	-24.1	5.1
5057	218.3443	67.5202	2013.466	13.116	16.247	22.943	68.6	-68.2	40.6	5.7	-70	42.2	5.7
5058	218.4185	32.5362	2013.8195	15.821	16.014	5.848	116.2	-80.1	-33	5.9	-82.6	-36.5	5.8
5059	218.4257	-8.2604	2013.7265	14.214	15.329	7.622	209	-75.4	-6.2	5.6	-75.2	-4.4	5.6

5060	218.4727	79.9622	2013.7605	13.478	13.876	16.466	149.9	-70.7	43	5.9	-69.6	41.1	5.9
5061	218.5428	-3.5783	2013.668	13.57	16.614	10.719	285.4	65.1	-29.4	5.5	66.1	-27.7	5.7
5062	218.5845	21.9063	2013.641	14.935	17.01	8.521	62.3	61.4	-54.1	5.9	60.7	-58.1	6
5063	218.5915	25.5943	2013.756	14.799	15.863	7.898	262	-95.2	18.2	5.5	-95.4	17.7	5.6
5064	218.6551	5.3963	2013.723	16.781	18.023	26.42	135.7	-69.7	-2.1	6	-70.3	-1	7
5065	218.7052	31.6892	2013.735	16.288	17.407	31.206	100.5	-97.3	-36	5.9	-93.5	-39.9	5.9
5066	218.7279	50.2288	2013.7315	13.068	14.798	7.833	270.1	-77.8	63.8	5.9	-78.7	62.3	5.9
5067	218.7346	5.3691	2013.8605	12.979	14.757	16.446	122.4	-14.3	-116.1	5.9	-14.5	-114.2	5.9
5068	218.7465	20.1081	2013.588	16.265	18.458	58.22	298	-91.6	-25.9	5	-88.1	-23.8	5.9
5069	218.8412	33.8156	2013.832	9.649	14.655	11.584	157.7	-91.4	125	5.8	-91.8	127.4	5.8
5070	218.8798	-8.7669	2013.7455	13.375	15.559	25.331	65.2	-113.3	-27.6	5.5	-112.3	-30	5.8
5071	218.9414	4.9692	2013.8395	16.155	16.256	51.811	335.4	42.9	-175.9	6	43.2	-169.9	6.1
5072	218.9593	0.7092	2013.8205	14.872	16.38	49.835	138.1	82.2	-87	6	78.2	-88.1	6.1
5073	218.9966	32.5784	2013.8305	15.402	16.025	6.273	192.5	-100.3	-2.9	5.9	-97.7	-4.8	5.8
5074	218.9976	34.6899	2013.8425	10.92	14.916	7.152	112.8	-78.8	70.9	5.9	-82.8	70.7	5.9
5075	218.9990	56.8113	2013.5755	16.889	17.164	5.775	37.9	21.2	-100.8	6	20	-103.5	6.1
5076	219.1123	42.1741	2013.4745	16.441	18.551	5.802	209.3	72	-62.7	5.6	74.3	-65.1	6.5
5077	219.1172	-5.6059	2013.649	11.661	16.409	13.955	301.8	-130.4	10.7	5.5	-134	9.4	5.9
5078	219.1290	52.1625	2013.734	11.341	12.699	25.576	238.9	-60.6	-20.1	5.6	-60.5	-20.3	5.6
5079	219.2286	9.9971	2014.053	14.81	18.157	7.019	277.2	-66.4	17.8	5.9	-62.3	27.9	7.3
5080	219.2425	74.1752	2013.505	15.629	15.63	8.886	233.3	-42.7	71.4	5.7	-45.9	72.4	5.7
5081	219.3709	7.2954	2013.8175	13.943	15.172	6.634	315.6	-13.6	-73.4	5.9	-9.4	-75.2	6
5082	219.4575	-3.1030	2013.7545	12.971	15.929	14.111	154.3	-46.1	108.5	5.5	-42.7	109.8	5.6
5083	219.5000	25.0263	2013.7475	15.361	16.317	4.683	343.4	-88	-33	5.6	-84.1	-38	5.6
5084	219.5255	49.4872	2013.6085	15.098	15.632	7.752	327.6	-88.4	57.6	5.7	-88.3	57.3	5.7
5085	219.6113	40.0823	2013.8575	13.85	15.354	6.208	273.2	-48.1	-63.8	5.8	-50.1	-64.4	5.8
5086	219.6624	29.5120	2013.8485	11.932	14.919	26.364	324	-64.2	-25.5	5.5	-63.7	-26.4	5.5
5087	219.6691	17.9952	2014.099	15.157	15.338	4.712	278.3	35.8	-134.7	5.8	38.4	-131.6	5.8
5088	219.7186	34.2133	2013.626	15.333	16.814	6.407	217.6	-113.1	11.9	5.9	-116.5	13.5	6
5089	219.7238	35.6200	2013.661	16.575	17.523	22.984	335.3	-71.6	70.8	5.9	-75.4	63.7	6.1
5090	219.7413	3.4133	2013.7235	13.551	15.301	14.548	69.5	-67.2	-15.4	5.9	-65.3	-17.7	6
5091	219.8239	50.4646	2013.7595	13.891	16.372	5.2	15.2	30.6	-98.2	5.7	21.1	-104.7	6
5092	219.8765	5.3530	2013.8025	13.574	13.788	10.261	227.6	-111.2	-54.8	6	-108.6	-54	6
5093	219.8791	-8.3580	2013.7985	15.21	17.409	38.698	189.5	-96.5	-11.1	5.6	-101.5	-2.6	6.3
5094	219.8849	51.9064	2013.67	14.378	15.378	13.039	102.7	-99.3	75.5	5.7	-97.9	78.4	5.7
5095	219.9052	-9.9216	2014.058	14.834	15.803	13.765	197.8	-65.7	29.6	5.5	-66.6	30.1	5.5
5096	219.9214	-3.2460	2013.7565	15.54	16.093	7.68	297.9	11.7	-61.8	6.4	9.2	-63.8	6.5
5097	219.9220	11.8348	2013.858	11.733	12.768	21.513	123	34.2	-89.5	5.9	38.9	-90.7	5.9
5098	220.0162	-1.5614	2013.7705	12.682	13.897	9.284	314	-92.6	6.2	6.4	-96.4	6.8	6.4
5099	220.0564	51.1509	2013.6465	11.612	13.196	5.208	289.4	60.3	-62.4	5.7	61.8	-65.1	5.7
5100	220.0695	48.7092	2013.7	7.916	10.85	7.219	102.4	-65	44.3	5.6	-64.7	44.3	5.7
5101	220.0990	18.9031	2013.575	17.724	18.072	5.764	102.4	-2.9	-60.5	6.1	3.2	-61.4	6.1
5102	220.1495	2.4777	2013.854	14.508	15.653	6.637	173.8	-74.2	-46.6	6	-71.4	-45.5	6
5103	220.1520	-7.3733	2013.9425	11.061	16.499	16.141	275	-79.3	2.6	5.5	-79.2	0	5.6
5104	220.1583	14.6969	2013.465	11.776	13.098	4.536	112.6	-42.2	-133.3	5.2	-42.9	-131.8	5.6
5105	220.1708	58.1930	2013.673	9.056	14.027	12.854	170.4	-30.2	-76.4	5.6	-33.9	-73.7	5.6

5106	220.2546	57.9572	2014.0705	7.328	7.95	7.481	50.2	142.3	-167.8	5.7	145.7	-177.1	5.9
5107	220.2656	33.7432	2013.7695	13.161	17.536	16.495	140.7	-102.1	39.8	5.5	-97.4	41.9	5.6
5108	220.3022	76.9008	2013.6175	11.644	12.713	3.905	303.5	-42.2	-113.4	5.6	-36.5	-110.8	5.9
5109	220.3345	33.6124	2013.755	14.237	15.392	20.112	28.5	-72.8	19.1	5.5	-74.3	19.7	5.5
5110	220.3496	11.6859	2013.7445	14.013	16.356	36.501	257.5	11.2	-102.9	6	16.1	-100.8	6.1
5111	220.3733	70.5749	2013.4885	11.417	12.574	15.493	177.2	-54.4	72.9	5.7	-51.2	72	5.7
5112	220.3788	9.7189	2014.0795	17.793	18.689	22.57	85.8	-94.9	-42.6	6.1	-97.6	-39.9	6.2
5113	220.3851	29.4259	2013.796	14.123	16.015	5.166	342	-96	86	5.6	-95.1	86.2	5.7
5114	220.4057	15.0042	2013.66	16.109	16.255	30.962	5.1	-3.6	-76.2	5.9	4.1	-74.4	5.9
5115	220.4375	-6.7039	2013.709	15.788	17.148	7.687	199.3	-69.8	-25.8	5.5	-71.4	-30.2	5.8
5116	220.4481	21.4966	2013.7615	11.786	16.493	16.18	34.2	-99.6	-44.8	4.9	-100.8	-45	5
5117	220.4692	-5.7754	2013.658	16.364	16.511	4.428	240.6	12.1	-91.1	6.9	21.3	-90.8	7.1
5118	220.4781	-10.9571	2014.2575	10.191	11.489	7.192	237	-37	-66.1	5.4	-42.4	-60.7	5.4
5119	220.5115	6.4944	2013.679	11.769	16.642	35.135	230.1	-61.5	-18.9	5.9	-61	-18.8	6.1
5120	220.5127	-4.8671	2013.4745	16.072	17.019	9.55	52.8	-77.9	8	6.6	-81.2	6.8	6.8
5121	220.5170	16.8932	2013.747	15.387	16.213	32.206	266.1	-72.3	13.2	5.2	-75	13.3	5.2
5122	220.5267	0.6522	2013.812	16.453	17.54	6.785	103.7	-42.2	-69.7	6	-39.7	-74	6.6
5123	220.5422	13.5378	2013.792	12.231	15.463	20.408	296.3	-121.7	-6.9	5.2	-122	-10.4	5.3
5124	220.5651	29.6513	2013.8615	10.842	14.423	10.243	9.3	-72.4	102.8	5.9	-74.5	101.6	5.9
5125	220.5793	26.2450	2013.643	14.938	16.143	13.866	316.5	64.2	-116.6	6	62	-117.5	6
5126	220.6347	20.7022	2013.7355	15.659	16.179	6.956	311.5	-103.9	-2.2	6	-106.3	-1.3	6
5127	220.6560	52.6943	2013.7385	11.679	15.802	7.426	108.9	-88.4	37.6	5.6	-86.9	37.5	5.5
5128	220.6851	-0.6656	2013.811	9.831	15.683	11.847	49.7	0.2	-74.2	5.5	2.9	-72.3	5.7
5129	220.6992	26.7809	2013.579	12.562	13.094	5.118	235.1	-67.1	-7	6	-68	-7.3	6.1
5130	220.7057	-9.6203	2014.0495	13.137	15.505	18.628	169.1	-116.1	1.6	5.4	-116.8	-0.3	5.5
5131	220.7557	61.4267	2013.913	11.689	16.588	7.082	169.7	-65	16.3	5.6	-63.5	27.2	5.6
5132	220.7595	13.1658	2013.904	10.202	10.336	6.639	137.1	-6.4	71.4	5.1	-4.9	71	5.2
5133	220.7747	15.8253	2013.855	11.196	12.233	11.797	250.9	-62.7	-38.2	5.2	-62.3	-38.3	5.2
5134	220.8138	17.1370	2013.367	11.054	11.122	4.166	332.5	-93	-8.7	5.4	-89.3	-12.5	5.4
5135	220.8378	-7.3780	2014.0405	11.622	15.367	5.999	250.3	-70	-110.1	5.5	-71.3	-111.1	6.2
5136	220.9105	25.9792	2013.603	15.692	17.329	8.205	182.5	-62.7	-0.8	6	-63.1	-4.3	6.1
5137	220.9424	-0.1447	2013.8445	13.078	13.527	12.042	113	32.4	-104.9	5.5	32.6	-105.6	5.5
5138	220.9722	48.0052	2013.633	16.072	17.224	34.698	259.2	-65.5	-34.6	6	-65.3	-36.2	6
5139	220.9856	15.2715	2013.58	15.386	17.587	24.135	105.6	-62	1.7	5.2	-63	2.1	5.4
5140	221.0672	6.9081	2013.795	11.448	15.476	22.647	221.7	-124.4	61.3	6	-129.2	63.4	6
5141	221.0918	45.6601	2013.59	15.615	16.472	7.382	196.5	-72.4	40	5.3	-71.1	39.8	5.3
5142	221.0976	16.6532	2013.664	16.308	17.138	19.335	48.1	5.2	-81.1	5.3	3.9	-84.1	5.3
5143	221.1219	-12.0801	2013.965	15.006	15.125	35.382	91.2	-36.2	-60.4	6.3	-33.3	-60.8	6.5
5144	221.2359	11.7760	2013.815	12.77	15.085	5.64	321.2	-91.5	-30.6	6.3	-87.3	-34.4	6.4
5145	221.2739	36.4885	2013.755	10.792	13.646	8.968	5.9	-66	-4.8	5.2	-63	-3.4	5.2
5146	221.3734	42.3827	2013.699	7.307	11.609	55.245	288.4	-78.4	65.6	5.3	-74.3	67.4	5.3
5147	221.3796	13.3206	2013.6955	14.05	16.76	17.351	118.1	63.6	-64.9	5.9	60	-65.8	6
5148	221.4704	45.6990	2013.657	15.25	18.29	30.787	106.5	11.7	-77.4	5.3	19.1	-79.7	5.4
5149	221.5142	44.1112	2013.5885	13.365	16.861	21.042	175.7	-83.8	38.8	5.3	-85.3	43.6	5.3
5150	221.5153	49.6935	2013.5995	12.212	17.433	29.876	191.3	-67.2	6.9	5.3	-70.4	8.4	5.4
5151	221.5351	57.9855	2013.6995	13.558	14.232	5.234	33.3	0.8	64.6	5.6	-2.9	61.6	5.6

5152	221.5355	10.9833	2013.7735	11.925	15.667	20.668	86.4	82.3	-165	6	81.4	-165.3	6.1
5153	221.5739	-3.5999	2013.538	15.809	17.158	5.905	259.9	-72.8	38.4	5.5	-72	34.6	6.2
5154	221.5988	67.8895	2013.679	15.4	16.002	3.709	200.4	-93.1	22.2	5.7	-89.8	26.1	5.8
5155	221.6237	6.8772	2013.197	18.341	18.818	9.826	354.7	-62.6	5.1	6.8	-63.5	5.2	7.7
5156	221.6433	75.6564	2013.585	10.19	15.304	10.506	325.6	126.1	-23.5	5.6	125.8	-20.4	5.7
5157	221.7276	-8.2549	2013.817	12.33	15.567	22.696	207.6	-70.6	-9.7	5.5	-72.4	-8.3	5.6
5158	221.7707	12.1700	2013.922	13.349	15.711	5.149	162.1	-69.4	24.7	5.2	-70.4	28.6	5.9
5159	221.8447	18.0719	2013.753	14.169	15.963	6.118	160.2	-72.3	21.9	4.9	-70.6	23	5
5160	221.8853	37.6179	2013.7665	12.268	17.918	20.792	343.1	12.9	-73.1	5.2	15.4	-76	5.3
5161	221.8990	11.5698	2013.826	12.493	15.173	38.594	171.8	-37	-171.3	6	-37.9	-170.2	6
5162	221.9023	-6.8085	2013.53	14.078	17.44	32.929	288.8	-123.8	-40.7	5.6	-128.4	-38.7	6.1
5163	221.9092	3.8860	2013.798	14.475	14.804	5.49	65.6	78	-19	6	77.9	-25.3	6
5164	221.9240	-9.5203	2013.6925	14.997	15.725	9.072	230	8.8	-82.9	5.6	7.8	-82.5	5.7
5165	221.9722	19.4916	2013.695	18.125	18.261	19.073	104.8	13	-65.1	5.1	11.1	-66.2	6.1
5166	222.0109	46.9664	2013.5115	15.931	17.029	8.598	290.9	38.4	-79.7	5.3	36.4	-82.3	5.4
5167	222.1301	47.8470	2013.478	15.141	17.219	5.152	210.7	-71.9	13.1	5.3	-70.7	18.1	5.5
5168	222.1890	32.5398	2013.8075	11.415	12.742	10.665	42.1	-64.6	27.1	5.2	-61.6	27.3	5.2
5169	222.2252	-7.1593	2013.8665	11.854	14.483	25.031	350.9	-60.5	-0.8	5.5	-60.1	1.2	5.5
5170	222.2334	14.3192	2013.7965	12.919	14.024	5.175	188.1	-31.5	62	5.9	-32.4	62.5	5.9
5171	222.2432	-4.9961	2013.3175	16.659	17.782	23.378	208.5	-68	45.3	5.8	-70.1	42.9	6.5
5172	222.2473	17.6162	2013.592	10.677	13.771	11.624	287	-22.5	-98.2	5.3	-23.3	-101.6	5.3
5173	222.2856	7.3383	2013.7315	12.327	14.954	28.47	42.3	10.8	-72.3	5.9	10.1	-70.2	6
5174	222.3664	45.2806	2013.7175	10.296	11.601	7.319	180.7	-161.4	-7.3	5.9	-155.7	-1.2	5.9
5175	222.4269	-4.1015	2013.993	11.334	15.145	8.966	47.8	170.5	-136.2	5.5	164.9	-144.6	5.5
5176	222.4700	-1.8525	2013.837	12.32	15.017	53.487	189.6	-75.3	15.1	5.5	-77.8	11.6	5.6
5177	222.4926	0.0224	2013.7635	12.077	12.63	10.44	309.7	-135.3	-50.8	5.5	-134.6	-50.5	5.6
5178	222.5274	55.6469	2013.611	9.358	15.309	11.337	73.2	61	-124.6	6	58.2	-123.6	6.1
5179	222.5784	12.1900	2013.5975	10.979	18.006	34.367	127.9	9.7	-66.3	5.2	10.5	-65.4	5.5
5180	222.7309	67.3200	2013.548	10.447	12.202	11.092	6	-116.1	100.6	5.7	-112	109.9	5.7
5181	222.8057	5.8055	2013.8075	13.845	13.9	25.547	192.5	18.2	-76.9	6	15.5	-75.3	6
5182	222.8235	53.1356	2013.75	13.034	18.375	44.27	318.8	-120	25.1	5.3	-123.4	20.1	5.4
5183	222.8415	8.7740	2013.71	14.9	15.57	14.059	144.7	-141.2	-49.9	6	-141.1	-52	6
5184	222.9058	-9.5101	2013.857	15.704	16.434	12.73	12.7	-16	-85.9	5.5	-18.2	-88.6	5.7
5185	222.9463	-5.5010	2013.951	11.123	11.903	49.549	221.7	-62.7	-1.7	5.5	-61.6	-3	5.4
5186	222.9823	42.3620	2013.5725	14.975	16.258	6.402	103.9	-72.9	-64	6	-67.9	-67.4	6.1
5187	223.0528	31.0355	2013.769	11.873	13.44	12.038	187.6	-139.5	-65.3	5.2	-139.8	-65.9	5.2
5188	223.1003	32.6302	2013.8515	15.859	16.522	5.119	44.3	-38.1	-144.4	5.2	-39.5	-142.3	5.2
5189	223.1065	-3.7179	2014.167	15.323	15.603	3.611	304.7	-87.3	63.9	5.5	-86.6	63.2	5.6
5190	223.1610	35.6071	2013.742	11.359	15.883	41.122	186.4	-82.5	55.2	5.2	-82.9	51.1	5.2
5191	223.1774	-11.8732	2014.4125	9.928	10.896	16.647	108.1	-178.5	34.6	5.4	-175.5	40.6	5.3
5192	223.1806	9.3335	2013.6535	12.074	14.721	15.917	139.3	-3.5	-104.2	6	-4.6	-107.4	6
5193	223.2122	55.9454	2013.5845	10.976	15.553	35.164	359.4	15.4	-67.6	5.6	19.8	-66.4	5.7
5194	223.2578	5.0018	2013.777	13.932	17.119	8.114	29.6	18.4	-68	5.9	13.2	-72	6.2
5195	223.3224	-2.5521	2013.6425	14.618	17.855	20.731	162.8	-70.2	2.7	5.5	-71.9	-2.9	5.8
5196	223.3500	32.1862	2013.7885	11.309	15.947	51.656	356.7	-66.8	-34	5.2	-66.4	-35.7	5.2
5197	223.3842	42.0701	2013.712	10.661	12.084	48.269	357.2	-75.7	-9.8	6	-75.8	-8.2	6

5198	223.3889	17.5563	2013.362	18.243	18.646	19.445	84.4	1.9	-63.1	6.4	-8.1	-61	6.6
5199	223.4137	7.3573	2013.682	14.776	15.246	23.963	180.3	-62.2	-31.5	6	-62.4	-32.9	6
5200	223.4322	17.7994	2013.839	11.853	13.342	4.545	314.6	-118.6	-14.3	6	-115.2	-14.1	5.9
5201	223.4324	18.5648	2013.0735	18.477	18.599	4.584	281.3	-82.5	-1.9	6.7	-80.2	-1.7	6.9
5202	223.5032	35.9184	2013.832	14.214	15.191	5.062	332.4	-63.5	-39.5	5.1	-64.7	-38.1	5.2
5203	223.5574	6.3765	2013.789	16.004	16.127	4.071	104	-22.1	-68.8	6.1	-23.3	-70.9	6.1
5204	223.5811	15.9024	2013.707	13.25	15.123	9.955	95	-75	-59.1	6	-71.9	-62.9	6
5205	223.5820	82.4061	2013.892	13.1	16.223	14.437	160.5	-82.9	16.9	5.5	-86	17.6	5.6
5206	223.6165	25.1577	2013.732	14.935	15.251	5.078	217.8	10.8	-84.7	5.6	9.9	-88.1	5.6
5207	223.6299	46.3319	2013.7125	12.902	15.707	22.796	35.1	-4.5	-74.9	6	-8.2	-77	6
5208	223.6728	58.7469	2013.789	14.607	15.491	4.388	287.3	-114.4	6.6	5.6	-112.3	8.3	5.6
5209	223.6860	19.3992	2013.593	17.956	18.715	21.969	265.6	-62.6	14.2	5.1	-61.7	20.2	6.3
5210	223.6868	-4.4106	2013.8795	14.131	14.733	25.147	112.8	-67.3	-36.3	5.5	-67.5	-34.4	5.5
5211	223.7295	23.1135	2013.604	15.924	17.159	6.638	37.2	-128.5	91.7	5	-127.7	89.5	5.2
5212	223.7897	67.6314	2013.4935	12.525	15.312	52.097	196.9	-79.6	36.3	5.7	-87.8	26.7	5.7
5213	223.9182	-14.4401	2014.45	15.542	15.543	45.369	128.3	-139	-105.1	5.8	-132.8	-106.7	6.7
5214	223.9848	25.9919	2013.7775	14.674	15.008	3.46	36	-94.9	-33.2	5.8	-94.5	-38.7	5.6
5215	224.0279	64.2294	2013.411	17.014	17.258	4.148	32.1	-64	25.2	6.1	-61.1	29.9	6.3
5216	224.2375	14.2483	2013.6135	16.18	17.798	21.517	209.9	20.8	-99.1	5.9	23.3	-100.3	6.1
5217	224.2568	45.1807	2013.789	13.208	16.312	4.984	184.4	15.7	-65.8	6	21.3	-67.4	6.6
5218	224.2585	4.0072	2013.751	12.462	14.519	19.803	247.7	-67.2	20.9	6	-66.9	19.3	5.9
5219	224.2873	60.9703	2013.7415	13.371	14.035	4.859	48.7	-39.5	-82.3	6	-38.8	-80.3	6
5220	224.3490	50.5025	2013.845	11.904	15.543	7.36	7.4	-100.1	44.3	6	-96.5	49	5.9
5221	224.3543	85.4905	2013.964	9.801	11.522	8.263	99.4	-113.2	70.4	5.5	-111.7	69.3	5.5
5222	224.4103	21.1528	2013.3625	11.398	16.407	12.124	167.3	-136.8	-70.4	6.2	-137	-67.4	6.3
5223	224.4135	7.4991	2013.6335	12.731	15.687	17.5	153.1	-70.6	-75.8	6	-70.3	-75.5	6.1
5224	224.4363	-2.9992	2013.878	13.497	13.939	30.82	187.7	-50.1	-76.4	5.5	-49.3	-77.1	5.5
5225	224.5160	75.8237	2013.5075	14.214	16.068	23.32	18.8	-7.4	76.2	6	-10.8	78.1	6.1
5226	224.5523	68.9509	2013.542	13.025	13.231	33.218	69	-63.4	35.6	5.7	-63.4	36.2	5.7
5227	224.6407	-5.0623	2013.8055	10.612	12.225	6.778	187.8	-57.2	-61.9	5.5	-54.2	-62.5	5.5
5228	224.6408	41.1481	2013.55	12.113	17.994	10.874	27.6	-43.9	82.8	5.2	-42.5	87.2	5.5
5229	224.6586	3.9218	2013.6775	9.751	13.945	38.812	65.7	-87.8	-11	6	-85.7	-11.9	6.1
5230	224.6632	45.1410	2013.6985	14.181	16.758	5.857	288	-79	-44.9	6	-82.3	-45.2	6
5231	224.6940	19.2943	2013.505	9.576	16.777	38.288	212.2	-98	-45	5	-93.5	-46.6	5
5232	224.7228	-11.1443	2013.9615	5.476	9.812	19.709	38.4	-107.7	-66.6	5.6	-110.1	-64.4	5.6
5233	224.7395	40.0325	2013.699	17.557	17.772	39.862	282.1	-95.4	-75	5.4	-101	-76.9	5.3
5234	224.7682	17.2412	2013.635	9.81	15.891	10.559	247	-89.6	8.8	5.2	-90.1	1.3	5.4
5235	224.7982	6.8081	2013.672	11.016	11.328	17.171	104.1	-31.2	-70.5	6	-31.9	-69.5	6.1
5236	224.8668	28.1435	2013.713	14.612	16.885	10.447	105.6	-77.3	-16.3	5.6	-77.8	-13.7	5.7
5237	224.8717	16.4326	2013.707	13.588	16.071	9.326	198.4	45.7	-146.1	5.2	45.2	-149.2	5.3
5238	224.9302	-5.9184	2013.757	9.958	12.731	8.25	21.2	-89.8	15.7	5.5	-89.5	15	5.6
5239	224.9315	53.8541	2013.6	16.303	16.531	6.874	22.5	-97.1	3.2	6	-98.3	4.8	6
5240	225.0109	68.2687	2013.4965	11.286	14.383	5.814	280.8	-95	19.2	5.7	-98.3	26.2	5.9
5241	225.0819	5.0059	2013.549	12.665	17.16	8.955	230.6	-126.3	17.2	6.1	-127.4	24	6.3
5242	225.1041	72.5262	2013.5875	11.89	11.964	25.181	16.8	-66.1	46.7	6	-65.7	46.1	6
5243	225.1126	47.2302	2013.742	11.157	13.865	19.898	330.9	-67.4	-4.9	5.9	-70.1	-3.8	5.9

5244	225.1174	65.4299	2013.6095	11.78	14.485	12.075	250.8	-0.1	-79.6	5.6	3.6	-83	5.6
5245	225.1837	40.1747	2013.712	16.399	17.166	10.888	73.5	14.8	-87.1	6	16.3	-86.3	6.1
5246	225.1973	23.1078	2013.741	14.591	15.415	26.211	338.9	-165.6	124.7	6	-165.1	127.5	6
5247	225.2162	13.2459	2013.725	13.96	14.177	19.246	234.5	26.5	-157.8	5.3	34.9	-158.8	5.3
5248	225.2734	33.4746	2013.8515	12.236	15.493	7.09	116.6	17.2	-77.3	5.1	15.4	-75.3	5.2
5249	225.2935	-2.4041	2013.7775	13.539	17.525	23.243	20	-37.3	-118.2	5.5	-36.8	-123.5	6
5250	225.3275	6.3585	2013.403	17.487	18.493	23.745	46.6	-2.1	-70.1	6.4	7.7	-68.5	6.8
5251	225.4303	53.5893	2013.666	15.477	15.703	4.082	268.1	-89.6	44.4	6	-92.5	44.6	6
5252	225.4923	-3.0996	2013.876	15.631	17.219	9.341	113.1	41.7	-122.5	5.5	37.3	-123.2	5.7
5253	225.5047	60.9528	2013.666	14.694	14.706	26.836	341	37.4	-89.4	5.6	39	-88.8	5.6
5254	225.5311	12.1691	2013.9045	9.757	12.232	8.545	26.7	37.7	-66	5.2	40.8	-66.2	5.2
5255	225.6451	47.0921	2013.204	14.221	19.359	51.95	255.3	33.9	-86.7	5.6	24.8	-90.7	7.4
5256	225.7079	66.9867	2013.513	12.065	12.641	10.412	318.2	-71	39.6	5.7	-70.2	40.6	5.7
5257	225.8065	39.4540	2013.715	15.666	17.388	9.228	115.8	-80.5	-65.9	6	-78.1	-65.2	6.1
5258	225.8142	18.0561	2013.8385	12.111	15.794	5.821	269	-16.9	-64.3	5	-18.6	-64.8	5.1
5259	225.8158	23.9962	2013.678	15.855	17.809	8.076	173.4	-75.4	-45.1	5.6	-78.7	-47.6	5.7
5260	225.8392	13.3018	2013.724	9.869	17.038	25.714	318.1	-77.1	-105.1	5.2	-75.4	-105	5.4
5261	226.0037	42.5601	2013.5155	17.118	17.539	53.578	173.4	27	-75.9	5.6	27.6	-74.6	5.7
5262	226.0183	-10.4973	2013.945	13.398	15.702	14.442	178.7	-127.1	0.2	5.5	-124.3	1.7	5.5
5263	226.0387	19.3670	2013.3935	17.438	17.866	45.682	133.3	-64.8	-15.9	5.1	-63.6	-20.8	5.7
5264	226.0654	18.0572	2013.506	12.066	15.231	16.238	183.3	-85.9	-88.2	5	-86.1	-85.5	5
5265	226.1049	71.7652	2013.49	13.046	16.679	6.913	59.3	-74.6	40	5.7	-78	33.5	5.9
5266	226.1549	22.7421	2013.588	16.948	19.079	10.907	132.7	-23.1	-91.1	5.1	-20.2	-88.9	6.1
5267	226.1761	77.0560	2013.5815	14.466	14.844	4.134	350.6	-71.4	27.8	5.7	-74.4	28.7	5.8
5268	226.1872	34.7259	2013.801	15.381	15.84	47.513	150.1	-109.1	60.9	5.2	-111.6	63.4	5.2
5269	226.2349	13.7231	2013.862	10.169	11.425	9.977	244.5	15.5	-104.9	5.5	21.4	-107	5.5
5270	226.2952	8.0107	2013.7125	10.421	14.606	38.539	149.6	-66.1	15.3	6	-63.6	20.4	6
5271	226.3096	20.9372	2013.3495	14.34	16.133	5.182	340.4	7.8	-92.1	6.1	5.9	-92.5	6.3
5272	226.3235	41.5522	2013.5675	16.471	16.885	14.79	189.4	-98.4	14.9	5.3	-98.1	12	5.3
5273	226.3418	18.1352	2013.4325	12.597	13.707	5.296	81.6	-79.4	-9	5	-82.5	-6.8	5
5274	226.3501	21.0284	2013.4145	14.21	18.006	47.099	222.6	-63.2	17.9	5	-65.3	21.2	5.1
5275	226.3639	3.5022	2013.6945	11.962	13.719	7.745	10.2	-26.2	-73.2	6	-23.6	-71.5	6
5276	226.4797	58.1369	2013.5515	13.479	13.937	20.743	150.4	82.3	-71.6	5.6	79.9	-74.5	5.6
5277	226.4818	10.4936	2013.265	16.592	18.388	11.53	77.6	-40.8	-96.8	6.1	-39	-101.5	6.8
5278	226.5521	-6.4242	2013.9285	13.998	16.475	6.553	20.6	-76.5	-8.7	5.5	-79.3	-9.8	5.6
5279	226.5636	28.3143	2013.731	16.843	18.525	10.246	4.4	65.6	-95	5.8	71.8	-94.4	5.9
5280	226.5785	32.0499	2013.733	12.835	16.966	8.297	242.5	-52.7	-68.5	5.2	-51	-65.7	5.4
5281	226.6074	21.7459	2013.5045	11.573	11.989	11.206	277.1	-91.4	-20.9	5	-90.9	-16.4	5
5282	226.7191	59.0531	2013.6385	11.521	13.865	8.886	218.9	-69.2	54.7	5.6	-68.7	57.2	5.6
5283	226.7807	14.7473	2013.6495	12.148	15.161	48.949	3.3	-71.6	17.3	5.5	-69.1	18.8	5.6
5284	226.8879	70.4606	2013.532	10.383	13.377	12.114	346.4	-62.1	43.1	5.7	-66.7	42.3	5.7
5285	227.0035	4.6068	2013.534	13.784	15.921	9.237	160.8	-22.9	73.1	6	-26.5	71.2	6.2
5286	227.0225	-2.7888	2013.898	13.802	13.852	5.313	0.1	-26.2	-64.3	5.5	-27.9	-62.3	5.5
5287	227.0363	61.4385	2013.5065	16.44	17.067	12.165	308.2	-0.4	-73.9	5.7	0.7	-71.9	5.7
5288	227.0377	73.7021	2013.62	11.545	12.25	8.158	117.9	174.8	-38.3	5.7	175.8	-40.7	5.7
5289	227.0447	33.1660	2013.775	10.798	16.277	43.592	306.1	-42.3	74.9	5.2	-43	73.5	5.2

5290	227.0801	-0.6649	2013.693	15.167	16.885	9.388	124.3	-100.2	-65	5.5	-98.5	-64	5.7
5291	227.1066	47.0697	2013.681	14.119	17.086	7.341	77.2	10.7	-62.3	5.6	10	-64.2	5.6
5292	227.1066	38.2406	2013.7565	12.092	16.097	20.334	276.1	-96.8	33.7	5.2	-97.2	31.5	5.2
5293	227.1310	-6.1820	2013.872	14.213	16.302	8.967	338.5	-52.7	-62.4	5.5	-46.1	-64.7	5.6
5294	227.2117	63.1636	2013.51	14.36	16.026	14.561	82.1	-51.4	76.9	5.7	-49.4	78.4	5.7
5295	227.2845	59.0736	2013.5895	11.413	14.099	9.021	285.9	104.8	-59.7	5.6	105	-51.5	5.6
5296	227.2925	29.5460	2013.6315	13.255	17.33	13.757	340.9	-60.3	36.7	5.6	-62.1	37.9	5.7
5297	227.3414	14.6284	2013.656	10.844	17.51	41.67	104.4	-10.9	-96.8	5.6	-13.3	-101.3	5.6
5298	227.3623	50.9456	2013.714	15.442	15.982	6.81	101.3	-60.7	9.8	6	-61.8	7.8	5.9
5299	227.3770	8.7949	2013.5985	14.593	15.691	4.429	187	-109.6	-11.5	6.1	-108.5	-8	6.6
5300	227.4194	27.6510	2013.557	17.256	17.445	7.807	167.1	-125.7	-36	5.8	-128.3	-37.1	5.7
5301	227.4233	49.6961	2013.722	11.006	11.429	25.13	65.9	38.9	-122.4	6	33.7	-122.7	6
5302	227.4621	31.5019	2013.6995	16.137	17.138	5.527	158.5	-82	-59.5	5.2	-80.2	-57.5	5.3
5303	227.4650	49.1852	2013.7565	16.956	18.388	38.007	234.6	-66.8	-21.2	6	-62.9	-24.8	6.3
5304	227.4706	40.6797	2013.865	15.77	17.316	4.872	157.8	-62.4	18.4	5.2	-62.2	28	5.7
5305	227.4753	38.9759	2013.79	8.143	9.887	9.729	331.6	-74.8	24.1	5.2	-75	21.2	5.2
5306	227.4958	-7.2994	2013.85	15.391	16.61	57.616	208.8	-29.1	-63.7	5.5	-26.4	-62.6	5.6
5307	227.5104	22.3904	2013.1785	17.375	17.703	9.326	351.8	36.4	-84.8	5.2	26.1	-84.1	5.2
5308	227.5660	30.7356	2013.857	11.991	12.439	10.83	349.8	-63.7	62.6	5.1	-62.9	63.3	5.2
5309	227.5958	56.7987	2013.5435	11.951	16.574	44.986	151.9	-108.2	79.5	5.6	-106.5	80.8	5.7
5310	227.6426	26.8192	2013.682	11.638	15.319	28.526	270	-64.6	-30.2	5.6	-63.5	-29.2	5.7
5311	227.6776	38.9452	2013.814	9.004	16.01	11.476	250.8	-110.5	-13.5	5.2	-110.8	-16.1	5.3
5312	227.6909	53.2294	2013.612	12.859	14.02	5.881	321.4	-22.4	-69.8	6	-23.7	-71.4	6
5313	227.7081	46.8486	2013.793	9.563	9.691	18.654	289.5	-96.2	26.2	5.6	-96	28.1	5.6
5314	227.7245	9.4673	2013.517	13.47	16.737	17.383	193.1	-64.3	23	6	-63.2	29.5	6.2
5315	227.7736	11.5687	2013.6235	12.069	16.378	27.105	250.5	-79.6	-20.7	6	-80.3	-18.8	6.1
5316	227.7863	-8.3058	2013.663	13.265	17.706	52.033	177.1	-25.6	-77.1	5.5	-36.4	-68.6	5.8
5317	227.7919	1.8956	2013.768	11.627	16.112	9.408	187.2	-78.3	-77.1	6	-75.3	-75.9	6.1
5318	227.7950	68.8493	2013.492	14.505	15.631	6.244	9.6	98.5	5.4	5.7	100.2	7.6	5.7
5319	227.8250	27.9416	2013.7305	12.01	13.65	11.075	184.5	-50.8	82.9	5.6	-52	85.1	5.6
5320	227.8258	52.2623	2013.7725	12.925	15.514	7.263	191.1	-69.1	7.8	6	-72	7.9	5.9
5321	227.8572	29.9130	2013.7145	10.846	18.499	51.823	298.6	-70.3	48.3	5.2	-65.9	56.1	5.4
5322	227.9265	62.0895	2013.524	15.328	16.009	12.724	26	-71.1	-71.5	5.7	-74.1	-74.2	5.7
5323	227.9585	61.8574	2013.741	7.967	8.533	16.047	353.6	-155.5	96.8	5.6	-152	97.2	5.6
5324	227.9758	39.8277	2013.718	15.508	16.228	15.356	86.6	-98.3	18.4	5.2	-98.2	18.1	5.3
5325	228.0294	29.8684	2013.7905	13.989	14.192	28.386	326.9	-87.9	29.7	5.6	-87	30.6	5.6
5326	228.0721	1.0568	2013.62	13.691	15.285	12.968	195.4	-44.9	-76.6	6	-45.8	-76	6.1
5327	228.0867	12.0516	2013.389	16.255	17.512	7.691	45.3	16.4	-64.7	5.2	16	-65.9	5.4
5328	228.1316	9.7101	2013.052	17.353	18.328	32.491	65.7	-11.2	-111	6.4	-2.8	-108.9	6.6
5329	228.1472	5.2567	2013.655	10.38	13.804	12.112	68.1	-91.6	58.2	6	-96.3	60.1	6.1
5330	228.1537	22.6895	2013.6075	13.795	14.855	4.095	252.9	-96.4	-35.3	5	-93.1	-30.5	5.2
5331	228.1658	24.0932	2013.659	14.252	16.024	7.953	305.3	-50.3	-119.6	5.7	-49.7	-114.8	5.7
5332	228.2357	32.3346	2013.7045	14.027	16.347	28.917	34.2	-70.8	32.2	5.2	-74.8	25.5	5.2
5333	228.2411	31.4568	2013.7495	9.882	16.16	18.88	23.1	86.5	-109.8	5.2	81.9	-115.5	5.3
5334	228.3399	8.6389	2013.6065	13.975	16.31	24.785	152.8	-123.1	-10.7	6	-122.3	-10.8	6.2
5335	228.3679	38.9973	2013.772	13.125	13.408	9.701	272.6	0.1	-73.9	5.2	0.2	-73.7	5.2

5336	228.4147	42.2463	2013.71	11.543	12.366	30.693	232.7	-24.6	-108.6	5.6	-27.6	-108.8	5.6
5337	228.4362	-8.6205	2013.998	9.308	13.488	14.813	121.8	-35.8	-140.9	5.4	-36.5	-138.2	5.5
5338	228.4641	72.4810	2013.5365	15.102	15.362	7.271	120.1	14.6	-91.3	5.7	15.9	-88.9	5.7
5339	228.5180	37.9662	2013.723	11.347	16.167	20.071	194.1	35.4	-73.6	5.2	35	-78.1	5.2
5340	228.5252	5.1521	2013.3795	13.784	17.372	9.311	179.5	-11.6	-83.2	6.1	-14.6	-83.8	6.5
5341	228.5429	56.5249	2013.6195	13.493	17.413	42.5	302.6	-68.4	23.8	5.6	-68.3	31.1	5.6
5342	228.6117	82.2100	2013.802	12.604	15.006	47.931	206.6	20.6	-78.5	5.5	23.6	-75.7	5.6
5343	228.6824	40.5284	2013.749	11.649	16.924	9.481	114.5	-38.1	-70.7	5.2	-45.6	-66.6	5.2
5344	228.8168	0.4072	2013.6115	16.27	16.872	20.815	331.3	-71.1	-37.8	6.1	-74.9	-39	6.1
5345	228.8350	35.0466	2013.618	16.054	16.361	6.824	324.7	-51	63.2	5.3	-51.8	63.3	5.3
5346	228.8630	22.5495	2013.756	8.246	11.12	11.093	345.5	-111.3	55.7	5	-111.4	47.7	5
5347	228.8709	20.3190	2013.571	14.734	18.24	52.434	337.5	37.5	-84.4	5	30.6	-87	5.2
5348	228.8788	12.7723	2013.6445	12.158	15.354	8.783	232.5	-26.7	-87.6	6	-26.8	-88.9	6.1
5349	228.8906	32.2516	2013.693	15.052	16.484	11.844	121.9	-13.6	-62.3	5.2	-11.3	-64.3	5.2
5350	228.9226	37.1121	2013.8445	16.088	18.017	6.73	120.6	-64	-129.5	5.2	-64.9	-125.4	5.4
5351	228.9518	49.8578	2013.65	12.948	18.409	10.763	92.6	-79.8	79.2	5.3	-71.6	86.7	5.7
5352	228.9579	13.7572	2013.7335	11.642	12.596	22.823	249.9	-209.3	-51.9	6.1	-211.1	-49.9	6.1
5353	228.9771	-3.1284	2013.7515	16.156	16.18	7.4	83.5	-65	5.7	5.6	-65.3	10.2	5.6
5354	229.0056	77.2663	2013.5715	13.867	16.221	24.185	299	-96.3	4	5.7	-95.9	6.6	5.7
5355	229.0407	25.2849	2013.56	16.842	17.102	5.176	115.5	-81.3	-28.1	6	-84.6	-31	6
5356	229.0569	65.9241	2013.566	11.808	15.64	7.742	266.9	-64.1	5.2	5.6	-66.2	4.5	5.7
5357	229.0799	32.7424	2013.822	8.976	14.773	53.921	116	-67.3	66.2	5.1	-62.1	68.1	5.2
5358	229.1495	-1.8002	2013.7285	15.508	16.378	37.485	119.3	-108.6	15.4	5.6	-108.4	13.4	5.6
5359	229.2507	19.1941	2013.802	11.341	12.054	26.295	77.6	-29.3	71	4.9	-31.6	68.3	4.9
5360	229.3112	20.8029	2013.674	15.057	19.037	58.877	249.2	-100.7	-22.8	5	-101.7	-24	5.8
5361	229.3202	47.5236	2013.676	15.412	15.488	18.941	158.5	-51.6	118.7	5.7	-51.9	113.6	5.7
5362	229.3310	25.8615	2013.6575	12.326	12.453	9.402	299.7	-71.1	54.4	5.9	-70.2	54.5	5.9
5363	229.3342	67.5747	2013.5035	13.988	14.982	9.88	79.6	-0.9	-109.5	5.7	-0.8	-110.1	5.7
5364	229.3448	9.0667	2013.39	16.069	17.3	18.167	337.8	-91.2	-126.2	6.2	-98.2	-125.5	6.5
5365	229.3449	35.9367	2013.71	14.623	15.853	10.377	26.7	-100.2	-88.8	5.5	-97.6	-84.9	5.5
5366	229.3512	43.2257	2013.7085	14.734	15.798	6.964	30.6	1.7	-63.7	5.6	3.5	-65.2	5.6
5367	229.3971	1.1598	2013.686	14.451	15.821	7.397	280.5	-48.8	-137.1	6.1	-49.2	-134.7	6.1
5368	229.4007	45.2520	2013.4675	18.368	18.426	35.416	24.2	-20.6	-62.5	5.9	-13.5	-66.4	5.8
5369	229.4019	15.2873	2013.566	13.947	16.941	12.661	21.9	-188	31.6	6	-192.1	35.4	6.2
5370	229.4209	29.1722	2013.674	13.457	16.236	10.265	223.3	-109.6	31.6	5.9	-111.6	34.3	6
5371	229.4534	66.6311	2013.4065	13.334	17.304	20.566	132.4	-78.2	29.8	5.7	-79.8	32.8	5.8
5372	229.5002	55.7752	2013.6255	12.302	16.87	29.247	263.9	2.6	-71.1	5.6	-2	-68.6	5.6
5373	229.5235	16.9508	2013.904	10.055	14.221	7.315	50.6	-75.7	28.5	5.8	-80.5	26	5.9
5374	229.5679	18.8552	2013.809	8.532	15.762	18.197	108.3	-92.2	63.4	4.9	-97.3	56.2	5
5375	229.5794	24.9973	2013.742	9.543	11.106	16.984	247.7	-122.6	11.5	5.9	-127.8	17.8	5.9
5376	229.5932	70.7302	2013.5215	14.195	15.746	42.854	266.7	-71.9	87.8	5.7	-72.8	84.1	5.7
5377	229.6021	18.4199	2013.621	12.031	18.026	21.271	313	-86.1	14.4	5	-87.4	15.7	5.1
5378	229.6109	51.8134	2013.5785	13.989	15.12	4.609	274.3	55.9	-182.7	5.3	58.1	-186.1	5.5
5379	229.6162	2.0592	2013.541	14.314	18.519	54.116	319.4	6.2	-65.2	6.4	-6	-64.2	7.4
5380	229.6326	6.9582	2013.342	17.338	17.517	5.883	57.8	9.8	-86	6.3	2.1	-83.8	6.4
5381	229.6923	43.2308	2013.727	12.809	13.117	16.187	27.7	-75.2	-67.4	5.6	-76.7	-68	5.6

5382	229.7221	19.0615	2013.666	13.638	15.039	11.699	354.3	-0.9	-66	5	-6.6	-65.7	5
5383	229.7597	8.5563	2013.615	12.788	16.402	8.929	52.6	-7.4	-105	6	-7.1	-103.4	6.2
5384	229.7629	34.0661	2013.799	10.443	13.559	7.77	106.9	29.9	-182.7	5.6	31.1	-176.3	5.7
5385	229.7935	14.4027	2013.735	12.559	13.79	54.386	313.9	-79.3	-38.5	5.9	-76.6	-39.7	5.9
5386	229.8402	6.4681	2013.8935	11.264	13.722	6.426	7	-78	25.2	6	-77.6	31.1	5.9
5387	229.8996	37.0680	2013.693	13.441	18.087	11.462	178.9	-19.4	-110	5.5	-15	-114.1	5.7
5388	229.9127	-3.0679	2013.6955	14.732	17.696	7.356	229.3	-96.6	90.1	5.6	-98	88.2	5.9
5389	229.9384	-0.5350	2013.681	15.38	15.457	10.83	218.9	-73.4	25.2	5.6	-73.8	25.6	5.6
5390	229.9646	-12.2649	2013.501	16.29	16.631	21.072	88.1	-75.4	-4.3	5.9	-77.9	2.5	6.2
5391	230.0455	57.4485	2013.628	10.095	13.316	9.286	107	38.7	99.5	5.6	29.7	98	5.6
5392	230.0462	32.7386	2013.7025	10.978	16.673	19.81	76.6	18.4	-139.9	5.6	18.3	-136.9	5.7
5393	230.0629	0.2586	2013.75	10.468	16.214	45.141	240.2	-61.1	-9.4	5.9	-61.6	-9.5	6.1
5394	230.0777	44.9545	2013.6845	12.966	16.289	43.105	148.2	28.1	-66	5.2	33.8	-65.8	5.3
5395	230.1151	20.2111	2013.799	11.91	12.103	27.272	259.5	-8	-80	4.9	-8.7	-78.2	4.9
5396	230.1529	55.1373	2013.6155	13.012	16.242	17.18	101.3	-81.6	79.1	5.6	-79.7	78.9	5.7
5397	230.1579	17.1099	2013.7525	12.812	13.246	48.807	167.5	-86.6	-35	5.9	-89.7	-32.6	5.9
5398	230.1737	26.6317	2013.59	11.289	16.055	54.858	233.9	67.7	-85.5	5.9	70.6	-86.7	6
5399	230.1807	72.2391	2013.7425	14.317	14.381	4.656	22	-84.7	92.6	5.6	-84.2	93.2	5.7
5400	230.2115	39.5298	2013.7675	12.447	16.252	9.71	51.2	-97.3	-19.9	5.5	-96.3	-20	5.5
5401	230.2446	50.7748	2013.733	10.574	12.634	17.257	292	-67.7	54.6	5.3	-68.7	54.1	5.3
5402	230.2779	18.1162	2013.308	17.314	18.579	17.609	160.2	-44.7	-77	5.2	-40.1	-79.2	5.6
5403	230.3051	-1.2236	2013.69	13.495	16.552	6.319	144.3	-79.5	-8.5	5.5	-77.9	-4.3	5.7
5404	230.3058	27.7476	2013.557	16.768	17.033	59.57	184.4	-137.5	-120.3	6.1	-135.4	-120.6	6.2
5405	230.3262	24.3452	2013.879	11.797	15.568	6.828	79.1	-86.7	13.5	5.9	-88.8	14.6	6.1
5406	230.3595	53.2037	2013.653	13.996	14.111	10.937	137	-65	74	5.3	-62.3	75.6	5.3
5407	230.3954	39.3665	2013.7335	12.528	14.932	24.942	40.6	-80.3	45.6	5.6	-82.2	41.2	5.6
5408	230.4381	-2.0143	2013.8405	8.598	10.811	14.101	165.8	36.1	-66.8	5.5	34.4	-67.9	5.5
5409	230.4563	25.8915	2013.6625	13.876	14.825	4.707	29.9	35	-91.5	5.9	34.5	-92.7	6
5410	230.5411	45.1315	2013.798	9.274	9.821	25.695	4.8	-76.7	-28.7	5.2	-74.1	-26.5	5.2
5411	230.5440	-2.6843	2013.74	13.342	18.033	18.173	244.6	-92.2	-24.1	5.5	-88.1	-27.5	5.9
5412	230.6785	34.8789	2013.767	16.615	16.929	5.97	82.6	-0.7	-68.4	5.2	1.9	-68	5.2
5413	230.6986	59.7149	2013.4415	17.379	18.483	14.455	114	1.8	-175.8	5.8	4.1	-168.1	6.4
5414	230.7919	16.2258	2013.573	14.288	17.99	17.222	44.3	-190.7	38.1	6.1	-192.2	37.6	6.5
5415	230.7961	-10.4877	2013.9365	8.157	16.041	22.864	317.6	-21.8	-148.3	5.5	-20	-149.5	5.6
5416	230.8539	11.1752	2013.65	13.759	13.838	23.5	54.2	-31.7	-80.5	6	-29.8	-84.2	6
5417	230.8585	43.3344	2013.654	13.077	16.505	11.324	184.7	-96.8	53.2	5.2	-100.5	49.6	5.4
5418	230.8661	-13.0987	2014.3265	12.064	12.138	12.115	179.3	-81.1	54.2	5.4	-81.4	55.2	5.4
5419	230.9392	44.7989	2013.6765	12.74	15.825	26.32	85.9	-71.2	-47.5	5.3	-67.9	-48.3	5.3
5420	230.9713	20.1607	2013.635	14.912	17.176	9.956	241.1	11.8	-65.4	5.9	11.2	-66	6.1
5421	231.0747	39.7068	2013.693	14.103	15.947	6.644	138.4	-41.1	179	5.7	-44.2	176.1	5.7
5422	231.1184	25.4465	2013.5015	14.588	17.62	29.938	115.4	-208.5	-47.3	5.8	-211.7	-44.7	6.1
5423	231.2248	13.9818	2013.883	9.492	9.953	11.479	95.8	-52.5	-98.4	5.9	-48.2	-97.8	6
5424	231.2249	75.3341	2013.42	16.889	16.893	30.486	137.4	70	-2.7	6.2	66.6	-9	6.2
5425	231.2480	16.1827	2013.786	10.74	13.1	35.783	300.5	-85.6	11.5	6	-85.8	12.3	6
5426	231.2647	63.3786	2013.4335	16.932	17.488	8.117	268.3	-85.9	-8.9	5.7	-81.4	-11.1	5.8
5427	231.2796	57.5018	2013.5375	15.502	17.44	32.901	348.5	38.9	-60.5	5.6	41.5	-60.4	5.7

5428	231.2808	38.2602	2013.707	12.024	18.048	13.788	328.8	-61.4	-27	5.6	-62.4	-19.7	5.8
5429	231.2820	37.5810	2013.732	17.913	18.261	20.879	58.2	-61	11.8	5.7	-64.4	0.6	5.8
5430	231.3428	16.7020	2013.51	15.658	17.428	5.952	322.9	-42.3	-73.8	5.5	-43.6	-74.5	5.7
5431	231.4596	31.2802	2013.767	12.883	13.289	5.827	48.7	-68.5	29.7	5.9	-66.5	29.6	5.9
5432	231.4628	3.6626	2013.4855	12.456	15.803	24.894	36.2	-24	-69	6.1	-25.5	-68.2	6.2
5433	231.4670	12.5345	2013.465	14.741	16.479	46.676	31.9	0.1	-76.1	5.6	-0.6	-72.9	5.7
5434	231.4931	13.1657	2013.5275	11.072	13.546	5.893	227.6	-60.8	74.4	5.6	-60.8	73.9	5.8
5435	231.5161	2.0690	2013.5035	14.617	17.936	13.219	109.8	-96.7	-8.6	6.1	-94.1	-10.1	6.3
5436	231.5453	-2.8574	2013.7405	14.616	14.711	21.363	222.6	-19.7	-99.5	5.5	-20.5	-101.3	5.5
5437	231.5561	32.5649	2013.767	12.269	14.613	9.912	265	15	-65.9	5.9	15.3	-68.9	5.9
5438	231.5675	18.0342	2013.509	11.795	17.028	7.248	35.9	-70.2	-40.6	4.9	-60.4	-52.3	7.8
5439	231.6495	18.0839	2013.7125	14.939	15.299	12.516	138.5	-105.6	5.5	5	-104.6	8	5
5440	231.6820	-11.6299	2013.6445	10.494	17.484	18.584	199.8	-72.6	-59.3	5.5	-75.8	-61	6.4
5441	231.7079	35.6696	2013.627	16.328	17.317	5.187	173.9	1.8	-83.1	6	-1.1	-84.4	6.1
5442	231.7527	31.3843	2013.768	13.637	16.21	8.054	6.2	-73.9	19.3	5.9	-72	19.9	5.9
5443	231.8100	23.4310	2013.7135	14.143	16.286	5.487	245.3	-67.8	-16.7	5.9	-62.6	-22.7	6.1
5444	231.8513	-1.4967	2013.6145	12.848	16.128	9.96	8.1	1.9	-149.6	5.6	0.8	-146.1	5.7
5445	231.8545	13.2395	2013.64	13.198	15.857	8.344	195.9	-2.2	-104.3	5.6	-3.3	-102.9	5.7
5446	231.8622	-6.0630	2013.746	14.078	16.762	15.959	71	-84.3	-85.9	5.5	-79.8	-89.5	5.7
5447	231.8701	-9.4963	2013.963	8.763	15.061	14.024	60.8	-196.1	-83.9	5.5	-194.9	-84.9	5.6
5448	231.9459	50.9939	2013.6755	12.778	14.371	4.911	35.3	-22.3	-75.6	5.6	-23.3	-78.6	5.7
5449	231.9465	65.7976	2013.57	10.791	11.713	7.78	359.9	-64	168.6	5.7	-69.2	174.9	5.7
5450	231.9481	28.4872	2013.419	10.649	10.763	4.086	249	26.3	-76.5	6	35.9	-72.9	6.7
5451	231.9665	49.1485	2013.6265	14.953	15.744	55.107	344.9	-56.8	60.3	5.7	-56.9	61.5	5.7
5452	231.9966	14.7068	2013.8495	10.229	10.236	9.166	280.4	-9.3	125.7	5.5	-10.2	125.7	5.5
5453	232.0248	33.8723	2013.765	13.737	15.35	6.371	288	-94.9	66.1	5.2	-94.7	66.3	5.2
5454	232.0778	76.0622	2013.7185	15.938	16.175	3.622	217.7	-60.9	65	5.7	-54.2	68.1	5.8
5455	232.1651	-9.5228	2013.7695	12.859	14.292	6.04	247.8	-65.3	-8.7	5.6	-67.3	-9.9	5.5
5456	232.2561	66.9979	2013.555	10.83	15.727	33.208	49.1	-40.1	78.1	6.1	-43.3	76.1	6.1
5457	232.2628	-13.4667	2014.38	11.653	15.238	8.025	117.9	-106.1	-12.7	5	-108.6	-15.2	7.8
5458	232.3503	17.5742	2013.5545	14.122	17.561	48.785	72.4	-89.2	-57.9	5.5	-93.4	-56.4	5.9
5459	232.4070	5.6873	2013.362	15.804	17.477	16.789	162.2	-48.9	-73.7	6.2	-48.3	-76.7	6.3
5460	232.4174	25.7749	2013.506	16.033	16.287	43.221	338.6	-71.9	-56.1	5	-72	-56.2	5
5461	232.4520	18.2455	2013.5435	16.566	16.969	4.216	202.3	-66.9	36.1	5	-66.4	38	5
5462	232.4644	82.2635	2013.7935	7.107	16.04	19.352	213.8	64.8	-24.4	5.5	64.6	-14.3	5.8
5463	232.5272	5.6105	2013.188	17.67	17.771	3.263	334.3	-60.3	0.2	7	-60.3	-12.7	6.5
5464	232.6200	5.7959	2013.418	14.702	17.838	57.178	85.7	34.8	-90.7	6.1	42.9	-88.7	6.3
5465	232.6826	32.3639	2013.718	15.982	16.24	4.028	92.6	-27.5	-130.9	5.2	-33.8	-125.4	5.4
5466	232.7204	51.2799	2013.794	13.637	17.851	7.486	242.2	-81.6	-45.3	5.6	-85.4	-42.6	5.8
5467	232.8072	14.6516	2013.538	12.517	15.929	50.915	37.5	12.2	-90.8	5.5	13.7	-88	5.7
5468	232.8445	-11.7753	2013.8235	14.716	15.138	5.671	263.4	-62	-81.8	5.6	-54.8	-81.8	5.9
5469	232.8606	25.9257	2013.4845	16.4	17.634	10.154	274.2	-102.7	75.2	5.1	-108.6	72.3	5.1
5470	232.8804	-0.0600	2013.3955	16.014	18.314	9.659	345.6	-71.3	-41.4	5.7	-69.7	-44.7	6.1
5471	232.9904	13.3573	2013.6785	13.427	15.302	13.7	186.9	-121.2	-39.8	5.5	-119.7	-38.5	5.6
5472	233.0024	50.1722	2013.6785	11.215	15.53	13.063	317.7	-23.9	-114.9	5.6	-24.1	-118	5.7
5473	233.0870	4.2902	2013.6285	10.866	13.497	31.043	47	-77.7	31.4	6	-77.3	24	6.1

5474	233.0898	56.9991	2013.5145	16.952	17.345	8.624	251.9	-61.5	-58	5.7	-61.2	-59.1	5.8
5475	233.1146	56.3881	2013.6225	12.143	13.579	6.626	288	15.5	-88.2	5.6	14	-87.5	5.6
5476	233.1258	8.5349	2013.653	12.209	13.245	14.439	310.3	-90	-189.2	6.1	-92.6	-186.2	6.1
5477	233.1541	-12.1895	2013.9475	13.269	16.72	30.999	101	22.3	-86.9	5.1	23.8	-88.6	5.4
5478	233.1621	39.1180	2013.7815	12.397	16.777	6.78	36.7	-74.4	37.5	5.2	-74.8	37.6	6.2
5479	233.1635	28.9078	2013.511	17.382	17.478	4.678	294.5	-77.8	50.4	5.2	-77.7	50.2	5.1
5480	233.2514	0.3583	2013.576	13.278	17.832	10.465	26.1	-83.6	-32.5	6.1	-79.5	-32.7	6.2
5481	233.3395	11.7571	2013.593	14.293	16.108	6.628	296.2	-93.5	16.1	6	-91.1	14.1	6.1
5482	233.4874	43.0076	2013.757	10.473	13.226	10.431	133.5	-76.8	34.1	5.2	-75.3	35.5	5.3
5483	233.4935	20.9577	2013.817	9.769	12.291	52.84	184.9	-178	48	4.9	-178.6	51.9	5
5484	233.5121	-3.0957	2013.52	17.585	18.045	8.804	255.8	-56.5	-87	5.7	-49.5	-87.7	6.1
5485	233.5384	40.7897	2013.7215	13.32	16.412	11.379	294	94.3	-130	5.5	97	-130.1	5.7
5486	233.6443	-0.1091	2013.693	14.239	15.078	4.585	72.1	31.5	-73.3	6.5	32.1	-69.5	6.6
5487	233.6467	55.6007	2013.583	14.141	17.332	8.939	166.9	-28.4	-80.7	5.6	-29.1	-80	5.7
5488	233.6642	2.2416	2013.471	17.409	18.786	13.165	247	21.7	-60.2	6.3	12.9	-63.8	7.2
5489	233.6943	36.0100	2013.668	17.336	17.575	19.995	186.5	-77.6	-57.2	5.6	-72.6	-59.3	5.6
5490	233.7447	20.5912	2013.4975	15.922	17.095	7.444	110	-67	-38.8	6.1	-65.4	-36.9	6.2
5491	233.7550	-4.8870	2013.8125	10.655	15.568	9.061	7	14.2	-98	5.5	13.7	-102.8	5.9
5492	233.7819	-5.0605	2013.6475	12.969	17.368	13.322	154	-71.4	69.5	5.6	-69.1	69.7	5.7
5493	233.8376	-0.2979	2013.5355	11.279	14.972	30.117	220	48.1	-87.9	5.6	47.6	-92.1	5.7
5494	233.8554	-8.1839	2013.6135	14.762	17.117	9.094	83.1	-48.1	68.7	5.6	-44.2	71.8	5.8
5495	233.8630	61.6451	2013.542	12.118	12.835	54.386	228.4	-71.8	37.3	5.7	-73.5	35.6	5.7
5496	233.8751	-8.7805	2013.7755	13.731	17.002	31.585	86.9	-169.7	-92.7	5.6	-172.1	-90.8	5.8
5497	233.8812	52.8659	2013.7175	10.149	15.135	34.848	164.2	38.3	-140.6	5.6	41.6	-137.4	5.7
5498	233.9539	-10.8042	2013.686	16.852	17.011	6.336	94.6	-145.2	-124.5	6.2	-155.7	-125	6.2
5499	234.0303	16.6585	2013.7905	11.06	13.123	17.638	114.9	-50.5	89.5	5.5	-48.8	93	5.5
5500	234.0641	7.0806	2013.7525	10.039	14.969	28.711	280.1	-89.6	-92	6	-85.2	-98.7	6.1
5501	234.0753	42.8056	2013.649	11.531	17.238	17.147	351.1	-40.7	-70.1	5.2	-38.7	-68.7	5.4
5502	234.0809	35.4300	2013.7305	12.599	16.691	57.887	357	13.5	-62.9	5.2	14.8	-62.4	5.2
5503	234.1221	16.1191	2013.8335	6.233	12.799	35.49	62.4	73.6	-4.2	5.5	71.4	0.2	5.5
5504	234.1239	5.0644	2013.4705	12.684	16.821	27.487	233.4	-32.9	-63.8	6.1	-36	-65.1	6.3
5505	234.2230	29.9910	2013.6425	6.738	17.06	36.142	84.6	92.5	-45.5	5.2	94.7	-35.5	5.4
5506	234.2952	4.0784	2013.864	14.647	15.991	4.41	196	2.9	-86.9	6.1	0.7	-91.1	7.3
5507	234.3377	-9.8839	2013.9155	16.082	18.248	29.261	94	-95.8	-11.1	5.5	-93.9	-13	6.4
5508	234.3885	-1.7877	2013.628	12.214	13.592	20.219	33.5	45.4	-128.5	5.6	45.1	-131	5.6
5509	234.5058	17.2479	2013.8285	8.612	14.422	9.857	154.2	-69.1	44.8	5.5	-69.7	44.9	5.7
5510	234.5656	-3.2537	2013.6195	13.698	18.234	36.822	79.2	-66.9	14.9	5.5	-63.3	19.3	5.9
5511	234.6663	-6.7936	2013.7335	11.737	17.002	15.808	73.7	114.9	-60	5.5	115	-61.5	5.7
5512	234.6847	51.8353	2013.569	15.962	17.233	25.914	292.2	-98.2	28.7	6	-97.8	27.9	6
5513	234.6883	38.0704	2013.708	14.105	15.552	8.748	356.4	-70.4	-14.3	6	-67	-14.5	6
5514	234.6913	7.2081	2013.025	18.197	18.576	16.942	287.8	-31	-65.8	7	-43.3	-61.1	7.9
5515	234.7026	34.4502	2013.8565	14.407	15.745	4.993	264.2	-77	0.2	5.2	-77.3	-1.7	5.2
5516	234.7467	1.6109	2013.4695	13.725	18.231	19.478	329.9	-61.9	16.6	6.1	-62.6	17.9	6.6
5517	234.7662	17.4368	2013.917	13.696	15.483	4.933	313.2	-87	27.9	5.2	-84.7	29.1	5.3
5518	234.8879	33.3437	2013.7605	10.759	13.548	17.425	41.7	-108.4	53.1	5.2	-107.1	51.9	5.2
5519	234.9239	-5.9741	2013.615	14.006	15.978	25.444	345.3	-74.4	-25.9	5.6	-73.6	-29.1	5.6

5520	235.0189	44.1165	2013.827	13.005	15.753	5.45	337.5	70.5	58.1	5.5	66	62.7	5.6
5521	235.0790	27.8426	2013.6085	15.668	16.45	5.563	60.6	-1	-117.6	6	-1.4	-118.4	6
5522	235.0926	2.2815	2013.357	15.417	17.124	35.044	78.5	-70.8	-44.9	6.2	-70.5	-46.1	6.3
5523	235.1378	31.4726	2013.809	8.253	17.005	23.505	317.7	-61.2	52.2	5.1	-64.4	54.6	5.3
5524	235.1527	22.0800	2013.786	9.604	17.262	28.681	342	-160.3	-13.2	5.2	-164	-12.3	5.4
5525	235.1630	23.6441	2013.65	13.777	14.63	41.112	169.9	-96.1	-57.6	5.3	-93.7	-57.2	5.3
5526	235.2002	65.8350	2013.4715	15.443	15.758	7.687	295.6	44.6	-118.3	6.1	45.4	-116.9	6.1
5527	235.2237	34.9397	2013.645	11.793	15.956	16.026	26.7	-93.8	11.6	5.2	-92.5	10.4	5.3
5528	235.2273	33.4747	2013.7475	10.483	14.715	9.611	32.7	18	-80.8	5.2	16.8	-78.8	5.2
5529	235.2639	-11.8071	2013.9055	12.644	14.841	17.525	287	198.2	-157.8	5.6	199.8	-156.9	5.7
5530	235.3631	32.4057	2013.5795	16.233	17.498	45.826	356	-71.6	-163.2	5.3	-74.9	-163.6	5.5
5531	235.4508	62.1559	2013.5145	13.426	15.225	5.281	242.7	-18.5	90.4	6	-19.2	88.9	6.1
5532	235.4570	26.7850	2013.5695	15.24	17.477	13.27	33.3	-85.4	-13.3	5.7	-88	-11.6	5.8
5533	235.4655	25.2804	2013.6825	12.946	16.542	6.161	281.1	-81.2	88.3	5.7	-80	95.9	5.8
5534	235.5002	-11.1411	2013.8835	9.038	9.16	23.106	52.1	61.8	-25.2	5.6	64	-23.2	5.5
5535	235.5450	6.4977	2013.654	13.839	14.156	6.174	288.1	-16.5	-73.3	6.1	-15.5	-73.8	6
5536	235.5459	59.8529	2013.597	13.517	14.564	44.056	296.7	-61.7	-19.6	5.6	-61.4	-16.9	5.6
5537	235.5464	-6.5728	2013.6475	13.471	14.949	27.373	110.8	17.5	-67.4	5.6	16.8	-65.3	5.6
5538	235.5575	12.7323	2013.591	14.432	16.94	8.276	214.1	-67.9	-52.3	5.6	-69.6	-47.6	5.8
5539	235.5894	11.0015	2013.488	16.037	16.149	5.736	208.4	53.5	-65.9	6.1	53	-65.9	6.1
5540	235.6341	26.3824	2013.684	13.402	16.104	5.692	63.3	24.6	-87.7	5.6	22	-85.9	5.8
5541	235.7198	23.9171	2013.7185	10.98	11.519	14.384	107.7	2.5	-61.8	4.9	4.1	-63.3	4.9
5542	235.7593	26.2762	2013.783	9.877	11.138	44.872	147.6	107.7	-132.2	5.6	108	-126.8	5.7
5543	235.7691	53.3915	2013.6155	13.873	15.866	6.706	308.3	-94.7	-52.5	5.3	-95.6	-52.5	5.3
5544	235.7858	18.7646	2013.811	15.713	16.114	8.52	216.9	-87.6	90.9	5.9	-87.7	91	5.9
5545	235.8305	40.4023	2013.5195	14.175	16.354	36.848	315.7	5.1	-142.1	6.1	7.3	-140.6	6.1
5546	235.8909	53.9265	2013.6195	15.026	16.037	17.751	56.7	-26.2	72.6	5.3	-25.9	69.3	5.4
5547	235.9051	19.8593	2013.6445	17.485	18.058	10.388	302.9	-36	-133.8	6.1	-34.1	-127.8	6.2
5548	235.9500	39.5739	2013.481	16.86	17.613	56.387	12.3	-18.9	-73.9	6.1	-30.2	-68.1	6.5
5549	236.1004	42.3897	2013.751	11.81	12.059	8.322	293.7	-91	139.1	5.6	-93.4	137.9	5.5
5550	236.1384	45.3159	2013.7275	12.59	13.248	5.72	118.5	65.5	-107	5.6	66.3	-107.2	5.6
5551	236.2409	-13.7513	2014.4335	14.004	14.773	5.925	119.1	-8.1	-120.2	5.8	-9	-119.5	5.5
5552	236.2485	12.3789	2013.5745	15.346	16.112	5.181	71.7	-133.6	-48.9	5.3	-134	-49	5.4
5553	236.2739	9.3878	2013.592	15.439	16.845	53.754	137.8	-93.3	7.2	6	-90.7	14	6.2
5554	236.3263	-4.1297	2013.622	15.813	16.756	10.86	33.5	44.7	-82.6	5.6	45.7	-86.6	5.6
5555	236.3973	18.9915	2013.8435	15.82	16.594	11.576	288.2	-86.9	-19.5	6	-86.9	-15.9	6
5556	236.4011	56.9839	2013.51	15.754	16.497	8.612	30.9	15.4	-66.6	5.7	20.4	-64.7	5.7
5557	236.4534	7.7080	2013.1955	16.931	17.149	29.398	289.8	-64.9	8.2	6.5	-67.7	4.4	6.5
5558	236.4759	-6.8606	2013.6195	13.96	15.567	5.5	180.5	-25.4	-99.1	5.5	-24.8	-98.4	5.8
5559	236.4812	1.8060	2013.344	15.639	17.51	8.786	1.8	-114.5	-12.5	6.2	-118	-13.1	6.4
5560	236.4875	17.1488	2013.881	11.263	16.525	7.447	199.8	-71.1	25	5.2	-72.6	30.2	5.5
5561	236.4924	3.5475	2013.4475	11.149	16.804	14.517	101.1	-9.3	-63.2	6	-8.1	-66.1	6.3
5562	236.5053	65.5803	2013.6445	8.18	10.882	13.052	333	-50.7	70	5.6	-58.6	63.2	5.7
5563	236.5910	35.5247	2013.68	11.366	17.48	27.975	209.6	-60.6	-116.5	5.9	-71.4	-114.1	5.4
5564	236.6016	36.4462	2013.8925	7.535	8.608	30.528	321.3	-101.3	61.4	5.2	-101.7	58.8	5.9
5565	236.6277	24.4670	2013.666	14.25	15.532	6.616	252.6	-9.6	-142	6	-7.5	-141.8	6

5566	236.6726	-2.0519	2013.572	14.268	16.463	17.291	45.3	-107.6	-45.7	5.6	-109.6	-44.9	5.7
5567	236.7110	-5.2564	2013.7695	9.714	13.937	11.999	156.8	-150.5	35.5	5.5	-147.6	32.8	5.6
5568	236.7274	64.0321	2013.2935	16.908	17.214	24.585	139.3	-64.8	26.6	5.8	-65.8	24.5	5.8
5569	236.7514	47.2982	2013.62	10.9	16.878	18.362	352.7	-13.5	-125.6	5.5	-12.4	-123.5	5.7
5570	236.8385	1.9072	2013.3055	15.449	16.447	7.729	105.3	-27.5	-68	6.2	-28.9	-68.1	6.3
5571	236.8493	53.4089	2013.654	14.916	17.65	11.114	192	-16.2	62.8	5.3	-17	62.8	5.4
5572	236.9097	77.3712	2013.671	14.41	17.285	5.989	248.4	-60.6	79.4	5.6	-55.7	82.3	5.9
5573	236.9761	38.1202	2013.499	14.376	19.036	12.795	49.5	-66	38.1	5.2	-67.7	27.2	7.7
5574	237.0167	34.1751	2013.7325	14.031	18.519	13.648	137	-112.5	-65	5.2	-116.7	-69	5.5
5575	237.0388	1.5711	2013.8035	7.229	11.596	17.789	352.5	-172.2	-174.4	6.1	-172.3	-180.5	6.2
5576	237.0593	-5.9232	2013.6065	17.174	18.679	45.154	330	29.3	-70	5.6	29.4	-73.3	6.6
5577	237.0734	37.0977	2013.627	14.652	16.133	37.835	109.2	18.9	-63.2	5.2	19.7	-62.8	5.3
5578	237.0738	4.7259	2013.4575	15.51	16.015	10.931	59.6	0.5	-68.6	6.1	5.9	-69.7	6.2
5579	237.1722	25.4081	2013.6125	10.723	14.226	7.63	228.7	-75.6	49.2	6	-72.2	46.5	6.1
5580	237.1728	50.8654	2013.4775	14.666	15.416	5.497	154.5	-93.9	52.7	5.3	-93.5	51.6	5.4
5581	237.2011	36.3549	2013.7175	13.147	13.576	10.311	138.1	-6.2	-77.8	5.2	-4.3	-78	5.2
5582	237.2103	17.8491	2013.815	11.74	12.855	29.168	257.8	-203.1	117.7	5.6	-202.6	116.9	5.6
5583	237.2240	8.9882	2013.3845	17.115	17.667	6.039	133.9	14.1	-69.3	6.3	15.9	-68.3	6.5
5584	237.2466	48.4619	2013.6585	10.06	17.884	18.321	351.2	-0.2	-62.6	5.3	0.3	-64.1	5.4
5585	237.2836	4.0566	2013.7585	8.684	15.56	27.176	194.2	-10.7	-68.7	5.9	-4	-68.3	6.1
5586	237.3033	45.0276	2013.5935	14.802	17.629	10.383	310.9	-5.6	-67	5.6	-6.5	-64.7	5.7
5587	237.3909	66.3009	2013.5425	12.404	14.244	38.03	341.2	-53	79.6	5.7	-53.5	79.7	5.7
5588	237.4504	0.6622	2013.622	11.56	14.514	44.229	354.1	-124.9	-24.6	6.1	-122.2	-26.1	6.1
5589	237.4623	43.8452	2013.8125	10.081	10.555	4.909	2.4	-94.2	28.8	5.6	-92.3	32.8	5.5
5590	237.4786	50.3670	2013.5785	9.622	16.516	37.563	305.3	-61.7	29.9	6	-64.1	19.4	6.2
5591	237.5305	-2.3701	2013.758	9.435	14.134	11.177	212.9	-63.8	-74.6	5.5	-64.6	-74.9	5.6
5592	237.5326	18.3660	2013.6865	13.7	16.601	54.504	139.5	-22.5	-70.2	5.9	-15.4	-74.8	6
5593	237.5769	34.6192	2013.774	11.595	12.683	4.576	349.6	-137.7	-152.3	5.2	-136.6	-144.6	5.3
5594	237.5865	28.0120	2013.5225	15.63	16.443	5.183	59.7	7.1	-63.9	6	5.6	-61.9	6.1
5595	237.6163	0.9136	2013.588	10.221	14.001	5.914	192.7	-21.3	-151.5	6	-23	-153.1	6.5
5596	237.6348	37.8355	2013.7335	8.338	11.734	28.152	172.5	-75.7	6	5.2	-77.8	6.6	5.2
5597	237.6771	3.0285	2013.4945	14.532	16.209	24.663	349.6	-39.6	-82.3	6.1	-40.5	-81.1	6.2
5598	237.7852	5.6301	2013.5965	12.464	13.045	16.553	178.1	12.2	-99.9	6.1	10.1	-99.1	6.1
5599	237.8676	50.9850	2013.662	11.227	12.477	33.744	316.3	11.3	-62.8	5.3	14.3	-63	5.3
5600	237.9330	17.1385	2013.8235	15.173	15.577	5.051	327.7	-107.2	55.1	6	-106.2	59.8	6
5601	237.9494	30.1584	2013.6755	16.879	18.05	7.124	97.8	13.4	-62.8	5.3	15.1	-60.6	5.3
5602	237.9752	4.3658	2014.004	14.886	14.909	4.968	256.5	93.6	-97.1	5.9	95.5	-95	6
5603	238.0337	31.1750	2013.7725	9.918	16.14	12.454	340.9	-87.6	38.2	5.2	-91.5	38.5	5.2
5604	238.0411	31.9170	2013.625	11.703	17.525	12.303	358	-68.2	22.6	5.2	-70.5	20.6	5.3
5605	238.0688	41.4884	2013.5955	14.537	14.777	19.691	135.3	-14.7	68.2	5.3	-14.2	68.4	5.3
5606	238.0756	34.2491	2013.7115	11.561	12.705	32.567	54.1	-78.4	186.8	5.3	-77.9	187.9	5.3
5607	238.1091	32.6723	2013.673	12.932	15.038	9.178	107.7	-36.9	64	5.2	-40.7	63.7	5.2
5608	238.2324	30.9503	2013.5795	12.159	17.003	16.824	355.7	-0.5	-63.3	5.2	-2.1	-63.6	5.3
5609	238.2570	6.6973	2013.5645	12.173	18.541	14.976	28	76.5	-92	6.1	69	-91.2	7.2
5610	238.2585	-4.2245	2013.6985	11.473	15.257	10.252	195.9	-69.2	54.1	5.5	-73.1	54.1	5.6
5611	238.3218	46.8079	2013.6325	13.829	14.942	20.625	300.5	20.1	88	5.6	16.5	88.5	5.6

5612	238.3634	50.6622	2013.6055	12.747	14.185	22.779	231	-65	5.1	5.3	-66.4	3.9	5.3
5613	238.5163	44.4293	2013.662	14.945	15.141	14.033	188.8	-90.2	97.5	5.6	-90	96.2	5.7
5614	238.5439	16.8202	2013.6665	17.179	17.714	7.642	189.6	-61.1	-4.1	5.2	-60.7	0.1	5.2
5615	238.5465	-11.8728	2014.305	12.045	14.528	28.831	141.5	-39.5	-67.1	5.4	-39.1	-69.5	5.4
5616	238.6016	55.0996	2013.601	11.367	14.398	15.761	5.6	-11.5	-63	5.6	-9.7	-62.1	5.6
5617	238.6025	29.0459	2013.7375	11.28	12.892	8.569	152.2	-63.3	93.7	5.9	-67.2	92.1	5.9
5618	238.6435	65.2377	2013.3125	16.902	17.556	30.876	76.3	-27	69.6	5.8	-20.2	70.9	5.9
5619	238.6617	13.0903	2013.735	13.418	13.814	6.23	158.6	26	-108.2	5.2	25.7	-107.8	5.2
5620	238.6780	-4.2409	2013.7395	12.267	17.637	29.243	308	-62.6	-24	5.5	-64.9	-18.6	5.6
5621	238.6823	61.5274	2013.517	9.297	12.529	11.42	233.8	58.1	-79.9	5.7	58.3	-80	5.7
5622	238.6859	57.7889	2013.5385	10.836	12.996	13.63	241.7	-135.3	132.4	5.7	-135.2	130.2	5.7
5623	238.7553	50.1135	2013.6905	11.497	12.076	6.516	53.1	-62.9	-32.9	6	-65.4	-28.5	5.9
5624	238.7950	25.7308	2013.6525	10.667	12.979	5.721	251.7	-1.3	79.2	5.6	-1.5	78.7	5.7
5625	238.8705	5.8764	2013.6745	9.849	12.362	10.059	281.9	-48.4	-141.8	6	-47.9	-140	6.1
5626	238.8728	15.7090	2013.7615	15.591	16.531	55.978	188.7	-74.2	60.1	5.2	-74.6	58.9	5.2
5627	238.8828	49.0731	2013.6435	15.896	16.863	21.504	3.5	3	-86.7	6	2.1	-90.4	6
5628	238.8868	39.0608	2013.523	16.252	17.379	4.944	297.8	-15.2	-79	5.3	-19.4	-78.2	5.4
5629	239.0369	56.9271	2013.5475	13.54	17.01	7.797	166.2	-150.2	-36.7	5.7	-147.3	-42.9	5.8
5630	239.0489	46.0241	2013.7265	10.413	13.248	6.438	261.2	-50	-91.9	5.6	-54.1	-90.7	5.6
5631	239.0500	5.2965	2013.709	11.815	15.569	9.958	94.4	0.4	-66.3	6	-0.4	-64.6	6.1
5632	239.1120	11.7398	2013.891	10.444	14.747	6.994	217	58.6	-71.5	6	63.7	-73.1	6
5633	239.1182	3.4220	2013.868	10.592	10.669	8.084	67.5	3.1	-146.5	6	1.9	-144	6
5634	239.1206	13.8909	2013.8265	8.689	16.096	40.67	41.8	16.7	-154.3	6	18.6	-148.1	6
5635	239.1294	57.2831	2013.583	10.014	10.438	19.436	107.8	-51.3	89.3	5.6	-51.3	90.2	5.7
5636	239.2021	35.5046	2013.788	9.046	9.926	9.42	13.3	-181.2	84.3	6	-181.8	81.8	6
5637	239.3274	-5.9612	2013.97	9.425	13.153	19.264	85	-73.4	0.4	5.4	-70	-2.2	5.5
5638	239.5171	-9.1348	2013.859	12.979	15.309	10.907	113.4	-44.4	-130.8	5.5	-53.4	-130.6	5.6
5639	239.5384	32.5784	2013.617	16.23	16.637	6.391	305.7	-39.1	-89.8	5.3	-39.7	-89.3	5.3
5640	239.5395	24.0719	2013.6675	16.276	16.549	5.613	118.7	-103.5	65.4	5.7	-101.8	64	5.7
5641	239.5528	39.3224	2013.3325	11.559	15.511	6.109	289.6	-71.3	-132.7	5.3	-69.6	-133.1	5.6
5642	239.5634	34.1451	2013.687	14.954	15.442	43.755	352.6	-15.6	-94	5.2	-14.8	-89.7	5.2
5643	239.5737	-3.4676	2013.661	17.518	18.412	8.103	120.3	-84.8	51	5.9	-85.1	59.5	6.4
5644	239.5899	-2.4537	2013.613	12.867	18.257	15.336	359.7	32.8	-74.9	5.6	38.8	-68.4	6.1
5645	239.6011	2.1752	2013.36	16.692	17.596	8.555	119	2.8	-81	6.2	2	-79.3	6.4
5646	239.6084	22.9468	2013.7065	12.326	12.927	26.321	225.2	-62.3	-81.1	5.6	-63.8	-79.9	5.6
5647	239.6284	15.0917	2013.592	12.177	17.057	27.831	344.7	8.3	-126.3	5	10.6	-125.2	5.1
5648	239.6542	13.5957	2013.8145	10.588	12.288	17.485	173.3	-33.7	-100.5	4.9	-34.1	-105.8	5
5649	239.6546	78.3665	2013.6735	12.691	14.223	7.465	356.6	31	-92.4	5.6	30.7	-88.2	5.6
5650	239.6831	23.8581	2013.8085	9.76	12.465	18.236	127.8	-108.9	-55.7	5.6	-116.7	-45.2	5.6
5651	239.7021	-11.0026	2014.1095	12.438	17.049	24.931	59.8	36.8	-67.6	5.4	40.1	-66.7	5.7
5652	239.7363	12.0950	2013.61	13.534	17.394	6.9	87.8	-20.5	-75.7	6	-27.2	-71.7	6.4
5653	239.7580	19.2640	2013.8465	10.305	14.613	27.822	130.9	-85.6	22.2	5.5	-87.7	23.2	5.6
5654	239.7953	3.9887	2014.009	11.175	13.065	4.439	148.5	-125.6	-31	6	-126.4	-25.1	6.9
5655	239.8087	-3.5925	2013.698	13.725	16.061	25.121	256.5	41.6	-106.1	5.6	39.1	-106.9	5.7
5656	239.8170	14.1703	2013.632	13.089	17.328	10.18	307.4	6.2	-145.6	5	6.1	-147	5.1
5657	239.8768	67.9990	2013.548	11.08	13.544	26.177	90.1	-40.9	65.3	5.7	-41.9	66.9	5.7

5658	239.8900	18.8854	2013.833	11.553	14.265	14.922	242.6	-75.5	-44.1	5.6	-75.2	-41.6	5.6
5659	239.9234	19.2463	2013.712	15.613	16.135	7.331	349	27.3	-99.5	5.6	30.3	-94.9	5.7
5660	239.9440	24.5287	2013.592	13.671	15.31	13.045	327.9	-53.6	100.8	6	-52.3	102.7	6.1
5661	240.0861	8.9806	2013.5125	15.346	16.81	5.726	227.1	-68.3	-41.3	6.1	-63.5	-43.3	6.4
5662	240.0893	31.4141	2013.753	11.415	15.191	9.745	46.3	33.2	-61.7	5.9	31.5	-62.3	5.9
5663	240.0920	70.5288	2013.58	13.29	14.553	5.877	78.4	-31.8	85.4	5.7	-31.6	87.1	5.7
5664	240.1446	41.9727	2013.684	13.163	15.866	5.518	259.5	-4.3	-101	5.3	-6.7	-101.7	5.3
5665	240.1648	48.3972	2013.6865	12.52	13.272	4.171	45.9	-27.1	-67.5	5.3	-28.7	-69.9	6.1
5666	240.2238	1.7707	2013.5295	12.666	15.93	6.175	298.6	-159.1	-132.3	6.2	-158.5	-133.1	6.3
5667	240.2348	11.6561	2013.739	14.968	16.062	9.378	327.2	-62.3	13.2	6	-62.8	10.9	6.1
5668	240.2590	50.4749	2013.6065	12.044	15.7	9.939	198.2	-118	9.9	5.3	-117.5	11	5.3
5669	240.2882	-4.6298	2013.737	13.67	16.479	11.648	305	-9.9	-63.2	5.5	-14.4	-62.3	5.6
5670	240.3409	83.1155	2013.831	11.989	12.845	6.101	338.7	28.5	-64.2	5.6	24	-67.9	5.5
5671	240.3613	39.1590	2013.549	12.306	17.028	13.878	89.9	31	-145.3	5.3	30	-145.7	5.4
5672	240.4444	25.4967	2013.7325	15.175	15.258	4.41	277.5	3.5	-77.8	6	1.3	-80.1	6.1
5673	240.4552	25.3116	2013.7125	13.604	14.774	8.282	192.8	-181.2	-74.2	6	-181.7	-74.1	6.1
5674	240.4888	25.2523	2013.52	13.163	13.821	3.822	157.3	-72.8	-53.3	6	-75.2	-45.7	6.5
5675	240.4918	27.0334	2013.685	16.918	17.612	12.775	287.4	43.1	-75.5	5.7	45.9	-73.4	6
5676	240.4950	12.2970	2013.6965	13.94	17.402	31.392	143.6	-107.4	-57.3	5.5	-107.4	-57.3	5.7
5677	240.5287	44.4727	2013.6215	12.778	16.565	9.02	294.3	-15.1	-84.4	6.1	-14.6	-82.5	6.1
5678	240.9434	70.9153	2013.617	12.469	12.952	6.149	204.6	-27.2	88.1	5.7	-24.8	88.4	5.7
5679	240.9636	29.5889	2013.7065	11.452	13.659	28.747	318.1	-26.8	74.4	5.6	-27.5	76.6	5.6
5680	240.9865	12.5466	2013.7205	14.992	16.096	30.692	213.3	-92	27.7	5	-91.1	27.5	5
5681	241.0443	35.5709	2013.7665	15.269	15.642	4.072	308.3	-8.2	-71.1	5.2	-4.8	-69.3	5.2
5682	241.0501	46.3375	2013.6295	8.804	12.531	6.448	58.1	-160.7	-76.3	5.7	-165.8	-75.7	5.8
5683	241.1558	-10.2426	2013.9175	12.135	12.905	50.2	342.7	-11.8	-199.3	5.6	-7.2	-197.4	5.6
5684	241.2200	42.6419	2013.581	14.557	15.894	7.475	34.1	-75.3	-55.2	5.7	-72.7	-53.2	5.7
5685	241.2684	54.4270	2013.629	9.481	11.847	5.461	11.6	-38.4	106.4	5.6	-37.2	103.7	5.6
5686	241.2991	22.4497	2013.764	14.733	15.965	4.467	250.5	-71.2	-41.2	5.9	-69.5	-46	6.5
5687	241.3805	19.9644	2013.7005	13.272	15.573	4.873	0.5	-66.6	-29.6	5.8	-67.4	-35.8	6.4
5688	241.3805	29.9237	2013.6085	15.087	16.503	8.324	140.8	-16.7	63.7	5.7	-15.6	63.7	5.7
5689	241.4271	13.5389	2013.814	12.584	13.493	27.266	348.4	53.7	-77.6	5.1	53.9	-80.2	5.2
5690	241.4294	-2.0354	2013.5545	15.678	17.123	5.393	260.1	38	-67	5.7	36.4	-65.9	5.6
5691	241.5531	57.3930	2013.3825	16.923	17.819	6.104	301	-5.1	72.3	5.7	-0.8	70.9	5.8
5692	241.6501	42.2622	2013.606	14.57	16.243	5.115	266.2	17.6	-62.2	5.6	18.9	-62	5.7
5693	241.6668	47.5041	2013.6455	9.203	13.137	11.353	264.9	13.9	93.5	5.6	13.4	98.1	5.7
5694	241.6947	18.3440	2013.729	9.592	17.23	40.819	323.1	-76.9	102.7	5.9	-88.8	101.3	6.1
5695	241.7550	68.2318	2013.5885	11.558	12.507	22.814	273	-37.4	66.1	5.7	-38.5	64.8	5.7
5696	241.8056	19.5272	2013.764	10.023	14.456	8.972	359.2	-83	-8	5.9	-84.6	-9.4	5.9
5697	241.8391	61.9204	2013.5075	13.663	15.303	6.384	339.4	-30	82.6	5.7	-29.5	83.2	5.7
5698	241.8413	39.0802	2013.1755	17.846	18.718	57.578	149.3	-17.9	-64.8	5.4	-14	-66.8	6.4
5699	241.9867	-12.1210	2014.164	11.421	11.78	7.62	44.9	-66.2	-85.2	5.2	-66.7	-84.7	5.2
5700	242.1177	31.4743	2013.867	15.191	15.607	4.333	43.8	-28.2	-68.5	5.2	-29	-69	5.2
5701	242.2032	36.5756	2013.65	14.065	15.456	17.11	341.9	-25.5	116.3	6	-26.2	116.9	6
5702	242.2534	6.8568	2013.483	12.521	13.439	4.994	189.5	-36.4	-107.6	6.2	-34.7	-107.4	6.3
5703	242.2711	17.2179	2013.668	16.596	17.952	18.595	198.4	63.2	-27.9	5.6	60.6	-28.4	5.6

5704	242.4289	52.1786	2013.6045	14.232	16.242	11.468	236.8	-61.5	-1	5.3	-60.9	-0.8	5.3
5705	242.4291	-12.3099	2014.245	15.16	16.327	7.448	178.2	-114.7	-6.7	5.2	-118.4	-9	6.2
5706	242.5581	-4.5284	2013.979	12.117	13.147	3.913	99.4	-18.9	-107.6	5.6	-31.6	-105.6	7.8
5707	242.5929	30.3350	2013.762	13.624	13.792	8.486	74.2	52.9	-69.9	5.2	54.5	-65.7	5.2
5708	242.5982	8.4523	2013.7425	10.824	13.515	7.494	314.3	-8.6	91.7	6	-8.1	91.8	6
5709	242.6175	-8.9427	2013.815	14.794	15.823	14.798	333.3	-76.2	5.2	6	-75.4	3	6.1
5710	242.6382	2.8022	2013.4375	13.797	17.253	34.025	9.5	67.8	-26.7	6.1	67.3	-31.6	6.4
5711	242.6398	16.5284	2013.8315	10.223	14.339	7.165	11.8	-170.7	97.5	5.2	-171.3	103.7	5.3
5712	242.7375	-4.8295	2014.033	12.783	14.533	3.98	225.2	174	-140.3	5.7	180.6	-143.5	7.7
5713	242.8168	19.4863	2013.7385	12.076	15.239	45.056	56.8	-64.8	-1.9	6	-64	-4.1	6
5714	242.8578	43.0923	2013.4775	16.09	16.431	15.768	145.7	26.1	-61.2	5.7	28.1	-63.6	5.7
5715	242.8604	45.3211	2013.6245	10.479	16.425	11.997	25.6	46.7	-94.5	5.6	44.6	-92.3	5.7
5716	242.8994	82.9691	2013.854	12.828	14.905	13.409	192.7	-51.8	61.2	5.5	-52.4	62.6	5.5
5717	242.9257	10.3736	2013.7245	12.176	15.967	15.154	90.9	48.9	-147.2	6	42.3	-143.2	6.1
5718	242.9761	7.3921	2013.3115	17.515	18.004	15.017	290.4	-13.2	-63.4	6.4	-18.8	-63.3	6.5
5719	242.9854	28.3169	2013.7395	12.137	15.354	7.7	268.7	-9.9	-126.2	5.6	-9.9	-126.9	5.6
5720	243.0153	17.3138	2013.739	13.821	14.41	6.114	21.6	-73.1	-22.2	5.2	-73.1	-21.6	5.2
5721	243.0154	18.0081	2013.675	15.51	16.695	5.703	295.3	-76.1	-22	5.9	-79.8	-14.7	6
5722	243.0243	26.7621	2013.5945	15.912	16.629	6.991	152.3	-4.3	-67.5	5.7	1.8	-66.2	5.7
5723	243.0385	13.6265	2013.8195	13.865	15.92	6	30.4	-70.6	-70.3	5.2	-72.1	-68.1	5.2
5724	243.0688	41.2433	2013.736	9.815	13.026	46.745	175.8	91.2	-115.9	5.2	89	-115.7	5.3
5725	243.1583	76.6978	2013.501	14.005	14.736	16.295	135.7	-36.5	89.5	5.7	-36.7	90.5	5.7
5726	243.1743	-13.3159	2014.308	10.504	13.831	8.641	130.6	-96	-75.9	5.4	-98.8	-80.2	5.7
5727	243.1919	14.6630	2013.607	16.46	17.505	58.576	190.4	-64.3	6.4	5.3	-61.5	12.8	5.3
5728	243.2172	36.7545	2013.75	11.065	14.788	11.281	136.7	-80.9	-66.4	5.9	-80.8	-63.8	6
5729	243.2485	26.4324	2013.7825	10.316	16.678	12.966	184.5	-133.1	-3.8	5.6	-131.4	0.8	5.7
5730	243.2961	-2.3341	2013.632	13.519	15.237	43.151	75.3	-54.4	-176.3	5.7	-55.8	-175.6	5.6
5731	243.3387	21.2512	2013.7865	10.564	10.826	7.105	352.3	0.5	-86.6	5.9	1.4	-89.6	5.9
5732	243.4187	39.5821	2013.6405	13.831	16.777	28.608	155	-60.4	34.8	5.2	-60.6	33.7	5.3
5733	243.5143	53.8247	2013.5505	12.814	16.124	26.54	322.4	56.6	-102.7	5.3	56.3	-98.3	5.4
5734	243.5665	42.2584	2013.596	14.358	16.241	24.669	69.8	-103.6	73.2	5.6	-105.1	69.2	5.7
5735	243.5693	-10.4146	2014.239	7.215	10.323	22.34	153.1	78.8	-15.8	5.4	76.1	-19.3	5.4
5736	243.5999	-14.7337	2014.283	11.462	13.488	6.152	178.3	-149.7	-159.8	5.4	-149.1	-160.6	6.3
5737	243.6586	15.5818	2013.6695	9.808	16.271	11.51	87.4	-51.6	-80.8	5.5	-59.2	-78.1	5.7
5738	243.6645	17.9981	2013.6775	12.676	18.358	29.793	226.8	-94.3	29.5	5.3	-92.2	33.1	5.5
5739	243.6906	38.0056	2013.6945	13.944	15.637	9.113	296.9	-44.1	109.7	5.2	-44.6	113.4	5.3
5740	243.7228	3.5066	2013.3955	15.172	16.039	23.252	247.9	-91	-90.3	6.3	-99.7	-77.9	6.3
5741	243.8374	29.0968	2013.7535	10.657	14.227	13.258	107.5	-62.6	-24.9	5.6	-60.9	-29	5.6
5742	243.8500	49.1331	2013.605	13.804	15.135	45.086	198.1	-72.9	-5.1	5.3	-72.9	-4.3	5.3
5743	243.8565	68.1927	2013.4835	13.662	15.582	10.371	108.4	-99.4	77.6	5.7	-97.9	78.1	5.8
5744	243.9681	55.2245	2013.424	15.135	17.919	14.666	158.1	66.3	2.6	5.7	63.7	3.6	5.8
5745	244.0008	38.3023	2013.566	14.215	16.996	5.326	135.6	-71.6	48.2	5.2	-72.2	48.2	5.9
5746	244.0104	-6.7288	2013.492	14.906	17.75	8.676	178.8	2.2	-74	5.6	2.6	-73.9	6.2
5747	244.0297	35.9374	2013.6715	13.532	15.358	6.682	285.3	0.7	-83.3	5.2	2.4	-82.3	5.3
5748	244.0406	14.2051	2013.856	14.952	15.21	3.665	165	-65.5	-115.4	5.5	-65.2	-114.3	5.6
5749	244.0812	-1.6480	2013.733	7.158	10.238	20.614	229.7	-123.8	41.4	5.6	-125.9	33.2	5.6

5750	244.0871	47.9842	2013.676	15.057	17.821	5.907	223.6	-13.4	63.4	5.3	-16.4	63.2	5.4
5751	244.0948	-4.5779	2013.5695	12.706	14.575	13.581	95.9	-71	-6.5	5.6	-69.6	-5.8	5.6
5752	244.1107	41.6949	2013.5285	12.598	17.1	11.686	201.3	-60.7	7.7	5.2	-61.9	5.3	5.4
5753	244.1269	-8.3226	2013.5165	13.729	15.304	7.044	119.7	-9.1	-102.8	5.7	-10.5	-105.2	5.8
5754	244.1730	22.7803	2013.673	14.064	17.686	9.355	103	38.7	-71.2	5.2	45.9	-70.1	5.4
5755	244.1846	-3.3843	2013.7085	9.606	14.47	11.33	286.7	137.4	-125	5.7	136.7	-130.9	5.7
5756	244.2681	15.3061	2013.346	14.069	18.821	53.944	335.3	-85.5	13.2	6.1	-86.2	18.4	6.1
5757	244.2790	3.3197	2013.4705	14.167	15.689	7.502	92.9	-76.4	-82.6	6.2	-76.9	-83	6.3
5758	244.2890	44.1134	2013.5765	12.943	15.487	56.816	16.1	-41.9	94	5.6	-42.8	94.3	5.7
5759	244.3896	-8.7514	2013.302	17.651	18.1	56.166	359.1	-26.4	-78.3	7.1	-19.8	-81.2	7.4
5760	244.3975	3.8276	2013.068	15.93	16.823	43.597	10.7	-85.7	-15.7	6.5	-88.4	-13.8	6.6
5761	244.4072	18.8282	2013.6695	14.51	16.469	21.479	155.8	-136.7	-62.8	5.3	-136.7	-63.3	5.3
5762	244.4575	4.2504	2013.112	16.458	18.286	14.747	15.3	22.1	-107.3	6.4	30.4	-104.8	7
5763	244.5012	23.0785	2013.7245	12.704	15.486	7.666	142.9	-81.3	6.5	6	-80.1	5.3	6
5764	244.5161	51.4232	2013.5165	18.405	18.516	27.235	160.5	1.9	-61.5	5.5	-4.8	-62	5.6
5765	244.5689	37.5331	2013.5935	10.766	17.412	10.332	285.3	15	-60.5	5.2	17.3	-63.2	5.7
5766	244.5949	62.0629	2013.569	15.019	15.436	4.031	222.1	-31.6	93.6	5.7	-33	89.4	5.7
5767	244.6302	10.9127	2013.6795	14.866	17.056	6.21	24.6	-42.6	-89.4	6	-46.2	-88	6.3
5768	244.6407	61.4759	2013.367	12.504	17.291	11.822	227.5	-132.1	106.8	5.8	-132	109.4	6
5769	244.6470	51.3077	2013.6765	10.18	12.498	23.253	250.2	-63.6	-16.7	5.3	-64	-15.7	5.3
5770	244.7264	45.1075	2013.624	16.411	16.551	4.58	3.7	-62.5	50.1	5.7	-60.6	50.5	5.7
5771	244.8658	0.9342	2013.4095	13.464	15.407	11.443	232.8	-20.3	-106.5	6.2	-19.7	-106.5	6.3
5772	244.9149	17.1588	2013.601	11.911	17.366	15.219	265.1	-83	22.5	5.6	-83.1	25.1	5.7
5773	244.9420	-3.7345	2013.5135	11.16	17.393	44.495	206	-39.1	-114.3	6.2	-37.6	-116.8	6.5
5774	245.0365	-1.1019	2013.4965	12.882	14.914	8.504	229	-3.9	-66.6	6.1	-0.1	-67.1	6.2
5775	245.0635	-11.9559	2014.3615	9.295	13.259	22.842	89.6	-11.7	-69.2	5.3	-13.6	-69.1	5.3
5776	245.0966	7.1684	2013.716	12.714	14.398	7.145	103.1	-93.3	37	6	-96.8	34.9	6.1
5777	245.1186	47.1756	2013.551	15.9	18.118	30.456	274.6	-2.8	-63	5.7	-11.5	-64	5.7
5778	245.1201	2.4958	2013.005	17.323	17.874	46.324	76.3	65	12	7.7	68.4	1.7	6.6
5779	245.1904	5.4643	2013.91	11.441	14.8	5.561	177	39.9	-80	6.1	38	-77.2	6.1
5780	245.3805	27.7398	2013.789	9.756	15.885	38.453	160.5	-74.1	44	5.9	-71.7	44.8	6
5781	245.3940	76.3344	2013.735	9.562	9.684	5.162	55.3	-10.9	63.1	6	-13.8	61	6.2
5782	245.4118	2.9155	2013.4375	13.286	15.728	32.1	135.8	-26.9	-186.6	6.2	-27.7	-187.4	6.3
5783	245.4144	13.2748	2013.774	11.49	15.105	7.729	48.2	-38.4	61.2	5.5	-37.8	62.2	5.5
5784	245.4796	5.1241	2013.6445	11.44	12.977	21.613	93.9	-104.3	15.3	6.1	-104.9	14	6.1
5785	245.5599	64.4531	2013.5245	11.67	11.826	10.483	173.9	8.4	-64.2	5.7	11	-65.1	5.6
5786	245.5948	31.1322	2013.587	16.347	16.49	6.439	216.4	-58.5	74.3	5.3	-58.5	73.9	5.3
5787	245.6347	9.5878	2013.6385	11.348	16.502	9.693	80.8	-39.5	-79.2	6	-39.3	-76.4	6.2
5788	245.6950	44.2520	2013.5675	9.242	17.187	17.803	339.6	-33.6	-164.5	5.3	-30.1	-163.9	5.5
5789	245.7254	38.2576	2014.0655	10.207	10.736	5.127	49.6	11.6	-97.5	5.2	8.2	-100.7	5.6
5790	245.7307	29.8233	2013.685	14.055	15.957	21.682	248.4	-70.2	98.5	6	-70.2	97.2	6
5791	245.7695	36.8426	2013.6605	10.234	15.831	54.92	71.8	-188.1	143.1	5.3	-188.1	142.2	5.4
5792	245.8025	68.7177	2013.48	10.254	14.806	10.897	355.2	-25.2	110.5	5.7	-23.9	113.2	5.7
5793	245.8048	-0.5404	2013.4995	9.905	16.823	14.724	204.9	-134.1	-23.4	5.6	-132.5	-15.5	5.9
5794	245.8328	46.7842	2013.7045	16.438	16.547	4.38	30.1	-102.2	53	5.3	-102.1	52.2	5.3
5795	245.8698	-1.3848	2013.339	15.016	18.497	56.122	155.4	-47.7	-64.6	5.7	-46.4	-62.3	6.5

5796	245.8855	67.3588	2013.399	13.789	16.451	9.116	273.1	-69.3	49	5.7	-68.4	49.7	5.8
5797	245.9791	20.5608	2013.689	9.681	16.658	26.742	291.6	-72	-16.7	5.9	-74.8	-16.2	6.1
5798	246.0062	39.7234	2013.74	12.392	13.164	6.068	357	9.9	126.5	5.2	9.6	125	5.2
5799	246.0184	22.7435	2013.69	12.616	14.634	11.417	48.4	-112.1	-121	6.1	-115.9	-118.2	6.1
5800	246.0429	37.0367	2013.811	7.915	8.493	8.202	340.4	-106	13.9	5.3	-103.9	6.4	5.2
5801	246.1114	32.8322	2013.667	14.045	15.328	10.302	287.6	-7.9	-97.9	5.2	-9.2	-96.1	5.2
5802	246.2138	52.2997	2013.6475	11.061	12.502	49.343	216.5	-107.3	171.8	5.4	-116.7	170.2	5.4
5803	246.2328	20.3174	2013.668	16.682	16.765	3.872	313.6	-10.6	-139.7	6.3	-13.9	-143.1	6.2
5804	246.2436	-5.1645	2013.5135	15.37	16.613	44.967	195.7	-22.3	-107.6	5.7	-17.8	-113.1	6.5
5805	246.2729	41.3600	2013.6525	13.222	15.331	12.616	345.5	78.9	-54.4	5.2	79.1	-53.7	5.3
5806	246.2824	42.0411	2013.7655	12.36	12.865	5.575	287.5	-38.4	-92.8	5.6	-41.5	-91.6	5.6
5807	246.2892	14.6126	2013.692	13.604	15.012	11.764	299.1	-18.9	-201.9	5.6	-20.1	-197.8	5.6
5808	246.3020	-9.9643	2013.9275	14.388	14.91	34.68	218	-21.9	-84.6	5.6	-24.6	-88	5.6
5809	246.3401	13.8783	2013.4005	17.199	18.586	12.275	54.4	-62	20.4	5.7	-63	10.6	6
5810	246.3814	29.4171	2013.547	14.336	17.413	35.091	97.8	-67.9	-11.1	5.9	-71.5	-7.2	6.1
5811	246.4698	48.3364	2013.666	9.063	11.467	30.743	151.6	68.4	-57.5	5.3	63.4	-63	5.3
5812	246.4906	65.5677	2013.437	15.941	17.024	5.584	100.3	-33.1	-83.9	6.1	-38.9	-84.8	6.1
5813	246.5113	40.8216	2013.74	13.73	14.803	5.292	32.9	-39.8	80	5.2	-39	82.2	5.2
5814	246.5777	3.9930	2013.208	14.819	17.587	8.474	0.3	8.5	-137.8	5.8	11.7	-140.8	6.2
5815	246.5807	2.7626	2013.669	9.902	13.335	8.187	222.2	-29.3	-147.7	5.6	-26.9	-145.3	5.7
5816	246.6356	2.5733	2013.6005	10.291	11.814	11.156	205.4	-46.5	-71.8	5.6	-48.7	-71.3	5.7
5817	246.6644	24.0903	2013.7775	10.007	13.925	20.355	242.9	-66.4	54.2	5.5	-66.3	54.8	5.6
5818	246.6788	42.4400	2013.6635	12.767	13.216	24.553	80.7	-33.7	82.5	5.6	-32.2	78.9	5.6
5819	246.6796	20.5424	2013.5695	15.95	17.349	5.639	260.3	61.2	-131.2	6.1	63.4	-134.5	6.2
5820	246.6997	-14.3871	2014.283	12.836	12.891	7.538	314.7	-124.6	-152.2	5.2	-126.3	-149.7	5.2
5821	246.7006	26.3601	2013.742	11.026	13.769	13.174	312.1	47.5	-145.6	5.6	48.5	-146.7	5.6
5822	246.7795	29.2558	2013.791	11.725	15.036	5.13	228.9	-77.2	98.8	5.9	-80.2	92.5	7.1
5823	246.8133	33.0065	2013.6265	13.273	17.754	32.215	319.6	-75	-54.8	5.2	-75.8	-54.1	5.4
5824	246.8142	42.6711	2013.619	13.68	14.377	11.997	312.5	-19.2	-62.9	5.6	-21.5	-62.2	5.6
5825	246.8250	24.6897	2013.734	14.164	15.377	8.339	98.3	63.4	-92.2	6	62.3	-95	6
5826	246.8864	8.5301	2013.6795	11.79	13.596	53.912	316	116.6	67.5	6.1	116.5	70.7	6.1
5827	246.8961	9.2719	2013.331	15.024	17.32	7.888	141.6	-79.4	6.6	6.2	-81.3	1.4	6.4
5828	246.9021	59.3701	2013.5565	10.116	12.264	12.744	322.2	-72.9	46.5	5.7	-72.7	50.2	5.7
5829	246.9256	31.7259	2013.6905	15.429	18.228	7.251	153.8	-33.9	-129.8	5.4	-32.3	-131.1	5.8
5830	246.9973	50.2537	2013.483	13.253	13.784	4.065	320.2	11.6	-71.2	5.4	14.3	-72.5	5.6
5831	246.9973	25.7273	2013.6095	14.867	15.584	10.524	73.6	-67.9	-68.7	5.7	-68.6	-63.9	5.7
5832	247.0406	-3.5214	2013.3015	16.59	17.641	16.4	279.2	-116.2	-52.4	5.8	-118	-45.3	6.1
5833	247.0492	-4.2501	2013.7495	11.277	12.712	8.57	77.1	-42	-70.3	5.6	-40.2	-71.9	5.6
5834	247.0592	-9.6855	2013.8385	15.096	15.906	54.94	318.3	-88.4	-197.1	5.9	-88.5	-197.9	5.8
5835	247.0724	13.4954	2013.6385	13.111	14.865	19.523	154.5	9.5	-63.9	5.6	7.6	-64.1	5.7
5836	247.1630	74.7174	2013.388	13.994	15.813	11.893	98.5	-133.5	2.6	6.1	-137.4	2	6.2
5837	247.2038	-8.1291	2014.2355	6.624	9.14	5.568	296.8	-63.3	-59.3	5.5	-60.1	-59.4	6.3
5838	247.2289	16.0138	2013.542	14.432	18.927	12.443	41.2	-14.8	-67.6	5.6	-21.1	-65.1	6.3
5839	247.2447	25.1522	2013.6855	11.982	17.199	19.468	144.1	20.6	-66.4	5.6	21.3	-65.5	5.7
5840	247.2807	5.2359	2013.7205	10.848	10.928	7.538	333.9	-84.1	-152.1	6.1	-82.4	-147.5	6.1
5841	247.3769	-8.7120	2013.5355	15.471	16.934	7.555	181	92.5	-58	5.8	95.8	-60.2	6.2

5842	247.5978	30.0804	2013.775	9.749	11.88	22.126	274	-4.2	-80.7	5.2	-11.2	-83.8	5.2
5843	247.6130	41.3668	2013.536	10.736	16.875	52.357	283.6	-28.5	116.3	5.6	-32.8	117.7	5.9
5844	247.6260	71.9699	2013.442	13.17	15.83	10.184	332.9	20	71.2	5.7	18.5	70	5.8
5845	247.6335	-14.5466	2014.283	8.823	9.798	9.516	228.5	-9.5	-96.4	5.1	-14.4	-93.1	5.1
5846	247.6470	-9.8017	2014.274	12.195	12.683	3.957	158.4	-118.5	-168.3	5.6	-115	-172.7	6.8
5847	247.8778	8.2944	2014.048	6.851	8.061	59.221	70.8	43.3	69.3	5.9	47.7	67.7	5.9
5848	247.9278	70.9334	2013.7165	8.001	9.932	39.943	312.2	-63.9	19.1	5.6	-67.2	16	5.7
5849	247.9542	38.0814	2013.5415	15.549	17.015	26.003	249.6	-3.7	80	5.7	0.1	76.5	5.8
5850	247.9594	64.4345	2013.33	15.643	17.041	12.232	300.6	-44.6	77.9	5.7	-43.7	76.2	5.8
5851	247.9641	30.7800	2013.652	12.382	16.876	39.234	105.4	43.9	-84.7	5.2	42.8	-82.3	5.3
5852	248.0605	10.6747	2013.7945	10.506	11.315	13.511	282.5	79.8	-152.8	6.1	86.8	-150.9	6.1
5853	248.1227	26.7103	2013.7755	10.319	11.986	5.776	170	-65.9	-95.6	5.9	-65.7	-97.7	6
5854	248.1354	4.6142	2013.2945	14.467	17.717	29.253	261.5	-27.9	-73.6	6.2	-32.9	-69.7	6.5
5855	248.1516	57.6178	2013.633	14.258	14.643	4.847	37.1	-8.6	-66.2	5.7	-9.5	-64.8	5.7
5856	248.1591	2.8775	2013.259	13.98	16.824	8.063	84.6	-106.7	35.3	6.2	-109.9	35.3	6.5
5857	248.2067	43.8950	2013.49	11.637	14.936	5.831	39.1	-5.1	121.4	5.3	-8	127.1	5.5
5858	248.3333	63.5244	2013.5415	11.819	15.361	26.509	336.2	-60.6	108.9	5.6	-64.8	106.5	5.7
5859	248.3747	46.0602	2013.6075	14.785	15.201	19.24	39.5	-34.2	69.3	6	-33.1	67.2	6
5860	248.4307	-10.3686	2014.332	14.331	14.616	15.78	166.8	-73.4	-25.5	5.6	-74	-26.2	5.5
5861	248.4507	38.6277	2013.5685	15.313	15.977	4.324	92.9	16.9	-60.8	6	8.6	-62.5	6.2
5862	248.4841	10.2869	2013.4245	9.355	18.797	22.201	193.5	-16.8	-147.8	6	-20.6	-145.5	7.7
5863	248.4953	63.7085	2013.5515	16.683	17.422	4.283	60.2	-54.6	120.6	5.7	-58.9	120	5.9
5864	248.5156	9.4536	2013.2745	17.501	17.961	12.756	79.6	35.1	-69.1	6.5	38.7	-70.9	6.3
5865	248.6067	14.6705	2013.944	14.676	15.3	3.353	180.2	-63.7	-29.8	5.6	-64.8	-20.1	5.6
5866	248.6544	24.4673	2013.697	10.122	15.795	50.535	63.9	-23.8	-84.2	5.9	-23.8	-85.8	6
5867	248.7578	20.6459	2013.638	11.465	15.48	25.239	311.2	-96.3	66.7	5	-93	65.1	5
5868	248.7687	64.2620	2013.5515	15.401	15.848	28.967	326.4	-76.6	31.1	5.7	-75.5	28.2	5.7
5869	248.8005	8.8250	2013.506	15.613	16.7	4.301	4.2	-74.3	10.9	6.1	-76.3	17.1	6.5
5870	248.8819	54.0531	2013.4475	16.199	16.64	9.583	57.5	-27.1	67.7	5.7	-30.1	67.7	5.7
5871	248.8911	40.3495	2013.7395	11.682	14.207	7.016	57.4	-23.5	-67.9	5.9	-23.7	-70.9	5.9
5872	248.9061	55.7148	2013.513	14.835	15.468	4.119	256.9	-61.2	79.4	5.7	-59.3	77.8	6
5873	248.9087	30.6130	2013.7555	10.806	15.283	27.794	30.3	13.8	-60.6	5.2	12.9	-61.1	5.2
5874	248.9111	18.3469	2013.679	11.614	12.775	4.307	338.9	-74.4	40.4	6	-72.3	39.9	6.1
5875	248.9623	41.3959	2013.6545	13.534	14.307	3.833	119.1	4	-85.2	6.4	-0.4	-82.1	6.2
5876	248.9735	14.9392	2013.608	14.674	16.467	5.759	54.9	-13.7	-64.3	5.6	-13.7	-65	5.6
5877	248.9751	29.5550	2013.8175	10.187	13.157	25.878	193.7	-145.8	-148.4	6	-144.7	-147.5	6
5878	249.0243	76.5938	2013.604	11.734	16.273	8.907	206.2	2.3	85.8	5.7	1.3	87.1	5.7
5879	249.0345	28.8512	2013.546	15.1	16.698	6.456	233.8	10.2	-95.3	6	12.6	-96.5	6.1
5880	249.0843	37.4586	2013.5995	14.127	16.836	21.21	342.6	30	65.2	5.2	27.3	68.2	5.3
5881	249.0934	85.3630	2013.818	12.584	14.307	8.849	143.7	66.1	-50.8	6	65.7	-55.5	5.9
5882	249.1028	45.7160	2013.5185	13.048	18.056	59.329	125.8	-10.6	-63.6	5.3	-16.9	-62.3	5.5
5883	249.1044	37.7952	2013.557	13.731	16.584	21.223	261.3	-105.4	1.6	5.3	-106.5	3.1	5.4
5884	249.1460	4.1472	2013.352	14.541	16.728	6.384	158.7	-73.5	32.7	6.2	-71.6	31	6.5
5885	249.1881	3.3382	2013.7015	9.613	14.605	25.579	319.2	-12.8	-65.3	6	-14.4	-66.5	6.3
5886	249.2269	47.7831	2013.5925	10.807	16.454	13.509	277.2	37	-96.5	5.3	36.3	-100	5.4
5887	249.2905	6.3646	2013.602	12.029	13.369	10.214	54.7	-89.7	-75.5	6.1	-90.8	-74.4	6.2

5888	249.3845	-5.5031	2014.3205	9.723	9.745	6.71	191.5	-102.4	20	5.4	-106.1	25.5	5.3
5889	249.4350	37.1997	2013.649	12.039	15.921	23.43	238.7	-169.2	49.1	5.3	-168.8	52.2	5.3
5890	249.4742	24.5158	2013.6025	13.584	17.92	40.73	253.6	18.9	-124.3	6	21.8	-127.1	6.2
5891	249.4765	0.8244	2013.596	14.037	16.203	11.046	65.1	-121.1	68.5	6.5	-124.7	57	6.2
5892	249.5015	31.4361	2013.5575	14.509	17.26	9.452	112.2	1.2	-64.4	5.2	0.2	-64.3	5.4
5893	249.5382	48.6416	2013.43	16.731	17.385	34.569	315	-49.1	64.7	5.4	-49.7	66.9	5.4
5894	249.5765	50.6735	2013.521	14.114	16.467	15.402	270	-32.7	-62.6	5.3	-35.9	-61.7	5.4
5895	249.5893	70.3378	2013.635	8.855	10.953	9.693	215.8	42.1	-180.6	5.7	38.2	-174.6	5.7
5896	249.5939	75.7155	2013.528	13.052	14.886	5.411	228.5	63.4	-36.1	5.7	66.1	-31.2	5.7
5897	249.6380	16.9695	2013.539	14.868	16.627	5.219	112.1	40	-162.3	5.1	39.8	-158.6	5.2
5898	249.6880	61.1842	2013.5435	15.538	15.829	6.994	354.2	31.7	-64.4	5.7	26.8	-68.8	5.7
5899	249.6906	39.3888	2013.7035	11.608	13.803	17.428	226.7	-1	-69.9	5.2	-0.3	-69.4	5.2
5900	249.7452	72.9122	2013.369	16.686	17.064	7.217	316.5	-135.6	192.1	5.9	-135	193.8	6
5901	249.7668	60.6994	2013.7305	8.24	9.409	8.23	4.7	26.7	-60.4	5.6	25.3	-60.6	5.6
5902	249.8066	52.6282	2013.705	9.277	10.565	5.534	228.1	-111.9	162.4	5.4	-112.5	152	5.3
5903	249.8435	25.2763	2013.6795	14.033	17.287	8.66	47.4	-141.3	-185.9	6.1	-147	-179.1	6.3
5904	249.9309	67.6331	2013.5515	10.789	11.046	15.787	149.5	-67.3	44.3	5.7	-69.4	45.3	5.7
5905	250.0756	34.0985	2013.68	12.997	16.031	17.056	253.6	-90.9	-36.8	5.2	-90.2	-41.9	5.3
5906	250.0849	46.6700	2013.3295	18.145	18.155	46.503	346.8	67	-33.1	6	66.6	-32.8	5.6
5907	250.1210	49.8108	2013.576	14.127	15.738	7.616	151.1	9.6	-62	5.3	7.7	-63.2	5.3
5908	250.1420	9.3211	2013.548	14.163	16.913	25.682	307.9	-31.7	-88.4	6	-29.4	-89.7	6.3
5909	250.1995	-9.3232	2013.9765	10.382	16.415	53.598	70.9	-84.3	128.5	5.5	-85.9	126.5	5.9
5910	250.3140	43.5814	2013.9215	9.802	11.517	4.813	256.2	16.4	-98.8	5.3	13.1	-100.6	5.1
5911	250.3425	31.5683	2013.767	13.24	13.753	18.518	45.7	-123.8	23.2	5.2	-125	17.1	5.2
5912	250.3849	37.9256	2013.634	12.964	15.287	10.96	24.1	23	-65.5	5.3	20.8	-66.8	5.3
5913	250.5933	32.9622	2013.725	14.796	14.946	5.863	88.9	-25.6	96.2	5.6	-24.5	97.3	5.6
5914	250.6065	19.1350	2013.4585	16.928	17.457	6.385	333.7	-82	10.6	5.1	-82	13.3	5
5915	250.6073	63.3508	2013.62	14.813	15.043	3.87	8.5	19.7	-125.3	6.2	17.6	-130.3	6.1
5916	250.6295	78.9244	2013.597	14.01	16.936	9.037	294.1	-26.2	63.3	6	-26.7	60.1	6.1
5917	250.6837	13.6076	2013.8185	8.797	9.229	12.791	333.9	-42.1	-161.6	5	-43.7	-158.5	5
5918	250.7284	-9.6662	2014.112	11.918	14.981	45.388	81.5	-62.3	38.2	5.4	-61.7	35.6	5.5
5919	250.7385	1.5981	2013.621	11.174	14.843	10.939	320.9	18.9	-140.7	6.1	20.5	-138.5	6.2
5920	250.7778	69.5006	2013.503	11.003	17.476	22.079	262.5	5	70	5.7	8.2	68.7	5.8
5921	250.8957	30.4090	2013.5905	13.391	18.888	14.057	213.2	-0.1	-122.7	5.6	-4.9	-116.9	6.5
5922	250.9114	55.3991	2013.335	15.113	18.528	8.212	121.9	-87.7	77	5.7	-88.8	78.6	6.2
5923	250.9272	-8.5686	2014.11	12.014	13.575	42.765	118.5	-64.2	-54.4	5.5	-63.8	-55.1	5.4
5924	250.9474	82.3415	2013.724	14.953	17.962	10.37	12.8	-62.6	112.6	5.9	-67.6	109.9	6.1
5925	250.9645	35.6101	2013.6085	15.286	17.265	5.797	329.9	-6	-66.3	5.6	-4.4	-69.3	5.7
5926	250.9689	38.5816	2013.5335	16.436	16.835	4.591	339.3	-1.2	-60.2	5.3	-3.4	-61.4	5.4
5927	250.9859	21.0102	2013.587	15.065	16.728	7.727	213.3	-32.2	-95.3	5.6	-29.2	-95	5.7
5928	251.0024	0.1891	2013.8485	11.932	12.128	9.276	170.2	-14.7	-64.6	6	-14	-64.6	6
5929	251.0160	1.5954	2013.325	17.991	18.523	44.192	195.4	-8.3	-67.3	6.5	-8.5	-68	7.1
5930	251.0251	36.1913	2013.667	13.087	14.163	5.527	115.8	-6.3	-61.5	5.2	-9.1	-62.5	5.2
5931	251.0765	26.6581	2013.5505	12.778	16.396	6.881	240.8	1.5	-85.4	5.7	-1.4	-84.6	5.8
5932	251.0831	68.9561	2013.363	17.406	18.34	29.355	209.8	-20.2	65.8	5.8	-26.4	63.2	6
5933	251.1154	1.6975	2013.469	15.122	15.965	11.075	164.5	13.9	84.7	6.2	14.5	86.7	6.3

5934	251.1455	51.8971	2013.6525	11.483	12.187	5.479	138.3	-5.2	-63.3	5.3	-6.5	-63.8	5.5
5935	251.1488	71.7574	2013.5705	8.184	12.41	27.725	113.3	-96.9	108.2	5.7	-100.8	105.3	5.7
5936	251.2134	17.0549	2013.625	14.263	18.145	16.381	10.4	-15.3	-97.7	6	-16.3	-98.3	6.3
5937	251.2472	36.4700	2013.52	13.813	16.605	43.303	301.2	-23.4	-72	5.2	-26	-69.5	5.4
5938	251.2579	34.2033	2013.6695	11.733	15.016	44.595	34.2	-71.1	-35.6	5.6	-68.6	-35.4	5.7
5939	251.3220	55.9876	2013.4885	13.679	15.477	13.171	27.3	3.4	91.4	5.7	1.1	91.8	5.7
5940	251.3222	3.5331	2013.5235	11.722	13.008	5.408	298.4	-23.8	98.2	6.1	-19.1	98.7	6.2
5941	251.3565	-3.6739	2013.9245	14.248	14.825	4.844	295.7	7.4	-130	5.5	14.4	-126.5	5.6
5942	251.3569	26.9159	2013.6605	10.641	13.112	14.654	162.6	33.6	-74.4	5.6	28.3	-78.2	5.7
5943	251.3787	50.4421	2013.6395	10.086	11.193	10.868	289.7	-36.1	-71.1	5.3	-34.5	-73.4	5.3
5944	251.3853	33.8295	2013.6745	8.921	12.858	8.083	146.7	168.9	-139.7	5.7	173.1	-146.6	5.9
5945	251.3974	25.3046	2013.605	15.154	15.317	13.996	347.1	29.7	-68.7	5.7	25.6	-68	5.7
5946	251.4380	34.2593	2013.732	14.809	14.92	4.19	220.9	-81.9	67.1	5.7	-86.3	68.4	5.6
5947	251.4845	58.2833	2013.402	16.467	17.317	25.496	81.6	-22.6	-74.7	5.7	-23	-77.5	5.8
5948	251.5169	2.4613	2013.393	12.969	17.202	34.1	94.8	21.3	-74.2	6.2	21.8	-72.6	6.4
5949	251.5540	19.7292	2013.6105	11.299	14.106	17.011	140.1	-56.2	-69.1	6.1	-57.6	-69.2	6.1
5950	251.5973	3.7993	2013.3595	14.167	18.098	11.604	340.3	-169.4	-64.5	6.3	-170.3	-68.6	6.8
5951	251.6191	47.9903	2013.6595	10.762	12.183	32.532	29.8	-49	88	5.3	-44.4	86	5.3
5952	251.6280	-8.6417	2014.1775	10.976	12.455	29.687	239.3	-79.1	-74.4	5.5	-78.1	-77.3	5.4
5953	251.6847	41.2955	2013.485	11.173	15.166	6.932	320.3	-57.5	-102.9	5.3	-54.7	-106.1	5.5
5954	251.7866	48.0601	2013.6415	14.879	16.134	5.656	285.5	-55.2	60.5	5.3	-55.6	61.1	5.4
5955	251.7937	15.0208	2013.685	13.936	18.246	12.575	355.7	-155.2	-26.5	6.1	-157.8	-23.4	6.4
5956	251.8080	3.2515	2013.2535	16.105	16.898	6.943	193.9	-98.5	-89.3	6.3	-95	-84.8	6.4
5957	251.8241	-13.3535	2014.334	11.905	15.263	10.197	50.3	-62.3	151.2	5.4	-59.2	160.4	5.7
5958	251.8577	62.8174	2013.513	9.742	15.146	17.505	166.5	67.8	-92.2	6.1	69.7	-89.2	6.2
5959	251.8838	25.3270	2013.7205	11.224	11.963	12.426	295.7	-132.2	-63.7	5.7	-129.5	-66.4	5.7
5960	251.9057	43.3096	2014.008	9.497	11.618	4.631	332.5	18.5	80.2	5.2	16.8	81.4	5.2
5961	251.9410	47.5418	2013.474	12.225	17.895	57.587	203.6	-19.8	-93.1	5.3	-20.8	-89.7	5.6
5962	251.9520	69.0765	2013.523	13.401	16.532	6.524	6.2	-55.6	113.3	5.7	-62.2	112.6	5.8
5963	251.9739	42.6071	2013.4525	15.204	17.158	5.994	354.9	-12.7	76.7	5.3	-14	79.7	5.5
5964	251.9804	7.6127	2013.6525	11.422	16.556	30.881	356.9	-30	-97.5	6.1	-37.1	-95.1	6.2
5965	251.9876	-4.9677	2013.6535	12.097	16.965	19.02	274.6	-21.9	-95.2	5.5	-18.7	-100.8	6
5966	251.9883	82.5952	2013.888	11.578	12.335	4.9	197.2	7.4	-66.8	5.6	6.2	-65.4	5.6
5967	252.1029	29.8481	2013.71	10.747	14.073	7.122	108.7	-25.8	-64.1	5.6	-25.8	-63.1	5.7
5968	252.1540	24.7177	2013.4295	17.287	17.358	45.535	95.9	64.8	-4.5	5.8	66.4	-3.4	5.8
5969	252.1631	68.0596	2013.5185	12.24	14.703	11.67	15.2	-95.7	22.9	5.7	-94.8	24.9	5.7
5970	252.1757	-11.6405	2013.706	14.587	17.072	43.861	61.8	-185.4	-60.6	5.6	-195	-53.8	6.3
5971	252.2045	16.8533	2013.602	13.292	15.771	10.979	286.6	-69.5	0.8	5.6	-69.7	2.1	5.6
5972	252.2122	-8.1237	2013.704	15.9	17.469	20.719	357	47.5	-107	5.7	44.3	-111	6
5973	252.2243	21.7171	2013.451	15.089	16.411	4.942	294	-21.3	-127.4	5.7	-20.6	-128.7	5.8
5974	252.2393	-3.0200	2013.746	14.963	17.448	23.633	117	-35.3	-188.5	5.6	-25	-186.9	5.9
5975	252.2522	11.3251	2013.8015	11.323	11.945	18.939	169.5	15.4	-92.4	6.1	14.8	-94.2	6.1
5976	252.2554	12.7760	2013.6015	17.557	18.071	54.505	189.9	13	-61.8	6.1	5.3	-66	6.2
5977	252.3613	25.9353	2013.689	12.08	14.202	36.604	165.8	-133.4	-50	5.7	-129.6	-49.8	5.7
5978	252.3700	20.9522	2013.6635	11.96	14.284	17.128	173.1	25.4	-92.9	5.6	25.4	-94.6	5.6
5979	252.5021	-8.1929	2014.0425	11.573	13.803	6.103	333.1	-69.1	-105.8	5.5	-66.4	-110	5.6

5980	252.6840	24.9047	2013.357	18.549	19.054	27.59	335.9	-60.9	-6.3	6.6	-60.9	-12.2	7.1
5981	252.7244	-7.2243	2014.005	12.595	14.1	7.126	20.8	80.6	-8.9	5.5	84	-10.2	5.5
5982	252.7303	28.2612	2013.59	12.09	13.204	4.032	231.3	-3.6	71.8	5.6	2.9	75.4	6.1
5983	252.7364	35.9334	2013.7155	10.281	13.77	12.577	244.7	-78.4	174.5	5.3	-72.6	173.4	5.3
5984	252.7427	27.7904	2013.4135	13.92	18.63	20.998	44.4	-124.9	116.6	5.7	-125	115.8	6.3
5985	252.7803	28.6007	2013.722	10.461	12.072	13.465	305.1	-23.1	114.6	5.6	-33	113.3	5.7
5986	252.8237	51.6496	2013.424	18.161	18.98	54.138	344.7	-63	5.2	5.5	-65.7	-1.5	7.2
5987	252.8837	61.5620	2013.3095	11.852	11.906	5.395	346.7	-17.5	87.9	6.2	-18.9	84.5	6.1
5988	252.9266	54.0858	2013.6985	15.046	15.426	4.076	74.7	-21.1	153.4	5.7	-22.2	146.6	5.7
5989	252.9846	20.3958	2013.5685	16.03	16.333	6.348	260.3	-5.1	-80.6	5	-1.2	-80.3	5
5990	253.0248	47.8432	2013.658	11.273	13.259	11.703	272.3	-83.7	2.3	5.3	-82.2	1.9	5.3
5991	253.0277	23.5222	2013.6435	13.268	14.731	6.602	189.6	87.8	-68.6	5	88	-69.7	5
5992	253.0529	-4.9072	2013.989	11.101	13.591	6.221	136	-76.7	-15.9	5.5	-76.9	-17.6	5.5
5993	253.0605	42.0910	2013.6455	10.755	14.692	50.22	153.7	53.8	-83.2	5.3	54.6	-83.6	5.3
5994	253.0688	58.2763	2013.492	14.701	16.653	22.294	242.6	-75.2	52.4	5.7	-72.3	52.6	5.7
5995	253.0732	54.6177	2013.689	11.02	12.733	4.716	214.8	-22.3	65.7	5.7	-29.1	64.9	5.6
5996	253.0788	3.2200	2013.514	13.183	16.674	7.043	147.7	84.7	-146.3	6.2	81.3	-144.3	6.4
5997	253.0848	70.5004	2013.3965	12.364	16.197	33.312	0.6	-98.3	136.4	5.8	-97.4	137.1	5.8
5998	253.1148	33.9576	2013.6855	10.318	15.83	12.719	318.1	-24.4	-64.8	5.9	-26.6	-64.4	6
5999	253.1401	17.7052	2013.6045	15.56	15.708	7.577	291.6	8	60.3	6.1	5.3	62.2	6.1
6000	253.1994	50.8015	2013.5665	13.232	15.079	11.638	242.8	-118.7	-45	5.3	-119.7	-40	5.4
6001	253.2159	42.9407	2013.624	14.408	16.929	5.888	245.5	-14	-69.8	5.3	-5.1	-69.6	5.4
6002	253.2761	40.3863	2013.467	13.873	18.488	33.129	303.1	20.2	-78.9	5.2	11.1	-77.2	5.6
6003	253.3438	49.0906	2013.682	9.602	12.234	13.25	125.8	-0.1	-151.2	5.3	-4.8	-154.2	5.3
6004	253.3587	29.4180	2013.8855	9.833	10.014	7.691	121.6	-81.4	-58.9	5.9	-78.9	-61.9	5.8
6005	253.3615	11.7694	2013.788	17.763	18.234	13.754	90	-10.5	-68.4	6.1	-13.9	-68.9	6.3
6006	253.3977	34.9409	2013.5315	10.986	17.249	15.669	151.7	-43.1	109.2	5.6	-44.7	108.1	5.8
6007	253.4001	73.7401	2013.6145	8.297	14.896	46.358	9.5	-21.7	113.4	5.6	-25.6	111.2	5.7
6008	253.4091	-2.3981	2013.8575	15.252	16.17	6.76	271.5	-70	-34	5.5	-70.9	-33.9	5.6
6009	253.4398	55.9318	2013.7395	13.798	14.225	4.733	33.9	27.2	188.1	6.1	24	189.8	6
6010	253.4627	26.0177	2013.6285	10.744	15.388	7.332	94.3	6.8	-69.7	5.9	2.6	-71	6.4
6011	253.4632	9.5650	2013.776	16.236	16.416	6.062	106.8	-8.4	-69.2	6	-10.7	-65.8	6.1
6012	253.5522	69.7186	2013.5185	14.274	14.485	3.69	187.1	-195.5	-17.9	6.1	-199	-14.7	6.5
6013	253.6586	38.2524	2013.709	11.611	13.05	6.333	253.1	-76.5	-8.7	5.2	-76.6	-8.9	5.2
6014	253.6642	50.8691	2013.4845	15.923	16.279	18.246	64.2	-44.6	101.3	5.4	-50.2	102.1	5.4
6015	253.8104	60.6711	2013.5155	13.354	17.951	54.081	125.9	-69.5	7.5	6	-66.8	8.3	6.2
6016	253.8202	5.8718	2013.502	15.255	17.493	49.488	3.4	-47.6	61.2	7.1	-38.6	63.5	6.3
6017	253.8372	13.3478	2013.8315	10.42	12.494	19.21	33.3	30.9	-78.3	5.9	31.2	-82.1	5.9
6018	253.8703	0.4893	2013.4745	18.322	18.387	48.063	285.7	-13.6	-66.7	6.6	-15.2	-64.3	6.5
6019	253.8775	35.9678	2013.69	12.771	14.27	8.66	284.2	-2.1	65.5	5.2	-1.7	64.8	5.2
6020	253.8922	57.2279	2013.5555	9.136	12.827	6.653	13	-0.7	86.7	5.6	-2.6	83.5	5.7
6021	253.9439	14.7582	2013.7485	14.92	15.815	17.16	301.4	-107.5	64.1	5	-109.7	62.8	5
6022	253.9621	68.7124	2013.3995	15.663	16.997	10.999	162.5	-18.6	67.7	6.1	-14.5	68.8	6.2
6023	253.9682	79.6321	2013.4665	17.292	18.543	46.427	54.5	-66.8	2.6	5.7	-63	12	6.1
6024	254.0255	54.3503	2013.814	14.128	14.2	4.349	14.6	51.7	-71.3	5.6	48.4	-76.3	5.6
6025	254.0416	52.4247	2013.5755	14.129	14.623	3.657	5.4	-72.1	24.4	5.3	-75.2	15.2	5.7

6026	254.1013	3.5081	2013.602	13.385	14.959	14.252	55.1	-143.5	9.6	6.1	-141.6	7.7	6.2
6027	254.1627	48.9190	2013.5035	11.605	15.66	8.892	69.6	-65	8	5.3	-67.7	12.2	5.4
6028	254.1879	27.5581	2013.829	10.051	10.61	4.745	216.1	-36	-122.1	6	-32	-116.9	6.2
6029	254.1908	30.2481	2013.5285	16.34	17.107	5.36	253.1	74.5	-13.8	5.7	72.5	-11.7	5.7
6030	254.1940	25.6832	2013.7735	10.109	10.171	8.523	265.7	136.8	-159.5	6	132.9	-171.2	6.1
6031	254.2249	-2.3958	2013.871	11.937	14.049	7.185	56.1	-23.1	-60.8	5.5	-23.2	-61.7	5.5
6032	254.2429	-4.1010	2013.626	16.815	17.66	31.065	152.5	49.7	-68.7	7.6	48.6	-60.9	6
6033	254.2781	25.6449	2013.633	15.602	16.524	5.446	69.7	-3.5	61.2	6	-4.4	61.5	6
6034	254.3167	33.9414	2013.796	11.483	13.871	5.32	111.1	-62.5	67.9	5.6	-60.8	69.1	5.6
6035	254.3624	0.7406	2013.6545	15.494	18.516	55.779	1.5	-18.5	-62	6	-16	-64.6	7.3
6036	254.4706	40.9309	2013.7035	9.646	14.123	12.011	326.7	-10.4	104.7	5.2	-9.4	101.7	5.3
6037	254.5149	6.2130	2013.5095	17.42	17.911	52.013	9.8	-23.7	-67.6	6.4	-13.6	-66.9	6.3
6038	254.5368	3.0471	2013.327	16.956	17.283	6.214	304	-2.2	-78.5	6.3	-6.7	-78.8	6.3
6039	254.5412	26.8738	2013.648	12.255	16.142	18.404	272.6	-85.1	-138.1	5.7	-83.5	-137.1	5.8
6040	254.5640	-11.6516	2014.086	12.17	13.754	4.583	33.9	-28.5	-70.4	5.5	-25.4	-70.3	6.5
6041	254.5802	36.1889	2013.51	12.429	16.965	50.606	38.2	-25.9	96.1	5.3	-30	93.4	5.4
6042	254.5964	59.7471	2013.562	15.107	15.83	4.352	266.9	-108.2	0	5.7	-106.8	-1.2	5.8
6043	254.6255	-6.7097	2013.9135	12.212	15.751	50.131	337.2	-3.8	-60.6	5.5	-3	-62.6	5.5
6044	254.6492	36.3241	2013.666	14.88	15.305	19.728	236.6	-7.5	98.8	5.3	-11.2	96.9	5.3
6045	254.7096	47.6919	2013.625	13.44	15.089	12.925	24.7	-110.3	95.2	5.3	-109.3	99.3	5.3
6046	254.7253	49.7143	2013.492	12.302	16.412	23.197	53.7	-34.7	79.5	5.3	-35	82.8	5.4
6047	254.7435	18.7751	2013.493	13.342	18.268	34.352	272.3	-7.7	-68.7	5	-0.6	-66.9	5.4
6048	254.8109	27.3769	2013.5535	10.631	17.221	49.332	305.4	-2.4	-101.5	5.6	-3.1	-103	5.9
6049	254.8204	62.4848	2013.885	11.38	14.075	5.298	224.2	-105.4	-5.8	6	-106.9	-7.4	5.8
6050	254.8512	32.5185	2013.5395	15.489	15.774	42.152	154.8	8.2	60.8	5.6	8.2	60.5	5.7
6051	254.8539	54.1315	2013.5015	13.645	16.298	12.286	297.6	-54.6	-74.7	5.7	-56.5	-74.4	5.7
6052	254.8644	6.1417	2013.6975	11.967	14.507	5.641	108	80.4	3.2	6	76.9	6.4	6.2
6053	254.9267	15.9610	2013.698	11.289	16.628	15.953	166.4	-23.4	-94.8	5.9	-20.4	-93.4	6
6054	254.9483	57.2111	2013.496	15.11	17.583	13.866	265.7	-8.7	65.2	5.7	-10.6	65.5	5.8
6055	255.0506	87.6216	2013.8155	11.468	15.787	28.492	23.2	-111	-24.8	5.6	-111.2	-22.2	5.5
6056	255.0940	11.2256	2013.833	11.334	12.587	5.545	210.2	-65	116.5	6	-61.1	120.3	6
6057	255.0976	39.7302	2013.546	10.353	15.706	12.633	131.2	2.2	80.6	5.3	0.3	82.4	5.3
6058	255.1075	28.2857	2013.589	15.44	16.713	18.573	120.5	-122.3	-1.4	5.7	-123.8	-0.1	5.8
6059	255.1082	23.7564	2013.56	14.64	16.245	17.957	259.1	-30	-62	5	-33.5	-62	5
6060	255.1218	29.2533	2013.641	12.847	14.589	4.583	66.7	-8.8	-75.1	5.6	-16.6	-74.9	6.2
6061	255.1259	33.2268	2013.599	14.655	15.376	27.126	289.3	-42	-78.2	5.7	-43.8	-73.2	5.7
6062	255.2204	23.3900	2013.688	14.078	16.586	7.629	101.8	-79	-4.5	6	-79.9	-5	6
6063	255.2400	73.4578	2013.5235	11.632	14.335	44.885	57.1	-57.8	80.8	5.7	-50.1	83.9	5.7
6064	255.2535	2.1016	2013.5615	13.186	17.295	13.661	224.7	11.1	-73.6	6	5	-74.4	7.2
6065	255.2584	0.8134	2013.7155	15.164	15.385	17.209	240.7	-0.7	-107.7	6.1	2.2	-105.3	6
6066	255.2748	84.9343	2013.8575	14.335	16.189	6.943	245.5	-19.2	144.9	5.5	-16.5	152.5	5.5
6067	255.2782	25.8412	2013.4085	16.607	17.643	8.271	238.4	-69.1	-40.4	6.1	-69.4	-36.2	6.2
6068	255.3022	38.4892	2013.568	14.584	17.035	6.743	48.4	19.7	-62.3	5.3	19.9	-64.4	5.3
6069	255.3516	1.9458	2013.4785	17.514	18.083	19.862	318	19.9	-64.8	6.2	17.3	-68.6	6.3
6070	255.4029	25.0583	2013.848	13.873	16.349	5.302	23.5	-44.3	-116	6	-37.9	-114.3	6.3
6071	255.4307	38.1157	2013.499	16.136	16.998	9.869	135.5	-108	-94.3	5.4	-109.8	-96.5	5.5

6072	255.4565	18.6734	2013.6645	14.669	14.844	4.11	206	47.7	-66.5	5	45	-63.6	5.1
6073	255.4627	43.7816	2013.566	10.337	15.609	14.889	161.6	-118.5	80.4	5.6	-118.6	80.8	5.8
6074	255.4847	-5.1787	2013.8255	12.511	17.198	11.805	277.9	-50.3	-79.2	5.5	-46	-78.1	5.7
6075	255.4966	45.1218	2013.5355	14.039	17.307	10.952	349.8	-56.1	122.5	5.7	-60.7	121.9	5.9
6076	255.5918	16.7918	2013.7035	14.151	14.908	9.279	331.1	-95.5	-124.9	5.6	-99.1	-126.7	5.6
6077	255.5923	44.9028	2013.4265	14.819	17.843	7.264	142.5	-30.7	-69.4	5.7	-32.6	-67	5.9
6078	255.6209	-3.0461	2013.908	10.875	15.414	12.393	224.7	-49.9	-110.2	5.5	-46.5	-112.8	5.6
6079	255.6651	2.1160	2013.5625	15.836	16.981	5.606	306.4	1.1	-83.8	6.1	-3	-84.2	6.3
6080	255.7139	9.2091	2013.8435	11.074	16.831	12.059	302.2	-88.3	9.5	6	-87.3	8.7	6.1
6081	255.7212	34.3873	2013.7085	14.621	15.335	22.348	233.1	-26.3	75.4	5.2	-27.4	76.7	5.2
6082	255.7230	84.4264	2013.733	15.63	17.53	12.582	334.9	-91.5	47.5	5.6	-85.6	49.5	5.7
6083	255.7279	0.7915	2013.5725	16.43	17.045	7.319	168.6	-24.1	-62.2	6.1	-23.4	-60.3	6.1
6084	255.8719	-9.3456	2014.266	10.329	11.265	5.576	60	59.9	-107.3	5.4	55.2	-115.8	5.4
6085	255.8781	26.3045	2013.6105	10.501	15.346	11.172	293.1	13.3	95.2	5.6	11.6	94	5.7
6086	256.0557	37.7340	2013.5765	11.247	17.125	25.022	191.8	-14.8	-76.2	5.2	-15.7	-78.4	5.4
6087	256.0821	3.7063	2013.5945	12.872	16.392	8.094	251.8	-30.9	-66.6	6	-31.8	-69.5	6.3
6088	256.0867	78.7126	2013.554	14.66	17.14	9.293	207.6	-4.7	-72.1	5.7	-6.6	-72.5	5.8
6089	256.0995	30.2900	2013.601	13.504	14.32	5.882	79	-77.6	-9.9	5.3	-78.1	-11.4	5.3
6090	256.1019	24.2940	2013.6595	10.529	14.861	6.979	182.1	-18.9	65.7	5.6	-17	63.8	5.8
6091	256.1186	13.2170	2013.8075	11.11	15.119	36.173	354.4	-89.1	33	6	-88.6	35.2	6
6092	256.1189	3.7395	2013.796	10.556	10.995	40.121	176	-219.4	-160.8	6.2	-220.6	-158.2	6.2
6093	256.1849	28.0868	2014.0425	9.245	12.822	5.874	272.2	-102.6	-95.5	5.6	-102.1	-97.5	5.8
6094	256.2009	-8.1990	2013.574	18.257	18.31	29.915	34.4	-13.1	-66.3	7.1	-8.3	-66.6	7
6095	256.2514	29.0337	2013.7455	12.631	13.008	5.745	100.2	-62.5	-1.8	5.6	-62.8	-2.2	5.6
6096	256.2514	-1.8645	2013.754	10.185	16.749	27.079	77.8	11.1	-64.2	5.5	4.7	-64.4	5.6
6097	256.2522	28.1495	2013.467	15.866	16.04	27.708	90.3	3.8	71.7	5.7	3.9	69.3	5.8
6098	256.2980	-0.3308	2013.9325	13.667	14.933	46.992	251.7	-17.8	-70.3	5.5	-17.9	-68.8	5.5
6099	256.3398	29.1361	2013.526	16.33	16.558	3.904	91.6	-16.5	-81.1	5.8	-24.5	-75.3	5.9
6100	256.3872	13.5692	2013.6345	16.498	17.657	34.1	76.4	-70.5	-31.6	6.1	73.2	-32	6.2
6101	256.4503	11.8777	2013.7325	12.825	15.03	7.622	1.3	-46.9	-94.8	6	-48.1	-92.8	6.1
6102	256.4574	4.9643	2013.827	9.825	12.808	28.843	205	-147.9	131.2	6.1	-143.4	132.3	6.1
6103	256.5009	22.6387	2013.897	15.365	15.588	3.324	185.5	-61.3	-1.4	5	-64.1	4.5	5.1
6104	256.5017	14.6077	2013.5815	15.406	16.356	5.612	333.5	-54.4	90.6	5.6	-52.9	90.7	5.7
6105	256.5099	-3.7223	2013.9295	12.554	15.662	17.345	357.5	9.7	64.7	5.5	6.6	64.4	5.5
6106	256.5577	36.8500	2013.662	13.366	14.41	8.203	9.5	-42.2	64.9	6	-42.8	61.8	6
6107	256.5952	7.7888	2013.6325	13.985	17.256	16.211	150.7	-44	-77.6	6	-45.5	-78.6	6.3
6108	256.6268	34.2301	2013.6145	14.61	18.1	10.453	181.6	17.6	-78.9	5.2	23.2	-77.3	5.5
6109	256.6592	12.8070	2013.831	9.127	15.312	11.667	124.3	-88	106.2	5.6	-88.8	99.1	5.6
6110	256.7088	1.6327	2013.6925	14.464	18.479	16.707	313.2	16.5	-74.9	6	23.6	-72.9	6.3
6111	256.7621	41.5872	2013.7045	10.682	14.57	6.846	4.8	-31	75.1	6	-30.4	75.6	6.1
6112	256.7668	-5.1609	2013.7985	11.377	16.669	18.548	304.6	-79.9	-54.4	5.5	-82.7	-49.9	5.7
6113	256.8944	30.6460	2013.6285	12.871	13.498	22.147	86.6	-64.1	74.9	5.3	-64.9	71.9	5.3
6114	256.9217	-11.5545	2014.141	12.883	13.284	13.173	215.2	-62.3	-38.7	5.4	-62	-34.2	5.4
6115	256.9543	50.5203	2013.3275	18.577	18.631	45.18	288.4	-74.3	-40.2	5.6	-69.4	-49.3	5.6
6116	257.1041	-12.0334	2014.3585	7.62	15.53	43.69	204.1	-60.2	-3.1	5.1	-61.1	-9	5.7
6117	257.1173	25.4938	2013.688	16.686	16.743	4.313	28.1	-62.4	-4.3	5.7	-61.1	-5.3	5.7

6118	257.1328	-10.8086	2014.0725	10.672	14.121	30.951	42.7	-27.5	64.9	5.4	-23.6	69.7	5.5
6119	257.1970	5.6209	2013.7555	12.532	14.409	12.139	138.6	-67.4	-4.4	6	-69.7	-3.6	6.1
6120	257.2004	35.9684	2013.6775	14.299	14.747	5.729	237.1	11.2	147.2	5.3	11.1	149.2	5.3
6121	257.2160	69.4191	2013.479	9.51	11.409	27.561	137.3	-10.3	-110.9	5.7	-11.3	-113	5.7
6122	257.2621	17.5159	2013.661	14.936	16.635	4.964	357.9	-162.4	12	5.7	-155.8	6.9	6
6123	257.3308	33.6253	2013.4665	17.438	17.794	46.526	21.7	-35.3	-79.9	5.4	-32.7	-80.4	5.5
6124	257.4150	15.5361	2013.665	11.608	14.418	7.641	46.3	23.7	-132.5	5.6	22.7	-133.7	5.6
6125	257.4289	17.3100	2013.5345	11.415	17.513	12.585	169.2	-22.9	-76.9	5.6	-26	-78.3	5.9
6126	257.4355	31.3103	2013.7005	13.716	15.986	7.349	25.8	-96.2	13.7	5.2	-97.6	12.5	5.3
6127	257.4483	45.0711	2013.855	12.648	13.301	5.171	58.7	-43	-100.7	5.7	-40.2	-105.6	5.5
6128	257.4759	12.4813	2013.699	16.324	16.925	5.626	251	-66.9	-103.1	6.1	-66.4	-108.9	6.2
6129	257.5026	18.1389	2013.696	14.917	15.617	5.934	102.1	-84.7	57.3	5.6	-85.1	56.8	5.6
6130	257.5418,	-6.5614	2013.685	16.319	17.153	24.092	164.7	-61.3	-99	5.7	-57.9	-106.3	5.7
6131	257.5443	54.4940	2013.7285	8.156	8.502	22.147	133.1	81.4	-99.5	5.6	83.5	-97	5.6
6132	257.5459	-10.7306	2014.054	14.36	15.724	8.303	356.8	-63.1	-98.1	5.5	-64.3	-98.8	5.5
6133	257.5499	41.6543	2013.719	9.824	11.966	22.609	331.5	-1.4	92.7	5.2	-4	92.4	5.2
6134	257.6062	27.9775	2013.55	11.158	15.158	48.1	128.8	17.7	-84.8	5.7	26.8	-86.3	5.7
6135	257.6610	39.9119	2013.5635	14.804	16.548	8.834	332	15.1	-85.2	5.3	16.4	-83.8	5.3
6136	257.6706	0.5815	2013.681	16.606	17.347	35.658	124.6	-66.6	-105.7	6.5	-71.7	-97.4	6.3
6137	257.7462	45.8736	2013.62	12.952	14.587	5.409	207.6	17.4	-70	5.7	17.1	-69.7	5.7
6138	257.8043	41.2869	2013.5465	10.808	16.188	12.2	30.3	-102.7	18.9	5.2	-106.4	19.3	5.4
6139	257.8157	27.2261	2013.443	16.974	17.428	42.456	283.6	7.9	-74.1	5.9	17.8	-75.9	5.8
6140	257.8933	-2.2384	2013.8175	10.784	14.682	33.68	123.7	-10.8	-76.8	5.5	-11.2	-74	5.6
6141	257.9064	2.7016	2013.7655	13.405	14.39	4.958	77.2	-52.6	-63.5	6.1	-55.4	-64.8	6.1
6142	257.9363	-0.5623	2013.773	12.605	15.177	15.28	272.6	115.5	-83.1	5.6	116.4	-81.5	5.6
6143	257.9454	-2.5382	2013.69	16.038	17.35	4.271	30.7	-140.8	-106.8	5.7	-143.8	-113.3	6.5
6144	257.9467	3.4785	2013.699	15.339	16.197	6.403	244.8	-93.5	-61.7	6.1	-92	-65.6	6.2
6145	258.0352	30.0161	2013.949	14.011	14.145	4.113	159.3	19.3	69.6	5.2	19.2	70.1	5.2
6146	258.0834	40.5953	2013.563	15.249	15.649	5.84	38.1	-64.4	-19.6	5.3	-66.4	-21	5.3
6147	258.0992	36.7918	2013.6355	13.508	16.242	8.347	286.3	38	-104.7	5.3	39.5	-102.5	5.3
6148	258.1405	16.1420	2013.4415	16.772	17.065	7.491	202.2	-109.4	62.3	5.8	-106.9	65.5	5.8
6149	258.1417	15.6164	2013.767	12.67	15.172	5.542	293.3	28.5	-120.8	5.6	31.5	-117.9	5.8
6150	258.1794	1.5347	2013.8225	12.279	14.57	22.454	145.2	-137.6	35.7	6.1	-138.7	33.6	6.1
6151	258.1851	17.6915	2013.637	13.762	14.517	10.683	321.1	34	103.6	5.6	31.6	102.3	5.6
6152	258.2240	40.7179	2013.4735	13.899	17.709	16.137	328.6	-54.1	102.7	5.3	-61.3	102.2	5.5
6153	258.2472	26.7678	2013.452	16.305	16.567	7.612	130.9	27.2	-71.5	5.8	28	-71.9	5.8
6154	258.2593	51.8663	2013.64	9.825	14.567	11.18	299.1	4.9	67	5.3	3.2	67.7	5.3
6155	258.3760	19.1653	2013.6265	10.722	15.038	15.364	326	-110.2	-154.8	6.1	-109.6	-158.1	6.2
6156	258.4113	-10.2675	2013.9895	14.706	16.092	21.723	299.5	-84	-79.5	5.5	-88.3	-79.2	5.8
6157	258.4344	68.1609	2013.4265	14.922	14.992	42.716	352.5	-8	-115	5.7	-9.1	-117.4	5.7
6158	258.5197	28.1185	2013.457	15.798	17.401	7.317	258.7	-48.6	82.1	5.7	-49.4	85.7	6
6159	258.5311	69.5890	2013.382	17.967	18.06	31.983	296.4	-69	6.5	5.8	-66.8	2.2	5.8
6160	258.5755	62.5890	2013.6255	9.996	14.055	10.674	140.3	10.1	-124.5	5.6	13	-127.2	5.7
6161	258.6503	33.6681	2013.543	12.724	16.096	12.634	22.7	-87.4	-17	5.3	-88.5	-16.7	5.3
6162	258.6806	2.8265	2013.4835	18.071	18.969	53.263	281.5	-18.9	-65.7	6.3	-14.6	-66	7.1
6163	258.7008	43.1961	2013.574	14.89	15.226	14.787	60.6	30.4	-131.1	6.1	31	-131.1	6.1

6164	258.8716	7.1989	2013.6715	13.094	17.521	13.192	29.7	4	-99.1	6	5.6	-98.3	6.3
6165	258.9102	-4.6129	2013.995	13.955	14.127	4.113	298.5	-22.5	-132.6	5.6	-21.3	-133.8	5.7
6166	258.9322	31.9257	2013.477	13.743	17.353	28.936	197.5	-49.6	-75	5.3	-49.5	-74.9	5.5
6167	258.9353	25.8101	2013.6105	10.917	18.341	49.919	160.7	9.8	-80.9	5.7	3.4	-80.9	6.2
6168	258.9730	8.4381	2013.657	13.266	14.859	19.669	278.2	69.1	-50.3	6	70.2	-51.5	6.1
6169	259.0236	63.1057	2013.5695	11.437	14.532	31.27	10.6	-3	-72.4	5.7	-2.7	-71	5.7
6170	259.0658	4.2314	2013.8455	11.449	11.875	20.125	318.5	66.6	189.7	6.1	68.9	193.3	6.1
6171	259.1014	74.7559	2013.391	15.17	15.927	30.036	245.9	-23	72.1	5.7	-23.3	71.4	5.8
6172	259.1280	34.8175	2013.764	10.449	12.066	5.498	322.3	-63.1	23.9	5.2	-62	23.4	5.2
6173	259.1287	-5.3938	2013.9375	15.867	16.076	6.685	87.7	91.5	-44.9	5.5	94.5	-47.5	5.6
6174	259.1610	4.6706	2013.8355	11.647	16.091	40.907	0.8	-87.5	-183.9	6.1	-91	-190.9	6.3
6175	259.2197	-2.4116	2013.8555	14.895	17.329	13.075	173.4	23	-118.5	5.5	25.2	-120.1	5.7
6176	259.2580	56.2055	2013.686	9.765	10.944	17.465	16	-26.6	72.7	5.6	-30.3	75.5	5.6
6177	259.2831	18.1156	2013.668	10.681	14.098	46.429	356.4	-61.5	-187.5	5.7	-63.6	-190	5.8
6178	259.3247	-5.2810	2014.064	13.545	14.847	4.723	3.1	-60.1	-64.6	5.4	-62.6	-66.9	5.5
6179	259.3410	71.8576	2013.573	12.406	17.23	7.919	15.5	-3.7	-65.7	5.7	-13.4	-65.7	5.7
6180	259.3468	42.3766	2013.4895	14.906	16.774	51.791	102.3	-36.4	-85.1	6	-36.4	-89.8	6.1
6181	259.3767	28.6887	2013.673	12.653	16.688	6.709	257.6	-112.8	-86.3	5.7	-111.8	-86.3	6.2
6182	259.4402	19.6461	2013.299	17.01	17.931	59.135	306.1	-60.6	25.2	5.8	-64	24.9	5.9
6183	259.4847	38.7942	2013.7185	10.181	14.813	19.275	171.7	-34.3	-112.1	5.3	-35.6	-112.1	5.3
6184	259.5202	-1.2855	2013.995	8.985	13.736	9.229	164.6	-126.8	-134.5	5.5	-123.8	-134.8	5.6
6185	259.5582	8.1239	2013.5115	15.717	15.9	9.678	87.3	-70.2	-60.5	6.2	-71.5	-59.5	6.1
6186	259.5632	4.9802	2013.8005	9.807	11.066	18.575	176.9	61.6	4.7	6	61.2	3.4	6
6187	259.5676	17.2800	2013.468	14.359	17.444	48.942	218.4	-69	36.4	5.2	-67.8	35.3	5.5
6188	259.6595	53.1546	2013.6295	14.326	15.879	6.243	207.1	-32.9	-64.7	5.3	-28.5	-64.6	5.4
6189	259.6995	47.5615	2013.612	11.812	14.595	30.623	216.1	-82.9	-37.3	6	-84.2	-39.6	6
6190	259.7268	55.8967	2013.568	12.404	13.403	6.952	147.3	-31.9	-67.9	6	-36.5	-64.4	6
6191	259.7524	17.6410	2013.656	10.559	15.033	16.175	212.4	55.2	-101	5.2	55.7	-100.7	5.3
6192	259.8543	16.1939	2013.6995	11.947	12.369	6.223	266.6	65.4	-33.8	5.2	-64.3	-34.6	5.2
6193	259.8757	8.5869	2013.689	11.177	15.356	9.945	239	0.8	-133.6	6	0	-138	6.2
6194	259.9235	-2.0732	2013.982	14.293	14.467	5.297	286.2	-24.6	-107.1	5.5	-23.7	-107.1	5.5
6195	259.9393	67.3827	2013.449	11.942	14.286	22.957	219.1	24.5	-62.5	5.7	24.3	-63.7	5.7
6196	259.9671	11.9192	2013.755	12.906	12.923	18.827	107.4	12.9	73.6	5.2	13.8	74.7	5.2
6197	259.9841	-8.0006	2013.446	16.602	17.972	11.886	187.3	-6.3	-63.9	5.8	-17.9	-64	6.1
6198	259.9959	84.4602	2013.8305	12.5	16.111	7.585	42.2	-37.5	68.5	5.5	-37.1	73	5.5
6199	260.0008	72.5841	2013.496	8.671	9.006	5.133	209.3	-37.6	89.3	5.6	-30.9	97	6.2
6200	260.0033	0.7137	2014.109	12.916	14.125	3.917	275.2	-14.8	-66.5	6	-12.3	-65.9	5.9
6201	260.0091	1.4841	2013.757	15.041	17.252	16.291	51.6	-61.4	-47.1	6.1	-61.2	-42.8	6.1
6202	260.0441	2.0328	2013.5865	18.185	18.355	58.332	217.9	-8.2	-61.7	6.5	-9.4	-62.5	6.3
6203	260.0763	2.0319	2013.7095	17.766	17.772	31.542	115.8	-3.6	-62.8	6.3	-15.5	-64.1	6.3
6204	260.1162	16.8898	2013.5985	11.949	16.335	11.056	231.5	-79.2	19.6	5.2	-80.3	18.3	5.3
6205	260.1379	-2.4239	2013.8225	8.498	12.64	7.075	100.8	-14	-63.3	5.4	-19.6	-61	5.7
6206	260.2332	5.2080	2013.6495	16.051	16.762	7.763	250.8	56.4	-78.3	6.1	55.1	-78.5	6.2
6207	260.2563	15.8003	2013.576	11.586	17.605	13.027	10.2	15.9	-103.8	5.2	8.3	-104.8	5.4
6208	260.2960	73.4864	2013.5305	10.218	13.15	10.562	51.9	-43.7	170.5	5.7	-41.8	171.7	5.7
6209	260.3690	-7.2559	2013.627	16.447	18.658	7.387	290.3	-75	-7.1	5.6	-76.5	1.3	7.9

6210	260.4344	12.1552	2013.7985	10.363	17.303	25.394	133.8	-7.1	-83.4	5.5	-2.7	-87.6	5.7
6211	260.4413	9.0866	2013.6235	12.708	17.301	12.427	59.8	-166.6	-107.9	6.1	-168.1	-111.4	7.2
6212	260.4661	70.5834	2013.356	14.96	17.214	23.858	232.8	21.9	79.2	5.7	21.4	77.4	5.8
6213	260.6459	29.5171	2013.507	13.305	17.201	31.587	342.2	-19.6	-176.8	6.1	-22.8	-177.1	6.4
6214	260.6786	71.7886	2013.4035	13.794	16.399	10.285	160.2	-2.8	89.7	5.7	-2.7	89.4	5.8
6215	260.7075	42.9149	2013.3815	9.708	17.886	28.496	43	-37.6	89.8	6	-39	85.4	6.4
6216	260.7116	21.9224	2013.6665	13.557	15.28	10.27	252.7	-76.6	110.5	5	-78.5	111.7	5.1
6217	260.7139	-2.3886	2014.0205	6.615	10.664	48.665	147.1	60.7	-107.4	5.5	60.1	-112.4	5.5
6218	260.7478	46.4118	2013.578	13.079	15.122	14.939	28.1	-30.8	67.3	6.1	-32	67.7	6.1
6219	260.7493	49.8910	2013.515	13.816	15.918	14.217	107.6	-73.2	23.8	5.3	-70.6	23.8	5.4
6220	260.8023	73.8298	2013.5075	11.222	13.955	44.66	358.4	102	117.9	5.8	100	120.7	5.8
6221	260.8259	62.1975	2013.5895	10.403	15.103	8.807	14.4	-43.8	-73.6	5.7	-45.9	-71.7	5.7
6222	260.9422	5.9178	2013.711	15.489	15.541	4.26	108.8	-18.8	-67.3	6.1	-18.4	-66.3	6.3
6223	261.0291	22.5887	2013.6955	14.365	15.216	4.155	79.8	-88	-13.2	5.3	-87.6	-19.8	5.9
6224	261.0341	20.1721	2013.702	14.201	14.74	4.605	196.9	-65.1	23.2	5.3	-65.6	28	5.2
6225	261.1263	28.9582	2013.588	12.85	17.082	11.822	5.9	29.6	-64.5	6	29.5	-63.5	6.1
6226	261.1614	-3.3013	2014.06	13.906	15.087	5.084	322.7	94.9	-32.4	5.5	97.6	-29.9	5.5
6227	261.2035	18.7166	2013.639	9.123	12.264	31.69	191.6	-64.5	-19.5	5.3	-61.6	-19.7	5.3
6228	261.2098	37.0543	2013.6325	10.862	15.141	11.283	39.9	18.2	123.2	5.3	18.8	123.1	5.3
6229	261.2263	50.4429	2013.616	11.6	12.775	9.147	198.2	32.5	110.2	5.3	24.6	110.4	5.3
6230	261.2327	10.5878	2013.729	12.735	14.398	10.691	148	-5.5	-131.2	6	-6.2	-134.5	6
6231	261.2347	43.6062	2013.473	15.21	16.562	14.877	2	-63.7	21.3	6.1	-65	22.8	6.1
6232	261.2440	3.0810	2013.59	18.026	18.353	34.05	317.9	20.1	65	6.4	31.4	63.4	7
6233	261.2730	3.8029	2013.967	9.277	9.627	7.928	4.3	5.2	-90.5	6	3.2	-87.1	6
6234	261.3140	2.6502	2013.773	14.243	15.791	33.445	215.4	-71.1	24.8	6	-73.9	24.2	6.1
6235	261.3274	49.4793	2013.591	10.449	15.03	20.592	38.6	-52.1	-105.4	5.3	-52.9	-103.2	5.4
6236	261.3679	4.9960	2013.618	17.008	18.555	59.762	72.6	-2.9	-63.7	6.2	-11.2	-63	7.1
6237	261.4827	50.8048	2013.416	15.498	18.152	34.709	44.6	-24.6	79.2	5.4	-26.9	80.8	5.6
6238	261.5171	1.2611	2013.802	17.781	18.61	55.57	277.2	72.9	20	6.4	76.9	9.2	7.3
6239	261.5423	7.3206	2013.63	15.144	16.672	6.054	125.4	-26.8	-62.6	6.1	-28	-62.1	6.2
6240	261.5432	5.7554	2013.6925	16.333	16.539	30.048	310	-12.3	-61.4	6.2	-11.5	-61.3	6.1
6241	261.5947	19.8312	2013.379	15.156	16.676	55.741	35.5	13.5	-70	5.7	13.8	-67.7	5.9
6242	261.6229	23.0502	2013.5825	11.512	14.799	26.064	80	-32.1	60.9	5.7	-28.8	62.6	5.7
6243	261.6546	31.4979	2013.6825	11.036	13.377	5.461	121.8	-35.4	-153.5	5.2	-37.2	-152.8	5.4
6244	261.7178	29.9485	2013.609	11.287	12.898	10.947	191.5	-48.4	90.8	5.3	-50.1	94.3	5.3
6245	261.7383	17.1767	2013.5345	13.712	17.972	43.165	135.7	-43.9	-89.6	5	-45.4	-93.3	5.3
6246	261.7417	30.9758	2013.787	15.178	15.484	4.026	213.1	20.2	-102.6	5.3	24.1	-103	5.3
6247	261.7518	33.1008	2013.576	13.465	16.592	39.941	311.9	6.3	-74	5.2	4	-75.6	5.3
6248	261.8665	61.5410	2013.555	11.823	14.333	12.662	197.1	-31.5	-130.4	5.7	-31.4	-130.4	5.7
6249	261.8847	4.4793	2013.8455	8.438	10.336	24.294	263.6	-61.8	68.9	6	-68	59.1	6
6250	261.9270	54.3993	2013.526	12.723	14.924	26.069	199.3	21.8	67.5	5.7	22	69.5	5.7
6251	261.9349	30.7250	2013.503	15.356	18.121	52.837	303.5	-15.6	-94.4	5.3	-18.9	-89.7	5.6
6252	261.9839	12.6103	2013.524	18.003	18.427	37.538	195.9	-3.4	-71.2	5.1	3.7	-69.3	5.3
6253	261.9949	17.1550	2013.6975	13.573	14.195	7.221	303.4	-70.7	-4.1	5	-71.4	-5	5
6254	262.0712	-6.3844	2013.5725	14.731	16.94	4.961	272.3	-42.9	-109.3	5.6	-35.7	-108.1	6.6
6255	262.1622	-5.8830	2013.7125	10.762	17.367	17.241	15.7	17.8	-72	5.4	14.7	-72	6.1

6256	262.1678	12.0728	2013.6035	17.684	18.427	20.246	174.2	35.5	-137.4	6.5	26.7	-136	6.8
6257	262.2318	72.1307	2013.5115	10.178	12.264	14.568	179	-0.4	92.2	5.7	-1.5	95.4	5.7
6258	262.3799	69.8778	2013.45	13.474	15.722	5.438	3.4	-47	60.9	5.7	-43.2	62.7	6.6
6259	262.3873	-4.8062	2013.5635	15.649	17.335	7.65	264.1	-8.5	-95.5	5.6	-9.3	-92.7	5.9
6260	262.4381	38.6957	2013.657	13.662	14.41	15.769	167.6	12.2	-61.8	6	11.9	-64.2	6
6261	262.4404	19.0288	2013.552	13.376	13.671	37.784	180.9	-127.3	90.3	5.1	-126.9	88.1	5.1
6262	262.4441	-5.3780	2013.4185	18.134	18.363	27.736	332.2	60.3	5.2	5.9	61.7	1.6	6.3
6263	262.4629	48.0549	2013.3005	16.325	18.298	5.808	344.8	-63.9	-3.1	5.4	-63.4	-2.2	5.9
6264	262.4643	7.7819	2013.8435	12.435	15.438	6.825	305.5	-23.5	-166.9	6.1	-23.1	-169.7	6.1
6265	262.4934	31.6388	2013.4645	15.241	16.264	8.525	38.2	-115	-128.6	5.4	-111.3	-131.4	5.4
6266	262.5278	4.9558	2013.756	9.505	13.563	43.406	240	69.3	-0.1	6	66.7	-5.5	6.1
6267	262.5374	10.3280	2013.944	8.968	9.492	49.923	211.3	-50.2	-148	5.9	-53.6	-149.3	6
6268	262.5555	49.9172	2013.5595	10.533	14.441	21.654	283.9	-19.2	90.7	5.3	-18.7	93.3	5.4
6269	262.5678	31.4865	2013.4545	16.136	17.528	17.137	141.7	-17.9	75	5.3	-17	73.1	5.4
6270	262.5741	5.2842	2013.6365	15.407	16.096	30.127	111.9	-76.8	-75.9	6.1	-77	-70.1	6.1
6271	262.5830	61.0021	2013.5415	11.34	14.681	9.438	357.1	-21.6	-65	5.7	-23.6	-67.8	5.7
6272	262.6307	21.4924	2013.5895	14.321	14.702	15.169	282.7	-207.7	-20.9	5.9	-206.9	-20.2	6
6273	262.6575	-11.1842	2014.362	12.224	14.429	13.566	123.2	81.4	-148.5	5.4	84.8	-148.7	5.5
6274	262.6627	-2.0693	2013.9245	11.014	12.519	44.386	296.1	6.4	-68.5	5.5	4.2	-67.6	5.5
6275	262.6870	6.6887	2013.77	12.716	15.494	37.94	152.6	-82	-114.5	6	-84.6	-116.5	6.5
6276	262.8163	9.8187	2013.7815	11.663	14.87	21.511	347.7	-45.9	62.9	6	-47.8	64.4	6
6277	262.8386	31.1291	2013.6155	13.616	15.713	21.024	216.2	-15.7	-69.2	5.2	-15.4	-70.3	5.3
6278	262.8944	-0.3859	2013.7345	13.937	17.084	48.169	110.6	-8.4	-77.3	5.6	-11	-80.3	5.7
6279	262.9256	41.1088	2013.5275	13.949	16.173	8.238	247.3	16.8	-133.3	5.3	18.1	-134.7	5.4
6280	262.9553	63.1210	2013.4455	13.331	15.145	9.616	199	-8.7	128.7	5.7	-10.4	122.9	5.7
6281	263.0312	55.1407	2013.4105	16.467	17.774	14.279	347.3	28.2	-78	5.7	23.7	-80.4	5.8
6282	263.0595	20.6543	2013.596	13.474	17.616	56.78	327.6	-40.3	-75.9	5.7	-36.1	-78.4	5.2
6283	263.0939	37.8725	2013.4155	16.559	18.161	15.356	140.6	24.2	-66.8	5.3	26.3	-66.5	5.5
6284	263.0982	70.7743	2013.497	11.905	14.883	15.329	64.9	1.3	62	5.7	2.2	63.4	5.7
6285	263.1041	13.8535	2013.4325	17.754	17.931	11.856	298.1	-12.8	-74.3	6.2	-21.4	-68.8	6.2
6286	263.1536	45.6289	2013.526	12.334	15.772	12.829	328.7	-24.1	-113.7	5.3	-26.7	-113.3	5.4
6287	263.1708	41.6131	2013.4675	15.456	16.584	29.076	139.7	85.8	34.5	5.3	86.6	36.7	5.4
6288	263.2953	4.5327	2013.6165	15.149	16.617	10.223	291.7	-75.2	116.7	6.1	-74.4	117.2	6.2
6289	263.3140	34.9021	2013.58	14.686	15.876	54.553	315	-22.9	-124.2	5.3	-22.2	-123.3	5.3
6290	263.3348	43.0224	2013.457	12.291	15.265	58.274	256.4	61.7	-100.1	5.4	60.3	-100.1	5.4
6291	263.3460	-6.2280	2013.661	14.961	16.208	5.338	328.9	-21.3	-71.3	5.6	-17.5	-74.6	6.1
6292	263.3619	65.7502	2013.571	14.186	14.704	4.31	294.3	-81.8	44.8	5.7	-81.2	43.7	5.7
6293	263.4221	24.3433	2013.635	15.685	15.853	34.22	175.9	-68.5	-26.7	6	-72.9	-16.2	6
6294	263.4243	-2.7613	2013.7535	10.257	16.615	11.847	113	-13.7	-161.3	5.5	-17.9	-164.1	5.9
6295	263.5127	10.6955	2013.6125	14.096	16.905	13.229	310.6	-109.8	-184.3	6.2	-112.1	-190.9	6.7
6296	263.5491	14.2780	2013.4595	17.377	18.099	7.678	155.1	-0.9	-64.5	6.1	-1.3	-61.9	6.4
6297	263.5608	5.9104	2013.4675	12.443	16.305	58.407	60.3	34.8	-93.1	6.1	28.1	-98.7	6.3
6298	263.6650	48.9687	2013.536	12.962	13.054	55.409	145.1	-56.9	152	5.4	-58	151.5	5.4
6299	263.7023	6.0236	2013.905	7.383	10.162	27.385	4.1	9.7	-78.3	6	11.2	-77.1	6
6300	263.8871	6.0529	2013.803	13.808	16.404	27.047	236.7	24	-103.9	6.1	22.2	-106.9	6.2
6301	263.9837	27.6976	2013.6005	13.665	15.562	8.764	204	35.2	-61.1	6	32.8	-64.6	6.1

6302	264.0241	44.1197	2013.5875	13.578	14.299	7.668	72.8	4.4	-75.4	5.3	7.4	-74.7	5.3
6303	264.0466	71.5960	2013.396	15.071	15.821	11.855	356.7	1.7	62.9	5.7	3.4	63	5.7
6304	264.1343	15.7807	2013.601	16.284	18.211	57.888	154.6	8.4	-75.8	6.1	-1	-76	6.2
6305	264.1362	7.1294	2013.7865	13.614	14.751	51.492	34.5	36.2	-103.7	6	36.9	-103.4	6
6306	264.1524	-6.8234	2013.7005	12.841	16.576	10.55	263.5	-8.4	-69.2	5.5	-6.1	-71.9	6.5
6307	264.1918	3.2208	2013.687	12.253	16.027	11.685	108	-46.2	-78.2	6.1	-48.9	-78.2	6.1
6308	264.2558	-1.3113	2013.758	14.808	16.266	4.614	343.1	-47	-133.4	5.6	-42	-130.2	5.7
6309	264.2693	46.4758	2013.4565	14.243	14.97	43.299	199.8	-13	-125.8	5.4	-16.1	-124.7	5.4
6310	264.3634	51.5366	2013.6	11.928	12.175	14.695	93.3	-79.8	14.8	6	-78.3	12.3	6
6311	264.3711	15.9812	2013.536	17.509	17.66	37.922	356.7	-69.3	21	6.1	-65	30.1	6.1
6312	264.3937	5.0017	2013.5225	17.212	17.361	50.707	53.7	-5.6	-70.2	6.2	-10.5	-69.3	6.3
6313	264.3965	55.1370	2013.567	12.122	12.84	9.528	122.9	-51.2	83.5	6	-51	86.9	6
6314	264.4004	67.7248	2013.4135	16.041	16.389	31.982	193.2	50.5	111.4	5.8	50.4	111.5	5.8
6315	264.4258	1.6337	2013.6605	11.285	16.778	48.846	344.9	-51.3	-92.8	6.1	-44.6	-93.6	6.2
6316	264.5087	35.1782	2013.5175	15.158	17.988	18.373	278.6	-16.5	69.1	5.3	-18.2	68.8	5.5
6317	264.5354	39.6548	2013.456	14.985	17.627	21.195	157.6	-52.9	136.6	5.7	-56.5	134.2	6.1
6318	264.5415	56.1203	2013.4855	14.281	15.872	6.257	60.9	-14.3	118.1	6.1	-19.2	119	6.1
6319	264.6140	30.9356	2013.175	14.044	18.734	40.793	201.3	-64.9	6.2	5.3	-67.1	-2.4	6.6
6320	264.6341	36.2066	2013.6825	11.598	14.576	24.563	158.1	-55	-105.5	5.6	-55.8	-105.8	5.6
6321	264.6444	-3.4554	2013.3815	18.283	19.167	28.911	18.6	-65.6	-21.8	6.5	-65.7	-19.5	7.1
6322	264.6798	-1.6607	2013.667	18.517	18.554	39.741	231.2	-14.6	-99	7.4	-17.5	-102.4	6.7
6323	264.6848	76.7543	2013.513	11.17	13.838	29.621	325.1	-10.6	200.4	5.8	-9.8	199.5	5.7
6324	264.7040	33.3268	2013.6525	12.781	15.509	44.925	180	15.5	-76.6	5.2	14.1	-74.5	5.3
6325	264.7299	51.9583	2013.4895	10.793	13.294	5.458	211.6	-11.3	-92.7	6	-13.5	-94.6	6.2
6326	264.7748	65.1522	2013.3985	14.972	16.421	17.467	109.1	-1.1	108.7	5.7	-5.4	111.1	5.8
6327	264.8137	46.7635	2013.4135	14.532	16.953	5.183	241.2	6.1	-93.3	5.4	2.6	-97.4	5.9
6328	264.9431	5.6242	2013.534	16.318	16.612	9.306	249.7	-136.3	-163.3	6.4	-134.1	-168.7	6.5
6329	265.0318	56.1297	2013.446	15.846	16.998	5.352	249.8	-7.2	72.2	5.7	-4.7	70.6	5.8
6330	265.1250	-3.2217	2013.59	14.342	17.345	33.422	86.1	10.9	-129.5	5.6	11.6	-132.7	5.9
6331	265.1899	5.8829	2013.7305	17.534	17.826	18.204	22.7	-61.4	-9.8	6.4	-63.9	2.8	6.4
6332	265.2582	23.0214	2013.4975	11.989	16.939	47.904	104.7	-1.9	-97.5	6	-4.2	-93.3	6.3
6333	265.3199	1.4588	2013.765	10.324	12.58	12.983	148.4	-98.2	-48	6.1	-97.3	-48.8	6.1
6334	265.3507	65.2748	2013.4645	10.422	13.654	8.268	19	-15.2	121.7	5.7	-13.1	123.5	5.7
6335	265.3518	39.0617	2013.645	13.178	13.404	11.929	222.2	5.7	-97.5	5.6	4.5	-93.1	5.6
6336	265.3520	41.4595	2013.724	9.969	10.913	7.603	181.6	-31.7	135.6	5.6	-28.4	137.6	5.6
6337	265.4959	70.0838	2013.4895	14.622	15.45	14.645	197.8	-90.2	91.5	5.7	-90.1	90.6	5.7
6338	265.5315	71.0315	2013.198	13.435	18.409	8.334	315.2	-56.8	79.6	5.7	-55.7	86.1	7
6339	265.5445	26.5104	2013.6065	15.567	16.416	7.295	81.9	-16.8	-66.7	6	-18.9	-64.3	6.1
6340	265.5524	23.0333	2013.638	11.154	14.267	24.316	139.1	-118	-38.3	5.7	-120.7	-43.3	5.7
6341	265.5941	36.0797	2013.762	9.125	10.715	26.756	195.3	-15.3	-141	5.6	-7.7	-142.9	5.7
6342	265.6065	27.9379	2013.5945	9.446	17.02	26.747	37.1	-54.1	150.1	6	-57	147.3	6.2
6343	265.6171	10.2830	2013.829	14.933	15.134	3.929	215.2	-26.7	-68.7	6.1	-25.6	-71.5	6.2
6344	265.7624	40.7442	2013.754	13.436	14.009	4.098	269.3	-38.7	-70.4	5.7	-36.3	-68.3	6
6345	265.7824	79.0767	2013.5635	12.428	14.002	5.905	100.5	8.3	-82.9	5.7	9	-81.3	5.7
6346	265.7971	57.0598	2013.4795	14.811	14.93	18.626	94.9	-25.2	205.9	5.8	-22.6	206.2	5.8
6347	265.8389	20.5268	2013.6095	12.066	14.232	23.606	41.4	-1.8	-71.1	5.7	-0.5	-70.9	5.7

6348	265.8401	-1.6425	2013.8565	14.128	14.388	9.405	120.9	12.6	97.3	5.6	13.3	95.2	5.5
6349	265.9078	-3.2147	2014.2785	8.338	11.838	7.343	252.5	7.2	-117.6	5.4	3.7	-115.4	5.4
6350	265.9388	46.2319	2013.499	13.507	15.855	7.306	282.1	-65.7	47.2	5.3	-63.4	47.8	5.4
6351	265.9456	26.7782	2013.3745	17.394	17.528	57.758	241.6	-26.2	-73.8	6.2	-18.4	-79.8	6.1
6352	266.1134	58.3106	2013.517	12.395	14.674	12.771	122.1	-41.2	-130.4	5.7	-41.7	-129.5	5.7
6353	266.1350	-0.6787	2013.5035	16.418	18.262	23.238	160.5	1.2	-90.3	5.7	-1.3	-94.8	6.1
6354	266.3052	62.3430	2013.497	12.349	14.824	10.325	113.6	-14.5	68.8	5.7	-12.5	69.3	5.7
6355	266.3401	-7.9355	2013.913	14.765	15.18	12.515	44.6	49.1	-66.1	5.6	47.7	-67	5.6
6356	266.3495	7.7618	2013.5875	16.168	17.87	18.179	220	-6.1	-60.2	6.1	-15.2	-60.8	6.3
6357	266.4020	34.6719	2013.563	17.589	17.617	4.099	38.2	23.9	-79.8	5.4	25	-81	5.5
6358	266.4573	17.9906	2013.6355	17.212	17.519	28.518	164.7	-60.7	15.7	5.7	-60.1	13.5	5.8
6359	266.5143	80.5303	2013.5955	13.927	17.658	10.459	176.5	-46.2	141.2	5.7	-45.8	144.5	5.8
6360	266.5151	55.4313	2013.482	14.907	15.887	7.392	151.1	-23.9	-71.8	5.7	-22.4	-70.3	5.7
6361	266.5888	70.9872	2013.499	13.426	14.19	10.179	1.5	-74.7	32.9	5.7	-76.4	32	5.7
6362	266.6205	6.6464	2013.708	14.456	16.281	26.585	331.2	-19.8	-69.1	6	-15.4	-71.4	6.1
6363	266.6330	9.3318	2013.801	11.564	14.891	16.171	0.1	33.7	-70.5	6	35.4	-70.3	6
6364	266.7333	64.9361	2013.4575	14.28	14.391	23.117	144.5	-85	101.7	5.7	-88.2	102	5.7
6365	266.7806	19.7482	2013.7195	11.792	15.139	27.525	258.2	24.2	-80.6	5.7	17.6	-84.4	5.7
6366	266.8367	64.7058	2013.4175	12.27	15.624	10.718	90.3	-40.7	-79.2	5.7	-37.5	-77.6	5.8
6367	266.8601	44.1453	2013.357	15.407	16.975	10.702	302.8	1.3	-80.1	5.4	2.6	-83.8	5.5
6368	266.9231	47.7050	2013.4545	14.891	16.516	4.841	227.5	2.3	61.9	5.3	0.7	64.1	5.5
6369	266.9571	32.9233	2013.4265	14.659	17.416	15.97	229.6	12.8	-150.4	5.3	14.6	-150.8	5.6
6370	266.9643	40.7892	2013.6515	9.886	14.338	23.073	336.9	-9.8	86.9	5.3	-11.5	85.3	5.3
6371	267.0165	-13.6121	2014.35	7.935	15.185	30.409	154.3	-122.9	-126.9	5.4	-115.6	-130	5.3
6372	267.0640	62.3479	2013.3655	12.994	16.657	12.535	176.7	-17.9	70.1	5.7	-19.9	69.5	5.8
6373	267.1422	9.9116	2013.85	12.968	16.159	7.053	311.3	-11.5	-73.2	6	-11.4	-76.4	6.1
6374	267.1462	46.0978	2013.444	14.694	16.882	19.793	34.3	-3.1	78	5.3	-2.6	76.7	5.4
6375	267.1628	35.6044	2014.068	9.042	13.843	6.65	230.2	95.9	-24.7	5.2	94.5	-19.1	5.2
6376	267.1652	38.5281	2013.6415	12.959	16.317	13.253	22.4	-3.4	-64.7	5.2	-0.9	-65 1	5.3
6377	267.1949	25.9540	2013.534	12.068	12.835	4.351	357.7	0.2	69.8	6	1.2	66.5	6.2
6378	267.2472	8.2974	2013.7085	12.957	13.447	6.413	240.5	-13	98.5	6	-8.6	99	6.1
6379	267.2655	62.7991	2013.5725	6.846	12.222	16.7	208.2	-70.3	68.9	5.7	-74.2	67.8	5.7
6380	267.2793	26.0155	2013.5605	15.438	16.371	16.364	349.1	0.1	-73.3	6	-4.2	-74.4	6.1
6381	267.2941	83.8743	2013.7585	11.925	17.759	14.979	256.7	-22	63.9	5.6	-15.8	66.4	5.7
6382	267.3293	8.5423	2013.472	16.368	18.495	19.177	224.4	-19.2	-61.8	6.4	-15.8	-64.6	6.5
6383	267.3388	49.0725	2013.8415	11.042	13.285	4.738	32.6	53.3	-94	6.1	44.1	-95.1	6.7
6384	267.3716	12.3784	2013.784	12.665	12.668	23.612	324.3	-34.8	-111.7	5.2	-35.9	-109.8	5.2
6385	267.3720	74.0624	2013.6065	17.715	18.803	57.245	189.5	15.6	72.7	5.7	12.1	72.8	7.6
6386	267.4499	18.1908	2013.7995	12.962	14.483	12.962	328	44.5	-112.8	5.9	45.3	-118.1	6
6387	267.4789	41.8554	2013.358	15.963	16.781	10.689	178	-31.8	92.3	5.4	-32.6	89.8	5.4
6388	267.5141	16.8399	2013.782	12.133	17.619	13.195	190.7	-47.2	-104.4	5.2	-48.5	-101.2	5.4
6389	267.5348	71.1281	2013.4275	16.205	16.55	4.834	71.9	-23.1	64.3	5.8	-22.2	64.6	5.7
6390	267.5492	8.8897	2013.7125	13.148	14.38	19.159	246	-129.5	-103.9	6.1	-129.1	-104.8	6.1
6391	267.5708	23.7938	2013.4905	17.387	19.164	57.474	139.8	-28	-66.9	6.1	-19.1	-71.3	7.4
6392	267.6394	23.7905	2013.6185	9.988	16.526	12	82.7	-67	-68.7	6	-65.4	-64.1	6.1
6393	267.7132	49.7146	2013.557	12.411	14.514	14.024	190.6	-32.2	66.4	6	-34.1	69	6

6394	267.7135	3.7911	2013.477	14.92	16.373	8.91	147.8	3.7	-90.9	6.2	6.1	-93.5	6.3
6395	267.7213	8.0951	2013.439	17.968	18.298	33.244	317.2	24	-60.3	6.2	12.8	-61.4	6.5
6396	267.7342	24.6205	2013.5135	14.267	16.991	10.55	309	24.4	63.5	6	25.5	63.3	6.2
6397	267.7540	57.1470	2013.4955	10.614	11.694	9.924	231.6	96.6	5	6.1	99.7	8.1	6.1
6398	267.7914	35.9191	2013.5065	14.932	16.999	9.69	186.9	72.2	-111.6	5.3	70.5	-108.7	5.5
6399	267.8403	15.0700	2013.899	10.159	14.402	53.502	130.7	-1.7	-96	5.9	-0.3	-99.8	5.9
6400	267.8529	43.5297	2013.447	14.232	17.011	15.74	253.3	-89.9	-60.2	5.3	-92.1	-61.1	5.5
6401	267.9170	4.4109	2013.384	16.137	18.233	8.124	91.9	63.8	25.9	6.2	63.9	29.6	6.8
6402	267.9455	14.3681	2013.808	11.893	14.157	23.01	223.6	-39.7	-80	5.9	-40.2	-78	5.9
6403	267.9947	25.9341	2013.683	13.856	16.129	6.49	110.7	42.5	-96.6	6	39.4	-96.2	6
6404	268.0167	21.4935	2013.6225	12.729	17.73	13.217	358.7	8	-115.2	6	7.9	-115.2	6.3
6405	268.0456	-0.1399	2013.268	16.96	17.775	37.937	110.6	-70.4	-20.8	6.7	-67.8	-29.9	6.3
6406	268.0492	76.4106	2013.387	16.893	18.216	15.504	347.1	-68.7	123.8	6.2	-65.7	130	6.4
6407	268.0741	33.9023	2013.39	16.009	17.74	5.378	324.4	-29.8	-82.8	5.3	-33.1	-80.9	5.6
6408	268.1157	-4.7178	2013.4095	16.446	17.994	22.799	45.6	39.2	-83.8	5.8	36	-84.8	6.2
6409	268.1505	45.5151	2013.628	11.955	12.757	24.298	94.4	5.7	-62.8	5.3	7.5	-62	5.3
6410	268.1559	14.4211	2013.7315	15.368	16.168	9.169	337.3	-35.3	-104.8	5.9	-34.1	-105.4	6
6411	268.2089	31.5335	2013.629	16.1	16.17	8.675	316.4	-47.6	66.1	5.3	-46	64	5.3
6412	268.2516	23.8648	2013.5935	14.205	14.511	35.88	243.3	4.2	-70.2	6	3.2	-72	6
6413	268.3153	58.7886	2013.413	13.645	15.659	15.232	184.8	-94.2	88.2	5.7	-93.2	87.7	5.8
6414	268.3979	2.9434	2013.055	17.113	17.483	31.858	255.1	61.4	-10.1	6.4	61.5	1.3	6.8
6415	268.4656	47.0167	2013.6375	9.317	13.264	8.222	214.9	-162.1	77.9	5.3	-169.2	70.7	5.4
6416	268.5210	27.3421	2013.6815	8.053	9.713	48.279	32.4	-38.5	-114.6	6	-41.2	-113.1	6
6417	268.6302	47.7112	2013.536	12.218	15.503	18.982	147.2	-7.5	71.3	5.3	-7.4	71.5	5.3
6418	268.6928	39.6294	2013.615	11.095	17.693	32.393	221.5	99.6	95	5.6	100.4	97.6	6
6419	268.7472	41.1094	2013.6365	11.747	14.85	11.959	130.5	-9.2	64.5	5.6	-5.9	64.6	5.6
6420	268.7796	31.5598	2013.608	9.799	14.252	11.248	107.2	-71	-101.6	5.3	-75.4	-99.4	5.3
6421	268.8408	43.5242	2013.676	10.598	11.818	10.812	194	-9.3	81.5	5.3	-7.9	81.2	5.3
6422	268.8467	19.4356	2013.738	11.649	13.461	17.626	198.9	-8.9	-146.6	5.7	-20	-145	5.7
6423	268.8670	2.2213	2013.29	16.993	17.875	57.331	5.4	61.6	-23.4	7.2	62.5	-20.2	6.9
6424	268.9075	-11.9309	2014.3465	10.009	10.516	11.085	265	-124.8	46.8	5.3	-120.1	56.4	5.3
6425	268.9120	1.4357	2012.8355	17.274	17.783	48.407	292	64.1	-56.5	6.7	72.1	-46.5	6.8
6426	269.0451	56.3672	2013.418	12.626	14.676	24.043	55.4	48.6	88.2	5.7	48.8	86.5	5.8
6427	269.1269	26.7532	2013.3545	18.16	18.364	9.847	220.1	-23.5	-64.4	5.9	-20.8	-67.3	5.8
6428	269.1334	33.1549	2013.596	14.219	14.377	5.803	335.5	-56.4	71	5.3	-53.7	68.9	5.3
6429	269.1695	8.1638	2013.958	10.923	16.025	8.243	305.5	-11.9	-82.3	6	-7.5	-84.3	6.2
6430	269.1854	45.9426	2013.4985	12.663	16.947	15.512	91.1	-87.3	94.6	5.4	-89.1	98.3	5.5
6431	269.2252	78.4765	2013.5385	13.386	13.774	53.802	276.4	7.9	-76.1	5.7	6.9	-75.1	5.7
6432	269.2352	-0.7288	2013.628	14.253	15.32	4.032	320.5	-67.7	-106.2	5.7	-66.7	-111.8	6.3
6433	269.2545	40.1533	2013.514	12.386	18.626	19.121	61.7	11.8	-122.2	5.6	14.1	-120.5	6.2
6434	269.2767	-3.4883	2013.9535	11.587	13.524	56.984	60.8	7.3	-88.7	5.5	10.2	-90.4	5.5
6435	269.2817	43.0571	2013.576	14.121	14.261	35.842	111.3	-8.9	-85.3	6.1	-9.1	-83.6	6.1
6436	269.2976	22.7848	2013.671	10.968	14.888	14.155	55.6	19.1	-80.4	5.7	15.1	-83.3	5.7
6437	269.3839	14.5974	2013.6925	11.124	16.987	53.702	185	59	-76.4	5.2	59.3	-72.8	5.3
6438	269.4127	25.9305	2013.5995	13.077	16.842	8.32	16.7	54.8	-79.4	5.6	52.3	-77.5	5.7
6439	269.4468	6.9477	2013.727	11.114	17.9	36.363	46.4	16.5	-71.6	6	17	-71.6	6.3

6440	269.5162	6.8485	2013.866	9.532	15.103	13.201	37.7	46.7	-70.3	6	35.8	-79.9	6
6441	269.5175	7.4123	2013.849	10.89	12.328	7.098	82.5	2.6	-75.8	6	3.8	-77.4	6
6442	269.5247	55.2154	2013.451	9.83	14.618	8.628	129.7	-11.4	70.3	5.7	-8.4	68.9	5.7
6443	269.5419	14.2398	2013.8405	14.045	14.126	8.152	130.3	-10.5	67.6	5.1	-9.4	68.6	5.1
6444	269.5456	41.6026	2013.767	13.164	13.228	5.569	258.6	-8.2	-81.3	5.5	-9.6	-84.5	5.6
6445	269.5706	20.7479	2013.476	15.443	16.899	22.173	342.5	-49.6	-62	6.1	-47	-66.1	6.2
6446	269.6061	80.7106	2013.6655	13.132	15.714	7.092	75.4	58.7	122.1	5.7	61.4	123	5.7
6447	269.6195	11.5934	2013.5945	17.517	18.095	39.314	275	-28	60.6	6.1	-25.9	62.1	6.1
6448	269.6431	23.8058	2013.6335	14.569	15.992	19.816	265.3	-7.6	-71.1	6	-9.1	-70.3	6
6449	269.6817	-6.3276	2013.944	10.691	13.847	8.731	66.9	21.8	-101	5.5	22.2	-101.6	5.6
6450	269.7175	44.6645	2013.4875	14.029	18.192	8.145	312.9	5.1	-133.4	5.3	2.2	-133	5.8
6451	269.7442	64.5766	2013.5255	15.605	17.094	10.118	153.7	-57	-112.2	5.7	-58.4	-105.5	5.8
6452	269.7493	12.8758	2013.7215	13.105	14.003	3.972	187.3	13.2	83.4	5.2	14.2	84.7	5.3
6453	269.7592	0.4465	2013.3475	15.163	18.362	38.578	198	0.8	-71.5	6.2	2.3	-74.8	7.3
6454	269.7745	2.5477	2013.6835	14.495	15.309	30.663	16.9	-45.8	77	6	-42.1	77.5	6.2
6455	269.7846	1.9772	2013.699	13.791	14.515	4.279	355.9	16	-96	6.1	18.5	-97.1	6.5
6456	269.8053	38.1098	2013.6465	12.897	13.453	6.366	85.7	-7.6	100.7	6	-5.4	96.8	6
6457	269.8729	53.9880	2013.3865	18.058	18.729	27.511	211.5	-21.6	-68.7	6.1	-32.7	-67.5	6.4
6458	269.8978	-8.3366	2013.887	10.346	15.462	23.569	127.3	102.2	-21.5	5.5	100.6	-22.8	5.7
6459	270.0158	80.0014	2013.801	6.291	6.632	18.725	51.9	48	134.2	5.7	43.2	139.4	5.7
6460	270.0704	46.8619	2013.501	11.384	16.587	9.375	219.8	-12.3	65.7	5.3	-12.7	66.3	5.5
6461	270.1504	68.5571	2013.571	9.158	11.129	30.606	171.2	-1.3	78.4	5.7	-1.8	82.3	5.7
6462	270.2202	0.8413	2013.6645	13.04	15.547	25.877	355.9	-30.9	-87.7	5.7	-30.9	-87.6	5.8
6463	270.2502	12.6334	2013.575	17.471	17.666	58.522	179.7	65.3	19.7	6.1	69	11.3	6.1
6464	270.2964	70.6643	2013.354	15.941	17.164	14.569	5.1	-15	-85.6	5.8	-8.9	-89.1	5.8
6465	270.3193	37.8533	2013.915	14.94	15.631	18.131	190.4	-29.7	-69.2	6	-26.2	-68	5.8
6466	270.3515	9.7005	2013.6175	15.6	15.984	4.361	2.6	-2.5	-64.4	6.2	-1.2	-62.4	6.3
6467	270.4818	17.9054	2013.725	14.297	14.785	4.991	92.1	-30.4	-204.5	6.1	-25.2	-202.8	6
6468	270.4989	40.9091	2013.613	9.84	15.849	23.637	242.6	-42.5	-71.8	5.2	-41.9	-75.8	5.4
6469	270.5381	45.1075	2013.6245	12.43	13.7	31.731	292.6	-68.1	-85.7	5.3	-67.7	-83.6	5.3
6470	270.6294	29.5150	2013.5495	14.529	15.229	16.548	265.7	-126.3	-92.6	5.7	-127.3	-93.6	5.7
6471	270.6539	56.6386	2013.41	14.88	16.72	9.897	346.3	8.4	184.6	6.1	8.2	189.7	6.3
6472	270.6547	41.6852	2013.4335	16.658	16.715	7.694	205	10.9	67.6	5.4	15.5	63.5	5.4
6473	270.6622	4.4798	2013.869	10.366	13.948	6.011	294.8	-54.8	-216.2	6.2	-51.5	-219.6	6.1
6474	270.7662	2.9204	2013.267	13.365	16.691	22.36	81.6	-21.8	-65.2	6.2	-20.9	-65.6	6.5
6475	270.8022	28.2167	2013.633	14.636	16.478	14.124	146	28.9	-83.6	5.6	31.2	-84.6	5.6
6476	270.8422	23.4074	2013.664	11.682	12.156	11.303	64.1	-2.3	-109.7	6	-1.6	-108.6	6
6477	270.9300	66.6841	2013.5135	9.395	13.561	30.241	80.5	-19	66.4	5.7	-17.2	68.8	5.7
6478	270.9678	50.6224	2013.293	17.981	18.701	6.937	223.1	-5.5	-60.8	5.8	-3.7	-61.3	6.2
6479	270.9841	4.7536	2013.294	17.472	18.244	37.872	294.6	-15.8	-65.1	6.4	-11.8	-69.4	6.5
6480	271.0133	2.4186	2012.8895	17.732	18.612	55.74	20.8	64.5	6.4	6.4	65.3	4.2	6.9
6481	271.0688	17.2938	2013.6035	16.003	17.596	6.871	41.3	-56.4	134.8	6.1	-56.5	130.5	6.6
6482	271.0898	13.9471	2013.824	9.826	13.119	7.249	147.5	-53.8	-76.8	6	-54.8	-81.4	6
6483	271.0945	8.8822	2013.6145	17.333	18.205	55.992	206.5	13.4	-63.9	6.2	24.5	-61.6	6.2
6484	271.1406	34.8621	2013.645	13.785	15.257	23.382	249.5	-5.3	-81.5	5.7	-5.7	-81.3	5.7
6485	271.1676	23.9417	2013.66	6.565	13.636	16.682	136.2	4.4	-69.4	5.2	8.1	-67.5	5.3

6486	271.1742	9.7376	2013.4445	17.382	17.916	14.623	335.8	-30.3	-63.9	6.2	-24.7	-69.2	6.2
6487	271.2227	30.3073	2013.4535	16.646	17.09	9.239	184	3.3	-61.7	5.4	3.8	-61.2	5.3
6488	271.2594	11.8926	2013.3945	17.681	17.845	54.216	187.9	64.4	-35	6.2	60.6	-34.6	6.2
6489	271.2917	23.8986	2013.432	17.273	17.431	5.004	130.5	9.7	-75.7	5.5	12.6	-76	5.5
6490	271.2967	34.4941	2013.861	10.309	12.18	11.238	245	-37.9	79.3	5.2	-40.6	75.5	5.2
6491	271.3077	45.9115	2013.5215	16.8	17.967	6.039	323.9	-22.3	77.5	5.7	-22.1	81.1	5.8
6492	271.3659	55.0351	2013.5295	12.052	14.234	7.629	133.2	-11.9	-75.5	6	-9.5	-77	6.1
6493	271.4229	74.4783	2013.4245	14.134	16.378	17.54	206.9	75.6	-12.8	5.7	74.2	-13.8	5.7
6494	271.4256	54.8625	2013.5965	11.176	15.28	6.701	51	-11.4	-67	6	-16.4	-64.7	6.1
6495	271.4746	0.5332	2013.1355	17.617	17.754	49.717	175.9	8.6	-69.4	6.6	1.6	-70.9	6.8
6496	271.6198	-6.2893	2013.295	16.604	17.24	6.396	52.7	-73.8	-140.5	5.9	-83.1	-140.9	6
6497	271.6608	10.2982	2013.696	15.399	15.487	3.644	213.8	-10.2	-78.8	6	-9.2	-76.8	6.1
6498	271.7413	8.3862	2013.658	12.347	16.173	7.681	132.9	89.8	-108.4	6	89	-108	6.1
6499	271.7701	30.5182	2013.3765	11.901	14.902	18.986	341.9	-36.3	-67.1	5.3	-34.4	-66.2	5.6
6500	272.0329	28.4952	2013.575	15.306	15.319	20.114	68.6	-5.8	-73	6	-4.2	-72	6
6501	272.0487	5.7491	2013.4225	16.928	17.173	3.768	344.4	-18.7	-66.4	5.9	-28.3	-65.5	6.4
6502	272.0632	11.4516	2013.7915	11.531	16.212	9.273	241.9	-9.6	-76.5	5.6	-7	-73.2	5.7
6503	272.0774	34.1058	2013.689	10.279	11.239	4.119	83.5	-2.3	-72.8	5.2	-1.5	-71.5	5.3
6504	272.0793	13.2926	2013.8615	10.639	16.048	29.022	10.2	2.9	-69.1	5.9	2.8	-69.8	6
6505	272.1521	-0.3677	2013.6125	10.631	16.232	14.982	132.3	55.7	-169.1	5.6	52.5	-169.6	5.9
6506	272.2369	10.4939	2013.624	16.616	16.798	5.738	208.6	12.9	-82.2	5.7	9.7	-84.6	5.9
6507	272.2448	24.1586	2013.74	7.288	11.251	51.248	119.8	91.8	94.1	6	80.4	97.5	6
6508	272.3850	12.2641	2013.7825	8.559	13.002	14.369	42.1	-67.5	-123.1	5.4	-73.3	-125.7	5.4
6509	272.3921	-12.7256	2014.3285	9.519	13.433	9.128	209.4	-65.7	-84.3	5.1	-64.5	-90.2	5.9
6510	272.4033	-7.9247	2013.774	11.821	13.203	25.1	44.4	-36.4	-176.1	5.7	-37.3	-176.8	5.7
6511	272.4363	1.4117	2013.295	16.568	16.738	12.716	323.2	30.1	-66.2	5.9	32.1	-63.6	6
6512	272.5192	22.1822	2013.477	15.469	17.442	12.113	315.4	-65.2	-82.7	6.1	-67.3	-83.1	6.3
6513	272.5284	18.7345	2013.5115	16.16	16.521	9.59	205.4	-48	-74.3	6.1	-49.7	-71.5	6.1
6514	272.5842	28.6439	2013.608	12.392	14.784	28.331	28.5	2	-72.2	6	7.3	-73.3	6
6515	272.5873	3.6878	2012.8575	17.783	18.526	50.253	166.6	-7.7	-65.5	6.5	-19.1	-63	7.7
6516	272.5895	0.2206	2012.9625	18.049	18.603	49.464	116.6	13	-64.2	6.8	19.5	-62.3	7.4
6517	272.7325	35.3274	2013.8325	6.38	12.377	24.862	4.4	16.6	-62.8	5.9	15.1	-64.1	5.9
6518	272.8326	-1.3686	2013.445	15.534	17.197	27.945	74.6	8.3	-61.9	5.4	4.2	-62.9	5.8
6519	272.9466	72.7954	2013.483	10.722	11.565	37.839	177.3	-25.3	80.2	5.7	-23.8	77.9	5.7
6520	272.9722	51.7498	2013.577	10.793	16.489	46.393	5.5	-52.6	-60.4	5.3	-54.9	-60.7	5.4
6521	272.9835	10.6161	2013.236	17.46	18.693	59.536	81.2	-27.8	-62.1	5.9	-25.3	-61.8	6.8
6522	273.0414	8.3488	2013.4995	17.828	18.274	57.027	149.6	-11.4	-67.3	6.5	-1.4	-67.3	6.3
6523	273.0817	66.5600	2013.348	13.848	15.385	16.458	117	-13.9	-76.7	5.8	-8.4	-78.9	5.8
6524	273.0834	6.1432	2013.329	17.407	17.579	47.083	242.6	41.8	-82.4	6.3	46.5	-74.6	7
6525	273.1381	8.3322	2013.4895	15.495	18.381	19.299	83.4	-10.4	-69.9	6.2	-6.4	-72	7.4
6526	273.1550	17.3879	2013.65	12.323	17.342	8.1	130.5	14.4	-94.2	5.2	8.2	-90.9	5.5
6527	273.1662	11.1356	2013.5515	17.264	17.588	34.396	277.9	-38.7	-61.5	6.2	-45.7	-60.3	6.2
6528	273.1761	-10.1540	2013.3545	15.382	16.883	39.469	306.2	-52.4	-98.5	7.7	-59.4	-92.5	7.6
6529	273.1808	28.5448	2013.189	13.093	13.114	15.671	130.9	-8.4	-90.7	6.2	-11.4	-90.2	6.4
6530	273.1967	67.6178	2013.4655	11.316	13.435	57.037	172.4	-37.9	-64.5	5.7	-36.3	-63.3	5.7
6531	273.2181	24.8121	2013.666	13.897	14.338	25.572	299.5	18.1	161.5	6.1	20.2	160.5	6.1

6532	273.2480	1.0899	2013.271	17.72	18.044	41.361	129.2	-22.4	-88.3	6.8	-16.1	-93.4	7.4
6533	273.2596	28.2470	2013.656	4.909	5.063	55.085	315.9	1.1	110.2	6	2.8	113.6	6
6534	273.2827	66.9244	2013.414	12.964	13.715	9.684	143.4	-24.8	-74.7	5.8	-24.2	-74.3	5.7
6535	273.2889	1.0597	2013.0795	17.123	18.486	21.461	141.3	64.6	27.9	6.7	68	24.7	6.9
6536	273.3084	25.2612	2013.464	13.921	15.829	4.11	14.2	-29.4	-84.6	5.6	-22.8	-84.4	5.4
6537	273.3954	7.7310	2013.591	12.873	17.591	21.122	168.8	-15.5	-62.6	6	-27.5	-66.2	6.3
6538	273.4961	7.3058	2013.815	13.323	15.306	5.032	51.4	18.6	-184	6.2	14.9	-186.5	7.1
6539	273.5199	42.6647	2013.588	12.466	13.93	5.091	26.8	6.5	-63.1	6	4	-63	6.1
6540	273.5980	21.4521	2013.815	8.738	9.575	16.322	109.9	-32.9	-86.5	6	-31.2	-88.7	5.9
6541	273.6009	68.8488	2013.4725	10.186	13.986	14.954	346	-181.5	-79	5.8	-180	-84.1	5.8
6542	273.6220	7.8286	2013.4115	17.454	17.635	49.912	60.1	-14	-69.3	6.2	-3.8	-67.5	6.4
6543	273.6668	1.9127	2013.447	10.422	17.821	14.565	338.8	-9.6	-61.1	5.7	-2.8	-60.9	6.3
6544	273.7350	11.3958	2013.5415	17.254	18.323	49.877	84.8	5.5	-65.5	5.9	6.5	-68.2	6
6545	273.7376	63.9198	2013.4785	17.545	18.17	32.825	155.6	7.9	-65.2	5.8	5.6	-67	5.8
6546	273.8279	4.0040	2013.492	12.466	16.025	13.552	223.9	78.7	-12.6	6.1	73.9	-21.6	6.4
6547	273.8959	66.4517	2013.8975	12.405	13.491	4.546	213.8	-23.3	68.7	5.7	-21.1	67.2	5.4
6548	273.9117	38.3309	2013.799	9.588	10.481	18.556	206.9	19.7	101.4	5.2	18.6	100	5.3
6549	273.9176	5.3443	2013.367	17.355	17.904	33.894	291.9	60.9	15.4	6.3	61.2	3.8	6.7
6550	273.9452	26.1012	2013.339	12.425	12.927	55.42	70.1	43.5	-89.6	5.5	39.1	-89.9	5.6
6551	273.9745	21.9487	2013.8025	10.035	14.9	7.792	234.1	-19.8	-70.1	5.5	-23.2	-68.2	5.6
6552	274.0138	42.8178	2013.866	11.943	13.611	4.79	245.8	21.6	85.4	5.3	25.5	87.8	5.3
6553	274.0377	29.0355	2013.52	8.562	12.38	22.774	342.2	-36.3	-94.1	5.4	-34	-94.8	5.5
6554	274.0674	80.4622	2013.765	10.104	11.466	5.645	108.7	-150.1	-74.3	5.6	-148.9	-70.3	5.6
6555	274.0895	7.0337	2013.8645	14.096	14.237	4.735	22.4	21.6	-135	6.1	19.8	-137.7	6
6556	274.1499	1.8850	2013.5555	12.017	14.032	8.824	65.5	-55.5	-116.7	5.8	-57.2	-116.3	5.8
6557	274.1939	52.5184	2013.463	17.735	18.709	50.03	89	28.6	-62.6	6.2	35.8	-60.2	6.7
6558	274.2215	48.9441	2013.5835	14.468	14.574	13.422	151.3	31.2	113.4	6.1	30.6	117.2	6.1
6559	274.2722	49.3443	2013.597	13.003	15.847	7.849	271.4	1.1	-64.7	6	1.4	-65.2	6.1
6560	274.3972	53.5471	2013.5965	13.045	15.202	28.577	121.7	6	93.8	6.1	3.2	97.7	6.1
6561	274.4253	41.3710	2013.609	11.059	16.445	56.372	134.7	65.5	150	5.3	63.9	151.4	5.4
6562	274.4455	61.5178	2013.311	16.286	17.278	5.008	30.6	-0.4	61.7	5.8	3.1	60.2	5.9
6563	274.4513	53.7014	2013.576	12.324	12.348	9.433	305.1	-22.9	128	5.4	-21	130.1	5.4
6564	274.4910	41.6115	2013.6325	13.612	13.937	7.118	271.2	-43.9	-73.3	5.3	-45.7	-74.6	5.3
6565	274.5082	48.3380	2013.683	13.099	16.05	6.479	141.7	-28.2	-104	5.3	-29.3	-106.3	5.3
6566	274.6118	48.0361	2013.542	13.384	14.527	9.165	155.3	-27.4	-62	5.3	-22.6	-66.3	5.4
6567	274.6278	58.7322	2013.701	15.306	15.379	5.366	264.3	9.7	84.5	6	12	82.8	6
6568	274.6943	27.5929	2013.2665	12.539	13.421	54.662	225.4	-14.7	-73.4	5.4	-23.5	-72.5	5.7
6569	274.6963	2.9199	2013.6825	10.753	12.099	6.403	36.4	13.1	64.8	5.7	10.8	64.5	5.7
6570	274.8609	2.1536	2013.219	16.05	17.375	46.71	62.9	4.2	-101.5	6	0.9	-97	6.1
6571	275.0315	49.0496	2013.7075	14.172	14.413	6.555	144.9	39.6	-69.8	5.3	41.5	-70.3	5.3
6572	275.0588	31.3791	2013.595	8.01	13.74	14.029	351.3	-10.5	186.2	5.3	-14.5	185.1	5.4
6573	275.0759	9.4085	2013.5995	17.53	17.679	32.065	221.6	61.7	0.6	5.9	64	-7.4	5.8
6574	275.0990	28.9807	2013.467	10.803	12.517	6.603	190.6	-65.5	-60.5	6	-67.5	-60.1	6.7
6575	275.1605	40.8099	2013.5725	14.047	14.84	36.831	179.6	-126.1	-180.9	6.2	-127.4	-183.4	6.2
6576	275.1861	-1.9145	2013.3595	15.639	18.284	20.298	144.6	20.5	-62.2	5.7	16.1	-60.3	6.3
6577	275.1998	45.9476	2013.9685	14.969	15.375	3.691	177.2	-26.8	-111.6	5.4	-28	-105.5	5.3

6578	275.2125	29.9759	2013.586	12.479	12.937	20.262	317.5	-10	-73.3	5.3	-14.2	-70.7	5.3
6579	275.4236	19.1582	2013.596	13.189	16.144	11.087	303.8	50.8	69.6	6	56.7	70	6.1
6580	275.4440	36.7924	2013.5375	9.276	12.443	10.47	79.8	-35.5	-80.2	5.3	-34.5	-82.2	5.4
6581	275.4483	32.1115	2013.671	8.225	11.433	11.007	264.8	-16.4	-109.4	5.2	-16.6	-113.3	5.3
6582	275.4914	31.2407	2013.372	12.358	12.86	6.037	33.8	-21.5	-60.1	5.3	-18	-61.6	5.5
6583	275.5430	64.8842	2013.426	11.81	12.331	5.886	306	4.4	-69.3	5.8	4.7	-68.9	5.7
6584	275.5569	50.8635	2013.5025	14.756	15.192	6.355	59	19.1	123	6.1	17.6	124.7	6.1
6585	275.5601	27.0597	2013.3	12.663	13.054	47.665	191.8	-3.2	-76.6	5.4	1.6	-78.8	5.5
6586	275.6189	20.9286	2013.6405	11.206	12.377	33.571	274.5	-16.8	74.8	6	-13.2	73.7	6
6587	275.6223	69.3948	2013.4815	10.81	13.425	23.829	264.1	-5.9	90.9	6.1	-4.8	91.8	6.1
6588	275.6227	-8.6042	2013.641	12.657	15.371	13.813	9.5	107.2	-67.8	5.7	113.3	-63.8	5.8
6589	275.6739	32.7903	2013.642	9.397	12.125	41.068	260.7	-3.8	-76.9	5.3	-4.4	-76.1	5.2
6590	275.7660	8.6227	2013.3305	16.966	18.102	44.615	261.6	-9.4	-72.1	5.9	-3.8	-74.1	6.3
6591	275.8378	-11.6166	2014.107	11.055	11.186	19.164	290.1	29.9	71.6	5.6	28.6	68.5	5.5
6592	275.8807	3.0884	2013.7785	8.031	11.191	19.455	129.5	-8	-124.7	5.7	-6.3	-123.3	5.8
6593	275.8911	19.3466	2013.512	14.765	16.503	5.318	130.4	-3.6	-63.1	6	-10.4	-60.8	6.2
6594	275.9036	61.3535	2013.4825	11.666	13.219	5.312	235.7	-24.6	118.1	5.8	-27.2	115.6	5.8
6595	275.9214	27.5065	2013.6015	9.246	9.334	6.285	185.9	-17	-85	5.3	-15.7	-85.9	5.4
6596	275.9367	20.5412	2013.6665	11.575	16.664	9.314	120.4	-22.3	-69.5	6	-22	-72.7	6.2
6597	275.9586	-9.5877	2013.2985	16.626	17.835	17.296	28.6	-76.8	65.2	7.6	-73.9	68.7	7.9
6598	275.9923	14.1758	2013.6215	17.528	17.806	27.824	241	60.2	-21.5	5.6	60.9	-21.2	5.8
6599	275.9934	7.0106	2013.439	17.986	18.161	37.605	238.7	-41.1	-63.1	7.4	-45.8	-63.4	6
6600	276.0182	2.2615	2012.963	16.352	17.873	51.24	20.6	-8.3	-63.1	6.1	3.8	-60.9	6.5
6601	276.0671	10.9923	2013.658	12.989	17.142	25.688	191.8	127.2	2.8	5.7	126.4	4	5.9
6602	276.1170	38.9184	2013.3925	16.119	16.94	55.425	274.8	-11.3	-64.3	5.4	-14.4	-64.2	5.4
6603	276.2106	15.0248	2013.588	12.737	17.402	20.596	42.3	13.3	-62.7	5.9	9.6	-60.5	6.1
6604	276.3207	5.7791	2013.2565	17.561	17.856	15.893	196.9	-14.8	-75.5	6.1	-5.2	-78.6	7.7
6605	276.4320	7.3836	2013.253	18.233	18.609	50.37	267.2	-19.9	64.2	6.7	-15.5	65.2	6.2
6606	276.4772	53.6282	2013.5125	14.238	17.101	16.189	84.3	-103.9	-129.1	5.4	-100.8	-127	5.5
6607	276.5403	57.8957	2013.4005	11.923	16.741	26.898	282.3	-11.8	71.1	6.1	-13.9	72.7	6.2
6608	276.5850	22.8325	2013.5565	9.712	13.122	31.391	103	-10.7	134	6.1	-11.6	134.6	6.2
6609	276.6896	14.8656	2013.672	14.079	14.254	6.76	294	-34.8	-87.9	6	-34.6	-83.3	6
6610	276.7592	34.9957	2013.7195	6.3	9.653	39.253	68.8	4.8	-74.2	6	5.7	-75.1	6
6611	276.7690	65.9993	2013.596	11.272	18.726	27.166	338	-45.3	-66.9	6.1	-53.5	-60.1	6.3
6612	276.7887	-5.3564	2013.5085	13.312	17.53	12.126	112	-3.5	-67.5	5.6	-0.3	-62.1	5.8
6613	276.8080	51.5943	2013.595	10.116	17.326	14.542	72.6	-10.6	-83.6	5.3	-11.3	-85	5.5
6614	276.8996	29.7615	2013.971	11.623	13.04	4.185	82.4	-11.8	-66.7	6.1	-4.1	-69.2	7.5
6615	276.9086	53.4089	2013.584	15.243	15.877	6.068	115.1	23.4	62.4	5.3	23.8	61.2	5.3
6616	276.9317	1.5725	2013.739	15.842	16.676	9.326	259.4	87.5	-52.9	6.7	85.5	-50.6	6.6
6617	276.9516	44.8609	2013.437	13.683	14.234	4.756	17.4	6.7	93.1	6.1	7.1	94	6.2
6618	276.9732	35.6080	2013.59	11.235	11.812	6.208	351.2	66.4	21.6	6	67.1	23.3	6.1
6619	276.9925	-9.1286	2013.3	16.449	17.466	7.272	240.8	2.3	-122.1	6.6	-8.4	-126	7.3
6620	277.0803	3.5606	2013.2695	17.809	17.919	56.751	81.2	-2.1	66.9	6.7	2.2	64.8	6.4
6621	277.0853	3.1024	2013.4485	10.091	18.972	35.178	70.4	-13.4	-70.6	6.1	-8.2	-71.5	7.6
6622	277.1038	32.2561	2013.0835	13.185	13.769	9.841	250.7	-67.1	-79.9	5.4	-55.7	-86.9	6.4
6623	277.1139	14.3448	2013.756	14.82	15.831	3.983	341.9	-4.4	-71.2	6	-4.7	-74.3	6.8

6624	277.2161	-6.5401	2013.322	17.567	17.685	30.879	144.5	10.3	-79.3	6.7	21.6	-73.8	5.9
6625	277.2249	22.3141	2013.372	13.277	13.59	39.139	204.5	1.5	-77.9	6.1	-0.8	-78.6	6.2
6626	277.2276	11.6865	2013.631	17.817	18.123	52.378	331.3	2.8	-69.8	6	5.1	-67.8	5.9
6627	277.2457	25.2252	2013.5745	8.696	9.528	6.784	144.8	51.6	-67.6	6.1	51.5	-67.7	6.2
6628	277.2629	32.5810	2013.4295	12.466	13.688	10.031	91.2	-21.1	-64.7	5.4	-22	-64.6	5.3
6629	277.2668	43.9399	2013.68	7.737	12.189	20.219	242.5	66.2	149.7	6	74.9	150.5	6.1
6630	277.2735	0.4401	2013.683	8.509	14.25	16.974	112	78.2	10.3	6.1	75.3	5.2	6.2
6631	277.2753	18.3891	2013.4175	16.15	17.523	5.426	63.9	84.1	-101.4	6.2	83.9	-102.4	6.4
6632	277.2816	59.3287	2013.4965	11.307	14.519	10.695	132.5	104.5	-1.7	6.1	106	0	6.1
6633	277.3407	4.3692	2013.4095	17.034	17.438	49.048	253.6	-0.2	-60.3	6.5	6.1	-60.5	6.7
6634	277.3492	26.4502	2013.5	11.782	12.504	50.221	287.5	4	-62.6	6.1	7.9	-62.1	6.2
6635	277.3678	33.1197	2013.6245	11.027	11.794	10.302	250.4	-25.1	71.6	5.2	-28.3	72.9	5.3
6636	277.4575	57.6841	2013.4185	11.56	15.06	7.346	197.6	-53.6	185.8	6.2	-53.4	186.5	6.2
6637	277.4811	37.2079	2013.5295	12.689	13.68	27.449	120.3	-9.4	68.5	5.4	-11.7	70.8	5.3
6638	277.5411	18.4677	2013.464	15.307	16.614	50.357	268	18.2	-65.1	5.7	9.2	-65.9	6
6639	277.5484	19.8103	2013.5205	16.918	19.275	47.16	34.4	60.2	-27.6	5.8	67.1	-17.7	6.3
6640	277.5825	52.5003	2013.5145	12.319	15.113	6.042	222.6	-25.4	-65.2	5.7	-23.8	-64.7	5.8
6641	277.5878	-2.1901	2013.3655	16.557	18.124	5.576	48.5	132.8	-34.3	5.9	125.9	-37.8	6.3
6642	277.5906	-6.4116	2013.375	16.414	17.707	34.128	264.6	-162.5	-148.2	7	-164.3	-151.1	6.7
6643	277.5981	47.9128	2013.809	10.195	15.487	8.182	194	-55.3	-107.5	5.3	-58.9	-105.2	5.4
6644	277.6030	30.6091	2013.598	12.544	13.475	4.759	24.1	-34.6	-98.8	5.5	-34.3	-104.1	5.3
6645	277.6143	36.5913	2013.643	8.116	11.828	24.301	222.8	22	109.4	5.3	20.5	107.3	5.3
6646	277.6670	7.1863	2013.449	17.496	17.856	54.308	125.7	18.6	-90.8	7.2	19	-89.8	6.2
6647	277.6740	11.6100	2013.306	16.599	17.817	59.893	131.1	-39	-63	6.9	-30.8	-67.5	6.2
6648	277.6811	63.7009	2013.5185	10.175	16.367	48.414	190.1	-0.3	64.1	5.7	-0.9	64	5.7
6649	277.7275	23.3280	2013.539	12.446	12.872	51.543	78.3	29.3	-94.2	6.5	17.6	-94.5	6.2
6650	277.7472	22.3752	2012.904	13.152	13.415	26.642	348.1	63.5	-29.7	6.4	63.4	-22.3	6.8
6651	277.7516	51.8075	2013.3785	14.394	17.355	8.486	328.3	12.3	-64.1	5.7	7.1	-65.7	5.9
6652	277.8264	43.9437	2013.6755	13.624	15.107	6.945	203.1	10.5	100.4	5.3	9.6	-103.6	5.3
6653	277.8973	-9.6661	2013.5195	15.254	16.975	50.994	254.3	-16.9	177.1	5.7	-13.7	176.7	7.5
6654	277.9591	20.6422	2013.679	9.802	12.825	46.414	190.4	65.2	-16.6	6	65.9	-21.7	6.1
6655	277.9794	84.2410	2013.674	16.938	17.042	6.006	101.7	13	113.8	5.8	12.9	115.5	6
6656	278.0011	-7.1722	2013.357	17.003	17.647	48.567	153.4	-22.3	-65.2	5.9	-25.3	-67.5	6.2
6657	278.0181	27.8205	2013.5125	8.994	12.778	15.086	136.1	109.6	-36.7	6.1	109.5	-32.5	6.2
6658	278.0495	-7.3216	2013.806	10.734	11.602	5.646	338.4	-18.3	-82	5.6	-15.5	-83.2	5.6
6659	278.0570	-6.9404	2013.411	16.175	16.996	36.472	95.2	201.3	10.9	6.1	202.4	22.8	6
6660	278.0794	-7.0641	2013.3695	17.766	17.804	9.26	255.1	18.2	-123.1	7.4	3.4	-130.5	7.6
6661	278.1089	72.7563	2013.4705	13.919	16.118	12.46	208.2	-36.9	-76.3	5.7	-37.4	-74.2	5.7
6662	278.1750	11.4021	2013.2275	17.914	18.229	12.814	344.8	170.6	-209.3	6	161.9	-217.6	6.6
6663	278.3442	-3.7668	2013.5175	12.134	12.408	7.615	214.8	-16.2	-87.9	5.7	-16	-89.8	5.7
6664	278.3887	-6.8958	2013.3605	16.968	18.025	28.678	179.3	-177.1	-157.8	6.2	-175.4	-150.3	7.1
6665	278.3918	21.9447	2013.684	6.549	12.129	15.414	178.7	84	10.3	6	81.5	5.7	6.1
6666	278.4551	70.0445	2013.507	10.015	10.121	10.526	135.1	78.3	3	5.7	76	3.5	5.7
6667	278.5972	48.0983	2013.588	13.215	16.383	26.637	72.7	-24.4	-73.5	5.3	-21.2	-72.5	5.4
6668	278.7177	6.1414	2013.1975	17.922	18.018	37.934	47.1	0.3	-66.1	6.4	-2.2	-66.1	6.7
6669	278.7479	28.6571	2013.5965	9.03	11.9	17.218	89.5	-23.2	-104	6	-22.9	-103.8	6.1

6670	278.7573	-0.1463	2013.325	17.795	18.964	25.838	99.9	-185.7	-191.7	5.9	-182.1	-200.4	6.1
6671	278.8333	30.4920	2013.601	8.724	13.258	35.228	149.8	0.5	-63.6	6	1.7	-64.6	6.1
6672	278.8535	-2.0369	2013.34	16.884	17.447	4.896	301.3	81.2	-12.5	7	81.5	-20.7	6.2
6673	278.9290	-23.8095	2013.9145	10.634	14.755	12.104	14.4	-57	-71.2	5.4	-57.2	-75.4	5.4
6674	278.9364	-4.6722	2013.381	17.314	17.88	24.537	90.2	112.5	-144.5	6.2	104.6	-157.1	6.9
6675	278.9607	45.4003	2013.5435	11.953	15.317	31.843	73.9	9.2	-82.9	5.3	9	-84.4	5.4
6676	278.9761	-7.9656	2013.3	16.626	17.135	35.534	56.9	67.3	52.2	6.4	76.1	42.6	6.7
6677	279.0295	-4.6531	2013.37	15.899	17.882	16.26	272.6	77.2	-74.7	6.1	80.2	-76.4	7.5
6678	279.0405	7.0093	2013.617	10.075	12.376	5.988	342.2	-48.4	-116	6.1	-48.5	-116.4	6.3
6679	279.1749	6.3601	2013.1175	15.682	18.35	45.199	23.3	39.6	-76.8	7.2	51.2	-72.3	6.9
6680	279.1797	-5.1796	2013.4065	14.957	17.273	44.314	60.6	-0.7	-74.9	6	8.3	-75.2	5.9
6681	279.2401	-3.0454	2013.3755	16.036	16.331	30.934	267.5	78.5	-71.9	5.7	82.4	-62.4	5.9
6682	279.2466	14.1439	2013.717	15.094	17.072	6.481	355.8	66.9	-116.7	5.6	65.8	-111.7	5.6
6683	279.2787	-1.5592	2013.288	16.373	17.245	41.411	124.9	132.1	104.1	5.7	125.7	113.8	6.4
6684	279.2901	8.3240	2013.283	17.529	17.721	34.944	108.1	68.3	-27.6	6.1	72.3	-15.3	6.5
6685	279.3234	17.2008	2013.5695	11.634	16.054	37.712	185.6	-22.4	-83.4	5.6	-22	-80.2	5.7
6686	279.3424	-4.2066	2013.4425	16.735	18.973	12.845	168.7	-47.5	-180.5	5.7	-37.4	-168.7	9.3
6687	279.3897	-4.1668	2013.365	16.425	16.492	8.939	357.1	87.7	-141.4	6.2	82.7	-151.4	5.8
6688	279.4303	21.9696	2013.602	7.729	13.575	14	161.8	-36.6	-86.6	5.3	-37	-86.1	5.5
6689	279.4560	-2.8327	2013.371	16.359	17.144	34.077	135.4	91.1	-28.2	5.8	92.9	-17.6	5.8
6690	279.4719	12.0522	2013.6025	12.038	13.988	5.741	9.8	-27.8	-84.7	5.7	-27.9	-83.2	5.8
6691	279.4808	-4.5445	2013.4245	15.342	17.983	43.07	135.7	17.1	-74.5	5.5	9.9	-79.4	5.9
6692	279.4849	18.2033	2013.492	17.059	17.227	3.96	139.9	-50.8	-65.3	5.6	-46.2	-64.2	5.6
6693	279.4895	41.8553	2013.577	12.78	12.81	11.888	118	24.6	87.5	5.6	25.1	87.8	5.6
6694	279.5012	36.8595	2013.665	6.774	8.423	9.391	171.5	4.9	61.1	5.6	2.7	64.1	5.6
6695	279.5168	63.6800	2013.3975	16.475	16.853	3.352	132.5	15.7	196.8	6.4	11.1	200.6	6.4
6696	279.5508	-7.3130	2013.401	17.279	17.315	27.23	150.5	89.9	-143.7	6.6	88	-141	6.1
6697	279.6124	23.6742	2012.908	12.973	13.262	28.377	39.8	65.9	12	7	65.7	11.3	7.2
6698	279.6409	-2.6256	2012.8705	16.182	16.565	17.083	254	82.9	33.8	5.9	87.6	32.4	5.9
6699	279.6459	19.7745	2013.3825	18.212	18.573	38.029	302.8	47.5	-93.4	6.7	39.5	-97.8	6.7
6700	279.6526	49.0119	2013.716	9.267	10.496	47.502	260.5	66.5	46.5	5.3	64.4	46	5.3
6701	279.6976	-3.6696	2013.348	15.56	17.533	36.616	310.4	57.9	-123.8	5.6	57.1	-131.5	6.6
6702	279.8107	37.8066	2012.965	13.929	15.885	38.923	79.2	-14.8	-67.7	5.7	-15.9	-68.6	7.2
6703	279.8553	-2.7047	2013.302	15.287	17.111	54.687	225.9	-32.6	-63.5	5.7	-21.6	-68.9	7.2
6704	279.8567	6.8947	2013.048	17.879	17.969	59.887	147.1	-4.5	-65.1	6.8	-6.8	-62.8	6.2
6705	279.8576	-7.1457	2013.4035	16.391	16.477	31.57	136.6	87.3	-209.6	7.2	73.7	-212.5	6.7
6706	279.8637	-2.2977	2013.3015	17.066	17.713	19.42	26.8	51.4	-69.7	6.9	63.9	-78.8	12.7
6707	279.8831	-17.2172	2014.162	12.274	14.994	22.675	136.3	-0.4	-106.6	5.3	-4.7	-107.3	5.2
6708	279.9000	-4.5215	2013.3795	16.716	18.049	28.385	225.2	-97.6	-197	5.6	-104.9	-204.3	6.1
6709	279.9512	-4.6625	2013.456	17.277	18.387	34.799	72	67.5	-178.6	7.1	76.9	-162.5	10.3
6710	279.9536	-8.7898	2013.3	16.438	16.896	22.661	76.6	103.8	-172.4	7.6	109.8	-178.5	6.4
6711	279.9859	-4.9808	2013.3305	16.564	16.86	30.204	6.6	-45.3	82.1	6.2	-35.5	83.7	7.5
6712	279.9889	-4.5843	2013.404	15.838	15.99	4.206	222.6	144.6	44.1	5.6	134.7	55.5	6.6
6713	279.9928	-5.2867	2013.4305	16.874	17.166	24.528	262.8	109.3	-90.9	5.7	108.8	-102.1	6
6714	279.9970	-4.8344	2013.392	16.519		22.164	357.7	-143.5	-191	6.1	-144.3	-198.5	7.5
6715	280.0078	19.9727	2013.5675	17.789	17.872	37.995	244.8	64.1	-52.2	5.4	76.4	-40.3	7.5

6716	280.0219	-7.0605	2013.4025	15.816	17.124	49.679	52.8	67.2	137.6	6.1	75.8	131.9	6.3
6717	280.0273	-4.5993	2013.3265	15.699	18.156	11.561	168.7	72.8	101.4	5.8	73.4	97.8	6.1
6718	280.0358	-2.1437	2013.421	16.666	17.638	57.924	336.1	-39.1	-82.9	6.8	-40.2	-77.4	6.4
6719	280.0845	-5.5008	2013.333	16.328	18.059	42.887	224.4	111.9	-136.7	5.9	122.2	-134.3	8.8
6720	280.1028	-5.8956	2013.3565	16.523	16.752	12.444	106.4	105.1	-105.3	6.2	116.3	-99.1	6.2
6721	280.1661	-8.4013	2013.3	16.698	17.561	26.33	279.9	67.6	-193.8	7.6	72.9	-194	6.8
6722	280.1729	-4.3006	2013.425	17.436	18.407	30.407	52	108.8	119.7	7	102	132.3	6.3
6723	280.1825	-2.2807	2013.0545	16.164	17.837	34.726	291.5	99.8	172.2	7.7	102.4	160.1	6.1
6724	280.1898	-4.5223	2013.4185	17.62	18.525	24.038	176.4	192.2	-71.4	6.6	207.5	-80.3	9.2
6725	280.1957	66.5479	2013.543	9.962	14.419	45.149	212.2	154.8	21.2	5.7	161.6	19.3	5.7
6726	280.2435	-7.0527	2013.4145	16.446	16.646	8.237	136.6	101.9	191.9	6.1	106	185.1	6.3
6727	280.3123	-6.3204	2013.354	15.698	16.793	15.941	111.8	149.2	138.6	6.8	143.9	134.5	6.5
6728	280.3231	-3.6368	2013.3945	17.659	18.34	47.161	56.4	115.8	188.2	6.2	124.8	181.3	6.8
6729	280.3296	-3.4768	2013.2545	17.301	17.454	45.731	23.3	-61.5	-79.2	6.2	-51.1	-85.5	5.9
6730	280.3426	33.0150	2013.6605	11.33	12.396	4.437	68	-0.5	-61.1	5.3	-6.5	-63.7	5.3
6731	280.3990	-6.2748	2013.5285	14.851	16	26.069	289.4	62.2	-72.4	5.9	60.6	-76.7	5.9
6732	280.3994	-1.1105	2013.3425	15.352	17.264	54.169	174.1	65.1	-151.7	5.8	63.7	-152.9	7.8
6733	280.4097	-3.2596	2013.15	14.094	15.834	33.97	69.4	71.8	-61.9	5.8	67.5	-63.2	7
6734	280.4158	-6.3277	2013.4005	16.5	17.473	49.355	106.2	185.4	-179.9	6.2	190.1	-188.6	6.1
6735	280.4170	-6.3030	2013.374	16.742	18.294	13.507	147.6	-13.9	-115.1	6.2	-7.5	-116.9	6.4
6736	280.4592	-7.2880	2013.3625	16.665	17.125	17.986	36.5	117.9	-166	6	127.2	-164.8	6.2
6737	280.4743	26.2877	2013.487	10.917	12.184	52.861	268.4	2.1	-62.4	6.1	-1.7	-63.5	6.2
6738	280.4789	-1.7301	2013.158	17.354	17.802	38.553	130.6	62.3	-19.8	6	61.6	-21.1	7.1
6739	280.4872	-7.6381	2013.3185	16.071	16.408	33.201	245.6	179.7	-27.4	6.5	176.7	-25.5	6.4
6740	280.4928	-7.1148	2013.3805	15.121	16.833	51.243	108.7	8.8	-98.5	6.2	3.7	-95.9	6.5
6741	280.4940	-12.3804	2014.1605	12.748	14.005	4.887	311.8	-32.8	-89.4	5.2	-37	-92.5	5.5
6742	280.5046	-6.8844	2013.445	15.034	17.715	25.816	68.4	71.5	-43.7	6.3	67.3	-44.2	7.8
6743	280.5068	-5.8958	2013.3245	16.811	17.753	16.932	259.5	80.2	124.3	6.2	65.1	129.8	12.2
6744	280.5145	-1.6905	2013.4225	17.9	18.418	28.497	204.3	70.9	-70.2	7.5	70.1	-74.6	6.9
6745	280.5173	24.8536	2012.9625	12.371	12.491	45.574	55.1	61.3	-16	6.3	61.1	-9.2	6.9
6746	280.5204	11.7606	2013.7715	7.706	12.823	14.86	79.1	-47.6	-95.7	5.9	-49.6	-100	6.1
6747	280.5381	-6.8114	2013.3785	17.295	17.432	20.789	40.1	-119.3	-101.5	6.8	-123.4	-97.8	7.7
6748	280.5861	-5.5531	2013.346	15.94	17.081	10.907	314.7	95.6	129.6	6.3	85.1	137.4	7.7
6749	280.6028	-5.4183	2013.4425	15.099	15.653	8.405	86.7	-2.4	-140.4	7	-11.7	-134.2	5.8
6750	280.6047	-6.2121	2013.3975	16.156	17.459	21.386	32.9	55.8	-146.9	6	46.6	-150.8	6.5
6751	280.6322	30.5814	2013.641	8.016	9.244	49.261	13.2	108.1	-40.6	5.3	109.5	-40.5	5.3
6752	280.6374	-1.6675	2013.339	16.316	18.152	32.788	119.5	67.7	-16.6	8.6	75.4	-20.4	6.3
6753	280.6768	-6.4335	2013.4215	15.282	15.691	58.357	297.7	-48.5	-78.9	5.9	-50.1	-82.1	6.1
6754	280.7042	-6.2789	2013.4405	16.69	17.144	7.329	84.2	-184.8	-146.3	6.1	-180	-155.9	7.1
6755	280.7177	35.3660	2013.5955	11.497	12.057	30.533	63.2	42.4	-78.8	5.3	45.5	-79.7	5.3
6756	280.7213	-2.6251	2013.348	17.135	18.354	44.502	25.3	-35.4	-183	6.4	-29.2	-177.9	7.4
6757	280.7244	-7.4997	2013.3435	14.783	16.814	10.291	153.3	74.3	40.9	6.3	76.2	41.4	6.3
6758	280.7245	-7.1767	2013.6815	13.79	14.369	26.04	89.4	-30.1	-114.2	6	-25.4	-121.2	5.8
6759	280.7306	-2.4118	2013.16	17.276	18.237	46.803	310	-194.9	-199.5	6.5	-200.8	-206.2	6.3
6760	280.7722	-6.3410	2013.3755	16.64	16.679	10.281	127.5	146	-171.7	6.8	146.4	-183.6	7.2
6761	280.7730	-2.6439	2013.456	17.75	17.937	14.551	73.6	-156.2	-135.2	6.9	-159.7	-147	7.5

6762	280.7762	34.3646	2013.711	10.758	11.795	33.928	212	7.2	-67.2	5.2	8.2	-63.9	5.2
6763	280.7917	-9.0865	2013.3	16.186	16.568	26.202	59.8	-4.4	-71.2	6.4	-15.6	-67.2	6.6
6764	280.7956	8.4446	2012.7545	17.56	18.359	16.325	185.5	75.6	-9.4	6.4	73	-9.4	6.7
6765	280.7986	-7.8427	2013.347	15.704	17.325	42.21	83.6	-27.9	-79.7	6.2	-37.5	-78.8	6.7
6766	280.8117	-4.8807	2013.37	16.185	16.592	25.45	355.3	17.3	-167.2	6.2	9.8	-163.7	5.9
6767	280.8257	-6.7201	2013.3295	15.201	15.747	12.921	357.9	-12.5	-69.4	6	-19.5	-75	7.4
6768	280.8462	-5.2907	2013.4385	16.062	17.126	33.79	15.7	95.2	-209	6.7	92.1	-212.9	6.7
6769	280.8627	-8.2651	2013.4695	15.507	16.452	50.198	60.9	74.1	71.9	6.7	65.9	79.2	7.2
6770	280.8918	-6.3636	2013.3755	16.499	16.631	15.856	131	-49.6	120.9	6	-46.1	124.6	6.2
6771	280.8924	-6.0502	2013.3885	16.063	16.965	13.575	48.2	66.7	-98.2	6.2	73.8	-86.4	7
6772	280.9357	52.7200	2013.348	18.467	18.53	51.08	337.4	73.3	39.2	5.6	75.1	37	5.8
6773	280.9401	-2.3484	2013.432	16.489	17.315	21.771	22.3	158.1	-141.4	7	164.9	-131.9	6.7
6774	280.9544	-5.9084	2013.4555	15.303	16.373	41.999	334.9	119.8	-71.9	6.5	117.3	-78.7	6.4
6775	280.9569	-7.6572	2013.418	14.14	15.494	33.838	318.7	61.5	62.2	6.4	58	64.4	6
6776	280.9581	26.7111	2013.0285	11.949	13.165	52.625	291.4	45.4	-90	6.3	36.3	-93.3	6.8
6777	281.0026	-4.3973	2013.3945	12.584	17.75	12.311	116	92.4	112.6	6.4	97	105.6	6.4
6778	281.0232	-6.6247	2013.38	15.468	16.216	27.309	252.4	75.3	42.4	6.8	69.7	53.4	6.1
6779	281.0427	71.4888	2013.4825	12.362	12.735	2.341	277	13.3	148.4	6.2	13.4	146	7.7
6780	281.0445	-4.6386	2013.399	16.824	17.393	39.727	309.9	-41.9	-182.5	5.9	-33.2	-178	6.9
6781	281.0525	-0.8109	2012.9395	17.072	17.657	16.423	236.7	-1.2	-69.4	6.1	-11.8	-68.1	6.1
6782	281.0633	-6.5803	2013.421	16.367	16.457	15.884	265.6	75.5	-38.5	6.4	71.5	-49.7	6.1
6783	281.0813	-6.3237	2013.4295	15.182	16.932	37.849	213.7	-94	-216	6.2	-83.4	-215.5	6
6784	281.0856	-5.7650	2013.4765	14.886	16.428	7.563	343.8	125.6	-61.2	5.8	125.7	-72.5	6.5
6785	281.0883	36.3187	2013.583	10.801	13.227	13.604	103.4	36.2	-126.6	6.1	35.5	-124.8	6.1
6786	281.0893	-1.9308	2013.3105	16.53	17.669	51.954	240.2	22.9	-64	6.1	11	-68.5	7.6
6787	281.1184	1.1050	2012.5595	18.135	19.063	25.728	224.5	-11.2	-89.8	7.7	-23.1	-83.4	6.7
6788	281.1265	-4.2499	2013.4585	17.85	18.503	23.187	266.7	158.4	71.5	7.7	151.8	80	7.3
6789	281.1626	-2.3974	2013.103	17.006	17.733	59.639	194.1	-41.5	-129	6.3	-44.3	-128.1	6.2
6790	281.1875	51.8963	2013.3345	14.013	16.403	5.315	41.5	77.5	-31	5.3	74.9	-40.3	6
6791	281.1906	32.6251	2013.5205	8.94	10.388	47.491	27.5	29.9	-118.3	5.3	29.6	-118.9	5.3
6792	281.2167	-5.0591	2013.434	16.713	16.798	36.814	306.1	-33.4	-69.6	6.1	-29.9	-72.6	6
6793	281.2178	59.2602	2013.496	12.35	14.695	24.105	240.3	67.1	24.4	5.7	66.9	21.8	5.7
6794	281.2322	-4.2598	2013.457	18.514	18.78	13.6	262.7	124.3	-143.6	7.8	118.5	-147.9	6.9
6795	281.2369	-3.5565	2013.457	17.691	18.37	15.094	41.2	165.3	-86.1	6.6	177	-73.6	6.8
6796	281.2491	15.8921	2013.624	13.83	14.305	7.192	65.1	-42.3	-98	5.3	-44	-99.7	5.3
6797	281.2642	-6.3408	2013.4165	16.049	16.466	21.388	6	-30.3	-104.5	6.1	-29.9	-103.7	7.2
6798	281.2692	-23.2525	2013.945	8.546	10.585	25.174	94.1	14.2	-120.3	5.6	14.7	-121.3	5.4
6799	281.2861	-5.7334	2013.541	15.8	16.38	6.881	127.1	30.1	-211.3	7.3	23.8	-208.8	6
6800	281.3213	-7.4915	2013.4365	15.245	15.502	30.972	208.3	72.7	-78.9	7.5	70.5	-75.7	6.6
6801	281.3227	26.4718	2013.1875	12.702	12.765	10.042	350.5	-30.1	-72.2	6.4	-29.9	-73	6.3
6802	281.3281	-4.3191	2013.458	17.823	18.428	30.883	341.4	-120.3	-202.9	6.5	-113.1	-198.2	7.9
6803	281.3327	40.0826	2013.6515	10.611	11.625	4.657	282.2	-14.9	-71.9	6	-12.1	-73.5	6
6804	281.3398	-2.6582	2013.269	16.641	18.421	17.872	121.3	149	127.2	6.3	151.1	130.4	8.2
6805	281.4155	-4.0245	2013.4575	17.719	18.564	5.521	187.7	44.7	-196.3	6.6	35.9	-204.2	6.6
6806	281.4348	-3.8397	2013.3345	15.503	16.968	29.173	160.1	-38.2	-70.2	5.9	-38.6	-66.5	6.1
6807	281.4532	-4.9693	2013.4125	16.835	16.866	15.153	200.1	38.2	-67.8	6.6	34.3	-67.5	6.7

6808	281.4562	13.3376	2013.3705	17.841	18.313	32.609	215.9	-13.4	-64.3	5.6	-14	-61.5	5.8
6809	281.4677	-5.9323	2013.436	15.505	15.788	22.229	230.7	64.8	-167.5	6	64.8	-175.5	6.1
6810	281.4835	-7.5865	2013.5435	13.807	16.232	28.907	165.3	112.7	21.1	6.5	111	16	6.9
6811	281.5013	-7.3192	2013.4335	14.348	15.585	22.564	112.5	-66.3	70.8	6.9	-65.7	73.9	6.4
6812	281.5239	-4.7211	2013.436	15.897	17.329	30.471	15.6	156.3	-89.9	5.9	169.1	-89.2	7.1
6813	281.5860	56.4965	2013.62	10.361	10.795	49.304	265	1.8	63.1	5.6	2.7	61.8	5.6
6814	281.6168	35.8368	2013.4575	12.897	13.634	51.723	31.2	-12	-60.1	5.3	-4.2	-60.4	5.3
6815	281.6284	-11.9895	2014.368	13.937	14.893	5.035	182.5	-37.7	-70.4	5.1	-37.7	-69.1	5.3
6816	281.6471	-6.3597	2013.3795	16.342	16.356	18.662	20.3	-111.2	-144.8	6.3	-113.5	-140	6.2
6817	281.6534	-4.6169	2013.4245	17.044	17.172	19.048	227	74.2	-24.6	7.9	74	-35.9	6.3
6818	281.7467	-4.6162	2013.394	16.254	17.711	37.69	260.9	-84.9	-102	6	-77.6	-100.3	7.4
6819	281.7587	46.8366	2013.3145	17.966	17.983	41.16	23.7	66.8	-12.9	5.6	62.6	-2.1	5.5
6820	281.8158	-5.4824	2013.3005	16.015	17.103	52.346	327.8	60.3	28.2	6.5	66.2	16.4	7.9
6821	281.8232	35.7643	2013.589	12.535	12.657	9.672	102.6	-25.4	-75.2	5.3	-34.6	-71	5.3
6822	281.8310	-8.2958	2013.4395	13.16	15.356	36.006	244.3	70.1	-9.3	6.9	72.4	0.7	6.6
6823	281.9441	-7.2172	2013.427	16.121	16.214	17.635	192.6	-15.2	-73.9	6.3	-9.7	-71.3	6.1
6824	281.9644	-6.0384	2013.4255	15.531	16.779	50.495	347.3	67.7	7.3	6	64	16.3	6
6825	282.0489	-5.4673	2013.6235	13.991	14.16	13.898	206.9	36.3	-63.7	5.6	38.4	-65.6	5.6
6826	282.0870	40.3190	2013.644	13.892	13.927	49.172	190.9	63	-24.7	6	64.5	-20	6
6827	282.1132	-0.5495	2013.007	16.9	17.515	27.24	238.8	20.7	-72.9	6.3	30.4	-69.2	6.5
6828	282.1192	45.2169	2013.594	11.576	12.326	24.01	312.6	-27.3	-68.8	5.3	-25.9	-67.2	5.3
6829	282.2532	-3.9368	2013.4575	17.103	17.257	11.464	133.6	142.5	154.9	6	142.1	160	6
6830	282.2980	-0.6880	2013.086	17.549	18.438	36.023	51.4	177.9	-190	6.3	183	-186.3	6
6831	282.3285	14.6709	2013.719	17.959	18.336	34.047	324	61.2	35.6	6	64.1	31.5	7
6832	282.3346	4.0467	2013.4725	13.088	16.017	10.412	321.4	38.9	-69.9	5.8	40.5	-67.7	5.9
6833	282.4686	-0.8933	2013.2975	17.17	17.753	17.874	216.5	-17.7	-65	6.4	-14.1	-65.7	6.1
6834	282.4849	30.4609	2013.3655	12.643	13.147	43.929	130.4	7.7	-76.2	5.3	13.9	-75.4	5.4
6835	282.5183	-13.2757	2014.0705	8.621	9.072	28.924	251.2	34.7	-151.1	5.5	34.8	-149.1	5.6
6836	282.5222	45.4344	2013.5735	15.396	15.425	34.996	266.5	-23.3	127	5.3	-22.4	126.4	5.3
6837	282.5403	-9.6117	2013.9515	12.074	12.998	5.504	177.3	-31.4	-72.1	6.1	-31	-75.7	6.1
6838	282.5480	22.6690	2013.136	12.468	12.905	24.327	142.4	-9.9	-70.4	5.1	-12.7	-68.7	5.2
6839	282.6234	-1.4153	2013.277	16.349	16.404	34.004	280.5	145.6	-9.8	6	146.4	-4.3	6
6840	282.6498	-3.7173	2013.4585	16.506	17.953	12.882	232.7	125	-33.9	6.1	129	-24	6.7
6841	282.6521	41.1130	2013.507	15.829	16.249	3.659	32.9	-89.3	-84.9	6.1	-82.8	-96	6.4
6842	282.7409	53.6566	2013.1865	18.303	18.597	40.373	204.3	73.4	-8.1	5.6	70.3	-2.5	5.6
6843	282.8136	-0.8078	2013.0815	17.164	17.269	18.615	266.8	35.8	-73.2	6	45.2	-69.7	5.9
6844	282.8159	-1.8450	2013.457	18.031	19.072	34.439	354.7	164.4	-131.2	7.1	167.8	-131.1	7.9
6845	282.8629	-0.7439	2013.229	17.895	19.292	11.931	7.2	-74.2	-188.4	6.7	-67.3	-175.6	14.7
6846	282.8868	-6.1749	2013.4755	16.413	16.697	16.315	312.9	64.3	49.6	6.6	60.9	52.3	6.1
6847	282.8869	1.2415	2012.766	17.853	18.516	33.138	11.6	-94.7	-108.2	6.4	-84.8	-108.3	6.6
6848	282.9061	-4.9469	2013.458	16.751	17.636	20.41	62.6	90.4	154.8	6.2	104.1	141	7.7
6849	282.9284	12.6704	2013.2305	17.433	17.541	40.987	12.4	113.4	-12.8	7.9	119	-8.2	7.6
6850	282.9461	-4.0667	2013.4575	17.797	17.939	35.087	32	189.1	15.8	6.5	189	25.3	6.7
6851	282.9603	30.8674	2013.343	6.983	12.917	43.069	340.2	-40	-75.4	5.4	-43.6	-76.5	5.4
6852	282.9780	-18.1433	2014.1005	12.161	15.633	19.94	231	-20.8	-60.3	5.2	-22.4	-60.9	5.2
6853	282.9999	-0.6153	2012.596	15.424	16.594	39.505	200.8	101.1	69.3	6.2	105.7	72.9	6.4

6854	283.0286	16.2921	2013.584	15.39	16.629	19.317	115.1	-18.4	-83	5.8	-24.6	-79.8	6
6855	283.0798	3.1939	2013.179	14.155	15.592	45.534	93.6	67.3	9.4	6	64.2	11.7	6
6856	283.1134	-0.9686	2013.3315	17.021	18.276	32.612	124.9	196.1	-78	5.9	185.7	-89.4	9.4
6857	283.1214	-0.0223	2013.3105	16.785	18.025	20.314	226.1	24.9	-205	5.9	16.8	-208.7	6.3
6858	283.1403	-1.3563	2013.356	17.786	18.175	10.338	68.3	-88	-184.1	6.3	-77.2	-188	6.1
6859	283.1408	-1.5426	2013.3695	16.065	18.49	17.633	158.2	-176.8	-65.7	5.8	-178.4	-70.9	6.8
6860	283.1751	-0.6960	2013.1275	16.575	17.725	55.873	25.7	73.3	-68.4	5.9	66.1	-77.9	6.1
6861	283.1826	36.9906	2013.6395	9.777	10.286	8.596	160.2	76.8	35.5	6.1	80.5	31.7	6.1
6862	283.1862	-1.0893	2013.307	16.397	17.716	28.367	156.9	115	-115.1	6.6	106.2	-110.4	6.5
6863	283.2036	-0.2859	2013.4065	17.384	17.859	9.981	38.5	12.6	-151.8	6.4	11	-149.3	6
6864	283.2408	-0.2581	2013.1125	17.005	18.028	14.759	234.7	39.5	-96.9	6	30.3	-104.2	6.2
6865	283.2450	30.9688	2013.384	7.024	9.777	10.452	356.5	-61.6	-191.9	5.5	-64.4	-191.2	5.4
6866	283.2614	53.0644	2013.5165	17.963	18.15	53.004	272.4	-0.6	-61.6	5.5	-11.6	-61.8	6.1
6867	283.2810	-2.4002	2013.456	16.096	18.177	21.699	273	-180.3	-160.1	7.6	-169.7	-157.6	6.1
6868	283.3151	-10.6156	2013.486	16.945	17.111	32.756	167.3	15.3	-63.3	6.4	21.4	-63.5	6.8
6869	283.3286	54.0700	2013.372	16.119	16.839	10.802	227.7	-28	-84.5	5.8	-24.7	-82	5.8
6870	283.3586	-0.5649	2013.521	14.619	17.657	29.899	291.8	-2.9	86.3	5.7	-5.2	86.8	7
6871	283.3773	74.3930	2013.344	16.756	18.619	7.208	125.6	32	77.6	6.2	27	80.3	6.4
6872	283.3883	-0.8877	2013.4735	15.182	16.375	42.139	297.1	34.6	-105	5.7	30	-103.8	7.1
6873	283.4036	-0.6047	2013.0745	16.373	17.141	18.184	58.1	112.2	139.1	6.9	113.1	150.6	6.1
6874	283.4180	-0.4183	2013.3845	17.239	17.295	17.103	347.3	190.9	-198.7	5.8	188.4	-202.6	6
6875	283.4397	45.8380	2013.5185	12.175	16.318	25.867	72.8	41.9	-93.7	5.3	40.4	-94.9	5.4
6876	283.4453	-0.1570	2013.379	17.547	17.697	35.925	336.9	84.5	-165.4	6.2	96.7	-163.9	6.5
6877	283.4819	-5.1277	2013.627	14.867	16.475	37.795	246.5	92.5	-111.7	5.7	104.8	-109	6.7
6878	283.5214	39.3713	2013.2905	10.595	13.696	31.248	344.6	55	-132.4	5.7	55.1	-135.7	7.1
6879	283.5723	20.4173	2013.445	13.786	16.459	10.111	290.3	-26.5	-109.8	5.9	-31.3	-109.5	6.1
6880	283.5925	9.9292	2012.8035	17.603	17.944	37.366	105.2	-3.4	-102.7	7.8	17.2	-113.9	21.7
6881	283.6170	-21.4448	2014.186	13.794	15.106	10.202	103.3	29.7	62.2	5.2	29	63.3	5.3
6882	283.6293	-4.1120	2013.5715	15.845	17.113	39.237	187.6	-76.5	-70.6	5.8	-75.1	-74	6.2
6883	283.6403	30.7251	2013.4245	8.502	12.39	5.657	84.9	55.7	110.6	5.4	56.1	112.1	5.4
6884	283.6595	19.2499	2013.3195	17.59	18.251	54.117	14.7	-6.4	-81.4	6	-6.3	-82.2	6.3
6885	283.6602	2.5564	2012.915	18.039	18.643	57.074	111.8	155.2	-191.9	7.3	154.8	-200.3	7.4
6886	283.6857	-1.6993	2013.195	15.887	16.419	24.522	14.7	91.2	67.5	6	93.6	68.1	6.3
6887	283.7129	32.3577	2013.453	10.539	12.23	6.306	122.6	10.5	-88.9	5.3	10.8	-91.5	5.3
6888	283.7433	-1.1260	2012.986	16.595	17.072	27.83	46.6	20.3	-191.1	6.2	20	-182.7	8.8
6889	283.7497	-0.6523	2013.0435	14.312	15.676	40.444	257.6	77.7	-69.8	5.9	74.7	-80.1	6.3
6890	283.7747	0.0209	2013.5015	15.541	16.133	48.635	329.7	71.2	-140.8	5.8	75.9	-138.4	7
6891	283.7853	43.2254	2013.0645	15.822	18.858	23.441	200	-19	-61.5	5.4	-16.7	-62.7	6.1
6892	283.7977	-0.3945	2013.206	16.741	17.324	29.255	197.2	70	-25.8	10.2	61.7	-16.2	5.9
6893	283.8210	-6.6004	2013.728	11.795	16.272	10.497	292.8	-55.8	-70.8	5.5	-54.6	-70.8	5.9
6894	283.8227	10.5120	2012.657	16.759	18.442	24.639	182	85.2	-66	6.4	90.8	-61.2	7.5
6895	283.8243	9.6698	2013.623	9.925	15.84	17.232	43.3	22.8	-75	5.7	20	-73.3	5.9
6896	283.9237	-7.1284	2013.4635	15.42	16.759	49.312	352.7	63.8	-48.5	6	64.8	-43.2	5.8
6897	283.9254	-1.8439	2013.1655	16.2	16.606	39.091	85.3	1.9	-72.4	6	11.5	-70.2	6
6898	283.9678	39.0622	2013.462	12.427	16.865	30.315	267.6	-67	124.4	5.3	-64.8	123.9	5.5
6899	283.9695	-0.0755	2013.347	16.599	17.955	24.953	331.4	142.5	-29.3	5.8	149.7	-30.5	7.3

6900	284.0295	1.4951	2013.147	15.256	18.361	33.284	286.2	-0.1	-65.4	5.9	-4.5	-62.3	6.4
6901	284.0372	-1.0689	2013.068	15.683	17.621	27.146	272.5	-13.6	-65	6.2	-14.2	-62.1	6.2
6902	284.0660	55.6062	2013.558	10.021	11.01	4.819	62.6	28.8	156.1	6.1	33.5	154.3	6.1
6903	284.0863	52.4256	2013.46	15.447	15.461	53.597	132.1	10.5	74.5	5.4	10.8	77.9	5.4
6904	284.0923	45.5073	2013.7525	8.358	9.164	29.319	341.9	17.2	60.6	5.3	18.5	62.5	5.3
6905	284.0952	-0.1120	2013.352	12.616	16.94	16.752	121.4	85.3	161.7	5.7	94.5	156.9	6.1
6906	284.1395	19.4436	2013.0695	16.284	17.081	53.499	205.9	87.1	39.3	5.2	89.7	34.4	5.5
6907	284.1470	-2.5765	2013.4575	15.956	17.789	50.499	19.1	189.3	11.6	7.1	198	17.7	6.2
6908	284.1548	10.6805	2013.5055	15.526	16.362	24.778	39.3	66.4	-86.2	7.1	64.5	-94	6
6909	284.1659	22.0800	2013.0795	14.811	15.963	30.116	73.5	-4	-65.5	5.2	-8.8	-65.3	5.5
6910	284.1941	40.6441	2013.6285	12.05	12.941	14.781	145.3	34.3	-118.4	5.3	35	-118	5.3
6911	284.2861	-0.8266	2013.261	16.499	17.681	24.511	311.5	108.5	-27.2	6.4	103.4	-29.3	6.5
6912	284.3001	22.0501	2012.95	15.495	16.098	27.399	274	-35.2	-64.7	5.3	-31.7	-69.7	5.8
6913	284.3049	20.3365	2013.13	17.223	17.749	42.955	303.2	-11.1	75	5.2	-16	77.5	5.3
6914	284.3356	70.8420	2013.383	17.661	17.713	32.567	358.3	12.6	-64.7	5.8	9.4	-62.9	5.8
6915	284.3378	3.0311	2013.504	12.948	13.408	19.807	179.7	70.9	43.8	5.9	72.2	47.7	5.9
6916	284.3662	20.8103	2013.824	11.461	15.112	9.096	304.3	0.4	-73.9	5	-0.6	-77.5	5
6917	284.3777	-2.7838	2013.392	16.871	17.422	51.401	134	21.1	84.3	6.5	17.6	85.3	6.5
6918	284.3795	-4.6711	2013.458	15.827	18.461	11.843	183.5	7.9	-67.3	6.5	14.5	-65.2	6.9
6919	284.3822	-0.1224	2013.4555	17.191	18.74	50.83	203	85.9	7	7	80.2	2.5	10.9
6920	284.4154	-24.2054	2014.0565	12.224	13.113	7.991	348.4	-56.8	-92.2	5.4	-57.2	-91.1	5.4
6921	284.4492	12.7228	2013.679	15.808	16.068	57.012	272.5	123.4	-47	5	119.6	-44.8	5.1
6922	284.4621	3.2537	2013.438	12.186	13.019	10.87	90.6	0.3	-70	5.8	-5.8	-71.8	5.9
6923	284.4687	11.4384	2013.7785	13.55	16.827	33.418	341.8	77.3	-46.4	5.7	73.2	-47.4	5.8
6924	284.4818	1.3159	2013.28	16.049	17.083	4.451	278.3	-6.3	-68.4	6	-0.1	-70	6.5
6925	284.5351	0.1579	2013.938	15.139	16.123	58.376	73.6	192.5	112.1	6.9	189	107.5	6
6926	284.6064	87.6657	2013.8235	13.536	14.588	23.819	141.7	36.8	142.4	6	36.7	146.4	6
6927	284.6411	12.2977	2013.6385	12.659	16.315	44.704	324.8	24.4	60.2	5.8	19.9	62.7	6.1
6928	284.6420	-23.5031	2014.245	12.043	15.396	6.856	27.1	-129	-92.2	5.7	-124.8	-97.8	5.7
6929	284.6474	73.7027	2013.378	16.709	17.343	9.835	322.1	-26.7	64.6	6.1	-28.8	63.8	6.2
6930	284.6622	-1.1232	2013.397	14.104	16.332	55.502	274.9	7.5	-62.8	6.2	15	-62.6	6.8
6931	284.7811	0.0828	2013.7355	10.615	13.061	24.976	138.9	21	-72.2	5.7	20	-72.3	5.8
6932	284.7831	25.7530	2013.3565	10.185	12.902	36.374	129.5	-18.6	-72.7	5.7	-30.1	-68.4	5.9
6933	284.7910	25.3518	2013.203	12.164	12.593	37.219	353	-10.6	-62.7	6.1	-4.4	-61.7	6
6934	284.8518	14.0908	2013.824	16.624	18.008	24.462	273	-5.2	-70.8	5.1	-0.2	-65.1	5.1
6935	284.8598	17.2942	2013.2955	16.724	17.771	55.915	143	29.2	-60.8	6.3	18	-62	6.3
6936	284.9268	16.2521	2013.639	7.124	14.743	18.419	207.9	-96.2	-118.3	6.2	-97.6	-123.4	6.2
6937	284.9853	27.5502	2013.5075	8.637	11.24	5.793	219.8	-21.7	-74	5.7	-24.4	-72.5	5.7
6938	285.0071	-2.6528	2013.349	15.619	15.844	46.487	230.7	-14.6	-86.8	6.6	-21.4	-82.7	6.3
6939	285.0075	56.9611	2013.649	8.417	12.532	6.879	278.1	-112.4	-153	6.2	-110.1	-155.1	6.1
6940	285.0228	35.8832	2013.479	12.76	13.34	52.585	317.9	61.6	27.5	5.3	62.2	31.2	6.1
6941	285.0358	19.8906	2013.3695	15.757	17.881	40.94	85.2	5.5	-80.9	6.3	11.4	-80.4	6
6942	285.0715	64.9851	2013.403	15.769	16.19	8.569	345.2	-96.2	-109.6	5.8	-93.9	-110.7	5.8
6943	285.1065	23.4557	2013.4445	11.06	13.577	9.999	12.4	12.2	-61.6	5.9	9.1	-61.2	6
6944	285.1148	5.8597	2013.4815	13.452	14.037	17.24	26.1	70.4	-183.9	6.2	69.9	-181	6.1
6945	285.1212	31.7020	2013.1905	12.605	13.026	12.214	245.2	-7.5	-69.6	5.4	-14.5	-71.9	6

6946	285.1734	17.6250	2013.6135	12.477	16.011	15.786	350.7	63.7	-71.7	5	64	-68.4	5.1
6947	285.2437	-6.1651	2013.946	11.162	11.182	9.228	23.6	40.2	-80	5.5	39.4	-79.5	5.5
6948	285.3076	22.8029	2013.119	12.601	13.356	13.458	111.7	67.3	-15.9	6	67.4	-23.5	6.5
6949	285.3116	29.1450	2013.609	10.175	10.541	23.388	138.3	-42.1	-125.7	5.5	-40.8	-121.7	5.5
6950	285.3216	19.9771	2013.478	17.47	17.617	4.808	105.3	76.7	-66.6	6.9	66.8	-72.8	6.2
6951	285.3749	7.2806	2012.9115	16.057	16.075	47.958	304	62.8	159.3	7	62.7	157.2	6.2
6952	285.3885	-24.1415	2013.985	9.031	10.585	31.117	83.7	-39.7	-80.6	5.6	-30.8	-84.1	5.7
6953	285.4229	-7.1032	2013.466	15.157	17.325	44.164	93	68.8	24.7	6.1	68.7	28.8	7.8
6954	285.4492	17.9960	2013.557	13.942	14.608	9.559	319.9	-10.8	-62	5.8	-11.8	-61	5.9
6955	285.4698	-3.6563	2013.4575	18.004	18.718	49.549	219.9	61.1	-44.2	7.1	70.4	-30.3	7.3
6956	285.5811	18.9014	2013.64	13.14	16.383	5.973	241.6	-18	-73.4	6	-20.2	-72.4	6.7
6957	285.6007	16.8941	2013.4825	10.255	17.313	25.741	283.1	-11.6	-71.8	5	-15.6	-73.2	5.2
6958	285.6122	0.3290	2012.8205	15.951	16.303	31.137	68.4	29.2	-106	7.2	27.9	-108.2	6.8
6959	285.6849	-0.7117	2014.0485	8.187	14.371	15.468	57	-28	-124.2	6.1	-31.4	-119.9	6.2
6960	285.7033	34.0570	2013.52	13.538	13.761	17.29	225.1	-34.8	-89.4	5.3	-34.2	-93	5.3
6961	285.7717	-7.5923	2013.483	16.54	17.127	5.949	106.8	-31.8	-72.1	5.9	-23	-71.7	6.1
6962	285.8286	47.3720	2013.362	14.316	17.364	8.568	200.3	82.8	-62	6.2	80.5	-63.1	6.3
6963	285.8951	9.4082	2013.2225	16.847	18.805	14.117	303.3	89.7	25.4	6.2	87.5	25.1	6.9
6964	285.9922	22.9712	2013.573	9.074	13.744	39.666	41.3	-5.6	-63.6	6	-4.6	-63.7	6.1
6965	286.0237	31.3494	2013.049	15.217	18.254	6.377	135.7	-0.3	-67	5.5	-8.2	-63.4	5.8
6966	286.0477	-7.4262	2013.5835	14.108	17.001	42.043	322.9	-4.5	-97.3	5.8	-13	-98.6	6
6967	286.0698	4.2450	2013.3855	12.329	14.409	8.911	321.1	33.4	-67	6	33.9	-66.9	6
6968	286.0757	28.3248	2013.301	12.776	13.066	49.32	74.9	6.4	-67.2	5.8	16	-63.5	6
6969	286.0819	-16.8735	2014.0375	12.732	12.797	13.976	264.2	2.3	-91.9	6	1.3	-91.3	6.1
6970	286.0999	8.7506	2013.677	15.976	16.371	11.574	318.1	-83.2	-62.6	5.8	-86	-64.9	5.9
6971	286.1003	7.2875	2013.3045	17.54	18.105	19.333	186	113.8	186.4	6	115.5	182.8	6.7
6972	286.1022	4.0942	2012.8645	16.038	16.991	43.239	21.7	-61.6	-63.2	6.2	-53.4	-73.9	6.2
6973	286.1298	-3.8045	2013.7095	12.295	16.622	11.876	242.8	-177.8	-136.1	6.4	-178.4	-133.9	6.5
6974	286.2284	32.0464	2013.5225	15.449	16.017	24.196	169.5	-2	-70	5.3	-3.4	-70.4	5.3
6975	286.2472	9.6071	2013.753	10.266	17.111	15.151	60.8	-38.1	-67.5	5.6	-42.6	-63	6
6976	286.2475	1.9588	2013.533	14.562	17.778	40.377	10.1	-2.6	-192.5	6	-10.8	-189.5	6.7
6977	286.2496	33.4095	2013.5585	12.435	13.53	22.711	223.1	-16.1	-61.9	5.3	-14.1	-63.4	5.3
6978	286.3016	19.7524	2013.171	15.699	17.418	44.132	245.5	-6.3	-66.8	6	-5	-64.7	6.9
6979	286.3218	3.4003	2012.97	16.704	17.207	41.091	249.8	67.7	-25.8	7.1	67.7	-28	7.5
6980	286.3344	3.5407	2013.5405	16.983	17.179	37.353	200.9	-132.5	-133	6.2	-120.3	-145.2	7
6981	286.3637	11.4845	2013.7535	15.952	17.824	11.471	27.8	-75.2	-82.6	5.6	-80.4	-83.8	7.6
6982	286.3651	-8.1572	2013.7045	14.443	17.366	6.933	223.5	-27.9	-80.1	5.6	-31.6	-78	6
6983	286.3848	0.2518	2014.023	9.97	13.688	52.492	182.2	32.9	-85	5.6	35.7	-87.7	5.7
6984	286.3976	35.0967	2013.5785	13.373	13.748	52.139	248.8	-30.8	-71.8	5.3	-28.3	-70.3	5.3
6985	286.4317	53.9143	2013.3575	17.693	18.8	44.602	5.7	65	62.5	6.3	59.4	67.8	7.4
6986	286.4609	3.8947	2013.301	15.646	16.655	16.431	308.2	-10.5	-69.9	5.9	-17.6	-69.9	6
6987	286.4761	14.0609	2013.9845	11.818	18.447	48.861	328.4	-11.6	-60.4	4.9	-2.5	-61.4	5.3
6988	286.4776	3.0289	2012.949	17.694	17.918	14.879	290.3	13.3	-65.7	6.2	3.4	-67.8	7.1
6989	286.4855	3.2084	2013.4005	15.798	17.101	27.236	266.1	61.6	-34.2	5.8	60.8	-41.2	7.5
6990	286.5042	-0.9783	2014.19	13.037	16.153	4.914	295.2	79.6	33.8	6	77.1	30.2	7.8
6991	286.5043	29.4727	2013.2625	12.817	13.185	13.439	53.7	33.1	71.6	6.1	36.9	70.8	6

6992	286.5393	58.5755	2013.427	8.868	13.93	36.512	227	66.3	91	6.1	65.3	89.5	6.2
6993	286.6502	3.4440	2013.056	13.757	17.466	32.067	195.9	66.5	10.2	5.8	70.1	0	7
6994	286.6602	24.2509	2013.8405	4.609	10.168	25.487	4.9	61	12.5	5.7	61.9	17.1	5.9
6995	286.6647	7.8217	2013.369	15.947	17.547	38.734	310.6	-168.7	-209.3	6.2	-163	-200.2	6
6996	286.7045	18.1600	2013.4465	14.311	18.245	7.391	140.6	73.2	-17.1	6.1	78.4	-8.2	6.7
6997	286.7196	0.8876	2013.41	17.883	18.421	14.183	18.1	-157.5	-85.1	6.1	-154.9	-94.8	7.4
6998	286.7377	27.4830	2012.9525	12.473	13.066	50.552	293.9	9.3	-66.1	6.4	17.5	-65.5	6.8
6999	286.7505	26.3728	2013.071	12.517	12.707	31.163	227.5	-4	-62.9	6.3	-13	-62.5	6.4
7000	286.7544	0.7923	2013.547	13.945	15.804	5.387	109.4	-17.5	-61.4	5.9	-15.2	-60.5	6.1
7001	286.7757	-14.0693	2014.368	8.452	10.105	20.734	317.8	94	27.9	5.8	95.8	34.1	5.9
7002	286.8139	1.9502	2013.4635	15.478	16.747	42.876	302.7	16.6	-65.5	6.3	8.5	-69.5	6.2
7003	286.8232	40.4199	2013.3725	16.227	16.511	7.303	141.6	-9.7	-76.8	5.4	-7.7	-77.3	5.4
7004	286.8548	5.4031	2013.6635	14.199	17.413	7.671	327.5	-26.7	-69.4	5.8	-33.5	-66.6	5.9
7005	286.8555	5.0353	2013.4965	17.386	17.837	41.284	4.8	129	-129.4	6.1	121.2	-138	6.3
7006	286.8923	-5.8366	2013.6985	14.641	16.307	40.678	350	-37.9	-62	6.1	-35.1	-62.7	6.5
7007	286.9240	-4.0297	2013.9585	11.538	12.878	11.116	26.7	1.6	-76	6	1.3	-75.2	6.1
7008	286.9859	51.7856	2013.556	10.371	15.051	17.103	126.5	20.4	101.9	6.1	18.2	105.4	6.1
7009	287.0416	13.6946	2013.479	15.929	17.395	40.009	131.7	-8.5	-89.2	5.4	-12.1	-89.5	5.6
7010	287.0498	-0.5396	2013.7625	14.881	15.474	4.417	66.5	-10.1	-67.2	6.4	-4.4	-66.4	7.3
7011	287.0679	67.2688	2013.4445	11.97	15.898	10.147	216.6	-6.7	77.4	5.7	-8	80.1	5.8
7012	287.1073	22.7602	2013.564	10.917	11.016	5.695	263.4	69.8	-2.1	5.4	66.7	-1.9	5.3
7013	287.1371	9.8269	2013.656	16.996	17.497	28.332	2.2	-53.4	-81.6	6	-55.1	-78.2	6.1
7014	287.1494	10.5036	2013.85	17.603	18.204	47.208	214.3	105.1	-21	6.3	104	-14.5	6.6
7015	287.1661	28.6087	2012.959	12.475	12.638	21.802	93.6	10.8	-65.1	6.4	7.6	-67.6	6.6
7016	287.1674	16.8509	2013.8095	6.772	15.735	38.455	129.2	-26.6	-94.1	4.9	-26.6	-95.3	5
7017	287.2016	11.0204	2014.1915	17.382	17.422	55.899	132.7	92.8	109.3	6	104.2	99.8	6
7018	287.2122	18.7268	2013.576	13.398	14.658	13.813	76.9	-33.2	-116.2	5.4	-29	-112.8	5.4
7019	287.2306	55.8480	2013.3505	15.518	16.132	21.557	175.7	11.5	109.4	6.2	10.3	107.6	6.2
7020	287.2808	11.8440	2013.6185	15.591	17.008	58.296	23.4	72.7	86.5	6.1	70.6	92.2	6.7
7021	287.3095	38.1007	2013.3775	13.429	14.619	7.475	316.1	19.5	-71.9	5.3	21.7	-70	5.5
7022	287.3270	41.3041	2013.346	16.783	16.947	5.816	23.5	-60	-72.6	5.4	-55.1	-75.3	5.5
7023	287.4104	37.6442	2013.537	8.883	13.715	27.499	28.8	78.7	118.3	5.4	74.6	117.2	5.4
7024	287.6010	24.2710	2013.525	12.25	12.403	53.751	50.2	63.4	-38.9	6.2	61.7	-47.8	7
7025	287.6458	12.1726	2013.9655	15.96	18.193	25.341	213	95.6	70.7	4.9	97.8	77.3	6.3
7026	287.6600	1.5355	2013.705	10.904	15.502	13.54	328.1	-160	-196.7	6.1	-160.7	-191.1	6.3
7027	287.6806	11.3281	2013.591	17.463	18.094	48.381	302.1	119.8	-73.6	6	116.7	-82	6.4
7028	287.6960	10.5547	2013.898	16.644	18.283	58.961	123.6	151.7	-7.8	6	157.7	-17.9	6
7029	287.7500	23.8045	2013.355	9.559	13.379	11.592	94.1	-44	-60.7	5.3	-47.3	-61.9	5.6
7030	287.7736	11.4714	2013.603	16.244	18.013	13.742	355.7	93.8	-183.9	6.3	89.5	-189.9	6.9
7031	287.7918	25.7402	2013.2185	8.92	13.602	59.424	153	17.3	-72.8	6.1	24.3	-68.9	7
7032	287.8042	12.6659	2013.7565	17.255	17.463	35.452	297.4	64.5	-15.6	5.2	64.2	-10.1	5
7033	287.8120	58.3940	2013.478	12.626	12.795	38.439	260.2	-16	-87.1	6.1	-17.5	-85	6.1
7034	287.8184	1.1018	2013.5345	15.464	16.588	47.395	260.5	76.1	58.2	6.9	64.4	65.1	6.8
7035	287.8340	25.5955	2013.3575	13.143	13.484	56.598	146.5	50.6	-76.2	6.4	58.5	-74.3	6.2
7036	287.8597	11.8470	2013.878	16.832	17.132	57.952	210.4	71.3	82.4	7.5	79.6	75.5	5.8
7037	287.8739	-11.8009	2014.2625	10.213	13.304	6.598	34.3	95.2	-2.7	5.3	98.4	-0.8	5.3

7038	287.8840	38.6020	2013.4765	11.302	17.148	23.254	295.9	-6.1	-62.9	5.3	-7.8	-61.2	5.4
7039	287.9251	0.6090	2013.319	16.004	17.882	49.139	146.6	-19.1	-83.9	6	-31.4	-81.5	6.8
7040	287.9718	22.7720	2013.4315	13.265	13.411	16.176	162.4	10.2	-186.6	5.5	9	-180.8	5.2
7041	287.9790	60.9991	2013.487	11.546	13.456	4.955	320.3	82.8	27.8	5.7	82.7	24.7	5.7
7042	287.9820	24.0942	2013.451	12.326	12.675	47.957	260.5	-19.6	-85.8	5.6	-21.7	-86.1	5
7043	288.0009	8.0367	2013.041	16.978	19.487	57.2	107.6	-213.1	-123.4	7.6	-223	-129	7.5
7044	288.0057	9.4086	2013.1785	18.176	19.398	45.207	16.4	-21.8	-66.3	6.4	-15.6	-66.2	7.7
7045	288.0465	-0.8918	2013.809	11.868	13.61	5.113	314	6.2	-82.2	5.4	6.5	-84.8	5.5
7046	288.0990	55.7896	2013.327	17.537	19.294	49.049	220.7	-0.5	-61.2	6.2	6.9	-62.4	7.2
7047	288.1118	28.3251	2013.3135	10.73	11.709	8.679	311.8	106.7	58.9	5.2	106.3	58.8	5.2
7048	288.1619	30.5799	2013.3265	15.088	15.563	6.944	50.9	76.4	-8	5.3	80.4	-2.5	5.4
7049	288.2046	7.3693	2013.6945	14.726	19.071	10.886	1.9	-24.8	-60.8	5.6	-17.8	-64.3	6.6
7050	288.2059	7.4976	2012.9625	17.795	18.372	12.59	332	-15.3	-67.7	6.1	-13.9	-67	10.1
7051	288.2060	7.5071	2013.767	14.371	16.98	5.656	199.8	-14.3	-60.5	5.6	-22.4	-62.9	5.9
7052	288.2064	7.5213	2013.683	15.738	17.065	6.819	94.4	-16.6	-62.5	5.7	-18.7	-60.5	5.8
7053	288.2067	7.4920	2013.3665	17.245	18.137	6.463	199	-25.6	-72.3	6	-29.1	-66.3	6.1
7054	288.2069	7.5320	2013.7995	16.167	17.14	6.064	239.1	-18	-67.3	5.7	-23.8	-70.7	5.7
7055	288.2075	7.5665	2014.055	18.58	19.576	18.83	59.8	-13.6	-60.1	6.1	-9.1	-67.8	7.1
7056	288.2081	7.3741	2013.694	18.362	19.071	13.642	240.9	-17.8	-71.1	5.9	-17.8	-64.3	6.6
7057	288.2083	7.3858	2013.2415	17.622	19.666	8.568	236.4	-29.3	-68.1	6	-41.8	-63.7	8
7058	288.2087	7.4853	2013.6415	17.161	17.072	14.571	31.3	-25.1	-66.8	5.9	-22.9	-62.3	5.8
7059	288.2095	7.5085	2013.543	14.889	18.815	11.138	197.3	-18.7	-79.7	5.6	-9.8	-76.5	8.1
7060	288.2109	7.4887	2013.6415	17.072	17.161	14.571	211.3	-22.9	-62.3	5.8	-25.1	-66.8	5.9
7061	288.2119	7.5867	2013.4215	17.942	18.912	12.32	54.1	-19.2	-65.3	5.9	-16.6	-74.1	6.1
7062	288.2129	7.4806	2013.685	16.466	18.052	8.123	222.6	-24.3	-61.1	5.7	-21	-68.1	5.9
7063	288.2132	7.5322	2013.741	16.093	17.319	14.514	331.1	-21.6	-71.7	5.7	-18.5	-65.5	5.8
7064	288.2139	7.4674	2013.604	15.409	17.72	27.706	309.1	-26.6	-70.5	5.7	-28.5	-64.3	5.9
7065	288.2147	7.5887	2013.286	18.912	19.022	7.069	30.5	-16.6	-74.1	6.1	-18.5	-68.1	7.2
7066	288.2154	7.5236	2013.3805	17.57	18.96	20.862	123.2	-26.6	-78.4	6.9	-32.6	-66.5	6.8
7067	288.2154	7.5857	2013.796	17.091	18.492	7.686	210.7	-20.4	-63.4	5.7	-10.6	-65.7	5.8
7068	288.2158	7.4232	2013.55	17.403	17.809	8.202	256	-25.6	-62	6	-27.4	-71.5	6
7069	288.2162	7.5822	2013.6725	18.249	18.492	8.875	310.6	-8	-67.3	6	-10.6	-65.7	5.8
7070	288.2163	7.5016	2013.0325	17.725	19.628	17.726	136.1	-11.1	-70.1	5.9	-21.3	-63.7	11.5
7071	288.2167	7.5533	2013.001	17.578	18.935	11.719	230.7	-18.7	-66.7	5.8	-19.7	-68	11.1
7072	288.2177	7.5327	2013.898	16.829	15.701	7.221	122.2	-14.4	-69.1	5.7	-20.1	-76.3	5.6
7073	288.2178	7.4853	2013.742	17.498	18.363	9.039	176.5	-15.9	-71.6	5.8	-23.6	-65.4	6.1
7074	288.2179	7.3700	2013.7745	15.483	17.761	11.939	60.5	-33.7	-62.1	5.6	-24.4	-68.8	5.9
7075	288.2184	7.5301	2013.8025	17.074	15.701	6.294	32	-19.7	-66.9	5.8	-20.1	-76.3	5.6
7076	288.2188	7.5815	2013.543	17.032	18.249	9.714	285.7	-12	-65.2	5.8	-8	-67.3	6
7077	288.2188	7.4136	2013.6325	15.639	17.969	12.259	34.7	-15.3	-64.5	5.7	-25.5	-60.5	6.1
7078	288.2194	7.5316	2013.8025	15.701	17.074	6.294	212	-20.1	-76.3	5.6	-19.7	-66.9	5.8
7079	288.2199	7.5885	2013.418	16.536	19.022	16.37	294.6	-14.5	-68	5.8	-18.5	-68.1	7.2
7080	288.2205	7.5742	2013.3085	18.298	18.691	10.322	63.3	-11.6	-61	5.8	-21.2	-67.5	7.5
7081	288.2213	7.5059	2013.3655	17.85	18.582	6.709	310.8	-20.6	-60.3	6.1	-28.9	-68.8	6.5
7082	288.2221	7.5445	2012.895	17.221	18.621	18.873	167.1	-16.9	-66.2	5.9	-11.4	-60.9	9.5
7083	288.2223	7.5137	2013.848	15.891	16.47	10.181	225.4	-25.8	-73.8	5.7	-35	-82.8	5.7

7084	288.2236	7.4197	2013.715	16.226	18.561	10.109	33.7	-20.7	-64.5	5.7	-13.6	-64.2	6.2
7085	288.2239	7.4707	2013.853	16.302	16.905	8.142	258.9	-24.3	-62	5.6	-22.2	-67.8	5.7
7086	288.2249	7.4117	2013.5515	17.993	18.61	13.225	177.1	-15.6	-66.5	5.9	-14.8	-66.2	6.5
7087	288.2252	7.4003	2013.669	17.412	17.854	9.946	269.8	-22.9	-65	5.7	-23.3	-62.6	5.9
7088	288.2257	7.5712	2013.6235	17.268	15.156	5.758	15.4	-19.9	-70.3	5.8	-13.9	-62.9	5.7
7089	288.2261	7.5727	2013.6235	15.156	17.268	5.758	195.4	-13.9	-62.9	5.7	-19.9	-70.3	5.8
7090	288.2267	7.5087	2013.057	18.108	19.428	12.308	291.9	-27.4	-72.6	6.3	-16.3	-80.1	10.6
7091	288.2273	7.4064	2013.589	17.446	18.61	9.779	305.7	-2.8	-69.7	5.8	-14.8	-66.2	6.5
7092	288.2275	7.3824	2013.701	17.521	18.453	14.869	315	-28.7	-60.3	5.8	-29.9	-61.9	7.1
7093	288.2280	7.4895	2013.734	16.107	18.027	7.86	132.7	-23	-63.9	5.7	-18.7	-65.6	6
7094	288.2282	7.4733	2013.6845	17.57	17.75	14.042	289.6	-29.2	-71.1	6.1	-20.2	-63.5	6
7095	288.2288	7.4114	2013.5795	17.904	17.993	13.93	274.4	-22.8	-72.1	5.8	-15.6	-66.5	5.9
7096	288.2295	7.3764	2013.3235	16.429	19.417	19.317	73.5	-25.5	-66.4	5.7	-21.5	-65.6	7.3
7097	288.2301	7.5097	2013.7555	16.206	18.108	12.623	253.8	-17.3	-64.7	5.7	-27.4	-72.6	6.3
7098	288.2303	7.4897	2013.6845	17.087	18.027	6.426	203.5	-21.6	-66.1	5.8	-18.7	-65.6	6
7099	288.2305	7.4633	2013.333	18.278	18.742	17.868	142.3	-21.6	-65.3	6.4	-26.1	-70.1	7.9
7100	288.2306	7.5695	2013.6725	17.12	18.859	9.219	328	-15.5	-61.5	5.7	-20.6	-62.3	6
7101	288.2307	7.4056	2013.733	15.182	16.231	9.056	43.8	-14.3	-64.5	5.7	-18.6	-69.6	5.7
7102	288.2315	7.5633	2013.5735	17.276	16.56	6.856	23.3	-14.9	-62.9	5.9	-17.8	-62.4	5.8
7103	288.2316	7.5582	2013.5935	12.188	17.276	18.25	358.4	-17.4	-68.8	5.7	-14.9	-62.9	5.9
7104	288.2322	7.5222	2014.1515	15.783	19.174	11.389	210.9	-16.7	-73.7	5.6	-18.7	-67.4	8.6
7105	288.2323	7.5650	2013.5735	16.56	17.276	6.856	203.3	-17.8	-62.4	5.8	-14.9	-62.9	5.9
7106	288.2325	7.4074	2013.82	16.231	17.458	5.593	94.7	-18.6	-69.6	5.7	-18.5	-66.7	5.7
7107	288.2334	7.4745	2013.4995	17.885	17.803	9.556	84.5	-25.5	-69.8	6.1	-18.1	-68.7	5.9
7108	288.2345	7.5448	2013.014	17.357	18.972	5.766	60.9	-19.3	-62.3	5.8	-19.5	-66.9	7.6
7109	288.2352	7.3878	2013.671	15.096	16.08	4.022	239	-24.1	-61.7	5.6	-18.8	-61.8	7.5
7110	288.2359	7.4639	2013.155	17.567	18.742	18.189	207.4	-14.3	-71	5.9	-26.1	-70.1	7.9
7111	288.2360	7.5169	2013.835	15.616	17.609	12.86	192.6	-21	-64.5	5.7	-30.8	-71	5.8
7112	288.2361	7.4747	2013.4995	17.803	17.885	9.556	264.5	-18.1	-68.7	5.9	-25.5	-69.8	6.1
7113	288.2367	7.5553	2013.9885	17.512		8.256	16.5	-13	-68.4	5.8	0.9	-66	12.1
7114	288.2367	7.5209	2013.478	17.233	18.356	12.374	65.4	-21	-60.9	5.8	-32.9	-65.8	6.9
7115	288.2369	7.5248	2013.4355	18.196	18.356	13.779	130.7	-23.6	-66.3	6	-32.9	-65.8	6.9
7116	288.2390	7.5119	2013.6945	16.044	18.359	8.756	82.8	-24.4	-63.1	5.7	-21.4	-68.3	6.9
7117	288.2394	7.4237	2013.6965	16.611	17.587	12.569	205	-15.3	-60.4	5.7	-25.8	-60.2	5.8
7118	288.2398	7.4761	2013.704	14.566	17.803	14.227	249.1	-21.9	-67	5.6	-18.1	-68.7	5.9
7119	288.2404	7.4122	2013.8345	18.545	19.363	10.169	142.8	-10.8	-73.5	6.2	-20	-75.5	6.1
7120	288.2416	7.5473	2013.028	14.171	18.972	21.096	253	-14.5	-61.2	5.7	-19.5	-66.9	7.6
7121	288.2416	7.4548	2013.453	16.132	18.25	27.901	268.8	-11.7	-60.5	5.7	-0.7	-66.3	7.9
7122	288.2419	7.5211	2013.525	17.401	18.356	8.474	300.7	-34.6	-68.1	5.7	-32.9	-65.8	6.9
7123	288.2423	7.4815	2013.494	16.113	17.771	7.065	234.3	-18.8	-65.8	5.8	-16.1	-61.2	6
7124	288.2430	7.5188	2013.588	16.812	18.386	5.742	39.4	-28.6	-64.4	5.7	-34.2	-75.9	6.5
7125	288.2437	7.4691	2013.9145	15.721	13.078	5.42	155.3	-23.9	-64.9	5.8	-22	-68.5	5.7
7126	288.2438	7.5628	2013.3	16.491	17.54	5.549	239.9	-12	-65.8	5.9	-8.6	-75.9	6
7127	288.2438	7.3734	2013.0765	18.42	18.488	8.369	169.2	-19	-65.8	7.8	-22.4	-68.4	8.1
7128	288.2442	7.3711	2013.0765	18.488	18.42	8.369	349.2	-22.4	-68.4	8.1	-19	-65.8	7.8
7129	288.2444	7.4678	2013.9145	13.078	15.721	5.42	335.3	-22	-68.5	5.7	-23.9	-64.9	5.8

7130	288.2450	7.4788	2013.7515	14.839	16.113	13.624	315.2	-22.7	-61.3	5.6	-18.8	-65.8	5.8
7131	288.2452	7.4060	2013.7835	16.867	18.394	8.662	64.3	-15.8	-71.5	5.7	-20.2	-73	6.1
7132	288.2453	7.5110	2013.726	18.297	18.359	14.425	288.2	-18.6	-69.9	5.9	-21.4	-68.3	6.9
7133	288.2462	7.5752	2013.7515	16.588	18.665	8.351	152.4	-15.1	-60.4	5.7	-20.9	-60.9	5.9
7134	288.2464	7.3946	2013.5505	17.125	18.031	14.129	12	-20.8	-68.5	5.8	-14	-72.8	7.4
7135	288.2466	7.4144	2013.956	14.383	14.79	4.004	60.3	-23.8	-66.3	5.7	-26.6	-65.6	5.8
7136	288.2469	7.4025	2013.755	17.972	18.032	9.491	102.1	-20	-69.5	5.8	-22.3	-61.7	6.2
7137	288.2470	7.5300	2013.4135	19.103	19.316	14.054	18.9	-26.1	-63.4	7	-33	-72.6	6.7
7138	288.2483	7.5240	2013.573	16.656	18.629	17.816	137.7	-22.1	-69.2	5.7	-16.9	-72.5	8.9
7139	288.2484	7.4496	2013.626	17.524	17.809	21.171	63.6	-15.3	-60.2	6.1	-26.6	-60.6	5.8
7140	288.2489	7.4776	2013.709	17.67	17.712	4.226	65.7	-19.6	-64.1	5.9	-29.9	-69.7	6.2
7141	288.2491	7.5471	2013.394	17.605	18.877	6.034	25.4	-11.1	-63.3	5.9	-23.5	-64.9	6.6
7142	288.2495	7.5668	2013.3365	18.657	18.665	24.175	340.5	-12.9	-61	6.7	-20.9	-60.9	5.9
7143	288.2499	7.4060	2013.8055	18.271	18.394	9.832	293	-21.8	-76.7	6.4	-20.2	-73	6.1
7144	288.2505	7.4113	2013.6115	16.636	17.976	9.666	52.9	-22.5	-79.7	5.7	-18.8	-74.9	6
7145	288.2506	7.5325	2013.6155	15.891	19.316	9.094	298.1	-26.8	-68.4	5.7	-33	-72.6	6.7
7146	288.2506	7.5947	2013.202	17.553	18.165	12.401	31.7	-8.4	-61.3	5.8	-4.3	-61.3	7.8
7147	288.2507	7.5156	2013.7365	11.807	17.649	11.024	205.7	-23.3	-65.8	5.6	-20.6	-63.2	6
7148	288.2509	7.4158	2013.633	15.781	18.378	6.078	205.7	-27.6	-62.8	5.8	-29.7	-69.5	6
7149	288.2510	7.4700	2013.5885	16.523	18.066	6.749	337.4	-25.3	-67	5.7	-34.9	-67.3	6.2
7150	288.2511	7.4029	2013.674	17.969	18.032	6.906	237.7	-13.7	-62.9	5.9	-22.3	-61.7	6.2
7151	288.2524	7.4899	2013.4905	16.392	18.574	6.832	302.4	-23.1	-72.3	5.7	-23.5	-76.5	6.6
7152	288.2533	7.4108	2013.6565	16.323	17.976	8.074	344.8	-23.3	-72.1	5.7	-18.8	-74.9	6
7153	288.2535	7.5070	2013.699	17.355	17.588	9.396	277.2	-24.9	-61.6	5.8	-28.9	-68.5	5.9
7154	288.2544	7.5479	2013.2055	18.341	18.877	16.299	278.3	-19.4	-60.9	5.9	-23.5	-64.9	6.6
7155	288.2554	7.4693	2013.656	16.804	17.503	11.497	153.1	-30.1	-62.7	5.7	-25.4	-63.8	5.8
7156	288.2555	7.4871	2013.7995	14.258	17.101	14.849	38.5	-26.1	-62.1	5.6	-29.3	-71.8	5.7
7157	288.2566	7.4033	2013.6885	18.507	18.521	11.205	336.3	-23	-63.1	6	-19.3	-74.6	6.4
7158	288.2584	7.5075	2013.9435	18.153	18.255	8.343	70.6	-17.8	-68	6	-21.7	-69.3	6.1
7159	288.2589	7.4839	2013.642	16.478	17.553	12.479	223.6	-29	-61.5	5.7	-30.4	-70.2	6
7160	288.2592	7.4182	2013.8115	16.075	16.474	10.644	188.4	-22.8	-63.9	5.7	-22.8	-65.2	5.7
7161	288.2599	7.4099	2013.812	14.523	15.936	11.642	94.7	-20.1	-65.1	5.7	-21.2	-74.4	5.7
7162	288.2605	7.4785	2013.555	17.538	17.061	8.136	30.2	-27.2	-60.1	6	-36.2	-64.2	5.7
7163	288.2615	7.4682	2013.3505	17.502	17.626	8.194	118.2	-24.2	-69.4	6.4	-23.1	-66.3	5.8
7164	288.2617	7.4804	2013.555	17.061	17.538	8.136	210.2	-36.2	-64.2	5.7	-27.2	-60.1	6
7165	288.2626	7.5128	2013.743	15.641	17.546	6.383	259	-21.3	-70	5.7	-23.7	-63	5.8
7166	288.2634	7.5625	2013.237	17.516	18.34	24.197	33.2	-21.3	-61.2	5.9	-14.4	-68.2	6.4
7167	288.2646	7.5452	2013.506	17.538	18.297	19.369	23.9	-17.2	-60.8	6	-20.4	-60.9	5.9
7168	288.2647	7.4645	2013.439	18.151	18.828	10.42	39.6	-30.5	-69.7	6.7	-24.5	-69.5	6.1
7169	288.2652	7.4161	2013.735	17.415	17.675	15.638	135.5	-19.6	-64.4	5.8	-16.8	-66.2	5.9
7170	288.2659	7.5219	2013.7995	16.181	17.116	16.126	319.4	-24.6	-64	5.7	-22.8	-60.7	5.7
7171	288.2660	7.4924	2013.3315	16.451	18.165	17.313	127.6	-25.5	-65.8	5.7	-19.7	-74.6	7.3
7172	288.2664	7.3989	2013.8245	15.798	18.207	10.399	165	-17.7	-73.7	5.7	-6.8	-62	8.1
7173	288.2671	7.3961	2013.7655	18.207	18.644	7.756	190.5	-6.8	-62	8.1	-15.3	-60.1	6.2
7174	288.2685	7.5036	2013.6505	16.785	18.638	7.229	52.2	-26.7	-61.7	5.9	-15	-68.3	6.1
7175	288.2692	7.3880	2013.379	17.476	19.101	6.64	255.7	-26.4	-70.7	5.8	-39.5	-68	8.6

7176	288.2693	7.3906	2013.73	17.181	17.476	9.174	182	-20.6	-66.6	5.8	-26.4	-70.7	5.8
7177	288.2701	7.4105	2013.6655	16.983	17.948	8.771	67.6	-15	-74.7	5.7	-19	-70.4	5.9
7178	288.2706	7.4999	2013.6385	18.174	18.321	6.471	132.1	-28.5	-67	6.2	-20.6	-69.4	6
7179	288.2711	7.4450	2013.732	14.255	17.697	8.957	31.8	-22.7	-64.6	5.7	-23.2	-69.4	5.8
7180	288.2722	7.4540	2013.6195	17.636	18.016	6.125	241.8	-25.6	-62.9	5.9	-26.4	-71.6	6.1
7181	288.2725	7.4516	2013.7385	15.013	17.636	8.835	352	-26.2	-72.5	5.7	-25.6	-62.9	5.9
7182	288.2727	7.4144	2013.721	17.41	17.948	10.717	187	-21.5	-67.9	5.7	-19	-70.4	5.9
7183	288.2729	7.4702	2013.963	15.331	16.285	4.847	53	-27.9	-63.8	5.7	-26	-62.5	5.6
7184	288.2729	7.5263	2013.7765	17.092	17.911	9.687	77.8	-28.9	-66.2	5.8	-33.7	-73.5	5.8
7185	288.2729	7.4572	2013.839	16.023	17.127	6.274	50.4	-24.9	-68.6	5.7	-22.5	-68.7	5.7
7186	288.2731	7.4900	2013.866	16.759	18.797	7.937	175.6	-25	-61.5	5.7	-24.7	-67	6.3
7187	288.2733	7.4841	2013.866	18.018	18.797	13.058	0.4	-21.8	-63.8	5.8	-24.7	-67	6.3
7188	288.2735	7.4749	2013.5905	17.016	18.234	5.677	306.2	-20.4	-67.9	5.7	-18.7	-65.6	6.7
7189	288.2737	7.4491	2013.7625	16.42	17.697	8.29	212.3	-26.2	-63.5	5.7	-23.2	-69.4	5.8
7190	288.2739	7.4710	2013.963	16.285	15.331	4.847	233	-26	-62.5	5.6	-27.9	-63.8	5.7
7191	288.2750	7.3793	2013.645	16.149	18.885	8.929	271.2	-16.2	-63.8	5.7	0.8	-61.1	11.8
7192	288.2751	7.4944	2013.5445	17.909	19.281	14.342	280.2	-27.3	-66.3	5.9	-32.4	-65.7	8.1
7193	288.2757	7.4209	2013.78	17.34	17.41	25.696	204.2	-18	-62.9	5.9	-21.5	-67.9	5.7
7194	288.2765	7.3944	2013.6685	17.921	18.514	6.202	219.3	-19.1	-68.2	6.4	-20.5	-64.2	5.9
7195	288.2770	7.4989	2013.686	16.745	16.766	8.462	70.8	-25.7	-66	5.8	-22.1	-66.7	5.8
7196	288.2774	7.5135	2013.664	17.109	17.367	6.171	190.1	-20.5	-68.5	5.9	-16.8	-78.6	5.9
7197	288.2785	7.4634	2013.8235	16.985	17.612	7.154	194.4	-25.7	-64	5.7	-27.3	-73.3	5.7
7198	288.2785	7.4884	2013.681	14.406	17.504	8.955	289.7	-27.1	-62.9	5.6	-25.5	-64.1	5.9
7199	288.2785	7.4758	2013.8055	17.046	18.076	16.863	197.7	-22.9	-60.1	5.6	-24.1	-68.3	6
7200	288.2791	7.4544	2013.0785	18.139	18.794	8.296	217.5	-27.1	-71.2	6	-26.6	-79.6	11.6
7201	288.2792	7.4996	2013.686	16.766	16.745	8.462	250.8	-22.1	-66.7	5.8	-25.7	-66	5.8
7202	288.2793	7.5307	2013.7525	16.688	17.335	16.254	82.3	-15	-61.1	5.7	-19.7	-62.6	5.9
7203	288.2798	7.4854	2013.8125	14.834	16.766	9.037	114.8	-20.3	-63.1	5.7	-25.5	-64.8	5.7
7204	288.2804	7.5208	2013.6985	16.527	16.605	6.727	142.1	-21.4	-65.1	5.7	-16.2	-65.7	5.8
7205	288.2805	7.4510	2012.8905	17.918	18.794	11.451	300	-31	-70.3	5.9	-26.6	-79.6	11.6
7206	288.2806	7.5102	2013.465	18.073	18.579	5.808	252.3	-23.5	-68.5	6.1	-20.2	-69.3	6.4
7207	288.2808	7.5263	2013.1885	17.385	19.413	15.563	162.4	-16.6	-67.5	5.9	-26.3	-71.6	8.4
7208	288.2815	7.5193	2013.6985	16.605	16.527	6.727	322.1	-16.2	-65.7	5.8	-21.4	-65.1	5.7
7209	288.2821	7.4843	2013.846	16.766	17.366	7.923	205.5	-25.5	-64.8	5.7	-19	-63.5	5.8
7210	288.2826	7.3981	2013.7795	17.563	16.337	9.908	65.7	-17.6	-75.8	5.8	-13.9	-73.2	5.7
7211	288.2827	7.4557	2013.853	13.957	18.139	13.415	249.6	-26.3	-68.5	5.6	-27.1	-71.2	6
7212	288.2832	7.4444	2013.887	17.797	18.152	7.318	314.7	-13.9	-74.9	5.7	-15.6	-68.5	6.1
7213	288.2835	7.4019	2013.926	17.549	18.976	6.565	82.5	-14.9	-68.1	5.8	-11.2	-71.1	6.4
7214	288.2839	7.4859	2013.7285	17.989	16.766	8.726	228.5	-28.7	-66.7	5.9	-25.5	-64.8	5.7
7215	288.2843	7.3874	2013.8585	18.126	18.423	9.785	81.5	-5.8	-70.7	6.1	-15.2	-64.4	6.2
7216	288.2851	20.5757	2013.7265	9.689	11.542	13.638	323.2	-4.7	-100.8	5	-4	-102	5
7217	288.2851	7.3993	2013.7795	16.337	17.563	9.908	245.7	-13.9	-73.2	5.7	-17.6	-75.8	5.8
7218	288.2854	7.3743	2013.721	14.424	16.975	7.278	217.5	-27.5	-72.8	5.6	-17.6	-70.5	5.8
7219	288.2880	7.5250	2013.675	15.578	16.962	4.912	144.4	-18.4	-66.6	5.7	-18.3	-60.6	5.8
7220	288.2887	7.3761	2013.5515	17.724	18.696	11.698	303.6	-25	-62.6	6	-29	-60.9	6.3
7221	288.2896	7.4695	2013.8005	16.287	16.752	10.996	82.6	-28.2	-65.7	5.7	-28.5	-68.3	5.7

7222	288.2898	7.4077	2013.775	17.341	17.755	6.573	158.8	-4.3	-61.6	5.7	-14.4	-68.6	5.8
7223	288.2900	7.4834	2013.827	16.78	17.413	8.159	129.2	-27.3	-68	5.7	-35.3	-62.3	5.8
7224	288.2907	7.4962	2013.6185	16.859	18.454	6.895	195.2	-29.8	-73.3	5.8	-33.8	-63.1	6.6
7225	288.2910	7.4877	2013.8335	15.164	16.78	16.06	192.3	-30.6	-63.8	5.6	-27.3	-68	5.7
7226	288.2910	7.4047	2013.896	17.108	17.755	5.186	336.4	-12.9	-73.8	5.7	-14.4	-68.6	5.8
7227	288.2911	7.5062	2013.8415	16.614	18.116	9.361	20.5	-23	-62.1	5.7	-30.2	-69.1	6
7228	288.2914	7.5111	2013.44	18.592	19.889	7.826	275	-20.8	-72.2	6.6	-11.8	-66.1	8.3
7229	288.2920	7.5086	2013.6595	18.116	18.592	9.061	346.6	-30.2	-69.1	6	-20.8	-72.2	6.6
7230	288.2924	7.3776	2013.89	17.31	18.24	6.13	29.7	-17.4	-70.9	5.7	-20.2	-63.7	6.7
7231	288.2926	7.4344	2013.2795	17.485	18.705	20.688	224.2	-27.3	-66.2	5.9	-20.4	-69	7.8
7232	288.2927	7.4699	2013.8455	16.752	18.514	6.871	155.2	-28.5	-68.3	5.7	-24.3	-70.1	6.9
7233	288.2927	7.4516	2013.6645	17.825	17.44	15.855	101.3	-25.4	-73.1	5.8	-27.4	-80.5	5.9
7234	288.2934	7.5267	2013.747	17.843	17.936	15.557	264.8	-23.7	-61.8	5.9	-21.9	-67.2	6
7235	288.2936	7.3849	2013.3455	15.83	16.544	4.124	78.8	-17.8	-65.5	5.7	-17	-65.4	13
7236	288.2938	7.4643	2013.7225	15.173	16.729	4.645	46.5	-28.7	-69.2	5.7	-32.6	-63.9	6.3
7237	288.2943	7.4836	2013.845	15.988	17.413	10.755	235.9	-24.6	-60.7	5.7	-35.3	-62.3	5.8
7238	288.2967	7.3779	2013.9115	12.366	18.24	12.96	288.2	-14.2	-62.7	5.6	-20.2	-63.7	6.7
7239	288.2968	7.5206	2013.9245	16.404	16.927	4.155	340.6	-26.3	-65	5.7	-23.3	-67.1	5.8
7240	288.2970	7.4508	2013.6645	17.44	17.825	15.855	281.3	-27.4	-80.5	5.9	-25.4	-73.1	5.8
7241	288.2976	7.4997	2013.654	17.424	18.336	9.332	195.3	-23.1	-66.1	5.7	-30.9	-76.2	7.1
7242	288.2978	7.5275	2013.678	17.27	17.843	16.069	259.7	-18.6	-64.4	5.8	-23.7	-61.8	5.9
7243	288.2984	7.5210	2013.5405	17.29	17.625	5.738	125.3	-28.9	-70.2	6	-31.9	-67.2	6
7244	288.2987	7.4938	2013.6725	16.666	16.982	11.089	157.2	-12.5	-61	7	-22.5	-61.8	5.8
7245	288.2988	7.3726	2013.6715	17.436	18.312	5.484	148.3	-22.7	-62.1	5.8	-14.8	-66.1	6.1
7246	288.2994	7.4551	2013.912	16.799	17.473	4.456	113.6	-22.3	-73.3	5.7	-27.9	-75.1	6
7247	288.2997	7.4019	2013.8105	17.956	17.021	7.508	87.4	-24.8	-70.6	6.2	-19.7	-65	5.8
7248	288.3006	7.4701	2013.6205	17.998	18.275	3.744	27.1	-30.3	-66.8	7.1	-24.5	-77.4	6.2
7249	288.3007	7.3584	2013.6135	16.833	18.276	25.19	117	-21.6	-69	5.7	-18	-63.4	7.1
7250	288.3011	7.4710	2013.6205	18.275	17.998	3.744	207.1	-24.5	-77.4	6.2	-30.3	-66.8	7.1
7251	288.3012	7.3712	2013.679	16.584	18.312	5.407	273.2	-19.1	-62.1	5.7	-14.8	-66.1	6.1
7252	288.3013	7.3681	2013.606	17.849	18.518	10.106	270.6	-8.7	-67.8	5.7	-19.6	-62.4	6.9
7253	288.3015	7.5081	2013.757	17.097	18.536	9.013	121	-21.9	-65.4	5.7	-32.8	-63.2	6.5
7254	288.3016	7.3866	2013.12	11.083	18.704	20.511	154.9	-12.8	-69.3	5.6	0.3	-81.3	9.2
7255	288.3018	7.4020	2013.8275	17.021	15.927	5.266	142.6	-19.7	-65	5.8	-12	-73.2	5.7
7256	288.3027	7.4008	2013.8275	15.927	17.021	5.266	322.6	-12	-73.2	5.7	-19.7	-65	5.8
7257	288.3027	7.4313	2013.4755	18.138	18.48	7.841	101.8	-12.8	-77.2	7	-20.9	-69.8	7.4
7258	288.3029	7.4534	2013.6895	18.219	17.622	8.678	54.4	-26.9	-71.1	6.1	-27.8	-76	5.9
7259	288.3047	7.4575	2013.926	16.295	16.666	4.201	359.1	-25	-72.7	5.6	-26	-67.8	5.7
7260	288.3047	7.4672	2013.8035	15.58	18.398	10.281	183.1	-33.2	-68.4	5.7	-29.4	-65.5	5.8
7261	288.3049	7.4548	2013.6895	17.622	18.219	8.678	234.4	-27.8	-76	5.9	-26.9	-71.1	6.1
7262	288.3054	10.6288	2014.0715	16.481	18.014	22.41	290.6	-78.3	103.9	6.1	-76.8	107.8	7.2
7263	288.3055	7.5119	2013.571	13.861	18.44	10.403	314.1	-30.3	-74.7	5.7	-32.9	-73.2	6.6
7264	288.3056	7.4332	2013.5935	18.201	18.48	8.758	197.6	-26.5	-61	6.1	-20.9	-69.8	7.4
7265	288.3062	7.3942	2013.543	17.356	17.851	4.907	91.7	-22.3	-66	5.9	-21	-73.9	6.1
7266	288.3069	7.4631	2013.8305	18.042	18.398	9.389	298.4	-29.7	-62.8	6.1	-29.4	-65.5	5.8
7267	288.3075	7.4600	2013.806	16.498	16.666	11	245	-27.8	-74.3	5.7	-26	-67.8	5.7

7268	288.3079	7.3709	2013.7305	15.665	17.967	19.327	282.5	-21.2	-70.3	5.7	-27.6	-62.7	5.8
7269	288.3088	7.3269	2013.7235	16.63	18.371	18.246	155.5	-11.9	-63.7	5.8	-10.4	-61.9	6
7270	288.3090	7.4516	2013.627	16.853	18.646	16.591	129.3	-26.8	-76	5.8	-21.7	-85.6	5.9
7271	288.3094	7.4048	2013.7865	15.864	18.624	21.475	234.6	-12	-63.7	5.7	-20.7	-74.7	6.2
7272	288.3102	7.3787	2014.0125	17.093	18.373	5.626	314.7	-16.3	-69.2	5.8	-15.2	-71.4	11.5
7273	288.3122	7.4242	2013.681	16.776	17.13	16.282	13.8	-14.5	-74.8	5.8	-13	-67.6	5.8
7274	288.3134	7.4748	2013.9045	14.859	16.75	4.476	236	-25.2	-67.2	5.7	-17	-67.4	6.4
7275	288.3140	7.4448	2013.6875	16.202	17.588	6.791	81.4	-35.7	-68.2	5.7	-27.6	-78.5	5.9
7276	288.3147	7.4538	2013.72	15.77	17.391	7.171	123.1	-29	-69.8	5.7	-18.5	-64.2	5.8
7277	288.3151	7.4113	2013.7235	16.716	17.235	8.113	45.4	-17.1	-62.9	5.8	-18.8	-61	5.9
7278	288.3164	7.4767	2014.011	15.422	16.242	8.436	224.2	-26.9	-63.6	5.7	-22.3	-60.3	5.8
7279	288.3167	7.3897	2013.527	17.102	18.032	10.848	155.4	-12.9	-73	6	-4.7	-68.5	6.1
7280	288.3175	7.4641	2013.723	16.835	17.272	4.699	215.6	-30.6	-66	5.8	-32.8	-70.3	5.9
7281	288.3176	7.5173	2013.6725	17.648	17.737	11.59	311.7	-32.6	-61.4	5.8	-29.2	-61.2	5.8
7282	288.3176	7.4379	2013.813	13.682	16.357	9.411	212.3	-22.2	-63.7	5.7	-26.4	-65.8	5.7
7283	288.3179	7.3777	2013.7575	16.15	16.838	12.903	46.2	-16.2	-63.4	5.7	-13.1	-71.9	5.7
7284	288.3180	7.4617	2013.767	16.544	17.272	6.892	317.1	-23.3	-67	5.7	-32.8	-70.3	5.9
7285	288.3194	7.4081	2013.927	16.42	16.716	19.181	307.6	-19.3	-63.5	5.7	-17.1	-62.9	5.8
7286	288.3198	7.4247	2013.815	16.584	16.836	8.636	338.2	-18.9	-63.9	5.7	-12.7	-67.7	5.7
7287	288.3203	7.3938	2013.7475	17.833	18.111	5.939	102.4	-8.1	-78.6	5.8	-18.6	-71.1	6.9
7288	288.3215	7.5085	2013.581	15.786	17.569	7.2	25.5	-23.9	-61.2	5.7	-33	-61	6.1
7289	288.3223	7.3815	2014.058	15.06	16.838	8.125	233.2	-19.1	-61.9	5.8	-13.1	-71.9	5.7
7290	288.3224	7.5104	2013.581	17.569	15.786	7.2	205.5	-33	-61	6.1	-23.9	-61.2	5.7
7291	288.3230	7.4947	2013.539	18.087	18.879	8.64	100.1	-18.4	-63.7	6.1	-23.9	-62	6.2
7292	288.3254	7.3930	2013.608	17.822	18.111	12.395	276.9	-20.9	-60.7	5.9	-18.6	-71.1	6.9
7293	288.3286	21.7203	2013.651	10.716	13.064	54.923	250.9	6.2	-62.5	5	1.5	-60.3	6.5
7294	288.3415	50.6560	2013.5005	15.793	15.885	8.595	263.5	-25.5	-87	6.1	-26.1	-84.7	6.1
7295	288.3607	80.2170	2013.6205	15.512	16.57	9.424	68	2.7	198.6	5.8	0.4	202.4	5.9
7296	288.3708	16.0310	2013.6275	15.839	17.693	35.598	239.4	-109	123.2	5	-112.8	127.6	5.3
7297	288.4136	13.6025	2013.88	11.384	14.473	17.772	332.4	16.4	-173	5	8.4	-170.4	5
7298	288.4461	7.0627	2013.7335	18.327	19.17	32.313	316.3	3.1	-70.7	6.1	-2.3	-61.5	6.3
7299	288.4861	14.9788	2013.969	16.384	18.158	30.706	352.9	47.9	-66.2	6.1	38.2	-76.8	7
7300	288.4863	6.2605	2013.554	15.521	17.487	59.503	352.8	-1.3	-64.5	5.7	-6.7	-64.2	6.1
7301	288.4925	4.9942	2013.6115	16.839	17.79	19.784	330.5	70.8	-59	5.8	69.2	-56.8	6.5
7302	288.5601	36.9927	2013.5235	10.749	12.673	7.01	339.9	16.2	-65	5.3	16.7	-66.5	5.3
7303	288.5612	8.4092	2013.4025	16.98	18.027	55.062	242.8	110.2	-60.9	5.7	116.7	-56.2	6.2
7304	288.5623	24.9199	2013.4615	8.068	9.955	4.066	259.1	112.7	3.8	6.6	115.6	7.3	5.1
7305	288.5629	15.8359	2013.168	17.04	17.641	36.835	296.6	70.6	-45.8	6.7	72.9	-43.1	7.4
7306	288.5947	6.5514	2013.781	12.729	15.176	9.822	110.8	-2.6	-82.6	5.7	-2.6	-83.8	5.7
7307	288.6112	6.1014	2014.015	16.669	17.483	35.387	349.9	-37.5	-109.4	6.2	-46.6	-108.5	6.1
7308	288.6112	15.5301	2014.0095	15.876	17.859	54.742	26.4	84.7	-80.5	6.9	71.8	-93	6.2
7309	288.6665	37.2891	2013.4365	11.973	15.226	47.806	64.8	28.6	-102.9	5.4	27.4	-100.9	5.4
7310	288.6669	10.1636	2013.744	18.281	19.046	58.569	156	-98.2	71.9	5.9	-98	69.2	6.6
7311	288.6843	12.7072	2013.727	18.103	18.496	6.649	72.3	74.1	23.4	7.6	78.6	16.9	7.1
7312	288.7066	23.7621	2013.543	11.226	15.063	17.267	265.4	-28.6	-130.5	5	-23	-132.2	5.3
7313	288.7363	75.3828	2013.451	16.101	16.709	16.905	276.5	10.3	91.1	6.1	14.9	92.4	6.1

7314	288.7536	16.2447	2013.8285	16.511	17.467	53.131	59.7	-23.7	87.5	6.3	-13.4	91.6	6.2
7315	288.7933	17.1581	2013.7305	10.909	15.498	25.128	139.1	37.7	-91.1	6	38.4	-96	6
7316	288.7990	8.3961	2013.4985	17.254	17.574	58.363	348.6	-59.6	-101.1	6	-65.9	-99.8	6.1
7317	288.8043	-8.4105	2014.084	11.128	13.218	6.27	265.3	-59.2	-130	5.6	-59.2	-130.8	5.6
7318	288.8313	12.9606	2013.799	15.702	16.907	43.494	24.9	16.3	-88.8	6	20.1	-88.9	6
7319	288.8687	10.6138	2013.9365	17.676	18.507	29.588	53.8	84.9	101.6	5.7	76.1	113.4	6.3
7320	288.8853	13.8838	2013.836	16.639	16.871	45.746	5.5	68.8	-15.8	6	68	-20.8	6.1
7321	288.9564	13.0320	2013.8735	17.059	17.553	47.499	301.4	0.5	-109.7	6.5	10.3	-104.1	7.5
7322	288.9685	8.8045	2013.5245	17.07	17.758	5.881	220.3	189.1	114.5	5.7	184.9	125	6.3
7323	288.9697	13.6492	2014.104	16.957	17.839	14.313	264.6	-106.3	-165	6	-103.9	-172.3	5.9
7324	289.0057	-4.4906	2013.5795	15.07	17.693	44.772	321.5	-72.7	-63.6	5.5	-60.5	-71.6	7.1
7325	289.0243	14.8265	2013.709	16.957	17.22	47.498	47.2	111.9	0.1	7.1	113.8	3.9	5.9
7326	289.0439	17.6529	2013.404	16.791	17.588	12.478	77.6	46.4	-68.9	6.8	59	-62	6.6
7327	289.0541	15.0661	2013.7605	16.895	17.596	17.328	79.1	65.9	-67.8	6.3	67.9	-62.3	6
7328	289.0603	15.5584	2013.9745	9.186	10.248	9.935	230	-18.1	-70.6	5.7	-16.9	-74.1	5.7
7329	289.0858	37.8909	2013.5285	13.653	14.164	10.908	74.4	66.3	172.9	5.4	65.3	169.3	5.4
7330	289.1510	29.2689	2013.4485	13.177	13.55	10.47	354.3	-114.4	-73.4	5.1	-117.6	-73.4	5.3
7331	289.1730	13.5045	2013.729	17.374	18.053	43.767	210.9	130.5	-136.8	6.6	140.8	-123.6	7.2
7332	289.1977	23.8982	2012.9965	15.641	16.545	33.857	220.9	-11.7	-72.6	5.3	-8.9	-73.7	5.3
7333	289.2008	32.2122	2013.456	17.511	18.571	34.273	271.3	-9.8	-63.3	5.3	-8.6	-64.7	5.9
7334	289.2208	37.2370	2013.4695	12.348	15.967	28.519	177.7	-33	88.7	5.3	-35.9	87.6	5.7
7335	289.2356	15.5798	2014.0665	17.751	19.148	56.35	194.5	-75.9	-96.8	7.1	-70.9	-105.2	7.2
7336	289.2997	14.2246	2013.935	17.837	18.172	16.89	250.6	206.6	-221.4	7.9	206	-220.6	6.2
7337	289.3096	16.4033	2013.869	10.358	15.03	14.837	270.2	75.1	-30	6	77.2	-34.1	6.1
7338	289.3322	4.8503	2013.3065	16.714	18.352	54.262	178.4	182.5	205	6.2	192.5	209.5	7.5
7339	289.3385	15.6055	2014.2395	16.515	16.521	15.426	352.1	80.6	-10.9	7.7	80.3	-17.6	6
7340	289.3574	47.4111	2013.4655	14.616	15.824	4.914	287.4	-99.5	-76	5.4	-100.1	-76.4	5.4
7341	289.3780	8.1558	2013.781	15.877	17.837	41.347	191.7	-25.4	71.6	5.6	-21.1	70.4	7.3
7342	289.3866	15.9909	2013.509	17.668	17.822	29.574	46.5	4.7	-63.6	6.1	-6.7	-66.7	6.9
7343	289.3943	45.1268	2013.487	14.322	17.936	14.228	39.3	-10.7	-66.1	5.3	-12.8	-65.8	5.5
7344	289.4262	28.3615	2013.2935	12.077	13.037	31.44	158	55.7	-93.1	5.1	67.7	-91	7.5
7345	289.4310	0.2425	2013.937	12.822	15.052	7.392	93.1	-1.7	60.3	5.7	-1.6	60.5	5.7
7346	289.4441	50.9293	2013.498	11.423	15.856	7.607	193.8	19.3	66.5	6.1	21.4	62.6	6.2
7347	289.4866	17.2030	2012.473	16.557	18.312	11.656	262.5	-16.5	-241.4	7.7	-3.9	-232.5	7.4
7348	289.5127	63.3514	2013.4285	12.876	17.197	33.38	167.6	-2.2	112.7	5.7	-0.5	109.9	5.8
7349	289.5145	15.3670	2013.8395	17.345	17.567	37.657	205.5	23.6	64.8	6.3	27.5	67	6.1
7350	289.5210	14.1109	2014.2075	16.306	18.056	41.098	118.4	23.1	66.4	7.9	21.5	69.1	6.6
7351	289.5245	31.9631	2013.4275	15.407	17.098	15.232	30.4	60.4	57	5.3	61	52.4	5.4
7352	289.5792	14.5222	2014.1955	17.685	17.973	30.059	323.5	-85.5	-117	6.7	-88.9	-109.8	6.6
7353	289.6487	22.2045	2013.318	16.247	17.692	59.335	58.3	8.9	-67.4	5.4	7.4	-64.5	5.1
7354	289.6529	16.2070	2013.8095	17.27	18.287	31.333	15.7	205	-109.5	6.5	214	-117	6.2
7355	289.7424	-0.7169	2013.829	15.403	15.543	8.071	158.3	-53.5	-100.3	5.6	-54.5	-103.7	5.8
7356	289.7434	26.5694	2013.495	11.956	12.937	16.402	241.5	-37.4	-67	5.1	-33.9	-65.7	5.2
7357	289.7725	40.5863	2013.3195	17.473	18.382	37.825	215.8	64.5	-16.8	6.2	69.4	-5.8	6.3
7358	289.7728	62.7524	2013.2825	16.876	17.548	26.854	59.6	87.1	112.4	5.8	91.9	113.1	5.9
7359	289.7854	8.9859	2013.831	16.508	18.06	47.412	206.6	-42.4	-62.2	6.1	-44	-65.6	6.6

7360	289.7929	15.0916	2013.882	18.292	18.475	44.576	146.9	42.1	75.1	7	41.5	69.4	8.1
7361	289.7936	8.6624	2013.4845	17.164	18.499	34.222	119.5	143	-173	7.3	145.2	-176.9	6.6
7362	289.8240	43.3509	2013.5515	12.919	15.849	48.616	91.6	54.7	-83.7	5.3	56	-84.7	5.4
7363	289.8394	-1.6551	2013.786	14.037	15.659	6.355	304.6	-58.4	-65.4	5.5	-57.6	-64.2	5.7
7364	289.9103	51.4231	2013.5	10.851	14.666	52.949	27.2	29.4	68.8	6.1	28.5	65.4	6.1
7365	289.9197	18.9201	2013.506	17.122	17.865	5.867	239.9	1.4	-117.9	6.2	4.7	-117.7	6.9
7366	289.9669	11.4694	2014.0995	16.256	16.628	42.557	340	7.7	-61.5	5.6	8.5	-60.3	5.6
7367	290.0608	10.8432	2013.886	17.124	18.508	31.857	32.4	192.7	-2.2	5.8	192.2	-11.4	7.2
7368	290.0624	55.0580	2013.278	17.113	17.786	31.219	140.6	-22.8	-100.6	6.3	-18.6	-97.8	6.3
7369	290.1899	14.5815	2013.6745	17.489	18.129	30.173	147.9	32.6	-83	6.3	38.7	-77.1	7.3
7370	290.1945	5.1068	2013.902	11.088	16.756	15.543	320.9	-63.4	-149.3	5.6	-62	-152.6	5.9
7371	290.2084	8.2862	2014.06	16.017	17.87	33.596	214.3	-13.4	-88.4	5.6	-15.2	-83.7	6.9
7372	290.2258	11.5101	2014.339	17.246	19.302	26.553	99.3	79.3	98.7	6.8	75.7	109.1	7.2
7373	290.2539	23.0715	2013.3365	13.143	17.697	39.056	27.9	-69.6	-64.3	5.2	-77	-60.7	5.2
7374	290.3597	49.9528	2013.229	18.065	18.787	28.932	236.7	26	-65.3	6.4	26.7	-61.7	6.9
7375	290.4304	30.1918	2013.4045	12.947	13.778	4.155	11.7	-66.9	-93.1	6.1	-65.4	-96.1	6.5
7376	290.4391	16.6166	2013.4025	16.131	18.199	36.185	258.2	91.3	-2.1	6.1	93.7	-1.1	7.1
7377	290.4981	-0.3008	2013.685	13.485	16.411	21.683	83.9	-10.3	-86.7	5.6	-14.9	-87.8	5.8
7378	290.5013	28.1203	2013.376	13.888	14.576	25.562	33.8	65.6	-20.4	5.5	64.4	-26	5.5
7379	290.5160	25.0732	2013.379	16.074	18.507	36.689	202.6	32.8	-67.8	5.6	27.7	-66.5	5.6
7380	290.5415	70.8457	2013.5165	11.384	13.273	53.82	172.6	35	-70.9	5.7	36	-71.6	5.7
7381	290.5573	1.0691	2013.5315	16.651	17.138	4.424	25.2	9.5	-67.5	6	11.5	-68.3	6.5
7382	290.5705	58.9674	2013.5625	12.364	12.939	6.955	246.7	-80.6	-112.1	6.1	-81.9	-116.1	6.1
7383	290.6214	20.5305	2013.535	13.273	14.056	5.562	241.6	85	24	5	85.3	25.5	5
7384	290.6439	16.0280	2013.6275	18.525	18.585	51.562	204.8	220	-56.8	7.1	221.3	-49.4	6.3
7385	290.6946	82.4331	2013.827	17.917	18.93	19.267	11.2	5.5	67.4	6.4	8	65.9	6.8
7386	290.7086	16.5599	2013.8265	17.511	18.984	29.271	206.8	116.3	-208.2	6.3	105	-215.6	7.4
7387	290.7264	0.0244	2013.7465	12.659	16.184	6.411	292.5	-43.4	-77.8	5.5	-43.4	-80.9	6.5
7388	290.7515	8.3962	2014.187	16.274	16.851	57.045	138	-39.3	-62.8	6	-32.4	-67	5.9
7389	290.7714	19.0142	2013.6415	17.305	17.709	19.413	183	-3.4	180.5	6.8	2.7	179.3	5.3
7390	290.7793	16.5432	2013.6665	18.099	18.64	43.38	318.5	95.9	-69.2	6.8	103.7	-66.8	6.8
7391	290.7866	58.2963	2013.377	13.184	17.664	10.171	314.7	9.5	-82	6.1	13.4	-81.9	6.3
7392	290.7903	7.4961	2013.926	18.387	18.806	41.883	29.4	61.2	25.7	7.5	61.9	16.8	6.8
7393	290.7911	16.9849	2013.423	18.19	18.62	34.874	198.3	-212.9	-228.8	6.3	-218.1	-215.6	7.6
7394	290.8053	11.9792	2013.9845	10.095	17.616	36.691	30.9	3.2	-63.3	5.9	2.2	-66.5	6
7395	290.8456	18.1398	2013.515	18.095	18.296	15.103	301.1	69.7	36.3	5.8	70.8	42.4	5.3
7396	290.8579	46.6561	2013.096	18.666	18.792	59.945	284.5	65	19.3	6	66.8	16.7	5.9
7397	290.8848	37.7084	2013.3115	16.335	18.07	12.042	19.8	-6.3	-86.1	5.4	-9.6	-87.4	5.6
7398	290.9938	4.7436	2013.815	15.418	16.906	7.441	341.6	2.2	-76.6	5.7	6.7	-74.2	5.9
7399	290.9949	28.0845	2013.655	10.301	12.494	38.149	235.6	80.6	80.3	5.4	81	81.2	5.5
7400	290.9955	60.7731	2013.3025	15.687	16.403	4.98	234.6	20.2	68.2	5.8	20.8	71.5	5.8
7401	291.0764	25.6528	2014.172	14.374	14.395	3.447	274.5	56.4	90.8	5.6	54.7	93.6	5.8
7402	291.0957	10.3788	2014.0045	15.368	16.859	28.744	235.6	-34.7	-82.7	5.6	-32.2	-85.6	5.7
7403	291.1861	-3.8804	2013.465	16.459	17.611	7.01	296.8	11	-66.8	6.5	15.9	-69.3	6.7
7404	291.1939	-3.8135	2013.6425	13.172	16.076	25.08	233.2	-54.6	-102.8	6.2	-66.7	-101.2	6.4
7405	291.1951	53.0885	2013.5515	10.902	15.869	14.373	138	-30.8	-64	6.1	-28.9	-66.4	6.2

7406	291.2272	25.3237	2013.28	18.393	19.154	49.888	301.3	64.8	-26.6	5.7	66.4	-19.5	5.9
7407	291.3024	52.6197	2013.4285	16.651	17.14	8.353	61.7	-52	-88.3	5.4	-49.4	-91.8	5.4
7408	291.3278	-11.5233	2013.47	11.387	12.24	6.745	343.8	-46.6	-103.5	6.3	-48.7	-104.9	6.2
7409	291.3729	18.6755	2013.505	18.216	18.47	54.691	178.5	48.7	-131.3	5.8	38.3	-141.2	5.3
7410	291.3894	17.3443	2013.6965	15.108	19.145	43.628	325.2	-31.3	-79.3	5.4	-42	-78.2	5.8
7411	291.4096	26.0844	2013.4975	17.731	18.141	50.747	300.3	62	23.6	5.8	62.3	20.5	5.5
7412	291.4317	14.8827	2013.7515	14.036	14.234	11.721	345.6	97.2	-1.3	5.4	94.7	-0.1	5.4
7413	291.4879	15.1678	2013.619	16.086	17.984	48.406	62.9	62.4	-100.2	5.6	64.7	-97	5.5
7414	291.4918	35.9148	2013.7705	8.818	10.946	6.016	106.4	-90.2	-160.7	5.4	-98.1	-156	5.3
7415	291.5553	37.5355	2013.203	18.588	19.665	45.31	103.6	11.8	-74.7	5.7	0.5	-75.6	6.8
7416	291.6257	20.6535	2013.444	12.735	17.307	22.046	47.8	12	-69.8	5	10.4	-72.5	5.1
7417	291.6388	40.0002	2013.5155	11.635	16.236	54.051	348	-7	-71.3	5.3	-12.3	-70.7	5.4
7418	291.6757	3.6407	2013.716	15.095	15.851	4.301	209.9	63.6	8.2	5.7	62.6	4.1	6.4
7419	291.7808	-2.7785	2013.8155	12.41	13.105	14.04	94.1	8.4	-66.2	6.1	3.9	-65.6	6.1
7420	291.8275	4.8762	2013.6885	13.267	16.292	5.067	349.7	17.8	-71.2	6	16.7	-73.3	7.5
7421	291.8344	42.5998	2013.4955	11.933	16.244	45.978	21.2	-4.2	-62.2	5.3	-5.5	-62.3	5.4
7422	291.8447	38.2359	2013.5825	13.45	14.573	6.64	26.4	-23.9	-118.3	5.8	-22.7	-118.5	5.9
7423	291.9726	28.4506	2013.744	9.834	15.766	12.198	80.5	-13.6	-71.6	5.4	-21.3	-68.8	5.4
7424	291.9885	12.5926	2013.515	16.715	17.83	34.018	250	185.4	-73.4	6.5	198.4	-66.9	6.7
7425	291.9926	16.7029	2013.7945	8.544	13.919	14.595	16	64.6	23.5	5.3	61.7	22.6	5.4
7426	291.9948	5.9720	2013.503	17.233	17.825	7.361	88.9	0.2	-71.8	6.4	-9.1	-70.5	7.1
7427	291.9958	-8.2882	2013.636	13.806	14.265	11.044	102.7	7.1	-62	5.7	8	-60.5	5.7
7428	292.0546	31.9145	2013.2535	17.378	17.942	49.513	349.6	-23.8	-63	5.3	-19.2	-61.6	5.4
7429	292.0770	19.4964	2013.3945	14.803	15.538	33.119	187.8	68.1	44.5	5.1	65.4	43.8	5.1
7430	292.0917	20.3186	2013.8275	9.189	9.985	5.513	155	-65.5	-87.2	5	-63.1	-83.7	5.1
7431	292.1091	29.0756	2013.495	10.584	12.377	4.881	209.2	52.1	-116.4	5.4	51.3	-121.6	5.6
7432	292.1199	19.9640	2013.783	17.3	18.956	58.887	131.1	88.4	-68	7.5	98	-57.2	5.3
7433	292.1241	71.7553	2013.46	10.975	15.573	8.76	296.4	-7.3	-82.7	5.7	-6.6	-79.9	5.8
7434	292.1291	27.8760	2013.16	17.197	17.605	54.712	69.2	64.8	50.9	7.3	64.6	50.5	5.7
7435	292.1604	26.2841	2013.389	16.267	18.274	31.302	39.9	-12.7	-73.3	5.6	-3.4	-73.8	5.8
7436	292.2165	15.4268	2013.6	18.309	18.746	32.718	48.1	78	-65.8	6	75.4	-67.3	6.1
7437	292.2250	43.8871	2013.4345	17.043	18.216	33.696	111.8	13.3	-61.6	5.4	22.8	-60.1	5.7
7438	292.2618	41.4432	2013.581	13.675	14.199	4.057	128.7	93.7	103.3	5.4	86.7	104.4	5.8
7439	292.2946	42.9831	2013.641	11.979	14.892	13.615	60.9	-11.4	-95.9	5.3	-11.9	-95.8	5.3
7440	292.3147	73.3213	2013.535	15.505	15.623	13.305	318.7	-56.8	-209.7	6.2	-50.5	-210.3	6.2
7441	292.3185	20.1692	2013.709	9.495	14.803	33.521	139.6	135.4	68.3	5	136.9	70.2	5.1
7442	292.3682	8.5728	2013.998	16.234	17.092	14.254	66.8	-17.8	-89.4	5.7	-20.8	-88.1	5.7
7443	292.3683	14.0565	2013.6065	18.389	19.136	41.889	33.1	-138.5	-177.2	6.2	-137.2	-170.8	7
7444	292.3811	21.5394	2013.62	11.74	16.319	43.208	74.5	-59.6	-134.3	5	-64.1	-136.9	5
7445	292.4087	47.3886	2013.5155	11.909	12.964	15.004	47.1	-7.4	-65.8	5.3	-7.6	-65.6	5.4
7446	292.4838	4.7863	2013.995	17.263	17.556	44.966	290.6	-37	-74.3	6.4	-30.2	-74.2	6.7
7447	292.5255	14.1365	2013.888	11.431	14.97	6.835	184	82.2	-17.9	6.1	79.3	-25.3	6
7448	292.5374	29.9779	2013.5685	12.262	13.902	14.298	316.4	64.9	2.2	5.1	64.6	3.3	5.1
7449	292.5584	16.5168	2013.3535	17.746	17.779	39.472	240.5	109.6	70.8	6.4	119	62.7	6.3
7450	292.5704	27.1818	2013.47	16.238	16.736	59.46	126.3	68.5	54.8	6.1	61.4	61.2	6.5
7451	292.6600	-1.2053	2013.9905	9.868	14.511	21.76	125.5	33.9	-125.8	5.4	34.8	-127.1	5.4

7452	292.6656	-5.2445	2014.0045	10.621	12.033	5.897	192.6	-33.6	-62.6	5.4	-29.6	-62.6	5.3
7453	292.6727	8.4525	2013.9415	17.608	17.976	17.789	48	62.8	-13.4	5.9	62.4	-12.4	5.7
7454	292.6901	19.0497	2013.3275	18.236	18.572	46.769	86.5	33.4	-78.7	5.2	41.2	-75.3	6.7
7455	292.7362	21.0336	2013.073	18.086	18.565	26.857	147.4	70	-24.5	5.2	73.2	-24.7	5.8
7456	292.7839	20.1974	2013.4235	18.014	18.129	35.068	119.6	141.9	186	5.7	153.6	184.9	6.3
7457	292.8017	36.1246	2013.6065	12.285	13.478	45.775	347.4	-132.6	-107.1	5.4	-132.8	-106.2	5.4
7458	292.8068	2.0469	2013.9125	11.468	14.95	6.672	279	64.6	1.3	5.7	61.9	0.1	5.9
7459	292.8080	22.9507	2013.7995	17.335	17.464	31.362	145	14.4	183	5.3	11.5	177.9	5.8
7460	292.8391	5.3000	2014.0775	9.643	13.849	42.696	53.4	6.9	-79.2	5.6	8.3	-79.1	5.6
7461	292.8502	-2.3398	2013.7	14.173	14.672	52.826	187.3	6.8	-63.9	5.4	-1.1	-64.5	5.4
7462	292.8792	17.8457	2013.7915	15.668	17.136	45.547	194.6	-197.8	154.4	7.6	-197.7	147.1	6
7463	292.8964	19.7337	2013.539	16.684	18.205	33.111	213.9	137.8	-88.8	5.1	139.2	-95.4	5.6
7464	292.9256	84.5729	2013.6245	16.709	17.926	55.81	23.2	4.6	62.1	5.8	8.2	62	6
7465	292.9358	17.5320	2013.9505	18.215	19.357	57.885	248.5	-8.1	-60.4	5.7	-13.9	-61.7	6.9
7466	292.9374	45.1070	2013.537	13.392	13.747	24.576	112.9	78.7	58.2	5.5	81.8	60.7	5.5
7467	292.9683	19.9079	2012.471	13.732	15.396	29.196	47.7	144.7	-16.2	6.1	148.9	-4.3	6.5
7468	292.9883	19.3092	2013.447	15.376	19.449	31.183	297.1	150.5	-19.8	7.2	159.2	-10.1	8.6
7469	293.0047	60.6330	2013.385	17.572	18.475	18.16	210.3	-1.1	-67.4	6	-0.8	-67.5	6
7470	293.0073	18.7774	2012.471	15.252	15.412	24.609	266.2	134.9	-98.2	6.6	144.5	-88.3	6.4
7471	293.0153	23.5071	2013.5835	16.585	17.188	32.121	261	-114.3	186.2	5.7	-111.6	184.7	5.1
7472	293.0340	19.4398	2012.9315	12.619	16.799	22.632	297.3	58.9	-89.1	5.9	66.4	-90.5	5.1
7473	293.0509	53.6827	2013.5605	9.28	16.644	57.022	139.2	-80.1	-128.7	5.3	-82.6	-125.7	5.4
7474	293.0749	18.4561	2012.471	8.987	13.844	19.492	285.9	84.9	-105	6.2	91.2	-99.4	6
7475	293.0858	32.7654	2013.472	11.625	13.369	4.894	316.9	-41.6	-70	5.3	-42	-71	5.4
7476	293.1074	-3.1911	2013.6075	13.578	16.928	8.359	83.1	70.2	-0.7	5.4	69.9	5.5	5.6
7477	293.1081	56.6070	2013.6185	10.109	12.325	14.649	334.3	102.9	-105.7	6.1	105.9	-110	6.2
7478	293.1212	-6.2684	2013.714	10.672	15.974	9.32	134.9	23	-72.9	5.7	17.1	-71.6	6.3
7479	293.1461	19.8110	2013.011	15.323	17.575	24.29	233.1	-126.1	-98.9	7.2	-127.4	-98.5	6.6
7480	293.1470	18.6322	2012.7295	15.227	18.549	39.632	125.3	73.7	-30.3	6.4	71	-30.3	5.9
7481	293.1600	20.2138	2013.0085	14.969	16.517	14.786	2.5	-93.4	-70.7	6.3	-98.9	-66.8	5
7482	293.1680	18.2557	2013.4455	10.147	18.723	19.14	57.5	-99.9	-89.6	5.8	-96.2	-84.3	6.1
7483	293.1691	53.7770	2013.375	16.741	17.909	43.334	112.4	13.7	-63.2	5.4	5.4	-63.4	5.5
7484	293.2067	18.6076	2012.471	15.921	16.085	29.746	241.5	-116.2	-184	7.1	-127.8	-177.4	7.3
7485	293.2362	19.6524	2012.471	14.643	15.211	24.937	154.3	4.4	-76.1	6.6	-3.7	-78.6	6.3
7486	293.2547	68.3280	2013.474	14.219	15.426	42.997	0.4	-50.9	-85.8	5.7	-51.9	-88.7	5.7
7487	293.2554	65.4474	2013.3815	16.377	16.472	17.527	233.2	-152.5	-103.9	5.8	-152	-106.8	5.8
7488	293.2555	2.3331	2014.055	13.635	15.103	48.216	1.5	-14.1	-87	5.7	-20.8	-85.8	5.6
7489	293.2567	18.1552	2012.9305	14.27	17.625	20.401	323.4	-27.6	-177.4	5.9	-27.2	-179.1	5.1
7490	293.2896	16.7213	2013.0065	12.133	19.059	39.262	286.3	111.3	2.4	7.7	112.1	-6	6.4
7491	293.3114	24.1259	2013.5325	15.65	16.428	45.138	136	-39.8	-102.4	6.8	-50.1	-92.3	6.9
7492	293.3332	16.2151	2013.5685	17.417	18.109	51.646	213.5	-80.1	-133.8	7.4	-93.3	-130	6
7493	293.3595	19.8677	2012.929	13.004	18.559	28.876	89.6	-40.6	145.3	5.9	-40.6	144.2	5.4
7494	293.3891	17.2870	2012.471	12.315	12.995	33.159	102.1	98.2	-95.3	7.5	95.9	-103	7.5
7495	293.3930	19.7758	2012.8595	15.041	16.839	44.612	306.4	104	-132.9	7.1	112	-124.8	5.2
7496	293.3935	29.0627	2013.341	16.193	17.831	34.398	197.6	82	127.7	6.4	84.8	125.8	5.9
7497	293.3994	4.8301	2014.0945	15.601	15.909	25.806	129	-29	-68	7	-18.1	-72.9	6.1

7498	293.4133	19.8662	2012.471	14.48	16.314	14.523	47.4	-102	-105.6	6.8	-121.4	-104.8	16.1
7499	293.4514	19.8904	2012.471	13.5	15.138	15.813	72.5	29.6	-143.2	6	23.8	-150.5	6.4
7500	293.4548	8.2261	2014	15.471	17.174	45.143	96.4	72.2	29.2	6.7	72.7	22.3	5.6
7501	293.4559	21.3254	2013.36	17.409	19.283	50.635	348.5	-97.9	112	6.1	-101.2	113.9	5.6
7502	293.4594	19.6321	2013.005	15.526	17.599	16.646	249.5	130.1	-129.6	6.5	122.9	-129	5.5
7503	293.4600	18.9051	2013.447	15.705	19.096	21.557	270.5	95.2	-118.9	6.5	85.2	-120.3	7.6
7504	293.4819	0.1495	2013.502	17.808	18.05	31.243	118.2	62.5	-10.5	7.1	61.7	-12	6.8
7505	293.4996	5.0528	2013.8765	12.165	15.448	7.313	319.5	116.2	-12.4	5.6	115.2	-17.7	5.8
7506	293.5228	21.8984	2012.8935	16.398	17.767	37.603	182.1	-26.1	-65.7	7.2	-21.3	-67.2	5.1
7507	293.5255	40.5692	2013.0795	17.982	18.461	50.22	187	1.8	-63.1	5.6	9.6	-63.6	5.6
7508	293.5361	46.4007	2013.3125	16.267	18.495	39.137	142.2	-34.5	-67.9	5.4	-24.8	-68.9	5.6
7509	293.5394	18.5045	2012.471	15.387	15.536	28.847	114	82	-197.5	6.2	90.6	-185.6	6.7
7510	293.5538	20.7792	2012.791	13.65	18.88	59.188	277.9	145.1	-95.8	6.1	138.8	-106.4	5.6
7511	293.5707	16.8701	2013.545	17.705	18.423	30.363	90.2	-152.1	-113.1	5.7	-149.4	-118.6	6.9
7512	293.5780	20.0037	2013.872	18.522	19.357	13.48	212.7	63.9	172.2	5.1	64.7	166.1	5.5
7513	293.6026	11.8598	2013.597	16.637	17.957	42.32	25.1	6.4	-104.3	6.4	13.1	-108	6.4
7514	293.6053	20.8374	2013.8025	18.574	19.321	41.15	202.7	-75.8	128	7.9	-83.5	123.1	6.8
7515	293.6101	20.4018	2012.471	14.814	15.7	43.55	136.4	187	12.9	6.5	180	17.8	7
7516	293.6118	34.1953	2013.1775	17.466	18.653	23.798	231.5	65.6	-1.7	5.5	66.9	8.7	5.6
7517	293.6126	-1.7928	2013.5695	15.172	15.78	9.585	53.4	-53.2	66.7	5.6	-52.4	65.9	5.5
7518	293.6216	11.2580	2014.0135	7.919	15.65	54.322	148.1	-61.1	-86.2	5.6	-50.6	-93.8	5.8
7519	293.6322	49.1593	2013.323	15.619	18.382	13.449	224.3	-10.1	-64.9	5.4	-13.1	-66.2	5.7
7520	293.6405	36.4288	2013.5085	13.095	15.11	18.059	328.8	34.7	-109.3	5.3	34.3	-109.3	5.4
7521	293.6454	39.2157	2012.9425	17.537	18.95	36.246	181.2	16.3	-90	5.5	6.2	-92.1	6.1
7522	293.6479	19.3658	2013.0205	15.41	18.14	7.506	250.8	-91.2	-170.2	7	-98.5	-175.3	5.2
7523	293.6716	20.2683	2013.4585	15.099	18.791	17.545	355.1	-13.3	-121.4	5	-4.5	-127.7	5.9
7524	293.6814	17.0517	2012.913	11.826	17.908	51.268	336.3	181.1	-120.4	7.3	183.1	-117.7	5.7
7525	293.6814	17.2068	2012.471	13.239	13.296	33.131	149.4	136.2	-6.5	7.4	138.8	3.3	7.7
7526	293.7069	18.4817	2013.4295	17.036	18.868	46.908	319.6	35.9	-77	5.1	30.4	-75.6	6.3
7527	293.7105	11.9788	2013.8845	16.312	16.938	58.136	124.4	-6.3	-71.8	6	-7.2	-75.1	5.7
7528	293.7136	17.1091	2012.471	11.979	13.489	21.442	336.2	152.2	187.4	7.2	154.8	174.5	7.7
7529	293.7396	19.7256	2012.471	10.972	14.211	26.458	94.1	81.9	-61.2	5.8	83.9	-58.7	6.3
7530	293.7477	39.7884	2013.5955	12.408	15.875	5.242	351.2	-5.1	-63.2	5.3	-4.8	-63.8	5.7
7531	293.7521	25.8484	2013.561	15.233	16.44	5.796	75.5	-22.6	-122.6	5.1	-21.9	-119.2	5.2
7532	293.7565	19.4716	2012.9445	15.448	18.177	19.29	198.3	2.2	-73.2	6.7	8.9	-70.8	6
7533	293.7650	37.5386	2013.4845	10.047	15.148	45.744	138.3	-3.9	-79.9	5.3	-3.7	-81.9	5.4
7534	293.7968	7.8876	2013.6705	17.706	18.116	54.706	50.3	-16.2	-62.1	6.4	-5.2	-66	5.9
7535	293.8696	41.9887	2013.393	17.333	18.005	15.224	347	-16.7	-69.8	5.4	-7.3	-68.5	5.5
7536	293.8830	19.5796	2012.471	13.898	15.868	20.508	9.5	-33.3	-121.8	5.9	-43.7	-113	6.8
7537	293.8867	11.3272	2013.969	16.75	16.921	41.368	127.5	-22.5	-82.3	5.8	-24.1	-83.2	6
7538	293.9156	6.2781	2014.0335	13.262	13.486	17.174	185.9	-146.6	-86	5.6	-146.2	-86.7	5.6
7539	293.9161	22.7886	2013.5565	14.56	16.288	47.757	1.5	72.9	62.2	5.3	70.8	68.1	5
7540	293.9185	27.1612	2013.371	16.111	17.709	15.162	281.2	70.2	-25.9	5.8	74.7	-22.7	5.2
7541	293.9417	20.6827	2012.471	14.755	15.382	38.431	3.8	185.5	-117.9	6.2	196.5	-104.8	7.4
7542	293.9577	35.2305	2013.0335	17.757	18.216	42.209	314.4	-16.6	-63.8	5.5	-11.1	-62.1	6.7
7543	293.9881	18.8602	2012.471	13.983	14.057	33.646	122.7	120.4	28.2	5.9	122.8	23	6.4

7544	294.0141	47.1647	2013.1535	17.811	19.026	30.2	184.8	75.1	9.5	5.5	78	0.5	7.3
7545	294.0278	20.7208	2012.468	15.205	15.341	34.584	63.6	-1	-135.8	6.6	-2.4	-139.5	7.2
7546	294.0520	11.4669	2013.929	17.2	17.781	55.036	304.7	160	-208.6	5.8	171	-199.8	5.9
7547	294.0665	23.1770	2013.2635	16.327	17.226	17.934	87.3	79.4	49	5.2	75.9	54	5.2
7548	294.0746	19.8781	2013.526	15.666	18.46	59.181	353.1	-102.8	-93.4	5	-112	-83.6	5.3
7549	294.1119	18.8128	2013.0495	14.944	17.722	5.295	93.8	-17.1	-96	7.2	-27.3	-97.8	5.2
7550	294.1160	19.2729	2013.467	16.982	18.067	46.354	286.5	183.9	-111	5.2	175.3	-120.2	5.1
7551	294.1251	9.3868	2014.0305	10.236	15.585	25.09	67.4	-24.9	-79.8	5.6	-28.9	-80.5	5.6
7552	294.1398	23.7640	2013.4235	17.558	18.545	34.901	230	2.9	-182.9	5.7	0.5	-188.5	5.6
7553	294.1459	9.5410	2013.999	8.723	15.841	23.516	172	142.2	10.4	5.6	142.2	13.6	5.7
7554	294.1531	24.0462	2013.163	17.271	19.007	21.378	297	94.5	-24.1	5.3	96.6	-31.4	6.1
7555	294.1644	18.1984	2012.956	15.382	18.309	35.184	114.4	3.9	133.6	6.5	5.4	134.3	5.4
7556	294.1834	23.2674	2013.321	17.713	17.981	26.069	195.6	117.9	12.7	5.4	121.5	5.8	6.7
7557	294.1948	18.0007	2013.6595	17.29	17.395	20.745	3.5	19.5	-81.4	5.8	13.2	-84.4	5.2
7558	294.2114	21.4106	2013.487	11.949	15.996	15.06	247.1	8.9	-61.7	5	8.7	-60.6	5.1
7559	294.2179	23.9518	2013.407	15.973	18.053	7.714	356.9	45.1	-67.4	5.6	38.2	-72.8	10.8
7560	294.2288	13.0906	2013.9085	16.919	16.928	49.818	276.7	-114.7	-183.7	5.5	-110.1	-186.6	5.6
7561	294.2322	20.1463	2012.9665	14.919	17.281	29.512	339.2	76.2	107.9	7.8	78.7	118.3	5.1
7562	294.2474	23.3557	2013.516	18.641	18.682	23.656	39.9	75.2	184.9	5.4	80.7	185.2	5.4
7563	294.2557	18.7369	2012.471	14.685	15.015	55.834	203.6	130.4	122.9	7.3	131.1	134.4	7.2
7564	294.2567	0.4996	2013.563	15.813	18.025	59.167	61.6	-34.7	-60.6	5.8	-29.2	-64.6	6.5
7565	294.2699	16.0662	2013.6995	17.136	17.684	48.458	53.5	62.3	81.9	6.7	62.2	79.2	5.6
7566	294.2914	-2.0310	2013.4775	16.016	17.82	17.531	305.4	40.7	-68	5.5	46.2	-62.1	5.8
7567	294.2943	19.9800	2013.0605	15.453	18.222	50.034	314.3	-112.4	-173.4	7.1	-114.7	-177.2	5
7568	294.3202	23.4651	2013.61	18.203	18.433	35.951	172.9	151.9	-156.2	5.4	152.8	-152.1	6.5
7569	294.3314	16.2556	2012.471	9.269	13.791	20.884	144.6	215.3	216.5	7.4	204.8	213.2	7.6
7570	294.3337	16.2032	2012.471	9.292	13.901	20.601	187.4	147.1	210.4	7.5	148.8	205.6	10.1
7571	294.3382	11.7748	2014.019	13.321	17.657	44.763	9.8	-15.7	-129.9	5.8	-25.1	-123.9	6.3
7572	294.3439	10.8036	2013.759	17.226	17.458	15.540	167.5	45.2	-107.1	5.9	40.7	-105	6
7573	294.3459	16.0525	2012.471	13.374	13.498	33.577	336.4	-38.1	212.8	7.8	-35	210.4	8.6
7574	294.3594	37.1459	2013.26	18.182	18.587	24.342	116.8	2.8	-61.5	6.4	6.5	-63.9	7.3
7575	294.3600	11.8721	2013.766	16.061	18.178	51.943	15.6	74	47.8	6	71.9	47.1	5.8
7576	294.3706	24.5034	2013.475	14.905	17.664	59.785	101.4	8.4	62.1	5.2	9.8	64.7	5.2
7577	294.3788	19.8107	2012.471	12.995	16.058	34.786	85.1	161.3	87.7	6.1	161.3	81	7.2
7578	294.4100	13.1260	2014.063	16.704	17.814	20.744	190.8	81.9	-7.6	5.3	79.4	-16.9	5.9
7579	294.4173	26.5957	2013.5095	13.505	17.904	55.506	257.7	23.5	-92.3	5.1	29.6	-88.2	6
7580	294.4216	20.3336	2012.4795	13.547	16.414	11.133	24	-7.2	-102.4	6.2	-25.3	-81.2	16.8
7581	294.4381	17.7100	2012.471	12.436	13.724	37.89	19.6	208.9	207.3	6.8	204.9	206.2	7.3
7582	294.4465	19.9132	2012.9765	14.267	18.62	53.245	296.4	-89.6	-180.8	5.9	-99.7	-186.3	5.3
7583	294.4665	29.6709	2013.5335	15.329	15.38	43.189	108.4	-9.4	-66.8	5.1	-3.1	-68.7	5.2
7584	294.4670	18.8887	2013.787	18.278	19.089	42.333	300.9	-8.4	-74.5	7.1	-3.5	-73.7	5.3
7585	294.4940	38.0988	2013.261	17.412	17.865	52.42	169	36.1	-62.7	6.2	29	-70.1	6.3
7586	294.4941	20.2411	2013.0185	15.33	17.761	11.645	125.2	93.8	74.7	6.4	97	77.8	5.1
7587	294.5300	23.6857	2013.4395	17.452	18.899	32.704	231.8	-49.9	-90.5	6.2	-55	-85.1	6.1
7588	294.5308	24.7459	2013.6915	14.12	14.205	4.542	105	-7.1	-67.9	5.1	-3.4	-66.7	5.5
7589	294.5474	2.5747	2013.837	14.647	14.912	51.112	153.9	77.1	-34.8	5.7	76.7	-39.5	5.9

171

7590	294.5516	19.7965	2013.7075	14.345	14.999	3.602	9.2	-33.8	-63.6	5	-29.1	-66.1	5.1
7591	294.5591	20.2552	2013.02	15.087	18.082	23.968	9.8	174.2	-15.1	6.4	166.8	-8.7	5.9
7592	294.5660	19.6179	2012.471	15.078	16.025	57.662	235.5	1.6	-95.1	6.4	2	-97.5	7.4
7593	294.5663	23.6137	2013.384	17.682	17.974	22.756	69.9	-49.2	-70.4	5.2	-41.1	-78.4	5.2
7594	294.5753	19.4671	2013.043	14.641	18.238	41.326	237.1	153.1	2.6	6.4	150.5	2.7	5.8
7595	294.5787	52.8304	2013.5475	14.22	18.493	41.857	174	84.5	38.3	5.3	82.8	41	6
7596	294.5811	37.0688	2013.5925	15.779	15.938	3.547	123.5	-4.2	-64.9	6.1	-11.2	-61.5	6.2
7597	294.6172	-13.0757	2013.47	13.688	14.82	11.107	292.6	30.1	-99.3	5.4	30	-100.2	5.5
7598	294.6300	20.0385	2013.4585	17.649	17.688	31.214	26	116.4	-88.9	5.3	121.7	-81.7	5.2
7599	294.6325	16.2671	2013.9275	16.319	17.441	17.751	328	-39.2	-63.3	5.8	-44.7	-62.3	5.7
7600	294.6405	20.2983	2013.327	16.923	18.019	22.177	214	8.3	-114.4	6.3	7.4	-111.8	5.9
7601	294.6848	9.1418	2014.0035	14.274	17.414	14.661	37.7	11.2	-62.1	5.6	15.8	-61.4	5.7
7602	294.7118	61.4983	2013.5555	15.777	15.923	4.69	135	117.9	67.6	5.8	119.3	76.4	5.8
7603	294.7563	23.4469	2013.3505	17.743	18.135	12.81	45.6	8.6	-97.6	6.5	1.2	-99.2	6.8
7604	294.7816	20.0567	2013.822	17.207	17.745	31.963	214.5	182.1	6.7	6.2	177.3	6.7	6.4
7605	294.7857	6.3656	2013.881	17.954	18.365	47.611	125.1	66.3	-37.5	6	65.5	-43	7.4
7606	294.7864	23.7960	2013.6365	17.816	18.006	59.153	118.4	60.6	-8.2	6.5	64	-3.3	7.5
7607	294.8056	8.9170	2013.7915	17.673	17.806	10.914	67.6	-52.2	-65.6	5.9	-46.8	-65.9	6.8
7608	294.8294	23.1853	2013.2115	17.702	18.381	17.854	257.3	-43.7	-87	6.3	-50.6	-79.3	6.6
7609	294.8456	30.9307	2013.405	15.144	17.959	48.523	158	9.4	-108.3	5.4	7.2	-109	5.5
7610	294.8533	37.1770	2013.1235	17.19	17.952	59.847	251.6	-152.7	-144.2	5.5	-161.1	-143.5	5.5
7611	294.8730	23.4455	2013.573	17.636	18.494	33.499	251.1	-114.8	-172.2	6.2	-121.9	-177.5	6.8
7612	294.8746	30.1794	2013.5005	17.587	18.081	21.266	344.9	72.9	9	5.5	72	2.2	6.2
7613	294.8893	26.0845	2013.51	17.718	18.198	49.625	134.5	71.2	174.5	5.6	68.8	181.8	5.9
7614	294.9266	22.3379	2013.5265	17.907	18.619	37.766	93.6	75.9	41.6	6.3	73.9	41.6	6.8
7615	294.9272	12.1200	2013.9015	16.025	17.035	20.797	141.8	-11.7	78.8	6.1	-2.4	77.7	6.1
7616	294.9487	32.5448	2013.459	13.076	14.326	4.463	359.8	25.7	-95.7	5.3	29.4	-98.8	5.3
7617	295.0378	23.6192	2013.65	16.862	17.517	24.52	120	9.1	-73	6.8	15.4	-72.5	5.1
7618	295.0468	19.7215	2012.471	14.081	14.437	27.725	174.5	155.9	196.8	5.9	165.5	198.1	6.2
7619	295.0499	14.6732	2013.9055	17.865	18.071	54.263	62.7	-17.3	-69.5	6.4	-26.3	-70.1	6.3
7620	295.0544	33.1045	2013.572	16.294	16.703	48.2	165.2	-14.6	196.2	5.3	-16.8	196	5.6
7621	295.0605	26.1979	2013.6335	13.335	14.061	16.659	250.6	25.4	92.7	5.1	24.4	95.4	5.1
7622	295.0811	30.9235	2013.3985	16.5	16.686	53.197	304.8	9.1	-66.3	5.4	15.5	-63.6	6.3
7623	295.0966	13.0454	2013.4855	17.01	18.828	30.146	177.5	60.4	8.7	7.3	60.6	5.2	6.9
7624	295.1171	13.1763	2013.9325	15.304	17.642	26.583	186.3	-42.8	-71	6	-44.7	-66.6	6.8
7625	295.1185	17.9920	2013.324	16.073	17.112	31.429	206.6	-80.3	-100.9	5.1	-74.1	-104.4	6.2
7626	295.1213	34.5391	2013.294	16.031	18.131	58.588	102.7	-3.9	194.8	5.6	-0.7	198.2	7.3
7627	295.1433	18.0186	2013.7385	16.928	17.917	34.24	149.2	69.2	-16	5.3	68.1	-8.9	5.4
7628	295.1735	13.4322	2013.882	16.237	18.366	12.641	26.2	17.3	-77.9	5.2	17.2	-74.6	5.7
7629	295.1885	54.7012	2013.5975	12.8	13.765	36.262	345.5	35.3	-115.8	6.1	33.4	-118.7	6.1
7630	295.1898	39.3563	2013.468	14.724	16.65	6.092	177.9	-58.2	-107.3	5.4	-58.7	-107.9	5.4
7631	295.2326	20.3049	2012.471	13.827	15.13	55.165	159.8	-51.4	-192	7	-60	-196.9	7.1
7632	295.2511	20.3341	2012.471	15.207	15.671	21.914	56.7	151.9	-171.2	6.5	165.3	-159.7	7.4
7633	295.2678	27.8818	2013.243	16.53	17.429	43.968	68.9	8.6	-60.4	5.4	17.2	-61.3	5.4
7634	295.2801	20.6329	2012.471	13.65	16.99	16.727	146.2	82.1	-7.1	5.9	85.9	-7.1	7.7
7635	295.3142	36.7717	2013.4235	12.614	16.471	5.381	241.4	-2.5	-96.4	5.4	0.4	-94.9	5.8

7636	295.3699	40.4641	2013.28	12.032	13.384	4.642	300.3	20.4	-94.1	5.3	20.1	-95.1	5.9
7637	295.3974	56.3009	2013.6185	10.703	13.702	16.502	336.5	-9.5	-67.4	6	-11.4	-69	6.1
7638	295.4023	31.3583	2013.519	17.225	17.674	36.905	71	-14.2	-70.7	5.5	-22.3	-65.9	5.3
7639	295.4446	36.3221	2013.0985	17.908	18.278	46.966	333	-15.7	-63	5.6	-20.4	-60.1	5.6
7640	295.5053	75.9923	2013.336	18.104	18.559	48.662	336.6	26.9	71.3	7.3	27.3	72.6	6
7641	295.5079	83.5770	2013.656	16.437	17.937	9.607	336.6	29.7	70.1	5.6	32.1	66.6	5.8
7642	295.5301	28.6710	2013.5955	11.92	13.542	57.042	24	-116.9	-153.6	5.2	-115.4	-150	5.2
7643	295.5406	34.1387	2013.4305	15.688	16.152	9.798	217.8	56.9	108.6	5.4	58.8	112.2	5.4
7644	295.5425	32.4900	2013.4635	18.163	18.224	31.784	352.1	-16.1	-61.5	5.5	-18.3	-66	5.4
7645	295.6106	16.3478	2013.164	9.725	15.996	56.925	64.7	34.3	-65.2	6.3	37	-61	5
7646	295.6432	57.2220	2013.5095	15.24	15.621	7.066	156.1	-29	-78.2	6.1	-28.5	-75.7	6.1
7647	295.6922	17.6680	2013.8395	13.766	15.237	7.628	281.3	5.5	-103.9	4.9	6.8	-105.1	5
7648	295.6950	20.7971	2013.0245	14.419	15.661	38.74	210.7	-41.7	-113.2	7.6	-39.4	-113.2	6.2
7649	295.7561	19.6210	2013.677	17.247	17.687	46.486	19.7	7.7	-64.3	6.8	11.3	-65.7	6.5
7650	295.7678	58.3561	2013.446	15.708	18.201	8.128	241.6	-25	-61.3	6.1	-22.4	-62.7	6.3
7651	295.8168	16.4919	2013.9845	16.568	18.272	27.388	193.1	-81.7	-69.4	5.5	-78.2	-80.7	6.1
7652	295.8321	54.5632	2013.584	12.25	13.042	36.759	102.9	82.2	163.7	6.1	80.9	163.1	6.2
7653	295.8504	38.8893	2013.2095	16.163	17.563	42.915	185.4	-2.8	-70.8	5.4	-5.5	-72.6	5.5
7654	295.8843	-6.9479	2013.8045	10.173	14.04	10.256	52.4	105.7	-84	5.7	108.1	-89.6	5.8
7655	295.8885	8.1302	2013.758	17.825	17.831	24.535	111.1	-23.6	-60.5	7	-10.7	-62.3	6.5
7656	295.8983	8.2641	2014.035	18.028	18.488	38.477	174.4	7.1	-64	7	1	-63.1	7.1
7657	295.9274	21.6474	2013.4905	16.325	17.371	21.117	143.2	109.4	-201.9	7.9	114.6	-204	6.3
7658	295.9638	38.1113	2013.633	10.488	14.075	6.422	75.9	-11.9	-74.4	5.3	-13.8	-73.1	5.3
7659	295.9807	21.3928	2013.7225	17.139	17.188	40.132	358	5.2	-66.7	6.2	17.2	-62.8	6.1
7660	295.9823	33.7634	2013.265	15.733	17.35	41.44	342.4	26.6	-67.9	6	23.3	-71.6	6
7661	296.0062	21.1240	2013.5885	15.12	17.051	58.725	95.1	81.6	29.6	6.1	87.6	19.3	6.1
7662	296.0166	26.3351	2013.4405	15.711	16.699	37.796	56.4	213.4	227.1	6.9	211.9	217.6	7.6
7663	296.0324	74.9022	2013.495	12.515	13.315	6.276	48.9	65.3	-53.3	5.7	70.8	-50.4	5.7
7664	296.0464	4.8751	2013.4925	17.734	10.04	31.653	169.1	16.3	-66.8	6.7	15.1	-65.1	7
7665	296.0544	1.9471	2013.4825	17.453	18.412	42.945	259.2	-5.4	-66.9	7.5	-9	-66.8	7.1
7666	296.0749	24.6623	2013.463	17.212	18.847	35.473	186.2	226.1	216	6.2	225.1	226.3	7
7667	296.0796	5.0079	2013.8685	17.388	17.847	59.285	342.9	4.6	-68.3	6.1	-0.8	-69.1	6.1
7668	296.0863	13.6895	2013.7805	16.926	17.557	44.651	40.2	-12.9	-65.5	5.6	-10.7	-65.1	5.1
7669	296.0916	26.1435	2013.6135	17.125	18.43	30.77	168.8	81.1	-11	7.5	78.3	-23.9	6.5
7670	296.0938	2.0222	2013.634	16.164	17.456	46.117	358.1	12.1	-61.1	6.2	15.7	-62	6.7
7671	296.1035	1.9785	2013.5345	17.793	17.888	35.075	338.8	-1.9	-65.9	6.7	-1.3	-64.4	7.3
7672	296.1123	2.0699	2013.523	17.099	17.451	29.666	334.3	10.7	-63.5	6.4	8	-73.9	6.6
7673	296.1338	2.1007	2013.5605	16.373	18.489	22.817	11.4	17.7	-62.8	6.2	13.4	-63.2	7.2
7674	296.1338	27.3150	2013.052	17.024	17.764	48.155	283.4	31.7	-66.2	6.2	39.9	-61.7	7.4
7675	296.1576	38.8817	2013.1395	17.315	18.525	13.849	200.8	24.8	-80.8	5.5	19.6	-84.7	5.7
7676	296.1587	22.1613	2013.9085	12.249	17.78	28.851	102.2	5.6	-67.9	6	6.8	-67.5	7.1
7677	296.1671	4.7597	2013.7455	15.782	17.157	12.689	51.6	-27	-64.9	6.2	-27.7	-62.1	6.3
7678	296.1698	43.6344	2013.2035	17.757	18.056	30.058	326.9	-12.1	-92.5	5.5	-3.1	-89.6	5.6
7679	296.1780	56.7914	2013.341	17.424	18.392	46.698	355.4	15.7	-62.1	6.4	22.3	-61.3	7
7680	296.1845	-2.8289	2013.571	13.164	14.061	4.173	258.2	19.9	-120.4	5.4	20.5	-117.7	6.6
7681	296.2090	-2.0661	2013.9375	12.755	13.727	4.719	308.3	-65.8	-119.9	5.5	-69.8	-113.4	5.3

7682	296.2247	18.7788	2013.8245	16.167	17.093	46.479	156	62.5	9.2	6.1	63	11.8	7
7683	296.2480	51.5956	2013.5625	10.087	16.077	12.549	184.8	126.1	136.1	5.3	120.6	143.5	5.4
7684	296.2517	5.2251	2013.9095	17.412	17.89	46.332	31.6	65.4	22	6.1	69.4	17.6	6.7
7685	296.2690	35.4208	2013.2655	17.28	18.898	22.296	314.7	-9.5	-69.7	5.8	-20.3	-70.6	5.6
7686	296.2856	25.4984	2013.4805	16.651	17.726	12.875	72.5	89.1	-185.1	5.2	95.2	-189.4	6.3
7687	296.3019	49.9482	2013.4045	16.905	17.51	21.884	132.1	24.4	-70.1	5.4	16.2	-70	5.4
7688	296.3536	34.8079	2013.505	17.387	18.65	40.075	148.7	-7.8	-67	6.3	2.9	-66.4	6.7
7689	296.3579	8.4391	2013.9945	11.568	14.733	5.931	350.4	40.8	-181	5.7	39.5	-182.9	5.8
7690	296.3635	22.9942	2013.4885	17.545	18.58	33.277	11	225.3	28.9	6.3	215.7	28.8	7.1
7691	296.3997	37.1991	2013.5705	9.789	14.034	6.696	26.1	-27	-108.4	5.3	-31.1	-104.5	5.4
7692	296.4314	22.6674	2013.4895	15.06	17.192	55.463	327	90.2	-46.2	6.3	96.5	-33.1	7.4
7693	296.4327	47.8123	2013.5675	12.132	13.024	19.246	262.5	73.1	39	5.3	71.1	42.7	5.3
7694	296.4644	16.6934	2014.1505	17.406	17.451	42.789	327.4	2.1	-69.4	5.1	-3.6	-66.1	6.7
7695	296.4647	11.8716	2013.9415	16.128	16.241	3.349	284.8	65	28.4	5.9	65.3	26.8	5.7
7696	296.4892	4.2484	2014.1215	7.073	10.401	10.171	180.6	75	-6.2	6.1	76.5	-14.7	6.1
7697	296.4910	11.2646	2013.905	12.095	15.244	13.882	201.2	14.7	-110.8	5.6	20.2	-113	5.7
7698	296.5294	1.5597	2013.455	16.714	17.644	44.329	193.3	-1.6	-68.5	7	-8.6	-71.1	7.2
7699	296.5382	14.4458	2013.868	16.824	17.315	33.608	166.1	-8.7	-69.3	5	-6.8	-72.9	6.7
7700	296.5987	27.7853	2013.2145	17.073	17.63	53.491	190.3	-18.2	-66	6.1	-27.3	-63.4	7.1
7701	296.6042	27.4648	2013.419	15.647	18.531	30.94	42.8	43.2	-82.1	6.6	44.1	-83.1	6.9
7702	296.6098	16.7682	2012.471	14.326	15.654	45.895	4.9	157.8	-59.9	6.5	163	-46.5	7.6
7703	296.7119	32.8474	2013.338	17.095	17.415	53.568	279	0.4	-64.1	5.6	0.2	-60.7	5.4
7704	296.7377	27.5117	2013.321	16.539	17.732	18.416	54.3	205.1	34.5	5.9	211.2	46.5	7.1
7705	296.7816	44.6953	2013.4485	15.856	16.599	15.082	123.4	-27.5	-94.9	5.4	-27.5	-95.9	5.4
7706	296.8043	25.2347	2013.5995	12.889	15.02	28.166	338.4	-27.7	-87	5.9	-27.7	-89.5	5.9
7707	296.8391	3.8066	2013.8295	15.537	17.332	21.489	153.3	82.7	-25.4	6.1	82.8	-32.6	6.5
7708	296.8414	21.5872	2013.826	16.949	17.814	58.141	134.5	69.5	-43.6	6.1	75.4	-32.7	6.3
7709	296.8524	11.2550	2013.739	13.617	16.273	58.909	219.7	-23.8	-60.7	6	-28.3	-61.3	7.9
7710	296.8954	28.3795	2013.4805	16.661	17.042	59.071	13.4	-9.9	-107.8	6.9	-6.4	-106.9	6
7711	296.9023	17.3115	2013.8955	12.222	15.207	5.622	94.2	105.9	31.8	6.2	108.1	33.7	6.3
7712	296.9125	44.6810	2013.3465	15.552	18.564	39.136	112.5	-5.4	-61	5.3	-6.2	-60.9	6.1
7713	296.9449	52.6627	2013.62	10.486	14.575	7.609	91.9	1.8	-72.1	5.3	5.4	-73.3	5.3
7714	296.9460	15.4995	2013.8885	11.872	16.234	55.469	34.8	-2.7	-62	5.7	-4.7	-64.6	5.9
7715	296.9491	-0.3253	2013.7175	10.487	14.91	11.815	179.8	41.3	-100.2	5.5	43.7	-94.6	5.5
7716	297.0162	25.3474	2013.6625	11.556	14.125	7.845	291.8	72.5	45.6	5.1	68.7	45	5.1
7717	297.0716	17.8311	2013.9695	17.021	17.8	30.939	260.9	-119.4	-75.3	6.7	-118.2	-67.4	7.2
7718	297.0760	27.4869	2013.43	15.423	17.561	52.633	267.8	95.8	-82.9	6	105	-78.5	5.5
7719	297.0836	38.6176	2013.4165	14.308	15.198	7.42	89	-32	-70.6	5.4	-32.7	-71.8	5.4
7720	297.0945	17.2503	2013.8285	16.219	18.064	8.947	272.6	-13.4	-72.7	5.9	-8.3	-71.1	5.9
7721	297.1194	34.2283	2013.3205	15.216	16.719	39.905	235.2	-61	-83.6	6	-54.5	-83.6	5.4
7722	297.2352	12.7096	2014.0185	14.516	17.916	31.681	109.1	62.9	12.4	5.8	65.6	6.2	6.4
7723	297.2596	44.2316	2013.6285	10.476	11.686	55.737	155.1	46.3	-66.7	5.3	47.5	-67.7	5.3
7724	297.2696	25.5454	2013.4405	16.963	18.188	57.732	284.5	74.2	90	7.4	61.3	102	6.4
7725	297.2819	27.8044	2013.411	17.054	17.178	23.072	61.6	6.5	-92	6.4	13	-93.3	6.2
7726	297.3223	40.0897	2013.5585	10.188	13.011	40.573	30.8	56.9	-90.1	5.3	58.9	-86.2	5.3
7727	297.3426	63.1107	2013.505	13.314	14.268	6.98	326.5	-26.5	-72.3	5.7	-28.5	-71.1	5.7

7728	297.3720	26.5397	2013.3815	17.339	18.678	30.512	94.4	-13.3	-122.4	5.3	-18.3	-127.7	7.1
7729	297.4021	2.5838	2014.0845	8.956	10.967	6.314	285.6	-84.9	-86.7	6	-74.4	-92.3	6
7730	297.4051	10.1752	2013.901	16.276	16.546	24.704	246.2	123.2	92.7	5.7	122.6	96.5	5.9
7731	297.4494	28.6014	2013.4	16.862	17.331	27.268	73.3	61.1	27.1	5.5	62.7	29.9	5.3
7732	297.5019	20.3658	2012.471	12.663	13.752	54.814	217.3	179.1	238.1	7.4	165.6	243.7	7.4
7733	297.5157	38.6433	2013.2565	18.104	18.35	53.464	168.1	10.5	-61.6	5.5	0.7	-65.4	5.6
7734	297.5159	27.3658	2013.366	17.074	17.934	25.541	50.1	54.8	-72	6.6	60.5	-63	7.8
7735	297.5663	28.6866	2013.4025	16.679	17.877	28.091	103.5	25.3	-68.9	5.2	32.5	-66.1	7.6
7736	297.5731	13.8822	2013.9635	17.493	17.888	6.914	39.5	-5.3	-60.2	5.1	2.1	-60.5	5.5
7737	297.5745	7.6927	2013.9555	16.115	16.476	33.097	246.8	0.6	-64.3	5.6	1.9	-64.6	5.7
7738	297.5793	23.6513	2013.503	13.391	13.781	54.824	7.9	-29.1	-71.1	6.1	-17.5	-75.6	6.1
7739	297.5821	5.5698	2013.815	16.494	18.145	33.891	98.6	21.8	-64.1	6.7	14.4	-64	6.9
7740	297.6755	47.2212	2013.6235	12.473	14.081	10.474	243.2	-99.3	-74.6	5.3	-100.2	-74.8	5.3
7741	297.6956	27.5433	2013.5065	17.823	17.994	36.043	49.6	-58.2	-103.9	5.4	-50.1	-109.5	5.7
7742	297.7586	-0.3428	2013.685	15.333	16.474	30.446	259.2	-41.2	-115.3	5.5	-37.8	-118.7	5.8
7743	297.8583	20.8299	2013.7315	17.834	18.828	35.874	109	-15.6	-77.5	6.8	-6.5	-77.5	7.4
7744	297.8788	34.6025	2013.352	15.24	17.913	48.96	151.3	-49.4	-75.9	5.9	-57.8	-69.4	6.1
7745	297.8835	32.2836	2013.5835	12.141	12.891	8.387	247.8	145.6	-5.6	5.3	141.7	-6.5	5.3
7746	297.9790	29.7154	2013.303	16.527	17.599	38.101	137.4	75.1	18.1	5.5	76.8	14.6	6.8
7747	297.9977	35.0469	2013.4965	17.767	18.974	23.217	23.1	-23.6	-62	5.5	-25.5	-62.9	7.9
7748	298.0018	38.0320	2013.5985	9.282	9.431	11.471	167.7	66.3	92.8	5.3	63.4	94.8	5.3
7749	298.0256	31.2506	2013.7	15.041	18.076	54.118	293.9	106.5	-74.6	6.2	101.4	-76.7	6.2
7750	298.0329	29.3779	2013.595	15.136	15.793	17.335	48.5	40.1	-133.6	5.6	40.1	-130.9	5.3
7751	298.0572	18.8378	2013.077	15.357	17.422	35.017	263.8	200.5	129.3	7.9	191.2	122.7	6.5
7752	298.1218	46.5519	2013.5395	13.741	14.157	12.736	122.4	78.7	97.6	5.4	79.3	99	5.4
7753	298.1222	29.0185	2013.559	13.764	14.807	18.762	70.2	17.6	-65.5	5.3	19.2	-62.9	5.3
7754	298.1312	26.1576	2013.6375	13.675	14.911	8.019	123.8	-30.6	-65.8	5.2	-20.8	-66.6	5.2
7755	298.1619	36.5684	2013.6905	16.179	16.614	57.044	322.4	-73.8	-194.1	7.1	-75.6	-188.8	5.3
7756	298.1906	85.0834	2013.704	15.303	16.207	9.902	259.9	-13.3	92.4	5.8	-11.2	94	5.8
7757	298.1927	67.5339	2013.549	13.154	13.225	29.169	85	-24	-63	6.1	-29.9	-63.2	6.1
7758	298.2301	43.2734	2013.487	9.852	11.898	5.057	257	118.3	105.3	5.4	120	100.4	5.5
7759	298.2693	28.8997	2013.41	17.865	18.565	47.768	123.3	-126	-122.1	6.2	-134.1	-121.2	6.2
7760	298.2696	32.8740	2013.487	16.936	18.054	50.056	25.9	62.2	-148.7	5.6	69.6	-152.7	5.7
7761	298.2826	27.7901	2013.229	16.69	18.243	18.888	57.8	15.4	69	5.4	14	68.8	5.6
7762	298.3577	12.1446	2014.0585	17.722	18.699	49.073	276.3	-40.4	-71.2	5.6	-49.6	-64.3	7.3
7763	298.4052	28.5520	2013.4085	16.234	17.621	26.535	277.7	-16.8	-79	5.3	-6.6	-78.5	5.9
7764	298.4110	29.4284	2013.5595	17.585	17.775	41.453	260.1	80.8	62	5.4	90.4	51.8	5.5
7765	298.4342	31.1376	2013.9825	15.36	16.779	16.274	261.8	0.4	-70.6	6.7	11.7	-69.6	5.5
7766	298.4478	13.9813	2013.8985	8.804	15.082	15.207	306.7	28.1	-84.1	5.2	26.3	-85.9	5.2
7767	298.4657	78.5553	2013.46	14.518	18.728	13.319	299.1	-3.7	61.9	5.7	-6.1	62.6	6.1
7768	298.4676	75.0275	2013.5575	11.107	12.276	16.036	225.5	-203.4	-145.8	5.8	-210.7	-144.9	5.8
7769	298.4711	26.9963	2013.9015	16.584	17.885	58.929	336	-82.3	-78.7	5.3	-86.6	-69	7.3
7770	298.5080	43.6822	2013.1925	17.765	18.456	52.557	46.4	63.1	-17.4	5.5	66.6	-11.3	6
7771	298.5222	31.4005	2013.6565	16.903	17.305	54.512	11.4	-59.5	-198	7.8	-59.1	-190.4	6.7
7772	298.5381	1.2316	2013.4735	16.948	17.813	50.265	305.2	65.4	-11.1	6.8	64.3	-12.6	6.8
7773	298.5676	31.1841	2013.6525	17.341	17.57	41.233	162.5	14.6	-64.3	7.2	7.5	-67.6	5.4

7774	298.5856	29.4144	2013.6835	16.862	17.821	51.22	154.1	15.5	-73.9	5.4	25.9	-72.7	5.4
7775	298.5904	60.6200	2013.0805	17.398	18.663	55.469	352.4	17.9	77.9	5.9	17.3	77.9	6.4
7776	298.6094	32.8258	2013.494	13.043	16.212	29.869	146.2	23.3	77.7	5.3	21.9	76.2	5.4
7777	298.6340	15.3440	2013.9505	17.866	18.232	18.728	178.1	-0.1	-67.8	5.2	-2.5	-63.4	5.5
7778	298.6366	6.5592	2013.8325	14.274	14.537	5.811	96.2	69.5	-33.7	6.1	67.8	-31.9	6.1
7779	298.6553	67.8096	2013.4515	16.603	18.651	8.306	25.1	40.9	116	6.2	36.8	115.8	7.5
7780	298.6726	75.6406	2013.535	12.199	13.995	5.542	170.8	44.9	88.3	5.7	46.9	89.5	5.7
7781	298.6834	83.4438	2013.7095	12.708	14.112	7.656	140.1	72.7	-36.7	5.6	74.7	-38.4	5.6
7782	298.7010	28.4295	2013.4845	16.631	18.53	34.529	16.9	-50.7	-154.6	6.2	-51.3	-149.7	7
7783	298.7068	70.3690	2013.497	12.708	15.772	11.37	358.5	14.9	60.2	6.1	16.3	62.3	6.1
7784	298.7128	53.3284	2013.705	12.007	12.192	5.898	309.1	84.7	62.1	6.1	83.8	63	6
7785	298.7480	44.4292	2013.1205	17.078	18.841	38.168	173.8	83.1	32.3	6.7	85.7	19.1	6.7
7786	298.7576	29.4732	2013.628	16.88	18.031	45.8	16.2	-8.8	-154.1	6.5	-1.4	-151.2	6.7
7787	298.7945	8.0461	2013.6985	17.501	17.607	46.621	125.5	62.5	23	6.5	61.6	17.9	6.4
7788	298.8059	49.3718	2013.49	14.983	16.702	10.372	311.7	63.5	60.8	6.1	63.9	63.9	6.2
7789	298.8065	54.1707	2013.6025	9.785	13.939	15.782	340.2	38.3	79.1	6.3	34.2	84.5	6.3
7790	298.8373	29.2412	2013.4865	17.111	17.238	14.361	10.9	70.4	50.7	6.5	68.6	49.7	6.2
7791	298.8882	31.2903	2013.5615	11.204	15.722	31.081	89.8	22.2	61.5	5.3	22.7	61.2	5.3
7792	298.8899	18.9130	2013.8075	11.056	15.865	10.826	245.4	16.8	-81.2	6	17.4	-84.4	6.1
7793	298.9296	-3.6742	2013.6435	14.13	15.7	28.638	118.6	-15.1	-103.9	5.8	-21.6	-100.3	5.9
7794	298.9789	35.9716	2013.4665	16.131	16.606	31.666	44.5	100.9	13.5	5.7	96.2	15.6	5.4
7795	299.0444	19.7407	2013.71	15.569	17.61	45.334	125.3	-1.1	-65.2	6.1	-3	-64.2	6.5
7796	299.1256	35.0819	2013.498	18.657	18.819	17.296	270.3	-10.6	-73.4	5.8	-12.7	-75.4	6.7
7797	299.1308	64.6171	2013.373	17.208	17.627	12.272	244.2	9	-63.2	5.8	3.2	-62.5	5.8
7798	299.1574	-12.6729	2014.445	11.049	11.358	19.137	157.2	-14.6	-120.1	5.7	-13.2	-122.2	5.6
7799	299.1897	16.3336	2013.778	15.846	17.536	42.34	26	-27.3	-65.2	5.5	-23.6	-64.7	5.2
7800	299.2415	29.8119	2013.0485	15.887	16.938	52.284	69.1	-30.6	-69.6	6.4	-37.4	-67.9	7.6
7801	299.2632	0.0542	2013.684	12.869	15.394	8.216	231.9	43.1	-85	5.4	44.2	-89.1	5.5
7802	299.2636	83.3076	2013.768	9.114	9.299	5.741	50.5	40.5	-60.5	5.7	33.2	-64.2	5.9
7803	299.2995	30.8827	2013.593	16.436	17.902	17.905	140.2	147.9	72.3	5.3	156.9	62.9	5.5
7804	299.3014	17.9855	2013.8415	17.595	17.62	26.274	72.1	-22.4	-65	6.3	-23.9	-64.3	7
7805	299.3152	0.9052	2013.577	14.739	17.326	8.308	90.3	-46.5	-76.7	6.5	-52.9	-75.6	6.6
7806	299.3184	-6.0552	2013.487	15.645	17.361	8.707	177.7	28.2	-81.7	5.6	31	-80	7.1
7807	299.3490	4.1425	2013.6465	16.486	18.395	36.187	31.1	-17.2	-70.2	7.2	-17.7	-72	6.8
7808	299.3552	45.1243	2013.6195	14.257	14.76	17.702	66.8	120.4	225.1	6.2	121.7	222.1	6.2
7809	299.3598	23.4794	2013.7065	11.67	13.733	8.092	173.8	27.9	-88.3	6	24	-86.6	6
7810	299.3633	-4.7427	2013.704	11.236	14.607	8.955	282.5	60.8	-98	5.5	58.1	-100.9	5.5
7811	299.4563	19.4666	2013.807	15.659	17.031	12.183	134.7	81.7	-21.3	6.4	81.4	-28.5	6.1
7812	299.4622	18.0608	2013.852	16.744	18.092	20.67	181.7	-136.9	-211.2	6.6	-132.9	-203.5	6.4
7813	299.4997	50.7367	2013.523	12.537	13.724	4.393	157.2	-21.1	-60.6	5.9	-22	-60.4	6
7814	299.5386	29.8784	2013.432	13.146	18.007	11.269	147.3	79.7	43.5	6.1	78.9	40.2	6.4
7815	299.5528	29.4364	2013.504	17.23	18.616	50.693	201.4	105.8	38.9	6.2	111.7	35.4	6.8
7816	299.5795	29.4513	2013.4835	17.489	17.639	55.054	218.3	4.4	-62.7	6.4	1.4	-61.2	7
7817	299.5856	21.5851	2013.4325	16.106	17.983	41.278	93.7	80.3	-17.4	6.6	80.4	-24.3	6.7
7818	299.6046	31.1727	2013.552	16.849	17.793	41.87	23	131.8	-160.2	5.3	131.2	-151.1	5.5
7819	299.6595	50.9513	2013.44	17.139	18.485	55.335	56.9	-1.8	-62.7	6.2	8.5	-62.1	6.4

7820	299.6934	29.8099	2013.5025	11.923	16.36	47.954	206.3	-24.6	-71.2	6.1	-15.5	-74.4	6.2
7821	299.8076	27.2078	2013.5555	17.151	18.424	23.825	47.6	-141.7	-78.7	5.5	-146.5	-72.5	7.2
7822	299.8285	47.3250	2013.1435	17.792	18.564	32.067	261.4	70.1	11	6.4	73.4	3.7	6.5
7823	299.8578	36.6129	2013.883	16.649	17.202	55.078	261.2	31.3	-181.9	5.7	30.8	-173.7	5.5
7824	299.8683	30.3200	2013.056	15.554	16.387	46.833	103.5	90	-2.4	6.9	87.8	-10.3	6.5
7825	299.8886	31.3725	2013.441	16.108	18.428	35.913	155.7	226.9	212.6	6.2	219.1	221.1	7.3
7826	299.8887	30.9200	2013.6015	17.928	18.436	26.352	32.9	76.8	144	6.3	79.8	145.8	6.4
7827	299.9453	34.5396	2013.4595	15.147	16.289	4.579	166	-51.4	-123.3	5.5	-47	-121.1	5.5
7828	299.9871	33.4113	2013.5105	18.039	18.403	33.611	347.3	196.9	167.3	5.4	192.1	168.2	6.2
7829	299.9956	5.8528	2013.856	9.973	16.138	23.038	209.7	63.2	53.3	5.7	64.3	53.9	5.8
7830	300.0031	38.8194	2012.896	17.872	18.067	30.537	189.1	124.6	207.3	6.4	120.4	201.3	7
7831	300.0332	32.7480	2013.2405	18.188	18.31	19.791	165	67.6	36.6	6.4	68.6	32.8	5.7
7832	300.0435	19.5390	2013.6075	17.559	18.071	59.236	237.1	74.1	-21.2	6.1	76.1	-21.7	6.2
7833	300.0575	15.3030	2013.87	15.115	17.442	30.369	339.6	2.2	-64.2	5.1	7.7	-64.1	5
7834	300.0591	47.6194	2013.6855	9.255	10.4	43.968	256.8	76.2	64.4	5.5	70.9	67.5	5.5
7835	300.1397	9.3528	2014.048	11.206	17.921	57.241	250.1	-186.6	-185.9	6.2	-196.1	-189.9	7.4
7836	300.1480	-4.3591	2013.8325	11.556	12.823	23.551	37.7	91.7	-40.7	5.5	89.1	-41.9	5.4
7837	300.1645	24.1955	2013.657	10.847	16.628	29.727	37.4	-15.3	-91.9	5.4	-18.5	-88	5.5
7838	300.1799	36.6983	2013.3585	15.216	17.904	8.707	149.7	-4.6	-63.5	6.2	-4.2	-65.6	6.6
7839	300.2222	45.0161	2013.605	13.968	14.377	40.445	281	87.4	24.1	5.5	84.7	22.9	5.5
7840	300.3138	2.7202	2013.8335	13.791	13.91	16.229	210	94.1	17.3	6.2	95.8	19.7	6.2
7841	300.3451	30.4817	2013.4035	15.961	16.781	21.469	340.1	-63.8	-93.7	5.7	-72.6	-87.8	5.7
7842	300.3740	4.0674	2013.5905	17.488	18.08	58.864	69.2	11.2	-68	7.3	22.6	-65.8	6.6
7843	300.3889	29.1049	2013.629	15.827	16.674	38.936	33.6	39.6	-189.4	5.8	42.9	-184.2	5.5
7844	300.4233	33.8468	2013.4835	10.724	16.766	12.088	186.7	-43.3	-77.1	5.3	-41.8	-74.7	5.5
7845	300.4902	30.3519	2013.4865	14.916	15.003	33.247	302.1	-45.2	-75.4	5.4	-54.1	-74.7	5.4
7846	300.4955	10.9856	2014.04	16.805	17.337	53.711	74.4	28.2	-61.2	6.1	20.6	-60.9	6
7847	300.5035	56.8736	2013.5815	10.042	16.787	58.132	39.2	87.9	53.7	5.7	85.8	51	5.7
7848	300.5609	31.8439	2013.4935	18.041	18.071	40.244	39.9	-53.3	-67 4	5.6	-45.9	-76.6	5.6
7849	300.6355	58.5114	2013.6485	9.88	10.542	18.137	182.9	62.2	19.7	5.6	64.8	16.2	5.7
7850	300.7008	4.6228	2013.71	14.455	17.258	14.301	29.4	-26.7	-69	6	-24.9	-72.6	7.9
7851	300.8093	23.0341	2013.62	16.838	18.093	42.836	222.8	66.8	4.7	6.7	65.5	7.8	6.8
7852	300.8414	23.2833	2013.6	11.734	16.284	51.477	277.7	64.1	41.5	5.9	61.3	42	5.9
7853	300.8622	31.5822	2013.5425	13.136	14.472	59.323	60	-2.6	-68.4	5.3	-6.3	-68.8	5.5
7854	300.8721	10.5013	2013.9575	17.438	18.334	25.642	37.4	-10	-61.6	6.4	1.1	-62.6	6.9
7855	300.8848	19.6428	2013.4285	13.557	17.1	56.343	90.1	94.4	20.6	5.8	98.3	14.8	6.7
7856	300.8898	30.1091	2013.6275	17.426	18.573	23.315	39.9	6.3	-61.8	6.3	3.5	-63.5	6.1
7857	300.9111	46.2944	2013.286	17.743	17.919	16.889	229.9	-7.8	-70.6	5.7	0.6	-68.4	6.2
7858	300.9294	39.2501	2013.204	17.789	17.924	49.742	278	12.4	-62.8	5.6	10	-61.4	7.6
7859	300.9299	5.2221	2013.702	11.89	13.734	5.722	61.3	24	-77.2	6.1	22.7	-78.5	6.3
7860	300.9312	-4.3759	2013.6745	8.614	15.817	36.714	220.3	-68.5	-154.1	6.1	-70.7	-146.8	5.8
7861	300.9982	32.5077	2013.553	10.346	16.644	21.084	87.7	78.6	20.4	5.3	77.9	22.9	5.4
7862	301.0188	49.0436	2013.356	17.428	18.171	16.878	145.2	-1.7	-61.6	6.2	8.7	-63.2	6.3
7863	301.0192	-12.7110	2014.445	13.697	14.681	33.894	341.6	87	-36.2	5.2	87.4	-33.8	5.3
7864	301.0204	18.0767	2013.53	16.256	17.99	25.425	260.1	-5.9	-109.7	6.2	-0.8	-113.3	6.9
7865	301.0372	32.8273	2013.5205	13.776	17.818	47.982	324.4	-60.7	-88.8	5.2	-67.1	-84.5	5.6

7866	301.0468	34.1532	2013.3175	18.079	19.003	28.703	217.9	128.2	-122.8	5.7	122.6	-123.4	5.7
7867	301.0524	31.6461	2013.659	14.717	18.325	13.036	62.9	91.1	-124.3	5.8	99.2	-123.4	6.2
7868	301.0527	7.6473	2013.851	15.52	16.447	54.433	106.3	-5.6	-65	5.9	-1.7	-63.7	6.3
7869	301.0886	37.3305	2013.3175	16.402	17.183	28.628	176.8	-7.6	-65.1	5.6	-13.4	-61.5	5.5
7870	301.1717	28.3289	2013.57	16.281	16.612	4.512	70.7	115.9	-49.4	5.1	110.9	-55.6	5.8
7871	301.1886	19.1346	2013.6435	16.762	17.277	27.637	328.9	98.9	-6.8	6.3	103.8	4.6	7.4
7872	301.2401	25.4269	2013.651	11.31	14.994	24.545	355.5	-137.8	-139.1	5.2	-137.8	-138.4	5.2
7873	301.2674	36.0732	2013.653	13.849	18.721	55.591	216.6	-3.2	-68.9	6.1	-8.1	-65.2	6.4
7874	301.3129	2.2594	2013.5315	17.632	17.964	31.782	32.4	-28	-60.9	6.3	-22.7	-60.8	6.8
7875	301.3736	8.5075	2013.9855	9.79	18.551	23.598	48.7	-52.8	-90.5	6	-55.7	-90.4	7.6
7876	301.3973	34.6455	2013.664	18.201	18.774	30.213	121.6	-48.1	-157.7	5.6	-41.6	-155.2	6.1
7877	301.4497	33.5940	2013.6795	13.807	14.275	5.478	72.7	-12.1	-70.4	5.3	-11.7	-67.1	5.2
7878	301.4787	10.5816	2013.953	11.154	14.114	49.167	4.9	-9.2	-60.9	5.9	-6.5	-63	5.9
7879	301.5516	-1.7192	2014.003	7.983	13.658	23.884	24.8	-25.1	-75.3	5.4	-30.9	-73.9	5.4
7880	301.5818	48.6717	2013.486	14.837	16.635	11.717	72.3	-24.2	-73.7	6.1	-27.3	-74.6	6.2
7881	301.6526	9.4979	2013.6785	11.734	17.639	37.611	18.6	-21.9	-63.8	6	-33.9	-60.5	6.2
7882	301.6814	15.6000	2013.7365	12.74	15.991	8.008	321.6	-17.8	-61.5	6.1	-15.8	-64.2	6.1
7883	301.6896	53.3291	2013.338	17.346	18.069	54.939	183.7	76.9	4.4	6.3	75.5	-6.7	7.2
7884	301.7462	50.3159	2013.4645	13.183	17.495	31.014	224.7	-68.5	-78.8	6.1	-62.5	-88.7	6.2
7885	301.7568	35.2897	2013.2785	15.588	18.531	55.038	302.1	79.5	-6.4	5.3	83	-3.7	7.5
7886	301.8072	56.4923	2013.522	14.641	15.862	12.161	11.9	102.8	54.2	5.7	101.1	52.8	5.8
7887	301.8258	-3.4684	2013.511	15.706	15.873	6.438	128.5	56.6	-91.5	5.5	59.4	-91.5	5.6
7888	301.8498	-2.0741	2013.791	11.122	13.686	8.858	248.7	41.2	-67.5	5.4	39.1	-68	5.4
7889	301.8689	33.7102	2013.428	16.933	18.084	7.678	176.4	18.8	-64	5.8	6.8	-69.3	6.8
7890	301.9137	23.4245	2013.642	15.258	16.195	8.619	303.6	-61.2	-114.1	6.1	-64.3	-119	6.1
7891	301.9840	51.7413	2013.6665	10.929	15.304	7.064	162	-30.1	-74.9	6.1	-33.8	-70.9	6.1
7892	301.9962	62.4160	2013.347	17.971	18.245	9.716	39.3	100.3	116.2	6	97	119.4	6.1
7893	302.0049	66.6336	2013.5475	14.671	15.877	53.424	223.9	-53.9	-70.7	5.7	-52.2	-74	5.7
7894	302.0319	62.4486	2013.6015	10.79	10.926	5.259	302.8	-62.6	-64.1	5.6	-60.7	-62.9	5.7
7895	302.1265	59.0008	2013.4815	14.588	16.769	6.124	38.6	-59.7	-80.5	5.7	-59.9	-76.5	5.9
7896	302.2324	41.9136	2013.5335	13.413	15.706	6.334	275.8	78.2	92.7	5.4	77.7	91.7	5.4
7897	302.2814	2.9805	2013.565	16.47	17.541	29.567	341	8.6	-75.8	6.3	10.1	-78.3	6.7
7898	302.3241	38.8352	2013.1495	18.381	18.446	32.536	285.1	-8.8	-62.8	5.5	-18.2	-60.2	5.8
7899	302.3511	-9.7651	2013.954	13.635	14.898	48.545	190.4	28.1	-119.9	5.8	33.1	-120.8	5.8
7900	302.3578	17.0737	2013.965	17.393	18.471	39.669	217	3.2	-71	5.6	5.4	-70.1	5.6
7901	302.3621	-5.3137	2013.5425	15.982	17.9	7.58	210.6	52.7	-102.5	5.6	48.3	-105	6
7902	302.3800	31.7958	2013.528	13.037	16.218	28.336	278.9	-32.4	-91	6.1	-33.2	-87	6.1
7903	302.4036	51.2016	2013.4765	16.588	17.247	42.218	15.6	-120.8	-143.9	6.3	-117.9	-153	6.3
7904	302.4762	8.3375	2013.7865	16.499	17.855	14.084	313.9	-2.6	-62	6	-4.2	-63.2	6.4
7905	302.4771	0.0338	2013.68	14.806	17.85	53.488	76.9	60.4	1.8	5.3	61.5	5.4	5.8
7906	302.4876	25.7290	2013.5925	17.759	18.065	30.607	48.5	-45.2	-71.5	5.1	-48.4	-71.4	5.9
7907	302.4880	48.5553	2013.4375	16.058	16.298	7.65	112.3	94.4	-43.4	6.2	93.4	-46.8	6.2
7908	302.5088	35.6341	2013.3395	17.586	17.831	35.173	281.8	114.7	173.8	7.1	111.6	167.1	5.4
7909	302.5151	53.4984	2013.326	18.559	18.597	47.867	176.1	2.5	-67.9	6.4	8.3	-64.2	6.4
7910	302.5322	8.4427	2014.093	6.739	9.825	7.52	80.6	-70.1	-185.2	6.6	-76.9	-190.5	6.1
7911	302.5356	38.5168	2013.4565	14.729	17.218	9.305	326.7	-12.2	-61.1	5.3	-12.6	-62.2	5.5

7912	302.5377	39.7439	2013.069	17.469	18.051	54.824	307.7	15	-72.9	6.8	12.8	-73.2	5.7
7913	302.6085	32.7560	2013.4345	16.559	17.168	46.715	304.9	148.4	0.7	7.8	144.5	1.9	5.4
7914	302.6270	35.2617	2013.5045	17.74	18.439	28.259	319.2	-4.4	-103.8	5.9	-6.4	-108.7	7.6
7915	302.6869	38.8836	2012.884	17.814	18.137	31.251	135.5	31.4	-60.4	5.7	20.1	-68.4	7.8
7916	302.7187	49.4240	2013.652	10.342	12.179	10.11	257.2	-16	-63.2	6.1	-16.9	-61.7	6
7917	302.7848	-7.1593	2013.751	13.039	14.383	7.894	216.3	81.9	-14	5.8	83.7	-15.5	5.8
7918	302.8085	56.6334	2013.5295	15.381	16.168	10.989	236.6	-45.3	-68.3	5.7	-42	-69.8	5.7
7919	302.8374	-5.6949	2013.638	13.376	17.019	51.186	157.7	-21.2	-65.2	5.3	-24.7	-64.2	7.4
7920	302.8638	52.0147	2013.385	15.404	18.191	54.989	60	83.1	28.3	6.1	87.8	25.4	6.3
7921	302.9294	18.3934	2013.594	16.845	17.294	7.158	242.7	63.8	-10.3	6.2	64.1	-9.6	6.1
7922	302.9766	10.5180	2013.977	15.207	16.678	22.698	222.9	-134.9	-116.8	6	-136	-113	6.3
7923	302.9924	39.3532	2013.172	16.889	18.384	59.56	100.6	-14.5	-62.5	5.5	-16.2	-64.2	5.8
7924	303.0343	42.3715	2013.456	13.694	15.028	22.949	141.9	-30.3	-73.3	5.5	-31.2	-71.4	5.5
7925	303.0548	52.2766	2013.5685	13.658	16.619	31.542	132.1	-50.7	-85	6.1	-56	-79.4	6.1
7926	303.0767	18.1092	2013.4625	16.447	18.355	45.017	302.6	69.8	28.7	6.2	72.3	20.2	6.5
7927	303.1096	14.1519	2013.889	10.289	14.704	32.68	217.3	18.9	-66.7	5.3	10.7	-66.7	5.3
7928	303.1590	37.4954	2013.59	15.126	15.719	6.282	264.5	-8.5	-68.1	5.3	-9.9	-68.1	5.4
7929	303.1895	16.4072	2013.7515	13.347	14.422	5.317	196.1	-2.4	-68.1	5.4	-5.7	-67.6	5.4
7930	303.2052	15.7975	2013.574	12.192	12.435	5.016	32.2	118.5	39.8	5.4	116.3	37.8	5.5
7931	303.2164	57.9977	2013.4545	10.743	18.391	17.963	173.2	-9.9	-65	5.7	-10.9	-64.9	5.8
7932	303.2296	23.5519	2013.8095	17.916	17.962	44.673	35.8	61.8	-36.7	6.3	62.3	-34.2	7.1
7933	303.2963	2.9390	2013.8875	9.504	10.601	32.491	123.9	107.2	21.2	6.1	105	17.3	6.2
7934	303.3058	-5.2394	2013.6605	11.753	15.809	13.316	316	31.3	-135.4	5.5	29.5	-132	5.5
7935	303.3061	36.2734	2013.5475	15.415	17.201	50.908	57.1	-1.8	-68.2	5.8	-0.2	-71	5.9
7936	303.3976	-11.6591	2013.9805	13.681	14.585	8.528	211.2	65.2	-138.7	5.9	65.4	-140.4	6
7937	303.4069	-5.0772	2013.8415	15.328	15.372	4.869	100	73.8	-6.3	6.8	73.9	-13.5	6.7
7938	303.4528	70.2459	2013.493	17.227	19.048	48.1	105.5	62.2	16.5	6	62.4	24.1	6.3
7939	303.4557	26.9113	2013.7535	8.659	16.985	18.882	253	75.1	97.5	5.9	67.4	97	6
7940	303.4746	27.8407	2013.5645	14.026	17.737	15.702	232.9	73	-85.5	5.9	-76.5	-80	6
7941	303.4897	33.8561	2013.336	16.524	17.073	6.351	315.5	61.3	33.1	6	60.9	27	6.1
7942	303.5236	34.3523	2013.3395	17.667	17.785	52.378	154.2	-60.5	-97.4	7.7	-51.4	-108.6	7.4
7943	303.5283	37.2120	2013.4055	15.85	17.106	34.252	260.3	100.3	74.9	6.1	97.2	70.2	5.8
7944	303.5389	7.7716	2014.0195	17.59	18.752	38.442	236.7	-26.1	-75.6	6	-29.2	-77.5	7.7
7945	303.5726	24.1517	2013.5715	15.792	16.795	5.615	100.8	37.3	-77.8	5.9	36.6	-81.1	5.9
7946	303.6345	20.8903	2013.368	17.916	18.861	35.779	171.2	10.6	-64.1	6.5	6	-67.1	6.7
7947	303.7941	-3.5038	2013.845	7.122	8.062	14.373	232.9	-51.9	-60.2	5.5	-51.2	-61.8	5.5
7948	303.7969	31.1866	2013.507	11.649	16.78	35.02	72.6	13.6	-63.9	5.8	11	-65.6	5.9
7949	303.8137	50.1635	2013.114	17.795	18.967	56.445	348.2	7	-84.6	6.3	10.2	-82.7	7.2
7950	303.9090	15.6917	2013.6765	16.568	17.827	54.334	110.2	10.1	-64.9	5.5	13.5	-66.4	5.6
7951	303.9206	13.0890	2013.844	16.81	17.472	5.146	60.2	-58.5	-69.7	5.5	-55.6	-70.1	5.8
7952	303.9481	11.0177	2013.896	11.405	12.932	8.518	269.5	90.2	81.4	6	88.1	85	6
7953	303.9782	16.9128	2013.47	13.021	13.849	15.05	14.4	-43.2	-77.8	5.5	-43.5	-76.7	5.5
7954	304.0238	-11.9775	2014.0495	12.26	12.388	38.355	174.7	-72.8	-150.8	5.3	-69	-147.9	5.3
7955	304.0519	19.4529	2013.6675	14.168	14.642	22.113	41.2	90.3	-120.4	5.7	92.5	-126.1	5.8
7956	304.1342	19.5274	2013.6215	12.54	17.421	9.934	18.1	72.8	51.4	5.7	76.3	46.2	5.8
7957	304.1722	17.4796	2013.5795	17.248	18.133	40.228	186.6	66.9	25.2	5.5	70.7	23.4	5.6

7958	304.1847	1.9738	2013.847	12.044	13.784	25.234	250.8	35	-166.3	6.2	36	-166	6.3
7959	304.2524	15.4427	2013.602	13.017	14.5	8.734	73.2	62.4	-13.1	5.4	60.7	-12.2	5.4
7960	304.4026	9.6735	2013.755	11.366	14.393	12.43	106.2	68.8	15.2	6.2	70.1	15.1	6.2
7961	304.4082	1.1092	2013.453	18.025	18.114	9.456	86	4	-64.7	7.3	1.2	-64	7.4
7962	304.5123	-5.6099	2013.459	17.367	17.706	24.078	224.6	-10.8	-83.1	6	-1.8	-80.6	6.1
7963	304.5285	11.6126	2013.759	12.506	16.854	14.002	314.5	17.7	-67.4	6.2	13.6	-69.1	6.2
7964	304.6269	14.0792	2013.8535	17.283	17.555	49.883	291.8	28	-63.5	5.5	22.6	-68.7	6
7965	304.6362	31.4210	2013.585	15.381	17.102	41.363	53.4	92	-52.9	6.1	87.1	-62	7.2
7966	304.7511	14.7817	2013.5665	13.264	13.304	7.479	60.9	102.5	139.1	5.5	101.4	138.9	5.6
7967	304.7903	10.2493	2013.8795	15.801	15.999	8.612	317.5	46.4	-92.7	6	45.1	-94.3	6.2
7968	304.8417	21.0230	2013.645	11.318	17.605	12.634	126.1	-20.6	-91.5	6.1	-21.1	-92.6	6.4
7969	304.9188	12.9032	2013.756	17.327	17.507	18.369	30.7	36	-62	5.7	30.3	-68	5.6
7970	304.9505	62.6096	2013.468	12.59	16.305	13.114	122.3	64	65.3	5.7	64	70.1	5.8
7971	305.0010	83.0932	2013.7425	11.687	16.59	7.668	145.8	92.1	199.2	5.7	86.3	207.1	6.1
7972	305.0215	15.6688	2013.6055	15.066	16.011	24.649	40.1	-8.7	-61.4	5.4	-11.1	-61	5.4
7973	305.0572	1.9970	2013.7195	10.947	12.604	7.356	331.6	-6.8	-61	6.2	-10.5	-60.6	6.3
7974	305.1498	22.7196	2013.356	14.953	15.582	29.978	105.1	3.3	-62.6	6.3	-0.7	-61.7	6.4
7975	305.1596	-6.7979	2013.6145	14.662	16.656	54.388	357.7	111.3	-75.2	6	104.2	-82.3	6.1
7976	305.1722	-2.8756	2013.629	14.095	15.678	56.089	228.9	-88.5	-137.7	5.6	-89.1	-145.7	5.6
7977	305.1950	43.9780	2013.4825	15.233	16.568	4.578	229.4	-41.7	-125.7	6.3	-37.1	-125.9	6.5
7978	305.2125	33.7648	2013.6255	11.897	13.3	5.188	188.9	6.4	-158.5	5.3	13.2	-158.5	5.5
7979	305.2251	44.8872	2013.4935	15.036	16.042	22.167	53.3	86.1	62.3	6.3	85.2	60.9	6.4
7980	305.2542	0.6019	2013.794	13.644	14.908	9.97	77.8	96.6	0.5	6.2	96.8	3.2	6.3
7981	305.2914	29.8729	2013.5505	12.739	17.572	32.134	8.8	-18.2	-72.2	5.3	-17.1	-71.4	5.4
7982	305.4559	19.8304	2013.582	18.562	18.933	56.714	152.3	77.1	74.5	7.4	77.2	76.2	6.7
7983	305.4795	16.3748	2013.505	16.672	16.721	4.551	13.9	164.9	-41.4	6.2	172.5	-43.7	6.1
7984	305.5404	22.3471	2013.6395	8.489	14.315	21.659	170.7	-138.6	-103.1	6.1	-132.1	-100.8	6.1
7985	305.5733	15.6839	2013.5255	17.911	18.086	51.454	129.8	-3.7	-70.3	7.3	3.8	-69.6	6.5
7986	305.6594	15.1814	2013.6685	13.974	16.44	19.382	24.9	-93.7	-68	5.9	-89.2	-71	5.9
7987	305.6632	7.8324	2013.4475	16.759	17.524	14.775	35.8	64.9	22.7	6.9	67.1	12.3	6.7
7988	305.6843	30.0596	2013.4745	16.776	18.189	13.089	28.4	-27.8	-66.5	5.4	-29.2	-68.3	5.7
7989	305.8141	12.7557	2013.4545	13.555	13.888	9.132	269.2	-54.6	-117	5.9	-50.4	-116.2	5.9
7990	305.9080	13.7817	2013.6355	14.543	17.773	27.094	65.4	79.4	98.2	5.9	80.4	96.3	6.5
7991	306.0109	33.1429	2013.513	10.866	12.829	4.766	13.4	144.2	71.5	5.3	143.1	75.6	5.5
7992	306.0780	44.3094	2013.561	13.813	13.935	4.176	211.4	-8.8	-91.5	5.5	-8.4	-91.9	5.6
7993	306.0806	-4.1248	2013.7115	11.032	11.255	25.085	36.4	122.2	-50.6	5.5	123.3	-44.6	5.5
7994	306.1082	53.9591	2013.49	10.534	17.873	35.358	235.7	-49	-75.6	6.3	-50.3	-71.3	6.5
7995	306.1576	-6.0400	2013.711	12.979	13.657	54.08	49.9	120.9	66.7	5.7	123.5	65.1	5.8
7996	306.2506	2.4326	2013.6525	12.79	15.252	44.57	246	209.4	45.1	6.4	209.4	35.4	6.4
7997	306.3010	17.6207	2013.7285	13.142	17.925	20.077	73.1	-57.5	-87.3	5.1	-57.2	-88.9	5.2
7998	306.3170	-8.4601	2013.5115	16.339	18.03	10.937	171.1	68	-13.4	5.9	64.9	-14.2	7.4
7999	306.3406	70.6461	2013.584	12.499	18.058	52.769	253	98.5	117.1	6.9	106	105.4	6.5
8000	306.3838	40.6234	2013.748	14.515	14.889	3.075	297.1	-60.8	-62	5.5	-59.8	-66.2	6
8001	306.4173	23.9567	2013.6775	8.994	15.773	55.37	344.8	75.1	96.5	5.1	74.2	97.4	5.1
8002	306.4298	37.5220	2013.6765	12.495	13.343	8.59	174	61.9	18.6	5.3	63.3	17.6	5.3
8003	306.4755	22.9911	2013.5405	11.847	12.646	32.603	24.8	15.6	-70.4	5.8	6.3	-60.1	5.7

8004	306.5605	31.9444	2013.468	15.167	15.222	17.214	60.1	110.1	103.6	5.7	113	108.7	5.8
8005	306.6519	17.5927	2013.6735	13.393	16.219	22.119	182.5	13.1	-74	5	12.3	-77.5	5.1
8006	306.7070	8.4330	2013.495	12.773	15.13	15.255	166.8	111.7	-62.5	6.2	111.3	-64.9	6.2
8007	306.7227	33.4749	2013.5435	13.543	15.458	10.048	284.8	-54.4	-91.4	5.7	-58.7	-93.2	5.7
8008	306.7294	0.7562	2013.5325	15.316	16.149	35.541	216.9	-1.5	-67.6	6.5	-2.5	-66.9	6.6
8009	306.7468	55.2322	2013.342	17.127	18.096	10.957	104.3	69.9	1.6	6.5	68.3	5.6	6.6
8010	306.7599	49.3836	2013.6005	6.726	11.81	55.599	146.2	78.8	62	6.2	75.7	62.3	6.4
8011	306.7874	-12.8355	2013.737	13.26	13.543	13.065	59.4	-1	-67.3	5.4	-4.3	-67.4	5.3
8012	306.8782	41.4693	2013.7525	8.704	14.617	9.116	123.5	100.5	0.8	5.3	96.2	0.7	6.1
8013	306.8802	18.2089	2013.573	14.892	15.729	6.063	157.2	11.7	-81.3	5.8	8.6	-78.6	6
8014	306.9247	61.1065	2013.2445	17.993	18.011	33.194	11.9	-7.3	-80.9	5.9	-11.7	-78.1	5.9
8015	306.9926	60.1103	2013.4895	17.504	18.487	34.427	299.7	60.7	-19.1	5.7	62.2	-14.1	6
8016	307.0004	-9.9810	2013.8585	10.69	14.246	33.872	205.3	-85.6	-107.3	5.7	-83.4	-107.2	5.6
8017	307.0298	8.5832	2013.5385	12.401	12.951	5.276	114.1	-25.2	-82.4	6.2	-22.9	-83.5	6.7
8018	307.0477	41.8596	2013.491	12.297	15.73	30.595	334	-70	-87.1	5.4	-71.6	-85	5.4
8019	307.0620	-7.0042	2013.8695	12.25	12.827	45.251	133.3	-5.8	-67.9	5.6	-9.6	-64.4	5.7
8020	307.0684	-13.8073	2014.447	8.592	10.876	30.918	337.6	1.4	-61.8	5.1	0.1	-62.6	5.2
8021	307.0956	81.1100	2013.589	11.717	16.488	12.755	352.4	-45.7	-66.9	6.3	-46.7	-70.4	6.3
8022	307.1487	50.6503	2013.7005	8.987	10.119	5.272	125.4	19.9	-74.6	6.1	23.5	-70.2	6
8023	307.1689	-4.3052	2013.564	15.725	16.292	12.226	295.8	44.6	-75.3	5.7	44.7	-76	5.7
8024	307.1938	48.2101	2013.361	13.509	13.779	4.708	79.8	152	3.3	6.2	148.7	7	6.4
8025	307.3167	-7.4771	2013.882	12.354	14.496	8.072	193.3	-27.3	-79.2	5.7	-28.4	-78	5.7
8026	307.3269	54.4707	2013.522	17.429	17.509	53.442	189.1	-28.8	-65.2	6.1	-21.8	-66.3	5.9
8027	307.3704	23.1909	2013.602	11.816	12.359	4.595	49.3	-141.2	-72.4	6.1	-141.4	-69.4	6.6
8028	307.4048	8.9578	2013.4885	12.561	18.084	51.172	53.3	-19.4	-68.5	6.1	-31.7	-67	7.8
8029	307.4054	-5.5176	2013.6365	15.037	17.382	36.023	84.9	-63.7	-79.8	5.4	-61.8	-75.5	6.6
8030	307.5097	7.5022	2013.738	15.031	16.19	39.399	132.8	174.2	-184.9	6.4	176.9	-190	6.3
8031	307.5295	45.4492	2013.6095	7.788	12.92	14.954	249.1	81.5	53.4	5.4	76.1	58.1	5.5
8032	307.6323	52.1398	2013.275	17.158	18.428	13.967	204.6	-38.1	-68.9	6.2	-39.5	-66.2	6.7
8033	307.6816	16.8060	2013.2325	14.708	16.647	18.636	27.5	9.3	-68.3	7	-1.2	-66.5	7.1
8034	307.6933	19.2406	2013.5585	17.96	18.226	53.553	91.6	20.4	-61.7	6.4	18.9	-65.3	6.3
8035	307.7382	56.2637	2013.5075	12.542	14.795	39.871	226.7	62.4	-9.7	5.8	60.9	-12.6	5.9
8036	307.7524	58.2364	2013.5335	14.006	14.853	10.2	209.2	-175.3	-88.3	5.9	-173.3	-96.6	5.9
8037	307.7619	11.9937	2013.4225	15.969	17.669	13.251	250.6	-138.9	-86.9	6.4	-139.3	-88.1	6.9
8038	307.7718	38.7831	2013.556	11.084	12.13	21.566	74.4	-23.5	-99.5	5.3	-22.8	-99.1	5.4
8039	307.7747	34.9337	2013.2025	17.153	18.369	52.118	265.4	12.2	-74.9	5.8	5.2	-72.5	7.1
8040	307.9102	38.4203	2013.494	13.397	13.422	4.286	350	-21.9	-82.2	5.5	-22.9	-83.5	6.6
8041	307.9219	35.2757	2013.568	10.926	16.659	16.627	3.7	90.3	44.8	5.7	89.1	52.5	5.7
8042	307.9222	-0.0649	2013.975	13.75	15.062	4.875	5.8	-10.7	-91.2	5.4	-10.4	-94.8	5.9
8043	307.9294	20.8244	2013.603	12.095	13.292	34.08	33.3	-68.9	-88.6	6.1	-64.2	-91	6.1
8044	307.9884	20.6910	2013.3295	17.003	17.648	30.081	69.2	22.9	-66.4	6.3	26.6	-62.9	6.3
8045	308.0376	8.7300	2013.341	17.271	17.498	33.13	75.3	-14.2	-60.1	6.4	-12.2	-61.1	6.9
8046	308.0429	0.1964	2013.605	11.245	16.984	27.769	69	69.6	-14.4	6.1	67.9	-11.7	6.4
8047	308.0847	-6.6395	2013.7655	13.137	14.263	9.769	56.5	77.1	-43.2	5.7	77.8	-43.3	5.9
8048	308.0979	36.3719	2013.5205	13.716	16.004	8.724	201.2	17	-150.7	5.4	18.2	-150	5.4
8049	308.1288	-8.4233	2013.6865	14.906	16.055	33.535	322.4	35.2	-116.4	5.9	40.1	-114.2	7.3

8050	308.2167	6.7798	2013.542	14.027	14.096	8.971	204.7	92.9	15.3	6.2	94	12.8	6.4
8051	308.2269	70.0973	2013.456	16.479	17.687	25.05	354	-44.8	-83.1	6.2	-40.2	-80.5	6.3
8052	308.2711	26.9405	2013.485	17.764	18.002	30.086	164.6	14.8	-65.2	6.2	23.6	-60.2	6.3
8053	308.3340	33.3904	2013.69	8.091	11.265	17.275	246.1	147.8	129.6	5.8	145.9	126.4	5.8
8054	308.3720	-13.5534	2014.447	9.229	13.168	24.951	17.4	-12.1	-83.5	5.1	-16.8	-81.6	5.1
8055	308.3870	51.5214	2013.5585	10.266	12.531	21.187	345.7	-84.3	-89.9	6.1	-82.5	-88.8	6.1
8056	308.4400	21.9499	2013.6335	10.912	13.866	17.949	24.4	-22	-98.1	6	-24.2	-97	6.1
8057	308.4721	26.3862	2013.7385	11.285	17.132	20.281	331.1	-2.8	-80	5.8	5.8	-83.1	6.3
8058	308.4813	22.4632	2013.645	7.828	8.619	19.639	82.9	11.9	-61	6	10.5	-61.2	6
8059	308.4920	15.7158	2013.5315	11.281	14.244	50.354	190.3	90.3	36.5	5.3	92.1	34.2	5.3
8060	308.5132	4.4892	2013.706	8.143	15.807	15.651	292.6	-84.8	-81.9	6.2	-84.6	-85.9	6.4
8061	308.5143	53.9315	2013.6435	14.535	15.878	5.082	4.9	83.8	37.8	5.7	86.5	34.4	5.7
8062	308.7240	24.4782	2013.521	12.829	13.277	14.882	215.4	-0.4	-64.4	5.9	-2.9	-62.8	6.1
8063	308.7270	-8.3652	2013.871	13.159	13.41	17.45	311.8	-9.7	-63.5	5.5	-11.4	-62.7	5.5
8064	308.7326	52.3506	2013.4025	17.147	17.485	54.418	93.7	60.4	20.5	6.2	62.1	23.7	6.3
8065	308.7421	19.9346	2013.55	17.385	17.633	59.064	123.3	11.5	-91.2	6.2	3.8	-88.1	6.2
8066	308.8666	-12.7835	2014.1525	12.383	13.136	13.957	315	-15.5	-72.8	5.5	-13.7	-73.5	5.7
8067	308.9363	27.2986	2013.604	17.783	18.567	41.78	311.1	-29.4	-60.9	6.2	-20	-63.9	6.8
8068	309.0304	9.1828	2013.3485	14.689	16.697	12.738	6.3	-5.1	-72.2	6.2	-6.5	-70.5	6.4
8069	309.0437	21.5325	2013.616	14.512	16.168	5.63	122.5	116.3	44	6.2	114.2	38.8	6.3
8070	309.0485	22.8650	2013.464	9.652	13.674	8.08	203.6	70	-2.4	6.1	72.1	-0.5	6.2
8071	309.0652	-3.0754	2013.7585	12.333	13.778	10.077	227.8	-52.8	-69.6	6.3	-54	-69.1	6.3
8072	309.0763	24.2797	2013.3035	12.893	13.143	57.152	314.6	1.2	-63.2	6.1	2.7	-62.1	6.9
8073	309.0778	28.6352	2013.3975	17.916	18.463	44.625	358.1	-18.5	-65.8	6.3	-7.2	-68.3	6.5
8074	309.1118	7.5549	2013.4385	12.171	13.89	8.582	84.1	1.6	-72.2	6.2	0.5	-73.6	6.2
8075	309.1494	56.1402	2013.221	17.91	18.277	23.463	36.4	24.4	-64.2	6	24.3	-65.8	6.4
8076	309.3171	1.5628	2013.6945	10.878	15.908	21.292	106.3	3.1	-99.1	6.1	3.2	-101.7	6.3
8077	309.3300	0.6839	2013.7175	12.619	13.815	17.77	284.2	-25.6	-73.9	6.2	-26.5	-75.3	6.2
8078	309.3385	9.6700	2013.213	13.998	17.645	8.979	102.4	62.1	2.3	6.3	63.2	1.3	6.4
8079	309.3464	6.6826	2013.3635	15.154	17.183	12.672	299.2	-18	-131.6	6.3	-23.3	-130.8	6.4
8080	309.4285	54.3570	2013.4635	17.389	18.808	57.767	179.6	-18.5	-72.8	5.8	-29.3	-69.6	7.7
8081	309.4345	53.5040	2013.4625	15.354	18.111	12.316	340.1	-14.9	-98.3	6.1	-10.5	-102.6	6.3
8082	309.4614	-7.3304	2013.574	15.525	15.551	3.099	197.3	74	31.2	7.7	66.6	44.2	5.8
8083	309.6001	-4.2601	2013.698	11.776	17.149	23.286	283.8	74.3	11.1	5.4	73.3	7.8	5.8
8084	309.6217	21.1415	2013.3535	14.759	15.704	22.306	337.5	6.5	-82.4	6.3	14.4	-82	6.2
8085	309.6516	44.6411	2013.6395	9.168	9.798	9.834	185	82.6	69.1	5.4	82.6	71.1	5.5
8086	309.6804	11.6517	2013.1215	14.904	16.338	5.75	165.8	-11	-109	6.3	-16.5	-108.5	6.4
8087	309.7162	1.4695	2013.6215	14.045	17.671	6.162	241.1	-51.7	-179.7	6.3	-44.2	-179.5	6.7
8088	309.7730	14.7977	2013.516	9.906	14.809	8.354	150.7	101.9	-144.1	6	94	-144.9	6.1
8089	309.8105	33.9024	2013.32	16.908	17.739	26.061	291	18.4	-65.9	6.1	21.1	-67	6.1
8090	309.8572	21.9253	2012.968	13.268	13.45	23.01	133.6	-26.7	-69.9	7.3	-36.6	-61.5	6.7
8091	309.8710	16.6275	2013.5125	12.399	15.449	6.921	234.5	-31.3	-70	5	-35	-72	5.1
8092	310.0453	15.0566	2013.588	10.907	15.136	52.099	290.7	-20.4	-85.6	5	-17.9	-84.7	5
8093	310.0562	29.2592	2013.6925	14.312	16.581	59.738	215.4	17.5	-186.8	6.1	14.1	-188.9	6.2
8094	310.0871	24.0962	2013.7205	12.322	12.447	29.979	217.4	-10.5	-64	6	-1.6	-64.7	6.1
8095	310.0915	-6.1703	2013.7475	14.523	16.255	35.08	73.4	9.8	-68.8	5.3	10.1	-71.8	5.8

8096	310.1598	48.9986	2013.581	9.284	12.42	8.55	285	-3.7	-73.7	6	-5.7	-74.7	6.1
8097	310.2094	43.1625	2013.573	12.991	15.905	5.837	65	126.1	43.4	5.5	119.5	42.7	5.7
8098	310.2645	4.7679	2013.4585	11.846	16.438	7.38	178.6	37.4	-79	6.3	44.1	-76.7	6.6
8099	310.2838	-2.0100	2013.6785	13.55	13.684	12.819	177.6	-31.9	-61.4	5.4	-35.8	-60.2	5.4
8100	310.2917	3.6383	2013.8155	9.339	10.605	6.457	33.2	-53.1	-88.1	6.5	-55.8	-87.8	6.7
8101	310.3200	22.6854	2013.6635	12.684	12.772	17.676	298.9	41.1	-104.4	7.9	44.7	-103.8	5.1
8102	310.3307	14.9602	2013.561	16.716	16.958	4.576	73.6	132.1	-17	5.3	127.6	-14.5	5.3
8103	310.3619	11.1722	2013.567	10.543	13.994	5.676	215.9	21.3	-60.5	6.2	24.2	-60.6	6.5
8104	310.3624	61.6404	2013.348	16.492	16.957	23.285	344	82.2	136.7	6	75.7	137.2	6
8105	310.4618	5.3226	2013.421	14.876	15.092	17.579	121.6	18.9	-75.8	6.3	17.9	-74.6	6.3
8106	310.4878	31.6880	2013.534	17.085	18.602	52.546	323.9	-22.4	-89.4	5.5	-12.2	-87.7	6.8
8107	310.5110	18.3508	2013.9025	8.974	10.551	5.483	81.4	67.6	41.8	5	63.7	46.2	5.1
8108	310.5881	4.3153	2013.3205	13.652	15.016	44.532	85.5	-14.2	-106.8	6.4	-14.2	-107.3	6.4
8109	310.6236	39.5734	2013.599	12.154	12.301	20.535	325.4	81.7	25.4	6.3	81.1	28.4	6.3
8110	310.6737	47.4728	2013.617	14.817	18.074	10.874	185.2	-0.6	-80.5	5.5	-7.4	-81.8	5.7
8111	310.7800	35.2437	2013.721	17.6	17.885	43.807	9.1	9.6	-69.1	7	7.9	-68.2	6.8
8112	310.7977	-13.7410	2014.447	11.213	13.892	42.619	27.2	137.1	-8.7	5.6	132.7	-7.7	5.6
8113	310.8487	-7.3387	2013.8515	14.147	15.121	6.252	12.8	5.6	-76.1	5.6	2.1	-74.3	6.4
8114	310.8986	45.2366	2013.6385	11.857	12.509	17.733	165.8	137.8	73.2	5.5	137.9	70.7	5.5
8115	310.9036	23.5047	2013.521	10.971	11.823	5.277	72.5	69.4	6.3	5	67.3	3.4	5.1
8116	310.9419	23.0935	2013.3145	12.696	12.703	19.35	125.6	-58.9	-79	5.3	-61.8	-73.4	5.2
8117	311.0311	-1.2680	2013.777	12.615	13.819	13.762	15.9	147.3	-12.2	5.4	150.5	-12.3	5.4
8118	311.0732	25.8269	2013.5685	12.155	12.698	45.713	11.2	-27.7	-93.3	6.2	-23.9	-98.2	6.2
8119	311.0911	30.7870	2013.683	18.329	18.787	41.462	285	14.4	-66.8	6.2	13.3	-64.6	7.2
8120	311.1288	8.9034	2013.609	10.092	11.002	15.094	344.3	205.4	121.1	6.5	212.8	121	6.7
8121	311.1662	-5.1913	2013.959	10.658	11.305	8.109	84.8	18.2	70.1	5.3	17.3	72.4	5.3
8122	311.1949	37.0714	2013.6	10.312	15.114	21.89	102.7	14	-64.5	5.3	11.5	-66.1	5.3
8123	311.1960	12.4264	2013.0665	17.713	18.338	31.978	235.1	13.2	-86.2	5.2	17.6	-88.2	6.1
8124	311.1992	12.2490	2013.189	15.971	17.663	6.087	221.8	12.8	-72.1	5.1	13.4	-71.6	5.5
8125	311.2344	82.0960	2013.678	11.492	15.514	22.711	231.1	108.7	81.5	6.1	108.7	82.9	6.1
8126	311.2550	-1.6828	2013.709	12.677	13.799	25.075	266	-35.1	-176.3	5.4	-35	-176.8	5.5
8127	311.2992	54.6456	2013.2505	18.278	18.379	51.305	318.7	23.7	61.8	6.6	29.4	60.6	6.6
8128	311.3016	15.8897	2013.306	17.996	18.193	18.306	18.4	-11.6	-71.4	5.4	-11.8	-68.1	6.1
8129	311.3084	23.3037	2013.2065	12.056	12.525	55.007	39.1	62	-25.4	5.5	64.3	-27.7	5.6
8130	311.3754	14.7995	2013.1195	10.981	16.644	32.238	228.7	-24.5	-121.1	5.2	-18.9	-120.6	5.3
8131	311.3766	2.0139	2013.8535	10.308	11.032	6.819	307.5	-35.2	-75.2	6.1	-33.6	-77.7	6
8132	311.3996	-8.8789	2013.7675	10.077	10.675	5.611	201.8	76.7	-91.2	5.6	74.5	-88.4	5.6
8133	311.4299	36.7763	2013.6245	9.583	9.605	6.642	207.1	71.2	-89.5	5.3	72.2	-89.5	5.4
8134	311.4630	53.1266	2013.7945	12.838	12.905	5.519	47.5	68	-53.9	5.7	65	-53.2	5.6
8135	311.4650	17.4739	2013.357	16.935	18.913	31.091	240.9	26.2	-68.5	5.2	18.4	-67.4	5.7
8136	311.4979	5.5475	2013.3555	14.756	16.079	5.232	110.6	57.2	-102.9	6.6	51.6	-100.5	6.8
8137	311.5372	38.9695	2013.609	13.179	16.923	10.753	272.2	17.5	110.7	5.3	21.9	109.8	5.4
8138	311.5583	8.3682	2013.8355	9.184	11.459	6.048	356.8	1.7	-99.5	6.2	-2.1	-97.3	6.2
8139	311.5726	61.5958	2013.3025	17.583	18.085	26.779	288.2	62.5	1.1	6.3	61.2	-8	6.5
8140	311.7541	31.8967	2013.2215	16.306	18.165	58.57	186.8	-24.2	-63.7	6.7	-22.3	-62.1	7.3
8141	311.8127	2.7844	2013.454	12.674	14.17	31.296	114.3	102.8	51.8	6.4	103.5	50.2	6.4

8142	311.8740	18.0073	2014.015	13.295	14.704	4.343	70.8	97.9	33.8	5.4	98	36.9	7.2
8143	311.9196	33.7176	2013.595	10.816	16.811	45.542	353.6	-144.9	-145	5.7	-148.6	-141.6	6
8144	311.9198	4.9032	2013.283	12.231	14.945	7.79	304.7	64.5	32.6	6.4	62.2	36.2	6.5
8145	311.9926	1.6830	2013.3915	17.092	17.247	24.653	222.5	10.5	-86.4	6.5	6.5	-83.1	6.9
8146	312.0401	46.5460	2013.49	14.218	16.864	17.24	108.1	44.4	-84.7	6.1	43.3	-85.3	6.2
8147	312.0639	75.3111	2013.4675	15.07	16.006	10.031	133	89.2	192.6	5.8	92.1	197.9	5.9
8148	312.1220	16.1026	2013.36	17.035	17.253	4.034	307.4	-30.5	-80.9	5.4	-34.6	-80.2	5.5
8149	312.1272	48.7257	2013.524	10.441	14.835	27.979	187.3	19.9	93.5	5.7	21	92.2	5.7
8150	312.2259	8.6456	2013.1205	12.825	15.971	11.356	339.9	66.3	36.6	6.5	64.9	32.4	6.5
8151	312.2478	31.7086	2013.443	17.089	17.194	47.821	136.4	-10.4	-64.2	5.7	0	-66.9	5.8
8152	312.2587	4.2330	2013.427	14.365	17.964	59.401	16.1	20.4	-65.6	6.3	19.2	-68.1	6.5
8153	312.3158	8.9213	2013.1965	17.657	18.428	55.61	298.5	61.6	-14.6	6.8	61	-19.7	6.9
8154	312.3616	28.6656	2013.664	10.267	14.657	53.689	317.4	20.9	62.2	5.8	20.6	65.4	5.9
8155	312.3967	-4.7999	2013.5995	16.357	16.7	6.866	317.5	25	-62.2	5.6	23.5	-61.4	5.9
8156	312.4155	-3.0790	2013.56	14.743	16.686	5.135	4.8	98.5	-135.3	5.6	105.5	-133.9	6.6
8157	312.4752	50.6026	2013.5795	9.626	14.434	25.015	1.5	106.3	88.9	5.7	109.4	94.3	5.8
8158	312.5240	8.0250	2013.2915	15.212	18.195	21.328	227.8	11.8	-96.9	6.4	6.7	-101.6	7.4
8159	312.5555	-6.6662	2013.588	15.565	16.174	4.224	332.6	-111.4	-131.4	6.1	-114.4	-127.5	7.8
8160	312.5652	-2.8124	2013.5295	15.112	17.721	9.526	340.7	62.4	-6.7	5.5	64	-7.3	6.6
8161	312.6675	46.5767	2013.717	18.136	18.32	30.916	212.4	-5	-62.8	6.4	4.8	-60.2	6.3
8162	312.7184	3.5826	2013.3705	14.743	14.929	7.459	318.3	90	60.8	6.7	94.6	60.4	6.5
8163	312.7201	19.8781	2013.579	13.626	13.68	11.914	82.8	9.4	-134.3	5.7	7.3	-136.6	5.7
8164	312.7654	-1.5663	2013.742	12.242	13.012	15.714	348.4	-108.2	-190	5.6	-112.6	-189.3	5.6
8165	312.7951	2.9548	2013.363	11.886	17.285	39.049	161.4	93.4	36.2	6.4	94	28.7	6.5
8166	312.8001	-4.0115	2013.7845	12.783	15.298	15.88	48.8	27.7	-63	5.4	25.7	-66	5.5
8167	312.9199	-7.5460	2013.8045	12.637	14.287	5.02	190.3	76.7	12.7	5.4	74.9	10.4	5.6
8168	313.0575	35.3630	2013.4455	17.919	18.281	10.224	240.8	76.7	-24.1	5.9	80.1	-23.5	6
8169	313.1000	12.0058	2013.221	14.707	17.551	11.116	251.7	3.9	-77.6	6.4	1.2	-78.4	6.6
8170	313.1130	43.2279	2013.6295	9.088	16.555	39.565	264.6	76.1	76.1	6.1	83.5	71.9	6.2
8171	313.1289	38.2118	2013.5225	11.407	16.17	39.174	55.1	-13.8	-65	5.3	-15.6	-64.9	5.4
8172	313.2350	10.2157	2013.5855	9.077	15.519	50.93	303.1	63.5	-56.7	6.2	61.9	-61	6.3
8173	313.2505	46.6959	2013.4255	18.062	18.319	27.99	10.9	77.4	-32.5	6.5	76.4	-38.7	6.6
8174	313.3328	14.9796	2013.213	17.457	17.879	35.728	109.7	89.7	-50.1	6.3	88.2	-47	6.7
8175	313.3374	5.8489	2013.3595	11.783	14.745	32.991	167.8	96.5	-53.8	6.3	95	-55.9	6.3
8176	313.4115	-2.8731	2013.766	10.5	13.059	11.277	287.7	116.7	-41.2	5.5	114.9	-40	5.6
8177	313.4553	10.5931	2013.1895	15.31	16.011	9.295	186.1	69.8	-75.3	6.4	66.7	-74.3	6.4
8178	313.4588	15.8757	2013.2005	12.879	17.273	14.854	213.9	123	-33.1	6.3	120.2	-37.9	6.5
8179	313.4694	54.4281	2013.3605	16.305	18.273	59.915	32.1	136.5	212.4	6.8	138.2	205.9	6.2
8180	313.4827	23.7064	2013.059	14.478	17.015	10.637	58.6	101.3	-55.8	6.5	107.3	-55.8	6.6
8181	313.5334	47.2561	2013.4605	16.093	18.48	41.728	149	132.6	18	6.7	137.4	22.8	6.6
8182	313.6400	30.7643	2013.6705	9.285	14.163	10.744	177.9	149.2	44.5	5.7	152.7	52.6	5.8
8183	313.6593	76.8880	2013.6095	12.246	14.762	6.62	81	-20.9	-63.6	5.7	-25.8	-60.1	5.7
8184	313.7085	-3.1084	2013.775	13.344	14.42	6.969	47	36.9	-60.4	5.4	37.3	-61.3	5.5
8185	313.7182	15.9581	2013.206	14.927	16.836	5.276	26.8	-99.4	-120.7	6.4	-91.7	-118.4	6.8
8186	313.7471	-13.2660	2014.447	11.192	15.185	39.935	351.9	40.1	-114.2	5.6	37.1	-117.9	6.8
8187	313.7902	28.1635	2013.517	17.446	17.748	57.051	228.4	37.8	-62.9	6.4	41.2	-60.5	6.6

8188	313.8691	80.1986	2013.454	15.969	16.994	15.772	314.3	69.5	70	6.2	72.3	69.7	6.3
8189	313.8909	33.8114	2013.417	17.552	18.479	39.514	288.8	-8.9	-78.5	6.4	-14.7	-77.9	6.2
8190	313.9371	2.1740	2013.2945	15.638	17.943	16.696	74.4	66.3	13.6	6.3	65.6	11.4	6.9
8191	313.9485	45.5718	2013.173	17.626	18.24	30.764	126.2	87.9	-6.3	7.2	90.4	-1.7	6.6
8192	313.9575	45.2189	2013.399	16.753	18.132	53.644	148.1	-40.3	-73	6.6	-49.9	-63.5	6.6
8193	313.9818	36.1593	2013.3605	17.393	17.68	36.638	226.2	-14.2	-63.3	5.5	-13.1	-64.1	6.6
8194	313.9921	7.1278	2013.3855	13.436	14.617	6.736	179.1	93.9	29.9	6.3	95.7	29.1	6.4
8195	314.0506	76.1756	2013.5865	11.287	13.452	22.333	299.8	97.6	88.3	6.4	97.1	90.8	6.4
8196	314.1006	7.3699	2013.0525	17.74	18.116	47.512	36	68.9	-18.8	6.6	68.6	-12.6	7.1
8197	314.1465	30.7981	2013.66	12.714	13.149	9.999	200.1	199.5	86	5.8	196.2	86.4	5.8
8198	314.2055	16.4557	2013.183	17.941	18.366	38.674	263.6	-26.6	-64.5	6.8	-28.7	-61.9	6.7
8199	314.2578	45.8517	2013.4375	16.053	17.777	6.195	344.9	-23.8	-67.9	6.2	-22.1	-67.3	6.4
8200	314.3212	44.7273	2013.501	13.102	13.92	15.658	118.9	64.5	5.3	6.2	62.3	6.8	6.2
8201	314.3355	32.2444	2013.623	10.809	13.11	9.869	28.2	26.3	74.2	5.7	24.9	76.3	5.7
8202	314.3432	20.1699	2013.0905	18.235	18.385	32.536	215.2	68.3	-12	6.5	64.7	-16	6.4
8203	314.4138	47.8633	2013.604	10.428	12.896	19.56	76	3.7	-81.6	6.1	0.4	-80.6	6.1
8204	314.4419	13.6463	2013.0655	16.837	16.969	12.987	290.2	154.7	-61.4	6.7	157.9	-60.8	6.7
8205	314.4976	4.3255	2013.323	14.713	15.313	22.393	204.6	-20.6	-63.9	6.3	-18.9	-63.1	6.3
8206	314.5918	34.1308	2013.5585	15.499	16.493	12.74	122.1	110.1	56.5	5.7	108.9	57.1	5.9
8207	314.6412	5.3915	2013.151	17.118	17.321	5.129	48.6	52.7	76.2	6.5	56.4	73.2	7.1
8208	314.6703	39.0212	2013.567	12.29	14.611	14.528	80.6	79	41.2	5.3	78.4	39.6	5.4
8209	314.6750	-1.2755	2013.195	12.986	16.597	9.219	311.6	68.7	-17.9	5.6	68	-19	5.9
8210	314.7034	6.0024	2013.576	11.015	12.551	40.445	30.9	21.3	-68.6	6.2	23.8	-70.2	6.2
8211	314.7682	4.2929	2013.917	6.49	7.163	10.502	66.9	-99.4	-156.4	7.1	-108.6	-144.1	6.4
8212	314.8123	18.2236	2013.657	9.812	15.072	10.453	21.1	-64.5	-82.4	6.1	-67.6	-77.2	6.1
8213	314.9435	42.4150	2013.383	13.391	17.765	38.193	347.3	167.6	22.6	6.2	166.3	22.8	6.6
8214	314.9468	77.3693	2013.448	13.927	17.112	29.178	174.9	51.1	112.9	6.2	45.8	118	6.3
8215	314.9682	17.6772	2013.438	12.889	14.285	10.037	129.4	63.9	48.9	6.2	63.7	50	6.2
8216	315.0289	-5.4990	2013.6045	11.703	13.476	25.126	324.7	-41.4	-87.8	5.6	-41	-89.9	5.6
8217	315.0731	14.7749	2013.1595	15.67	16.175	55.316	158.4	85.6	6.5	6.3	90	3.3	6.4
8218	315.0923	47.7370	2013.29	18.193	18.454	52.782	136.2	8.4	-62.8	6.4	10.3	-63.4	7.1
8219	315.1611	63.4877	2013.4435	14.182	16.477	8.434	76.7	-54.1	-69.7	5.9	-51.8	-70.4	6
8220	315.1661	5.9158	2013.4885	11.057	11.186	33.491	116	71	23.4	6.2	69.5	21.5	6.3
8221	315.1669	29.3358	2013.6265	10.442	13.948	34.76	323.8	-49.1	-119.5	5.4	-50.3	-118.7	5.5
8222	315.1822	21.5540	2013.366	13.231	16.931	16.315	66.5	70.2	64.5	6.2	67.5	64	6.3
8223	315.2059	-7.0735	2013.4485	11.77	13.539	39.182	52.5	88.1	12.4	5.5	90.2	16.2	5.5
8224	315.2137	10.6849	2013.3005	15.596	16.275	5.453	201.6	141.8	3.3	6.3	139.9	2.4	6.4
8225	315.2367	21.1205	2013.3535	12.193	17.022	17.315	255.3	90.9	6.9	6.4	93.9	7.4	6.5
8226	315.2792	37.1231	2013.2495	14.738	18.219	12.339	248.2	-0.1	-68.3	6.2	-4.2	-65.8	6.9
8227	315.2816	22.6574	2013.1545	17.079	17.93	50.367	74	11.3	-66.3	6.6	13.2	-64.9	6.9
8228	315.4206	45.0020	2013.111	17.164	18.406	27.868	88.5	-14.2	-68.1	6.4	-14.8	-65.5	8
8229	315.4833	45.9138	2013.0705	17.405	18.572	38.321	139.5	-20.5	-69.6	6.9	-29.8	-63.1	6.7
8230	315.4994	-0.4181	2012.8615	16.174	16.518	5.051	324.3	19.1	-92.6	5.8	17.1	-95.6	6.3
8231	315.5261	62.1162	2013.5445	11.209	16.093	25.717	94.9	109.3	137.5	6.3	111.4	140.7	6.5
8232	315.5654	47.0997	2013.3865	17.149	18.074	36.57	19.7	39.2	61.4	6.2	36.8	61.9	6.4
8233	315.5715	33.7392	2013.6935	7.723	12.889	10.841	4.3	44.1	72.7	5.2	43.4	73.4	5.4

8234	315.7054	9.7578	2012.829	16.883	17.174	5.109	180.1	59.7	64.3	6.7	56	66.6	6.6
8235	315.7287	17.5427	2013.13	17.668	17.948	51.252	321.7	1.4	-60.6	6.3	-6.1	-63.2	7.4
8236	315.7690	6.8445	2013.3595	11.622	12.221	10.462	287.1	62.7	40.2	6.2	63.7	41.8	6.2
8237	315.8046	62.2215	2013.75	8.488	14.091	8.914	347.7	0.5	-107.8	5.8	4.1	-104.4	5.8
8238	315.8678	59.3630	2013.584	11.372	14.163	5.978	175.3	133	98.3	5.9	129.9	98.9	5.9
8239	315.9418	35.7991	2013.5535	14.693	17.301	6.399	334.6	61.9	30.4	5.3	61.2	32.2	5.4
8240	316.0397	41.5533	2013.3975	15.286	16.662	23.665	232.3	97	5.7	6.4	97.8	6.2	6.5
8241	316.1032	22.3718	2013.5135	17.028	17.471	49.631	101.9	68.1	25.6	7.2	68.9	22.3	6.3
8242	316.1223	1.3601	2013.5355	14.786	15.411	5.456	359.4	77	18.2	6.5	77.3	21.5	6.4
8243	316.1544	0.2506	2013.308	13.787	15.762	9.691	118.7	118.2	13.8	6.6	118.9	16.2	6.9
8244	316.2180	35.0824	2013.4395	16.351	17.42	5.695	189.3	69.3	21.5	5.4	67.4	24.1	5.5
8245	316.2455	23.5266	2013.446	13.879	16.229	59.412	189.9	6.5	66.9	6.1	10.4	64.3	6.2
8246	316.2803	26.5597	2013.397	12.742	16.27	12.219	120.8	62.7	31.9	5.4	64.6	29.1	5.6
8247	316.3445	39.6195	2013.5745	10.921	13.395	29.265	351.8	65.7	27.6	5.3	67	27.6	5.3
8248	316.4392	23.8738	2013.413	13.63	15.141	13.403	252.4	70.4	-4.8	6.2	69.6	-0.4	6.2
8249	316.4396	50.2622	2013.5935	12.004	12.576	31.474	296.5	101.1	37.7	6.3	99.8	26.6	6.3
8250	316.4551	29.5236	2013.606	10.043	15.787	23.44	246.8	-54.2	-95.2	5.4	-51.3	-92.7	5.5
8251	316.4703	59.4032	2013.311	16.036	18.076	39.811	159.9	8.5	-88.5	5.9	8.3	-91.2	6.2
8252	316.5211	28.7122	2013.491	12.798	13.929	5.964	319.8	197.1	43.2	5.5	198.7	43.1	5.6
8253	316.5469	47.4047	2013.472	14.322	14.661	4.864	45.3	77.3	51.4	5.7	78.6	52.9	5.8
8254	316.6309	39.5266	2013.169	17.476	18.568	59.481	94.8	-21.7	-66.4	5.6	-18.1	-66.3	6.4
8255	316.6695	52.1344	2013.7205	11.113	13.731	5.713	161	93.4	49.9	6.3	93.3	49.6	6.3
8256	316.8740	69.8410	2013.4905	11.625	15.638	27.442	231.7	8.6	-64.3	5.8	7.6	-66.1	5.9
8257	316.9197	30.3696	2013.707	8.477	10.555	48.077	67.2	-4.8	-66.9	5.8	-7	-69	5.8
8258	316.9777	-7.4170	2013.618	10.669	10.696	17.976	190.1	-10.7	-63.4	5.8	-15.2	-64.5	5.8
8259	316.9799	45.0216	2013.485	12.148	16.45	16.51	255.8	-2.3	-106.6	6.1	0.6	-106.2	6.2
8260	317.0334	45.6152	2013.25	16.4	17.153	3.902	136.8	25.4	-77.9	6.2	22.3	-79.7	6.7
8261	317.0696	33.9212	2013.444	13.352	14.497	4.962	268.4	107.6	22.9	5.9	109.1	21.8	6
8262	317.1075	36.3003	2013.445	12.017	16.665	22.9	128.5	106.5	85.6	5.4	107.4	87	5.5
8263	317.1323	30.7647	2013.4905	14.246	14.798	4.693	305.8	109.7	23.5	5.9	110.4	23.4	6
8264	317.1425	25.0791	2013.566	12.071	13.42	9.986	332.1	96.2	74.5	5.5	90.2	75.6	5.5
8265	317.1466	9.6291	2014.0165	9.918	10.492	4.029	243.8	-35.1	-121.9	6.2	-30.5	-121.7	7.1
8266	317.1471	37.7998	2013.4215	11.804	15.753	19.149	140	67.1	-28.3	5.4	68.2	-25.7	5.4
8267	317.1864	31.6869	2013.371	16.716	17.738	42.811	282.9	13.3	-66.5	6.1	14.6	-64.9	6
8268	317.2539	5.3633	2013.204	13.531	17.334	14.623	159	74.2	-15.2	6.3	71.7	-13.6	7.1
8269	317.2843	32.5372	2013.5175	12.634	12.924	9.639	38	-25.5	-95.9	5.9	-22.8	-98.4	5.9
8270	317.4584	42.9556	2013.482	14.015	15.864	21.462	135.4	196.3	-18.7	5.5	200.8	-10.4	5.6
8271	317.4677	7.0498	2013.116	15.694	17.137	6.02	7.8	120.2	25.5	6.5	125.3	26.6	7.1
8272	317.4741	62.0196	2013.366	11.813	18.352	28.26	75.5	63.8	16.5	6.2	63.1	21.6	6.5
8273	317.4989	49.1306	2013.6435	11.146	15.651	8.654	255.1	160.6	28.7	6.2	159.3	32.3	6.2
8274	317.5412	3.5201	2013.3005	10.312	15.346	59.151	215.5	16.4	-67.9	6.3	15.5	-68.6	6.5
8275	317.7553	37.9512	2013.4155	11.531	17.325	15.231	93.9	10.5	-116.2	5.4	11.1	-117.3	5.7
8276	317.8417	68.8241	2013.5485	12.443	15.562	9.063	79.5	-33.2	-60.6	5.8	-32.5	-62.8	5.8
8277	317.8651	27.3087	2013.5145	14.263	14.357	7.24	210	-41.8	-211.8	6	-44.6	-211.4	6
8278	317.8849	2.2697	2013.397	12.634	14.137	22.173	106.4	177.5	-26.9	6.6	174.1	-29.7	6.6
8279	317.9383	57.0975	2013.5235	14.258	17.428	28.42	133.2	75.4	26.4	5.8	73.3	26.9	5.9

8280	317.9819	13.0365	2013.217	14.483	17.401	22.41	142.5	80.4	-33	5.8	78.6	-33.8	6.3
8281	318.0365	55.8347	2013.4225	18.104	18.14	6.004	37.8	-6.8	-66.8	6.1	-12.9	-69.1	6
8282	318.1453	6.7412	2013.3455	12.783	15.713	6.494	60.8	90.4	-197.5	6.6	86.9	-195.3	7.3
8283	318.2221	43.3651	2013.608	10.029	14.723	56.161	3.7	79.6	60.7	5.4	77.8	63.2	5.5
8284	318.2370	-0.7459	2013.014	15.73	15.801	14.001	210.8	62.4	10	5.9	62.9	6.7	6
8285	318.2383	59.6906	2013.612	11.124	14.828	8.744	188.3	-195.2	-66.8	5.7	-197.4	-69.8	5.8
8286	318.2414	11.1537	2013.2095	12.397	17.076	25.575	113	91.6	-13.4	6.3	87.1	-13.9	6.5
8287	318.3091	19.2558	2013.4245	11.242	15.544	18.341	146.6	29.9	-104.6	5.4	29.5	-100.1	5.5
8288	318.3196	-0.3961	2013.2315	12.871	13.607	7.925	264.8	179	30	5.9	179.5	28.2	5.9
8289	318.3538	35.7732	2013.499	11.531	14.432	9.546	264.2	97.7	38.8	5.4	98	36.3	5.4
8290	318.3640	45.9123	2013.717	15.874	18.096	5.553	309.8	-90.1	-199.7	5.6	-99.9	-196.9	6.4
8291	318.3913	30.6741	2013.5225	12.99	16.049	20.94	37.4	1.1	85	5.3	2.7	89.1	5.4
8292	318.4497	74.1494	2013.4315	14.342	15.514	32.712	167.9	27.9	77.6	5.9	25.5	76.9	5.9
8293	318.4614	19.7070	2013.206	16.751	17.333	7.636	184.3	71.1	6.2	5.6	70.1	11.4	5.7
8294	318.4648	59.4456	2013.491	14.469	16.506	16.224	197.8	-23.1	77.4	5.7	-25.1	75.1	5.8
8295	318.4879	76.7836	2013.512	12.899	16.23	15.145	34.6	84.4	68.6	5.9	85.5	70.7	5.9
8296	318.5610	35.2888	2013.5315	10.645	14.541	9.958	107	95	54.3	5.3	94.5	48	5.4
8297	318.5845	65.6349	2013.253	17.869	17.89	33.009	258	-9.4	-68.6	6.6	-20.7	-63.2	6.3
8298	318.7272	38.6485	2013.458	11.878	14.1	54.991	13.1	91.8	91.3	5.4	96.7	91.4	5.4
8299	318.7636	37.4976	2013.302	17.155	17.649	20.621	202.4	8.2	75.8	5.8	5.6	73.1	5.7
8300	318.8484	37.6226	2013.184	17.688	18.77	34.548	128.8	-3.7	-68.3	5.5	-7.8	-68.4	6.3
8301	318.9000	13.5766	2013.2705	15.377	16.924	5.589	310.3	-13.3	-95.1	6.3	-10.8	-95.3	6.6
8302	318.9214	52.6618	2013.492	13.306	17.277	9.552	136.1	58.3	63.4	5.5	56.3	68.1	5.6
8303	319.0321	29.1611	2013.5565	13.076	16.243	5.516	76.3	-21.2	-70.7	5.1	-20.3	-71.7	7.3
8304	319.0359	22.2381	2013.414	12.78	16.491	10.817	131.3	65.5	17.5	5.4	63.9	15.7	5.6
8305	319.1636	8.8433	2013.437	10.814	14.917	15.472	257.7	-13.5	-105.5	6.3	-10.1	-105.9	6.4
8306	319.1954	20.2470	2013.4175	9.367	12.894	6.73	254	-78.5	-86.5	5.4	-77.5	-81	5.6
8307	319.2136	3.7875	2013.3485	11.809	14.744	27.308	251.3	73.4	5.1	6.4	74	4.5	6.5
8308	319.3108	42.6795	2013.143	17.982	10.775	22.78	273.1	-0.1	-83.3	6	5.1	-84.8	6
8309	319.3147	82.1532	2013.6385	12.772	16.124	15.362	105.8	88.8	63.8	5.8	88.5	65.9	5.8
8310	319.3171	24.7091	2013.949	14.149	14.217	5.122	339.6	1.5	-73.4	5	4.9	-73.5	5.1
8311	319.3176	9.1960	2013.7035	13.947	15.598	5.079	144.6	64.7	-16.7	6.4	65.6	-16.1	6.7
8312	319.3354	65.1498	2013.5185	12.562	13.502	5.933	304.6	-31.6	70.2	6.1	-31.1	69.7	6.2
8313	319.3406	19.7319	2013.312	12.53	17.477	11.371	157.3	64.6	80.3	6	58.9	86.1	6.3
8314	319.3662	76.9698	2013.5655	9.848	9.889	8.606	305	-3.4	61.6	6.1	0.8	60.4	6.1
8315	319.4440	61.0668	2013.081	18.446	18.966	30.07	27	-0.2	94.1	6.6	-3.2	95.3	7.1
8316	319.4790	36.2230	2013.225	15.153	17.117	51.857	253.5	-1.8	62.3	6	-5.6	60.1	5.9
8317	319.5103	33.2234	2013.442	17.324	17.812	43.029	355.4	-7.2	-67.6	5.4	-3	-69.1	5.4
8318	319.6053	48.6944	2013.8745	14.837	15.249	3.608	116	69.2	27.5	5.4	70.7	30.3	6.4
8319	319.6104	12.0375	2012.93	17.191	17.9	4.503	296.8	-66.6	-107	5.6	-64.8	-107.1	5.6
8320	319.6127	28.2492	2013.491	11.429	14.745	7.582	299.5	100.4	-10.1	5.1	100.2	-8.9	5.2
8321	319.6249	43.7585	2013.467	10.633	14.231	6.11	86.4	62.7	46.6	5.4	60.7	48.4	5.6
8322	319.6929	37.6353	2013.253	14.982	17.715	43.117	223.1	-15.8	-68.7	6	-5.5	-70.3	6.1
8323	319.7282	14.9961	2013.513	9.877	13.339	16.005	333.3	84.1	23.2	5.5	88.1	22	5.6
8324	319.7705	59.8604	2013.016	18.531	18.549	46.575	281.5	44.2	-62.6	6.5	42.5	-61.5	7.8
8325	319.7723	28.2312	2013.4955	12.517	15.68	18.31	128.2	46.4	88.1	5.1	42.8	91.7	5.2

8326	319.7758	12.7299	2013.3455	11.156	11.84	34.892	64.9	103.7	29.3	5.7	99.5	28.1	5.6
8327	319.7950	-5.0143	2013.265	12.917	13.138	12.159	222.1	4.1	-76.4	6.6	3.5	-72.8	6.6
8328	319.8248	-4.6618	2013.1825	12.563	15.369	8.684	300.8	-80.2	-61	7	-80.8	-61.3	6.9
8329	319.8425	0.7310	2013.188	12.71	15.561	7.616	126.3	-9.5	-65.2	6.3	-5.8	-67.7	6.5
8330	319.8501	39.2189	2013.14	17.949	18.246	21.329	13.3	-6.8	-66	6.1	-6.6	-67.5	6.2
8331	319.9635	29.6887	2013.4895	14.471	16.342	7.196	59.4	86.1	76.5	5.1	87.8	78.8	5.2
8332	319.9930	36.3518	2013.303	12.426	17.184	14.556	189.3	139.1	19.6	5.9	134.2	21.7	6.1
8333	320.0573	-0.5733	2013.193	13.149	14.161	7.432	117.5	-93.8	-208.1	6.8	-101.3	-205.6	6.8
8334	320.0688	45.1749	2013.639	10.175	15.392	20.407	345	130.9	16.9	5.4	128	15.3	5.5
8335	320.1543	79.2844	2013.3325	15.955	17.486	5.308	287.2	107.5	63.9	5.9	105.1	66.9	6.4
8336	320.2041	45.9851	2013.482	15.019	17.065	8.137	211.3	14.1	-72.3	5.4	16.6	-73.3	5.5
8337	320.2183	40.1641	2013.514	12.129	12.638	54.744	109.2	85.2	56.1	5.9	84.7	60.9	5.9
8338	320.2338	20.8376	2013.369	11.831	16.683	25.528	326.9	63.8	11.6	5.9	67.1	7.8	6.1
8339	320.2858	22.7380	2013.3955	14.67	16.018	12.342	14.5	97.1	-46.7	6	97.4	-49.2	6
8340	320.2988	-7.1158	2012.864	11.899	13.302	8.351	332.4	-9	-75.1	6.8	-11	-75.6	6.9
8341	320.3334	29.5873	2013.42	16.24	17.655	7.056	218.6	11.3	88.2	5.2	11.2	86	5.3
8342	320.3814	-7.6710	2012.864	15.347	16.43	8.22	150.9	92.3	-7.3	6.9	92	-4.4	7
8343	320.4160	5.1913	2013.048	9.991	14.93	19.844	197	-83	-109	6.5	-86.5	-109.1	6.7
8344	320.4255	79.0252	2013.474	12.905	14.918	30.23	353	43.5	66	5.9	43.7	62.8	5.9
8345	320.4977	56.5113	2013.5135	11.313	15.529	12.884	340.9	89.6	-68.1	5.7	89.7	-68.8	5.8
8346	320.5221	19.8048	2013.9015	5.892	8.786	36.213	311.4	109.2	62.5	5.9	112.9	59.4	5.9
8347	320.5872	20.2216	2013.3185	15.893	16.778	5.939	72.4	-26.9	-93.7	6	-28.5	-91.3	6.1
8348	320.5981	33.3527	2013.3265	17.569	17.754	16.85	236.7	-9.2	67.5	5.6	-6.8	68.8	5.5
8349	320.6160	4.0554	2013.0065	14.439	15.119	29.175	71	6.4	-81.1	6.6	2.8	-80.1	6.6
8350	320.6613	-4.8709	2012.857	12.869	14.436	5.484	271.1	-100.5	-79.1	5.8	-96.4	-82.6	5.8
8351	320.6644	41.9800	2013.2765	17.395	17.84	21.226	333.8	-25.9	-65.8	6	-34.2	-63.5	6.1
8352	320.6769	-3.1685	2012.8555	14.113	17.014	8.213	261.1	86.7	14.3	5.7	87.5	15.4	6
8353	320.7022	61.9299	2013.572	8.472	12.907	21.683	181.4	-97.4	-113.6	5.9	-98.4	-112.3	5.9
8354	320.7087	24.3573	2013.3955	11.156	13.949	5.651	320	67.6	-16	5.1	67.4	-21.3	5.6
8355	320.7272	47.2836	2013.499	13.958	15.442	16.701	23.4	5.1	-69.2	5.4	5.7	-69.1	5.5
8356	320.7305	48.3113	2013.512	12.524	16.518	22.287	21.9	-52.8	-112.1	5.5	-49.5	-116.1	5.6
8357	320.8139	16.8254	2013.282	17.036	17.288	49.103	8.2	10.9	-61.9	5.3	6.5	-62.5	5.3
8358	320.8416	50.8188	2013.3835	17.776	17.83	13.208	239.3	-4.5	-64.2	6.8	-2.1	-68.6	6.4
8359	320.8556	41.5664	2012.9845	17.886	18.253	10.992	1.6	-18.9	-71.2	6.1	-25.3	-67.7	7.6
8360	320.8861	30.1961	2013.558	11.446	13.193	33.364	242.5	62.4	43.3	5.3	61.3	41.6	5.3
8361	320.9060	14.5127	2013.443	10.191	13.496	34.995	224.6	100.1	11	5.1	100	12	5.2
8362	320.9317	44.3252	2013.499	13.555	15.983	17.281	230.8	119	191.5	5.6	116.2	189.5	5.7
8363	320.9454	69.7342	2013.551	9.825	13.156	12.709	334.6	8.2	-60.4	5.8	5.9	-60.8	5.8
8364	320.9734	66.3197	2013.6695	13.918	14.653	3.81	282.9	150.7	161.7	5.9	155	158.7	5.9
8365	321.0430	21.5842	2013.3565	14.954	15.732	8.189	136	60.4	15	6	63.7	14.1	6
8366	321.1201	13.9647	2013.3275	13.191	14.848	30.65	67	145.3	86.7	6.4	149.7	80.8	6.4
8367	321.1235	34.9676	2013.496	14.168	15.588	16.548	123.2	-61.9	-96.6	5.4	-61.2	-93.1	5.4
8368	321.1967	11.6030	2013.32	13.427	17.17	11.153	181.1	14.1	-65.4	6.2	11.6	-65.5	6.6
8369	321.2099	50.2506	2013.7345	8.708	10.319	8.101	36.4	74.7	77.3	6.1	76.3	83.2	6.1
8370	321.2124	18.0613	2013.369	14.035	17.2	8.03	286.6	160.8	-51.9	6	156.3	-52.9	6.3
8371	321.2992	36.4809	2013.4645	13.586	15.146	12.918	56.4	65.7	48.6	6.4	63.6	52.9	6.4

8372	321.3408	46.3130	2013.62	10.269	17.014	10.882	51.8	107.6	37.4	6.1	104	39.8	6.5
8373	321.4001	34.1489	2013.096	17.304	17.602	3.618	96.7	-2.5	-75.3	5.6	-9.4	-75.4	5.9
8374	321.4469	43.0702	2013.532	14.106	15.888	31.094	103.5	112.5	18.8	6.1	110.9	19	6.2
8375	321.5289	-4.5244	2012.858	15.105	15.374	9.288	132.7	161.2	-25.6	5.8	167.9	-23.7	5.9
8376	321.5434	30.5374	2013.429	14.844	16.411	22.656	279.6	81.8	-7.3	5.5	82.9	-5.4	5.6
8377	321.6310	31.0275	2013.339	15.506	18.039	6.894	196.3	74.4	24.5	5.6	74	27.6	5.9
8378	321.6650	20.4422	2013.3555	14.121	15.331	43.839	334.3	76	9.4	6	73.7	10.1	6
8379	321.7375	53.7164	2013.4815	14.527	17.04	46.562	146	1	-64	5.5	-0.3	-63.2	5.6
8380	321.8330	-6.8665	2012.864	12.065	16.689	9.223	302.4	63.3	-21	6.1	65.9	-18.9	7.4
8381	321.8673	16.2969	2013.459	12.598	13.723	48.326	151.5	-31.3	-71.3	6.4	-29.5	-70.9	6.4
8382	321.9018	51.3979	2013.5285	16.501	17.921	28.506	313.2	30.5	-83	6.1	42	-77.1	6.4
8383	321.9155	72.7835	2013.3545	12.495	15.21	6.059	4.3	-117.3	-120.2	6.2	-114.8	-116.2	6.4
8384	321.9386	71.3224	2013.3755	12.791	17.172	11.451	115.7	31.7	93.8	5.9	32	92.9	6
8385	321.9615	25.1961	2013.4925	14.727	15.254	6.754	199.3	30.7	-66.8	5.2	29	-65.8	5.1
8386	322.0350	32.2256	2013.51	6.043	12.865	12.141	11.1	116.7	78.6	5.4	117.2	69.1	5.5
8387	322.0793	15.3216	2013.3585	12.408	14.668	10.986	175.3	2.1	-103.4	6.5	3.4	-102.4	6.5
8388	322.0830	36.6689	2013.442	14.464	15.441	25.745	45.9	59.9	101.5	6.4	61.8	101.8	6.4
8389	322.0838	22.9390	2013.663	9.125	15.789	44.782	84.7	125.3	14.2	5.9	121.1	18	5.9
8390	322.1380	40.4295	2013.443	13.284	16.57	8.242	59.2	0.1	-65.3	6.3	-2.1	-68.6	6.5
8391	322.1562	8.6367	2013.237	13.297	16.641	12.976	135.2	-15.6	-70.9	6.3	-22.3	-68.5	6.7
8392	322.1613	-7.5504	2012.864	11.818	14.299	6.525	96.3	9.7	-94.8	6.1	-1.6	-95.8	6.8
8393	322.1618	41.0308	2013.5195	12.406	12.857	48.342	199.5	63.1	-17.9	6.3	61.5	-12.3	6.3
8394	322.1806	59.8994	2013.4815	13.799	14.6	24.267	34.5	69.3	94.6	5.9	72.7	91.3	5.9
8395	322.1813	26.5701	2013.6865	7.693	13.666	9.099	218.8	66	80.2	5.1	73.1	74.4	5.1
8396	322.2133	29.0910	2013.5285	11.848	13.809	4.638	248.9	-28.1	60.7	5.2	-23.1	61.7	5.8
8397	322.2378	36.1980	2013.5325	10.251	12.345	7.362	335.9	112.1	71.1	6.1	110.3	62.5	6.2
8398	322.3818	36.6874	2013.452	11.549	13.844	18.739	351.4	37.8	76.7	6.2	36.1	75.9	6.2
8399	322.3874	51.1881	2013.7135	10.358	11.392	4.454	232.1	3.4	-96.9	5.4	2.3	-95.5	5.4
8400	322.4164	27.7978	2013.635	14.225	14.813	5.123	331.2	135.8	17.5	5.1	135.3	16.6	5.2
8401	322.4546	10.5094	2013.331	11.623	15.694	10.256	230.5	183.6	9.8	6.6	183.1	9	6.6
8402	322.4868	36.2730	2013.509	14.514	17.732	42.817	94.7	-6	-62.1	6.2	-12.2	-62.9	7.4
8403	322.4888	41.7849	2013.5135	10.415	13.467	23.543	136.3	69.3	-6.1	6.1	69.2	-8.9	6.2
8404	322.4989	65.0209	2013.184	18.224	18.455	59.584	193.2	31.6	-62.5	6.1	23.2	-63.1	6.4
8405	322.5325	9.9937	2012.9835	14.741	17.61	7.084	155.7	52.2	-97.5	6.6	46.9	-103.9	7
8406	322.6357	19.2428	2013.179	17.781	18.218	52.727	268.8	-19.5	-71	6.2	-10.7	-74.7	6.6
8407	322.6500	-0.1010	2013.1995	11.39	15.274	6.725	29.5	-45.4	-99.7	5.9	-45.6	-99.4	6.7
8408	322.9049	53.7638	2013.476	15.417	16.271	13.642	49	76.2	28.4	6.2	75.9	28.5	6.2
8409	322.9287	37.5097	2013.389	10.067	11.351	24.57	235.9	76.8	54.6	6.4	72.6	57.5	6.4
8410	322.9399	13.0804	2013.218	14.078	15.761	8.042	135.2	74.3	30	5.6	74.8	34.8	5.6
8411	322.9824	33.0774	2013.287	17.953	18.343	28.07	49.9	77.6	-5.8	5.5	72.7	-15	5.8
8412	322.9850	28.8382	2013.461	15.167	15.39	5.407	103.2	-7.4	-137.2	5.2	-7.3	-134.3	5.2
8413	323.0217	9.7240	2013.2955	10.958	16.325	10.878	234.1	63	29.8	6.2	65.7	24.4	6.3
8414	323.0316	24.8600	2013.6055	10.973	15.221	10.988	356.8	127.7	42.5	5.1	128	45.1	5.2
8415	323.0556	48.6328	2013.4695	16.736	17.998	57.31	201.6	33.5	65.2	7.9	34.2	62	6.4
8416	323.0858	-5.3477	2012.864	13.824	14.331	11.235	315.6	64.8	-24	5.8	63.6	-25.4	5.8
8417	323.1681	8.5276	2012.9715	16.1	17.817	32.392	278.3	-6.3	-62	6.5	2.4	-61	6.9

8418	323.1860	62.3282	2013.4345	13.906	15.559	5.31	327.7	-34.4	-65	5.9	-33.4	-62.4	5.9
8419	323.2670	69.4711	2013.3435	16.619	17.13	17.34	136.8	79.6	77.7	6.1	80.1	81.1	6
8420	323.2773	41.9031	2013.46	9.706	16.421	43.737	52.1	66.7	37.7	6.4	62.4	43.9	7.8
8421	323.2956	-3.9147	2012.864	14.427	16.941	23.542	65.7	80.1	-14.5	5.7	82.7	-13.7	6
8422	323.3299	61.0624	2013.4945	8.11	14.548	53.651	100.3	75.7	27.3	5.9	73.6	29.2	5.9
8423	323.3942	42.1494	2013.529	10.655	13.091	25.067	145.6	-39.9	-71.2	5.5	-42.9	-73.9	5.5
8424	323.5464	28.0626	2013.3925	14.634	15.31	4.07	275.8	-117.4	-111.3	5.6	-116.3	-111.3	6.6
8425	323.5699	59.7247	2013.236	16.417	18.527	47.84	49.9	75.4	-61.5	6.8	75.3	-55.8	6.3
8426	323.5781	49.8941	2013.559	14.062	14.878	4.43	102.3	117.4	83.4	6.2	110.1	81.9	6.2
8427	323.5850	1.0131	2013.2115	14.33	15.345	7.736	283.5	-53.9	-67.4	6.3	-56.3	-69.3	6.4
8428	323.6067	36.4160	2013.4695	12.729	16.621	31.779	182	149.7	22.1	6.4	148.1	22.3	6.6
8429	323.6241	9.4994	2013.45	7.98	9.785	7.121	239.4	56.6	-98	6.9	57.4	-94.3	6.3
8430	323.6532	54.7023	2013.0665	17.729	18.88	46.075	165.7	-80.6	62.7	7.7	-80	63.8	6.5
8431	323.6923	39.0496	2013.6015	9.762	15.013	19.923	168.4	-156.4	-123.4	6.3	-154.7	-116.7	6.4
8432	323.6970	13.0009	2013.076	13.333	15.885	7.177	79.4	59.2	-68.4	5.6	60.4	-66.7	5.7
8433	323.7555	4.5006	2012.9055	15.799	17.105	20.136	300.5	-12.8	-88.9	6.5	-11.4	-87.8	7
8434	323.7809	26.1221	2013.4825	11.025	14.288	15.079	320.1	67.6	18.5	6	70.7	17.5	6
8435	323.7862	-6.1723	2012.864	13.467	14.568	5.981	17	94.2	-21.2	5.8	90.8	-22.9	6
8436	323.8345	41.3678	2013.325	17.724	17.842	26.013	104	-18	-77.9	6.4	-28.6	-71.4	6.5
8437	323.8437	19.7970	2013.215	17.628	18.641	8.147	341.7	28.9	61.7	6.1	29.9	61.1	6.6
8438	323.8579	40.2132	2013.095	17.984	19.004	45.984	175.1	-3.9	-70.2	6.5	-4.7	-69.2	7.5
8439	323.8859	30.8304	2013.4545	15.518	15.776	3.653	339	78.9	34.4	5.4	79.7	27.5	5.7
8440	323.9392	46.1573	2013.513	15.771	16.151	5.324	200.4	-114.3	-90.3	5.6	-116.4	-92.4	5.7
8441	324.1468	2.5047	2013.2005	10.351	13.791	7.243	194	151.6	13.5	6.4	147.1	19.6	6.6
8442	324.1728	37.5648	2013.2375	17.967	18.089	46.73	301.6	-16.7	-61.3	6.6	-7.3	-64.6	6.5
8443	324.1946	17.9108	2013.126	12.152	14.443	5.254	64.1	-47	-76.6	6	-49.2	-79.6	6.3
8444	324.2199	57.5912	2013.324	17.255	18.243	11.612	217.7	-10.3	-71.5	6	0.5	-69.5	6.9
8445	324.2312	6.0619	2013.039	13.261	17.106	45.922	256.6	4.7	-81.4	6.4	12.2	-84.2	6.8
8446	324.3087	-4.8438	2012.8635	12.651	14.625	9.66	81.7	1.8	-118.6	5.8	0.1	-121	5.8
8447	324.3376	42.9629	2013.307	17.655	17.862	34.877	280.7	24.6	-64.7	5.7	20.4	-62.8	5.7
8448	324.3435	28.0613	2013.3855	15.401	15.994	42.706	254.6	77.6	62.4	6	77.3	64.4	6
8449	324.3909	24.6314	2013.4845	11.788	16.633	58.898	299.7	51	-95.8	5.9	52.2	-95.3	6
8450	324.3985	-7.7412	2012.864	9.817	12.017	21.906	128.7	-94.1	-78	6.5	-89	-78.4	5.8
8451	324.4201	-3.2089	2012.86	10.77	12.093	14.335	141	-12.5	-71.6	5.7	-13.4	-70.5	5.7
8452	324.4798	51.3791	2013.5305	13.168	13.521	10.241	180.5	68.1	61.7	6.3	68	63	6.3
8453	324.5501	26.3301	2013.506	11.989	14.58	24.901	317.7	74.5	0	5.1	76.4	2.1	5.9
8454	324.5803	56.9640	2013.439	16.02	16.835	12.001	10.8	-15.9	-91.6	6.3	-18.1	-87.1	6.4
8455	324.5819	-2.4582	2012.944	11.929	13.083	6.477	65.7	68.3	8.8	5.7	69.2	12.2	5.7
8456	324.5999	30.7199	2013.3825	12.232	13.23	17.443	213	-3.8	-109	6.2	-8.7	-109.2	6.2
8457	324.6260	66.6117	2013.404	12.711	13.975	25.178	257.9	112.1	79	5.9	114.8	78	5.9
8458	324.6355	19.0354	2013.357	9.974	16.172	16.133	90.5	218.1	-93.7	6	215.6	-92.6	6.4
8459	324.6416	45.7412	2013.506	13.649	13.973	9.29	58.3	87.9	21.4	6.3	88	20.5	6.3
8460	324.6934	37.3265	2013.405	14.009	14.018	11.606	341.7	8.6	94.4	6	7.1	94.9	6
8461	324.7027	16.4515	2013.483	12.853	17.862	52.164	59.1	195.1	25.7	5.6	191.9	24.3	5.8
8462	324.7164	54.6145	2013.5155	16.1	16.187	1.792	338.6	-21.7	64.7	6.9	-23.1	67.7	6.8
8463	324.7244	45.3734	2013.275	17.599	17.757	22.907	295.2	9.4	-60.8	6.3	0.2	-61.3	6.3

8464	324.7639	7.9983	2013.286	10.221	14.559	9.177	270.1	29.8	-121.4	6.4	38.1	-122.3	6.5
8465	324.7639	13.4624	2013.4815	10.136	11.603	9.126	127.7	34.4	-72.3	5.5	30.7	-76.5	5.5
8466	324.7754	29.9144	2013.451	14.166	16.671	7.291	85.1	4	-89.1	6.4	-0.5	-91.1	6.5
8467	324.7983	-5.5274	2012.864	12.016	15.328	48.565	359.1	-94.1	-104.2	5.7	-89	-109.4	5.8
8468	324.8008	38.8277	2013.3575	12.686	12.942	7.447	129.9	-62.8	-65	6	-60.8	-60.8	6
8469	324.8122	56.4568	2013.4115	18.243	18.25	34.474	62.7	35.4	-73.2	7.6	22.3	-76.3	6.6
8470	324.8215	41.6104	2013.447	10.814	17.713	33.709	205.1	84.3	28.3	6	78.6	32.7	6.4
8471	324.8223	43.2012	2013.5195	11.612	11.98	6.837	275.1	140.2	131.5	6.3	143.4	130.8	6.2
8472	324.8258	40.3449	2013.5095	8.106	15.288	28.852	186.9	64.7	-15.5	5.9	65.4	-13.6	6
8473	324.8943	40.8288	2013.5015	10.289	11.859	6.822	339.5	-61.3	-62.4	6	-60.4	-63.1	5.9
8474	324.9260	14.3848	2013.2715	15.043	16.322	18.387	319.4	-26.4	-120.9	5.6	-26.9	-117.9	5.7
8475	324.9619	66.5081	2013.3715	13.762	17.585	18.431	99	-27.7	167.2	6.2	-24.2	169.6	6.4
8476	325.0644	5.2736	2013.0275	12.232	14.951	14.906	22.9	-7.7	-69.1	6.5	-6	-67.7	6.5
8477	325.1425	-0.3032	2013.2965	11.74	12.711	16.577	83.3	66	1.8	5.5	65.6	1.2	5.5
8478	325.1857	57.2064	2013.4145	15.507	17.158	5.317	104.9	-40.6	-96.2	6.2	-40.7	-96	6.3
8479	325.1938	1.6918	2013.444	10.716	11.155	5.442	66.5	82.2	3.2	6.6	83.1	1.9	6.7
8480	325.1976	6.4447	2013.009	11.499	16.516	12.521	275.2	70.1	15.4	6.4	72.7	15.6	6.6
8481	325.2148	16.6083	2013.3335	10.332	12.738	7.708	46.9	122.2	57.7	5.6	119	54	5.6
8482	325.2302	22.8245	2013.417	10.266	14.413	14.878	171.2	81.7	8	6.2	78.6	11.1	6.2
8483	325.2399	3.3504	2013.0505	14.19	16.928	8.449	309.3	-70.4	-137.5	6.5	-69.9	-142.5	6.7
8484	325.2555	11.2626	2013.2405	9.156	11.685	17.258	283.1	-69.4	-130.7	6.5	-64.5	-126.5	6.5
8485	325.2615	35.9206	2013.151	17.926	18.045	50.859	259.2	6.4	-65.1	6.5	3.9	-67.5	7.7
8486	325.2712	72.0479	2013.5825	15.306	16.003	5.291	248.2	-73.9	-63.5	6.2	-78.2	-61.4	6.1
8487	325.2873	22.2191	2013.333	15.912	17.12	5.188	38.9	98.1	-23.8	6.4	95.6	-24.2	6.7
8488	325.2879	9.9586	2013.381	13.387	14.168	5.111	154.1	89.7	-17.1	6.3	88	-13.6	6.3
8489	325.2958	18.0026	2013.5375	12.068	16.372	9.72	33.7	-52.4	-69.1	6.2	-55.7	-67.7	6.2
8490	325.3144	32.4908	2013.385	14.976	15.603	6.326	324.4	8.9	-105	5.9	9.4	-104.3	5.9
8491	325.3351	23.9110	2013.374	13.53	18.337	11.356	163.8	35.1	-119.3	6.2	34.4	-115.7	7.5
8492	325.3364	5.2733	2012.9625	14.754	16.961	8.054	236.8	117.6	10.4	6.7	121.3	9.8	6.8
8493	325.3689	14.7264	2013.1895	11.511	15.36	7.941	241.3	-3.4	-91.5	6.3	-3.1	-93	6.4
8494	325.3887	37.0879	2013.3775	8.953	14.985	24.509	83.4	153.3	42.8	6.2	155.3	45.2	6.3
8495	325.4469	17.4181	2013.2745	14.44	16.617	6.717	13.6	63	3.9	5.5	60.7	4.8	5.7
8496	325.5292	43.5756	2013.3615	16.801	17.442	28.145	243	-25.6	65.1	5.9	-23.5	64	5.9
8497	325.6046	9.0320	2013.133	15.111	15.415	40.418	36.2	67.2	13.6	6.5	63.5	26.3	6.4
8498	325.6225	12.5548	2013.284	13.383	14.594	4.741	9.9	68.9	-4.3	5.6	68.3	-6.8	5.5
8499	325.6235	46.1462	2013.501	11.581	14.54	17.758	332.5	36.6	66.5	5.5	34.1	68.3	5.5
8500	325.6640	48.0554	2013.4265	12.541	17.379	10.647	114.3	-26	-104.8	6.2	-27.3	-105.4	6.4
8501	325.7539	2.8886	2012.846	16.697	17.167	5.853	160.7	-14.1	-65.8	6.7	-19.5	-61.2	6.9
8502	325.7659	-0.9916	2012.982	10.996	14.769	12.408	138.2	16.1	-84.4	5.7	17.2	-82.3	5.8
8503	325.7741	39.0763	2013.2975	12.003	17.483	10.191	75.6	63.1	-7.6	5.9	65.7	-2.6	6.2
8504	325.8127	48.4667	2013.521	10.311	13.046	46.592	116.6	66.9	30	6.3	68	24	6.3
8505	325.8282	54.0528	2013.486	15.557	16.781	4.562	206.4	92.5	27.5	6.2	94.8	32.9	6.2
8506	325.8362	51.0271	2013.455	18.201	18.458	53.526	204.9	72.9	10.6	6.5	70.5	7.1	6.9
8507	325.9305	24.6386	2013.4415	11.244	16.654	10.294	98.7	92.6	70.7	6	85.2	72.5	6.3
8508	326.0234	12.7343	2013.127	14.504	18.128	51.147	307.5	4.1	-74.5	5.6	2.1	-75.3	6.6
8509	326.0239	47.1224	2013.529	9.744	16.056	12.771	252.1	60.1	-30.2	5.4	60.3	-29.6	5.6

8510	326.0385	51.5917	2013.453	14.213	16.258	4.672	261.9	-10.9	-75.2	6.3	-12.3	-72.2	6.5
8511	326.0464	52.2885	2013.4535	14.997	15.685	15.779	0.4	-16.2	-98.7	6.3	-18.9	-100.2	6.4
8512	326.1330	39.9960	2013.268	17.118	18.08	34.246	233.8	14	-67.7	6.7	26.2	-64	6.6
8513	326.1761	33.5365	2013.3985	13.003	13.18	12.872	165	208.9	110.9	6.1	212	113.4	6.1
8514	326.2467	39.7458	2013.5465	13.199	15.668	5.158	170.6	90.7	82.3	6.5	91.5	86.2	6.6
8515	326.2728	0.3670	2012.847	16.701	17.689	18.957	143.3	-74.2	-62	6.8	-75.9	-60.3	7.1
8516	326.2908	52.8962	2013.606	8.875	11.428	7.539	221.9	-60.9	-70.8	6.2	-72.6	-65.1	6.1
8517	326.2931	34.5483	2013.3055	11.176	12.735	4.673	297.8	112	25.9	6	111.9	27.7	6.2
8518	326.3091	4.2248	2012.8425	16.685	17.183	26.75	253.5	-15.8	-85.1	6.9	-22.4	-85	6.9
8519	326.3514	4.1026	2012.927	15.53	15.894	19.996	174.6	71	-48.4	6.6	72	-47.5	6.7
8520	326.3568	61.3631	2013.031	17.781	18.524	54.884	352.5	4.2	-61.7	6.1	2	-62.8	6.4
8521	326.3623	13.2014	2013.1455	10.929	14.67	23.289	142.5	78	-6.7	5.6	77.1	-3.3	5.6
8522	326.3716	42.5285	2013.325	12.693	18.271	53.604	149.7	-51.6	-73.7	6.4	-52.7	-73.9	6.8
8523	326.3815	55.8378	2013.3945	13.113	17.6	13.376	45.8	50.5	-102.2	6.2	52.9	-97.2	6.4
8524	326.5255	55.3553	2013.402	14.044	18.16	37.034	107.3	67.2	17.6	6.1	67.4	23.6	6.4
8525	326.5333	80.0123	2013.5905	12.982	15.213	29.989	80.5	118.8	71.9	5.8	122.2	74.6	5.8
8526	326.7429	4.1328	2012.842	11.009	17.263	24.341	182	-20.6	-78.9	6.6	-30.5	-78	7.3
8527	326.7667	52.2546	2013.3805	18.181	18.404	54.413	292.6	-31.3	64.6	6.4	-27.6	62.5	6.7
8528	326.7873	53.1618	2013.4375	11.798	13.251	13.706	191.1	139	-179.8	6.3	138.3	-185.2	6.3
8529	326.8214	50.2317	2013.479	17.953	18.285	28.66	258.1	-23.6	-61.7	6.3	-18.8	-65.2	6.5
8530	326.8596	33.6358	2013.2985	12.304	16.54	45.063	0.4	9.7	-70.3	5.9	14.1	-70.4	6.1
8531	326.9030	12.6493	2013.3495	15.752	16.234	5.679	205.5	-19.3	-85.1	5.2	-18.8	-86.9	5.3
8532	326.9444	23.2621	2013.355	15.805	17.381	9.972	87.4	96.9	-11.5	5.5	101.5	-10	5.8
8533	326.9523	17.9084	2013.3045	15.262	16.544	32.714	66.3	96.8	-13.2	5.2	95.4	-11.9	5.4
8534	326.9897	47.8584	2013.159	17.673	18.805	48.788	59.4	-0.9	61.5	6.3	-4	63.2	6.7
8535	327.0059	8.8181	2013.248	13.806	15.647	35.767	192.9	85.6	-35.1	6.4	83.8	-34.4	6.4
8536	327.1123	23.8368	2013.392	13.967	14.839	22.181	130.9	67	-41.3	5.5	69	-40.9	5.5
8537	327.1862	8.6287	2013.0855	12.245	14.536	14.664	132.3	90.9	21.5	6.4	90	24.2	6.7
8538	327.2284	3.1468	2012.845	12.698	13.639	29.128	230.3	102.2	-14.3	6.6	101.2	-14.8	6.6
8539	327.2526	29.7061	2013.184	16.727	17.38	5.777	147.2	84.6	13.3	5.3	87.6	15.9	5.5
8540	327.2610	37.1711	2013.2055	15.712	17.621	7.73	357.2	118.1	70	6.3	120.4	69.1	6.6
8541	327.2971	26.6734	2013.3765	10.59	13.641	16.514	133.7	118.8	28.5	5.2	117.4	28	5.2
8542	327.3464	39.8968	2013.3985	10.768	15.931	32.608	270.8	40.8	-66.8	5.5	41	-65.3	5.5
8543	327.3540	51.1719	2013.388	10.3	14.999	57.132	354	99.8	56.1	6	98	54.3	6
8544	327.4200	32.6439	2013.301	17.484	18.427	9.344	280.3	64.7	-26	6.1	65.5	-21	6.4
8545	327.4460	1.1764	2012.8685	14.792	15.813	13.177	176.4	62.8	-6.3	6.6	64.2	-10.4	6.7
8546	327.4540	14.0380	2013.208	9.338	16.02	13.236	338.3	166.6	-43.4	5.3	164.8	-51.3	5.4
8547	327.4729	18.1389	2013.3775	13.149	14.311	4.729	98.3	-71.2	-81.7	5.5	-72.4	-78.8	6.4
8548	327.4733	63.1107	2013.3325	14.76	16.974	14.621	199.3	-79	76.5	5.9	-80.2	76.5	6.1
8549	327.4948	46.3624	2013.5835	15.97	16.898	23.041	344.9	50.4	68.4	7.2	50.5	63.1	6.5
8550	327.5059	46.2574	2013.4915	11.559	13.443	7.328	24.3	66.2	-15.8	6.1	62.9	-17.9	6.2
8551	327.5426	5.9704	2013.2005	13.385	13.501	11.996	252.8	135	-28.4	6.6	138.8	-27.7	6.6
8552	327.5681	49.1533	2013.488	11.53	14.022	58.558	258.2	80.9	-26.1	5.9	80	-19.8	5.9
8553	327.6040	57.3315	2013.416	14.998	17.19	8.244	57.8	67.4	47.1	5.9	67.6	50.7	6
8554	327.6194	44.9631	2013.302	17.554	18.658	25.9	31.9	4.7	-63.5	6.4	7	-65.4	6.7
8555	327.7476	16.6114	2013.293	12.1	17.128	27.989	91.5	12.5	-78.2	5.2	14.5	-79.9	5.3

8556	327.7777	16.0683	2013.2005	11.912	17.892	49.997	342.3	-48.1	-64.4	5.2	-44	-66.7	5.7
8557	327.8130	8.1282	2013.3225	9.748	11.129	5.893	247.1	80.5	-109.9	6.4	89.7	-105.8	6.4
8558	327.8239	20.6069	2013.4545	10.607	15.094	18.933	352.6	108	-45.9	5.8	102.6	-49.5	5.8
8559	327.8268	8.3472	2013.3215	8.349	10.76	14.538	354.7	107.5	13.5	6.6	103.5	17.5	6.4
8560	327.8394	-1.0100	2013.1735	10.16	11.74	24.695	36.3	-102.7	-75.8	5.7	-99.2	-73.8	5.7
8561	327.8491	10.8620	2013.216	13.755	14.205	41.191	93.3	146.4	25.2	6.5	149.9	22.7	6.4
8562	327.9230	57.6330	2013.2455	18.122	18.693	10.422	213.9	66.4	26.6	6.2	67.7	27.1	6.5
8563	328.0000	37.3625	2013.3385	13.568	18.139	12.193	289.4	97.2	18.5	5.5	98.7	23	5.9
8564	328.0416	48.9811	2013.532	15.136	15.439	4.578	342.1	109.3	45.4	5.9	106.2	43.5	5.9
8565	328.0672	23.7416	2013.3335	16.678	17.995	40.027	98.2	-13	-66.4	5.9	-9.4	-65.9	6.1
8566	328.0958	6.3890	2012.9405	9.608	12.154	7.063	253.1	103.1	-25.5	6.7	102.7	-35.4	6.7
8567	328.1052	49.6957	2013.5055	10.039	16.005	11.077	359.8	-35.5	-66.3	5.9	-33.9	-63.5	5.9
8568	328.1383	11.7957	2013.4485	13.806	13.877	4.922	71.8	61.7	-16.8	6.3	63.2	-8.3	6.7
8569	328.2413	38.6263	2013.282	12.023	17.468	20.796	179.2	68	17.3	6	66.9	21.2	6.2
8570	328.2539	54.3088	2013.1435	16.626	18.156	23.422	282.3	-10.1	-66.9	6.4	0.2	-69.4	6.5
8571	328.3820	15.8463	2013.367	10.576	13.52	51.244	137.9	-79.1	-107.4	5.8	-76.1	-105.3	5.8
8572	328.4167	81.2068	2013.613	10.431	12.883	26.148	166.8	141.8	109.7	5.9	140.6	107.6	5.8
8573	328.4238	9.8938	2013.2075	10.285	15.823	32.285	230.3	71.3	14.8	6.4	73.6	18	6.4
8574	328.4395	49.1036	2013.337	17.354	17.489	43.329	117.7	91.7	-5	6.1	88	-8.9	6.3
8575	328.4669	37.7199	2013.118	15.191	17.258	5.17	343.6	94.5	-4.9	6.3	95.2	-3.7	6.6
8576	328.5660	35.7856	2013.434	12.401	14.996	22.526	88.2	-3.4	-97.7	5.9	-8.9	-96.2	6
8577	328.6061	7.6376	2013.2005	16.237	16.932	13.077	73.1	-83.1	-102.4	6.7	-69.4	-104.9	7.1
8578	328.6598	-2.5554	2012.896	14.691	17.626	11.763	316.8	91.8	-66.9	5.7	90.5	-64.9	6.6
8579	328.6947	7.3298	2013.1685	12.5	12.7	7.327	357.6	68.7	6.9	6.5	69.4	4.4	6.5
8580	328.7353	51.0466	2013.294	17.372	17.707	45.3	14.4	13.7	-64.6	6.3	17.6	-62.7	6.3
8581	328.7499	53.9359	2013.6645	7.123	11.813	11.642	310.3	156.8	87.4	6.1	164.2	90.5	6.2
8582	328.7615	1.6653	2012.912	12.506	15.417	11.001	107.2	12.4	-125.1	6.8	9.8	-128.5	6.8
8583	328.8034	59.8769	2012.8985	18.478	18.712	22.224	206.1	-16.4	-61.7	7.2	-24.9	-62	7
8584	328.8163	59.0363	2013.0395	18.106	18.245	30.21	184.3	72.5	-3	6.1	74.5	-11.6	7
8585	328.8282	51.0608	2013.3645	15.759	17.224	55.397	353.5	12	-79.6	6.2	14.7	-77.2	6.3
8586	328.8289	7.1101	2013.1345	13.96	14.688	8.263	266.3	-69.6	-67.8	6.7	-70.3	-63.2	6.7
8587	328.8463	-5.7125	2012.866	12.867	14.942	6.942	274.4	22.5	84.7	5.8	25.3	84.7	5.8
8588	328.8697	16.9626	2013.545	14.137	14.195	5.166	297.2	-22.3	-97.1	5.9	-18.2	-97.1	5.7
8589	328.9022	51.6379	2013.526	10.282	15.236	9.614	61.5	12.2	-80.9	6.2	5.8	-83.9	6.1
8590	329.0242	81.2459	2013.578	10.263	16.122	26.145	359.1	62.9	63.3	5.8	62.6	62.8	5.9
8591	329.0303	22.8175	2013.495	12.831	16.927	7.779	244.5	121.5	-73.4	6	123.3	-76	6.5
8592	329.1513	34.9983	2013.3565	12.399	13.782	20.621	61.8	87.3	33	6.4	89	33.2	6.4
8593	329.2012	0.2834	2012.8815	15.62	17.426	19.541	248.6	-30.8	-64.1	6.7	-28.4	-64.9	7.2
8594	329.2112	42.6981	2013.4305	12.642	17.589	11.251	59.7	67.9	-38.9	5.9	67.5	-41.4	6
8595	329.2131	67.3572	2013.3915	15.131	15.165	5.556	265.3	-45.9	-65	6.2	-45.7	-66.3	6.3
8596	329.2450	-6.7292	2012.869	9.974	14.707	26.718	357.9	167.8	19.3	5.8	160.3	20.8	5.8
8597	329.2826	16.3182	2013.329	12.565	13.249	24.733	121.8	135.4	-4.1	5.6	139	-2.9	5.6
8598	329.2850	48.9544	2013.475	17.614	18.411	47.314	191.4	24.6	-110.9	7	28.6	-110.6	6.9
8599	329.3085	38.4812	2013.2335	12.174	17.517	29.979	258.4	65.1	90.1	6.3	69.2	86	6.5
8600	329.3974	29.3325	2013.387	9.772	14.101	21.105	43.6	119.3	-17.6	5.2	119.1	-11.3	5.3
8601	329.4005	-3.4698	2012.864	14.525	17.729	37.058	128.3	165	-146.4	5.7	159.2	-141.9	6.9

8602	329.4582	61.2196	2013.422	14.115	15.058	9.429	283.6	69.9	-8.5	5.9	66.4	-8.9	5.9
8603	329.4734	32.4558	2013.2125	16.746	18.252	10.684	64.5	-43.3	-65.5	5.9	-46.5	-64.5	6.2
8604	329.4941	27.9142	2013.326	12.056	13.308	14.18	113.6	-63.2	-159.6	5.3	-64.9	-166.1	5.3
8605	329.4962	65.2833	2013.0015	17.68	18.325	57.854	324.3	64.8	-1.9	7.1	62.4	7.4	6.7
8606	329.5743	2.8147	2012.912	11.021	13.87	5.748	347.6	63.1	-75	6.7	65.6	-77.9	7.7
8607	329.6026	36.7040	2013.3245	12.847	16.908	33.689	163.1	198	-29.6	6.3	193.9	-31.1	6.5
8608	329.6918	29.9185	2013.1765	13.385	16.768	13.231	255.3	-54.3	-73.8	5.8	-53.2	-75.1	6
8609	329.7669	5.6818	2012.841	13.427	14.162	34.719	202	66	-3.3	6.5	64.7	-4.7	6.5
8610	329.7878	2.4000	2013.0525	10.118	17.484	24.469	234.7	-32.6	-79.4	6.5	-31.4	-78.8	7
8611	329.7988	35.3614	2013.2475	16.086	16.508	57.948	44.9	77.6	-14.2	5.8	76	-10.8	5.9
8612	329.8233	30.0156	2013.2055	12.858	16.96	16.572	129.2	64.1	-17.7	5.8	63.5	-13.8	6
8613	329.8949	12.5753	2013.2265	9.621	13.317	27.22	14.2	65	5.5	6.5	62.4	10.4	6.5
8614	329.9129	31.4215	2013.328	14.202	14.255	10.848	255.4	9	-120.4	5.8	10.5	-120.7	5.8
8615	329.9300	60.8119	2013.352	15.778	17.1	13.334	245.9	98.4	35.7	5.9	99.2	32	6
8616	329.9856	41.5663	2013.421	11.547	14.965	11.591	317.6	-85.2	-109.1	6.2	-87.5	-105.8	6.3
8617	330.0626	31.6877	2013.1935	14.004	17.929	15.753	12.4	75.7	-6.6	5.8	76.2	-0.7	6.1
8618	330.1031	10.5004	2013.253	11.437	13.214	7.648	348.2	80.9	5.5	6.4	79.4	8.1	6.4
8619	330.1133	52.4303	2013.354	12.826	16.873	11.315	125.2	121.8	64.9	5.9	123.7	67.5	6
8620	330.1186	38.2457	2013.4205	9.802	14.377	9.95	113.9	72.3	77.2	6	71.5	71.7	6
8621	330.1467	29.7976	2013.1635	12.204	18.084	26.604	232.6	116.1	4	6	112.3	13.3	6.6
8622	330.1488	11.6169	2013.2255	13.743	15.47	35.64	227.5	-20.8	-71.1	6.4	-21.5	-70.8	6.4
8623	330.2631	15.7415	2013.333	10.757	16.426	11.44	227.6	67.9	121.2	6.1	67	124.3	6.1
8624	330.3202	-2.3536	2012.9305	14.387	15.094	12.662	299.3	48.7	-102.9	5.8	46.7	-102.4	5.8
8625	330.3274	12.3375	2013.119	14.984	16.402	8.883	126.2	99	20.8	6.2	99.8	26.9	6.2
8626	330.3914	45.4634	2013.501	10.72	14.492	7.866	292.9	149.8	105.5	6.2	150	102.7	6.2
8627	330.4141	46.3684	2013.446	17.514	17.959	45.873	272.6	64.7	-11.9	6.2	64.1	-8.5	6.3
8628	330.4277	20.1289	2013.3255	12.736	14.144	5.376	183.6	-54.2	-76.9	6	-55.6	-79.4	6.2
8629	330.4473	16.9142	2013.432	12.452	12.765	46.304	297.7	-47	-83.7	5.5	-45.8	-82.3	5.6
8630	330.4656	63.9436	2013.343	14.416	16.919	12.154	268.8	98.5	80.4	5.9	99.1	78.9	6
8631	330.4859	28.8183	2013.3115	12.092	17.592	10.686	112.7	11.8	-70.3	6	9.6	-70.7	6.1
8632	330.5180	-3.8545	2012.866	15.618	16.057	6.269	288.2	119.7	54.7	6.7	122.5	56.3	6.8
8633	330.5300	11.5681	2013.239	7.658	15.568	17.772	202.6	114.8	31	6.3	112.7	28.3	6.8
8634	330.5478	30.3340	2013.2915	11.546	16.634	21.983	311	96.9	67.3	5.5	96.4	71.4	5.6
8635	330.5536	49.0680	2013.398	13.515	17.959	47.865	103.6	-13.2	-70.7	5.9	-20.5	-69.5	6.3
8636	330.7199	16.7830	2013.436	16.496	16.69	3.694	29.5	67.9	2.3	5.7	70.3	5.7	5.7
8637	330.7754	11.5493	2013.1815	14.069	15.052	5.872	323	89.2	-2.5	6.4	89.5	-2.6	6.5
8638	330.7997	57.1276	2013.4295	15.736	15.742	10.352	95	191.8	8.1	5.9	195.4	11.8	6
8639	330.8001	1.9474	2013.088	10.164	10.295	8.354	3.3	6	-74.7	6.6	2.4	-76.2	6.7
8640	330.8503	36.3871	2013.3725	13.369	14.703	4.264	135.5	84.9	-31.1	6.3	85.9	-33.2	7.9
8641	330.8757	19.6534	2013.3875	13.017	13.93	4.524	112.8	-22.2	-68.1	6.3	-24.7	-66.1	6.3
8642	330.9019	43.9833	2013.439	13.424	16.967	13.278	86.3	146.9	42.9	6.4	147.1	43.2	6.5
8643	330.9226	27.8769	2013.3765	12.476	13.369	5.538	51.2	61.5	16.7	6.4	60.8	19.5	6.4
8644	330.9273	-3.5878	2012.865	13.328	16.071	13.769	14.9	-24.3	-71	6.7	-28.2	-70.4	6.7
8645	330.9403	5.4478	2013.2685	12.028	12.57	4.845	170.7	-41.4	-84.2	6.4	-40.7	-89.3	6.6
8646	331.0201	6.9574	2013.5255	11.589	12.9	14.602	9.2	127.8	-5.5	6.4	126.4	-7.3	6.3
8647	331.0266	34.5512	2013.5205	9.69	11.171	6.06	33.2	-3.5	-66	5.4	-6.9	-66.4	5.3

8648	331.0269	56.0876	2013.41	9.588	14.859	39.645	231.7	109.5	-13	6.2	106	-5.1	6.3
8649	331.0612	10.0693	2013.4295	10.678	14.69	12.33	285.8	-119.2	-92.2	6.5	-118.9	-99.8	6.5
8650	331.0728	9.8599	2013.4045	10.996	12.016	40.481	135.4	176	87.5	6.6	178.6	89.6	6.5
8651	331.0881	50.8156	2013.4515	9.94	12.125	37.108	314	65.5	35.1	5.8	66.4	36.7	5.9
8652	331.3019	79.7580	2013.462	16.769	17.084	14.54	353.9	8	-60.8	5.9	11	-60.2	5.9
8653	331.3259	24.8885	2013.3485	11.945	15.688	40.648	106.1	12.9	-69.1	6	17.5	-67.2	6
8654	331.3460	35.0743	2013.339	18.025	18.74	58.744	288.6	67.7	-5.9	5.6	65.3	-8.4	5.7
8655	331.3951	64.3660	2013.42	10.431	13.773	26.185	143.6	147.9	27.9	5.9	145.7	27	5.9
8656	331.4135	71.8502	2013.3915	10.757	12.005	22.916	52.4	8.7	82.2	6.2	7.3	85.8	6.2
8657	331.4613	18.1187	2013.322	14.324	15.247	6.27	0.1	-38	-68.8	6.1	-37.6	-69.7	6
8658	331.4885	37.4089	2013.38	11.155	14.639	33.001	79.8	149.5	-83.6	6.3	147.4	-79.5	6.3
8659	331.5573	17.3868	2013.4705	11.358	15.064	7.5	288.4	59	-75.1	5.7	67.2	-73.4	5.6
8660	331.5801	12.7902	2013.2115	15.079	15.811	5.845	93.5	14.5	-92.9	5.7	12.5	-93.3	5.7
8661	331.6224	46.6428	2013.4625	17.616	17.7	19.608	74.8	-15.4	65.3	6.4	-16.7	66.2	6.5
8662	331.6495	51.3135	2013.439	10.69	10.857	10.513	105.4	33	-80.6	5.9	39.8	-74.7	5.9
8663	331.6755	7.5727	2013.3845	10.527	13.471	12.668	47.5	92.3	-63.6	6.4	99	-60.7	6.4
8664	331.7045	70.9653	2013.355	10.003	10.732	8.398	339.9	107.8	13.2	6.2	110	9.1	6.2
8665	331.7991	52.5768	2013.137	18	18.586	53.039	336.4	-6.6	-66.2	6	-1.4	-67.2	6.4
8666	331.8805	48.2990	2013.4335	17.892	18.074	52.967	328.6	-23.7	100.7	7.4	-19	100.2	6
8667	331.9512	11.7648	2013.413	10.063	14.685	24.987	355	-47.3	-79.6	6.3	-44.5	-82.5	6.3
8668	332.0368	33.9250	2013.3255	16.221	16.56	8.579	81	67.8	56.3	5.4	67.9	54	5.5
8669	332.0469	45.5360	2013.399	17.215	17.984	46.713	62.3	-13	-60.9	6.5	-17.3	-61.3	6.5
8670	332.0775	50.3582	2013.368	11.897	13.299	4.471	311.1	69.2	23.9	5.8	67.4	23.4	6.5
8671	332.0920	15.3722	2013.294	15.554	17.021	10.826	76.4	72	18.3	5.7	76.3	15.7	5.7
8672	332.1346	2.1599	2012.8995	13.703	15.852	32.248	188	121.6	-174.7	6.7	120.9	-174.9	6.6
8673	332.1363	14.4209	2013.3415	11.665	14.662	55.747	67.1	21.7	125.7	6	18.6	128	6
8674	332.2654	48.6527	2013.31	17.686	17.698	20.402	120.1	-11.2	62.2	6	-8.4	63.1	6
8675	332.2855	25.7567	2013.3195	11.802	12.09	8.31	28.1	75.7	-17.6	5.2	77.1	-20.5	5.2
8676	332.3018	49.7300	2013.5045	15.538	16.877	4.885	317.3	-55.7	-75.6	5.9	-56.4	-75	6
8677	332.3817	-7.2907	2012.868	12.869	14.421	11.206	36.6	-73.2	-110.5	5.8	-68.4	-119.5	5.8
8678	332.4149	22.2421	2013.477	7.163	16.399	41.232	222	88.3	-21.9	5.9	87.9	-13.3	6
8679	332.4251	23.3919	2013.2895	12.251	16.187	15.755	239.8	-38.1	-70.6	6	-36.5	-68.7	6.1
8680	332.4451	8.0010	2013.3505	11.112	11.302	9.739	319.4	-34.3	61.3	6.3	-35.4	60.6	6.3
8681	332.5699	-1.4420	2013.1705	10.757	12.671	36.75	19.5	121.6	-50.5	6.1	118.7	-56.1	6.1
8682	332.6200	30.8102	2013.48	7.888	14.677	18.257	240.8	-65.1	-96.2	5.9	-72.9	-89.6	6.1
8683	332.6798	32.4048	2013.232	16.397	17.094	4.931	326.8	100.3	-28.5	6.1	95.3	-29.2	6.3
8684	332.6840	48.6220	2013.4775	14.082	16.819	7.34	10.9	68.8	31.5	5.9	71.2	28.4	5.9
8685	332.6969	64.6774	2013.194	16.027	18.173	7.888	135.3	76.2	0.2	5.9	74.8	0.5	6.3
8686	332.7474	10.3397	2013.423	12.067	12.149	42.034	155.4	127.7	-23.5	6.3	124.3	-23.9	6.3
8687	332.7593	32.3530	2013.4535	10.791	13.826	5.725	88.6	73.7	50.3	6	74.8	45.8	7.7
8688	332.7864	32.0615	2013.309	16.868	17.488	39.174	211.4	-23	-67.8	6.1	-34.5	-64.2	6
8689	332.8173	22.4402	2013.349	10.672	15.73	28.899	328.7	-37.5	-91.1	6	-42.2	-89.7	6.1
8690	332.8368	45.2033	2012.984	18.632	19.101	24.942	123.8	75.2	6.5	6.9	74.4	19.4	6.7
8691	332.8383	3.5439	2013.3505	12.51	16.047	7.646	21.6	-30.3	-73.8	6.8	-30.5	-74.9	6.7
8692	332.8607	63.0010	2013.0625	18.236	18.649	32.548	32.5	86.3	32.8	6.2	83.5	28.1	6.8
8693	332.9281	3.8421	2013.246	13.297	17.168	16.734	51.1	111.9	-34.6	6.9	113.8	-43.5	7

8694	332.9327	18.3201	2013.2465	13.164	13.613	17.322	171	-56.4	-135.1	6.1	-61.6	-135.9	6.1
8695	332.9501	2.3142	2012.859	16.311	16.461	43.776	230.9	30.2	-67.3	6.8	28.5	-64.9	6.8
8696	332.9543	45.1338	2013.2915	17.921	17.948	57.384	106.2	68.4	26.8	6.1	69.8	20.5	6.1
8697	333.0106	21.9389	2013.3135	14.071	17.207	20.085	191.1	-52.1	-92	6	-56.1	-89.8	6.2
8698	333.0527	9.5681	2013.49	15.48	15.511	7.017	45.8	25.3	-68.8	6.3	21.6	-67.8	6.3
8699	333.0743	27.7363	2013.3665	12.037	17.612	9.332	81	93.9	38.2	5.3	92.6	41.2	5.7
8700	333.1204	44.7932	2013.5585	15.609	15.666	5.041	160.9	77.8	22.1	5.5	78.8	18.3	5.4
8701	333.2514	5.6110	2013.462	14.959	17.383	24.938	351.3	90.2	11.6	6.4	89.8	7.9	6.6
8702	333.2860	23.4256	2013.314	11.267	14.931	25.568	151.7	63.9	-30.7	6	65.5	-28.1	6
8703	333.2972	70.2088	2013.363	12.139	15.882	11.568	359.2	93.3	-1.4	6.2	92.3	-1.1	6.3
8704	333.3425	13.3064	2013.332	12.783	13.422	4.991	339.5	62.4	-46.5	5.2	61.9	-49.8	5.3
8705	333.4573	52.3718	2013.353	16.076	17.058	50.389	117.9	18.8	98.3	6	21.5	98.7	6
8706	333.4634	27.2051	2013.281	11.304	13.549	47.551	290.9	67.7	74.3	5.3	69.5	74.2	5.3
8707	333.5029	6.2081	2013.387	13.46	13.573	11.571	30.5	93.8	-90	6.4	94.2	-91	6.4
8708	333.5329	70.1700	2013.343	15.393	17.174	5.1	28.2	61.7	9.5	6.2	62.7	2.3	6.3
8709	333.5398	63.1451	2013.366	10.596	18.871	49.656	179.5	-10.4	66.2	6.4	-1.4	66.4	6.4
8710	333.6329	18.7205	2013.3025	14.749	16.835	5.932	138.6	79.6	58.2	6.1	75.9	62	6.1
8711	333.6481	24.7744	2013.3515	12.052	14.003	16.839	108.5	72	-41.2	5.2	71.5	-37.7	5.2
8712	333.6695	84.4509	2013.75	13.034	16.007	10.285	320	72.6	3.6	5.8	70.3	3.7	5.8
8713	333.7049	42.1292	2013.327	15.335	18.512	41.201	266.7	62.7	-25	5.6	63.4	-18.9	5.9
8714	333.7635	77.1137	2013.231	15.917	17.706	9.86	194.1	74.7	18.6	6	70.7	20.1	6.1
8715	333.7721	52.9965	2013.33	16.021	17.703	40.975	25.1	-40.1	-61.9	6	-28.7	-71.1	6
8716	333.7772	14.5380	2013.4745	13.253	16.353	9.88	76.1	68.3	-38.8	5.1	65.8	-38.2	5.3
8717	333.8518	16.6174	2013.5175	11.744	16.242	11.988	330.3	-25.8	-65.4	5.1	-26	-63.7	5.1
8718	333.8562	17.1074	2013.456	10.043	10.809	20.34	211.4	-19.2	-76.3	5.1	-16.7	-79.1	5.1
8719	333.9568	0.9588	2012.863	17.055	17.524	7.321	304.6	-14.3	-146.6	7	-20.3	-139.8	7.5
8720	333.9694	50.0549	2013.418	13.125	13.954	11.829	189	106.1	37.9	5.9	105.5	36.1	5.9
8721	334.0315	65.5255	2013.3935	13.344	16.925	6.328	227	92.9	34.1	5.9	93.9	31.7	6
8722	334.0659	34.7652	2013.347	13.703	14.097	8.926	308.2	94.4	9.7	6.4	94.2	8.7	6.4
8723	334.0729	33.7322	2013.4505	7.303	10.813	12.963	162.9	95.5	1.2	6.4	95.8	-3.1	6.4
8724	334.1849	38.5690	2013.303	14.75	17.025	9.462	356.2	67.7	30.9	5.5	63.8	33.4	5.6
8725	334.1907	0.2541	2012.864	14.405	14.73	6.716	15.7	72	9	6.6	69.6	6	6.6
8726	334.2195	49.0479	2013.589	8.396	10.682	48.749	180.2	79.7	65.1	5.8	80.4	59.9	5.9
8727	334.2525	18.3920	2013.2585	9.842	10.933	30.43	324.4	-56.3	-69.1	6.1	-59.1	-64.9	6.1
8728	334.2545	42.9090	2013.4685	15.073	15.139	58.918	59	96.7	-21.3	5.5	95.1	-22.2	5.5
8729	334.3295	15.5376	2013.4105	17.598	17.613	6.238	138.8	97.2	15.6	5.8	99.5	15.6	5.4
8730	334.3356	6.2414	2013.4705	12.435	16.281	7.553	216.9	-56.9	-72.2	6.3	-59.8	-68.2	6.8
8731	334.3565	13.6616	2013.53	13.254	16.751	7.226	329.8	20.6	-141.7	5.2	16.8	-148.1	5.3
8732	334.4441	44.3792	2013.284	17.705	19.295	39.395	347.5	67.3	-10.8	5.6	69.6	-5.8	6.3
8733	334.4659	19.7629	2013.2235	13.266	14.196	57.237	0	63.7	48.3	6	62.5	50.3	6
8734	334.4671	34.7987	2013.3115	11.427	14.735	10.309	137.4	75.2	-6.7	5.4	75.9	-7.6	5.4
8735	334.4767	59.0605	2013.4255	12.169	13.622	4.836	280.2	49.1	62.1	6.2	48.2	62.7	6.2
8736	334.5108	8.4552	2013.4885	16.502	16.937	5.506	247.2	72.6	-56.1	6.5	83.4	-44.7	6.4
8737	334.5535	-2.6730	2012.8685	14.958	16.246	6.064	222	23.6	-78.8	5.7	22.1	-82	5.8
8738	334.5582	15.8158	2013.634	15.481	15.941	4.45	183.7	-57	-92.9	5.1	-56.3	-92.9	5.2
8739	334.5928	53.8804	2013.293	15.723	16.661	4.989	276.4	70.7	7.2	5.9	71	8.5	6

8740	334.5997	28.9763	2013.307	14.945	15.313	6.225	140.5	-14.8	-83.8	5.2	-16.2	-84.8	5.2
8741	334.6978	25.1292	2013.1475	15.018	16.001	7.485	140.7	54.1	-76.8	5.3	57.6	-77.8	5.3
8742	334.6992	32.3382	2013.226	14.838	17.936	15.078	123.7	65.5	-31.1	6	67	-26.3	6.3
8743	334.7129	-6.6965	2012.868	14.709	16.24	47.415	297.2	54.4	-95.9	5.8	52	-91.8	5.9
8744	334.7167	25.0966	2013.086	16.832	17.525	7.325	328.5	6.6	-60.8	5.5	1.5	-62.6	5.5
8745	334.7826	12.9863	2013.2035	15.107	18.654	9.967	134.9	86.1	-13.5	5.1	84.3	-11.3	6.1
8746	334.8010	10.1891	2013.4565	10.442	12.25	57.496	309.9	108.8	-61.4	6.4	108.6	-64.9	6.5
8747	334.8380	66.6648	2013.3	15.666	16.161	20.668	13.8	156.2	83.9	6	152.4	83.1	6
8748	334.8560	63.5715	2013.391	11.122	13.432	22.945	18.3	92	71.4	5.9	93.5	69.9	5.9
8749	334.8902	35.6377	2013.293	13.016	16.307	16.449	262.4	67.6	38.4	5.4	68.4	37	5.5
8750	334.9313	56.6112	2013.2915	15.958	17.008	17.326	98.5	61.3	51.9	6.4	63	48.7	6.5
8751	335.0568	28.7770	2013.2785	12.744	17.621	43.764	85.2	36.4	-93.7	5.2	34.3	-92.8	5.3
8752	335.0913	47.4087	2013.4675	13.607	14.653	11.98	302.2	102	22.4	5.5	101.2	23.2	5.5
8753	335.1020	2.8515	2012.868	10.902	13.369	26.107	146.3	100.9	-38.2	6.6	97.6	-32.5	6.6
8754	335.1053	5.3322	2013.253	15.492	16.944	31.551	271.3	-117.6	-119	6.8	-118.6	-119.5	7.2
8755	335.1071	73.4165	2013.268	18.326	18.441	37.883	84.8	-19.2	-60.2	6.3	-7.7	-60.9	6.6
8756	335.3783	-2.0382	2012.8735	14.585	16.744	10.679	165	-32.8	-77.8	5.7	-35.7	-77.6	5.9
8757	335.3970	55.6553	2013.3765	8.973	17.536	21.844	158.3	-94.8	-68.7	5.9	-94.6	-70.8	6.4
8758	335.4643	20.0311	2013.3415	10.99	18.118	16.765	113.7	-33.7	-72.7	6.3	-35.6	-75.6	6.8
8759	335.4750	2.4907	2012.865	14.295	15.1	15.728	351	113.8	13.7	6.6	111.5	6.8	6.6
8760	335.5309	-7.7911	2012.868	11.987	15.122	51.752	165.5	24.1	-73	5.8	24.8	-72.4	5.9
8761	335.6041	-5.8394	2012.868	12.385	13.172	10.316	189	-26.8	-106.9	5.7	-29.9	-110.1	5.7
8762	335.6066	21.8003	2013.2695	15.618	15.955	13.005	132.1	80.7	14.7	6.3	81.3	15.2	6.3
8763	335.6203	29.3699	2013.3275	10.231	15.749	36.117	13.5	65.6	-61	5.2	63.5	-61.3	5.3
8764	335.6427	7.4562	2013.4015	11.044	11.997	9.958	155.6	92.2	32.6	6.5	91.2	34	6.4
8765	335.7009	0.0555	2012.87	12.528	17.14	17.254	193.7	87.9	49.3	5.7	88.8	49.4	5.8
8766	335.7569	22.0252	2013.099	16.174	16.374	6.557	167.6	54.4	-77.6	6.2	60.6	-66.8	6.2
8767	335.8183	42.0308	2013.348	13.906	15.308	33.301	175.9	1.7	-90.5	5.5	1.4	-87.4	5.5
8768	335.8330	54.0037	2013.2395	17.783	18.094	42.937	321.3	140.2	85.9	6.8	146.5	73	7.8
8769	335.8628	26.3511	2013.34	10.16	13.202	10.444	110.1	73.6	-6.6	5.2	71.4	-7	5.2
8770	335.8806	31.9800	2013.2395	12.045	14.041	34.056	348.5	145.3	70.8	5.5	144.8	71.3	5.5
8771	335.9073	23.0302	2013.209	17.335	18.096	54.013	317.9	62.3	-8.4	6.4	62.9	-9.4	6.7
8772	335.9323	40.9025	2013.583	8.805	9.282	6.69	53.2	67.1	37.7	5.4	64.6	39.1	5.7
8773	335.9395	28.3895	2013.239	14.071	17.457	21.774	188.8	91.3	20.3	5.3	94.1	20.8	5.3
8774	335.9518	27.7776	2013.133	11.975	14.725	57.652	335.4	134.3	100.1	5.4	134.6	99.2	5.4
8775	336.0131	43.3995	2013.531	10.461	12.697	11.721	317.6	-15.3	-62.2	5.4	-15.9	-61.9	5.5
8776	336.0360	37.8375	2013.3595	14.006	17.144	58.337	139.5	-66.9	-69.1	6.2	-59.3	-73.5	6.3
8777	336.0520	39.4489	2013.3435	17.333	17.697	43.75	86.8	38.5	-63	6.2	40.9	-64.6	6.3
8778	336.0877	-4.7800	2012.871	17.054	17.261	6.195	321	57.6	-84.5	6	55.1	-87.6	6.2
8779	336.1880	-3.7452	2012.872	15.069	16.04	6.857	58.1	-66.6	-100.7	5.7	-66.2	-100.1	5.8
8780	336.2237	56.1173	2013.263	15.549	18.153	20.627	111.2	-5.5	-63.7	6.3	-13	-60.9	6.6
8781	336.2624	32.6166	2013.276	17	17.307	39.912	127.5	78.1	20.3	5.6	75.4	22.1	5.6
8782	336.2828	44.5634	2013.4975	11.707	14.361	16.061	203.5	102.9	10.6	5.5	105.3	7.1	5.5
8783	336.2861	55.6722	2013.3405	14.509	18.046	27.698	247.7	33	-65.5	6.4	30.1	-70.3	6.8
8784	336.3355	51.3184	2013.2855	14.223	18.969	9.859	325.8	105.7	111.4	6.5	104.5	107.5	7.3
8785	336.3466	20.8573	2013.174	11.824	13.893	14.179	122.7	-19	-85.1	6.4	-18.7	-84.4	6.3

8786	336.4082	48.2320	2013.535	9.043	14.816	45.5	17.9	105.1	97.3	6.3	110.8	101.3	6.4
8787	336.4763	27.3230	2013.25	11.434	15.795	36.43	129.1	90.4	49	5.3	90.2	50	5.3
8788	336.5242	34.6984	2013.2805	15.329	16.958	4.933	84.2	79.9	7.7	5.6	77.9	14.3	5.9
8789	336.5632	35.5844	2013.334	12.086	14.38	50.778	289	75.7	18.4	5.5	75.2	17.4	5.5
8790	336.5723	-3.0486	2012.873	13.665	16.371	8.287	327	-117.9	-104.1	5.7	-122	-103.9	5.8
8791	336.6561	54.0148	2013.4295	12.919	14.861	8.43	345.6	-75.3	-66.7	5.9	-74.1	-68.1	5.9
8792	336.7081	34.6870	2013.4355	15.088	18.799	19.265	58.5	70.3	-15.3	5.6	72.4	-18.9	6.1
8793	336.7871	35.8369	2013.319	15.199	16.642	4.91	309.1	70.4	6.1	5.5	72.5	7.5	5.7
8794	336.8193	15.3110	2013.2245	9.779	16.903	20.156	116.1	-42.8	-101.7	5.6	-45.5	-99.5	6
8795	336.8214	18.0835	2013.186	12.614	15.159	33.643	243.2	68.9	-28	5.2	65.7	-27.1	5.3
8796	336.9622	7.4060	2013.4325	12.76	15.738	8.22	84.6	7.6	-106.5	6.3	5.5	-106	6.5
8797	336.9927	15.2647	2013.1325	14.731	15.41	5.767	39.1	-14.3	-81.3	5.6	-17.1	-80.6	5.6
8798	337.1335	43.3283	2013.2585	14.431	17.692	5.915	317.4	75.6	25.1	5.5	73	25.7	5.9
8799	337.1341	38.4139	2013.304	11.357	13.729	7.831	111.1	72.1	48.5	6.1	73.5	49.5	6.1
8800	337.1794	44.1387	2013.3145	17.323	17.712	15.88	113.5	69.6	34.3	5.6	67.8	36.3	5.6
8801	337.1870	53.3202	2013.2875	17.402	18.336	53.209	44.2	-11.7	-60.1	5.7	-2.6	-61.7	5.6
8802	337.2509	9.6505	2013.527	14.631	16.701	8.136	166.6	124.6	-20.7	6.3	123.1	-20.4	6.5
8803	337.3195	-7.4425	2013.599	13.821	15.321	6.116	324.2	54.9	-65.3	5.8	49.6	-68.1	6.9
8804	337.3226	36.2173	2013.1895	10.644	17.019	50.025	184.4	112.2	24.5	5.6	110.3	24.5	5.8
8805	337.3991	17.7611	2013.212	13.666	16.552	7.288	356.1	87.2	-0.7	5.6	87.7	5	5.8
8806	337.4019	-7.2732	2013.288	14.702	16.802	34.398	70.3	20.6	-96.6	5.6	20.4	-95.6	5.7
8807	337.4216	6.3340	2013.4875	13.011	18.083	25.891	76	117.6	-39.9	6.5	115.2	-39.6	6.8
8808	337.4296	22.4428	2013.0855	14.2	16.105	8.746	352.4	111	10.3	5.3	108.2	13.7	5.4
8809	337.4462	-4.5347	2012.883	12.054	14.099	8.551	135.8	158.4	-83.2	5.7	156.7	-82.8	5.8
8810	337.4604	-9.4138	2013.599	11.346	12.654	7.608	146.6	-39.1	-104.7	5.5	-40.9	-103.4	5.5
8811	337.5178	32.5895	2013.2885	11.382	14.962	27.385	291.9	79.6	11.7	5.5	78.8	14.5	5.6
8812	337.5411	1.3022	2012.875	16.546	16.994	10.291	95.3	-8.6	-71.9	6.7	-8.8	-68.5	6.8
8813	337.6338	39.8813	2013.3775	14.273	16.524	11.322	281.4	53	-115.7	5.5	51.6	-118.1	5.5
8814	337.6546	-7.7692	2013.3365	11.721	12.449	12.949	209.3	83.5	-75.7	5.7	81.4	-75.9	5.7
8815	337.6929	48.2895	2013.394	17.751	18.44	44.422	253.1	6	-74.3	6.5	1.9	-72.2	5.9
8816	337.7020	48.2499	2013.3085	11.042	17.605	10.475	306.2	65.4	22.7	6.4	65.7	25	7.4
8817	337.7114	21.0545	2013.0315	17.184	17.201	29.173	8.1	-35.1	-71.7	5.3	-40.5	-72.4	5.4
8818	337.7837	68.3398	2013.352	12.176	18.3	43.521	163.5	96.3	45.5	5.9	91.2	52.7	6.1
8819	337.8118	36.7414	2013.3375	12.844	18.81	36.428	206.1	169.7	-148.7	5.7	178.8	-141.3	6
8820	337.8314	40.2602	2013.338	11.766	15.534	10.075	214.5	63.9	17.4	5.5	61.4	16	5.5
8821	337.8364	35.7610	2013.269	12.71	16.961	36.367	202.8	85.5	-39	6	85.3	-44.9	6.1
8822	337.8547	38.1753	2013.2625	14.918	15.413	6.237	22.4	-166.1	-123.6	5.7	-166.5	-123.7	5.7
8823	337.9364	13.3231	2013.3465	16.039	17.398	10.504	297.9	-15.3	-76.6	5.6	-8.7	-78.2	5.7
8824	338.0007	34.9182	2013.2785	11.439	11.564	46.364	253.3	132.2	81.4	6.5	131.5	78.5	6.5
8825	338.0446	32.3465	2013.275	12.093	12.517	5.392	106.7	-8.4	-114.7	6	-6.7	-115.2	6.1
8826	338.0514	69.3348	2013.5255	14.041	14.532	4.96	141.5	64.9	12.3	5.9	63.9	17.1	5.8
8827	338.0650	64.4116	2013.5085	10.128	13.953	6.332	329.3	149.6	75.9	5.9	152.5	80.2	5.9
8828	338.0715	41.6422	2013.384	13.346	13.864	5.713	104.9	97.1	-3.8	5.5	98.9	-7.3	5.5
8829	338.0901	56.7193	2013.371	13.665	17.429	8.866	157.8	83.1	-10.7	5.6	83.9	-15.4	5.7
8830	338.1183	19.3157	2013.493	14.014	14.532	4.163	327.3	81.7	-68.7	5.3	82.1	-70.4	5.5
8831	338.1476	47.9904	2013.2695	16.326	17.469	58.818	231.1	24.3	-61.5	6.8	20.1	-63.8	6.2

8832	338.1839	9.6749	2013.287	16.773	17.481	9.357	92	71.3	-2.1	6.5	70.2	-4.5	6.8
8833	338.2228	26.3400	2013.299	11.413	12.228	21.092	160.2	64	42.7	5.6	65	45.2	5.6
8834	338.2513	41.4020	2013.451	14.62	18.287	8.74	234.6	64.9	34.8	5.5	68.6	34.4	5.7
8835	338.2774	56.8690	2013.4645	10.526	11.843	7.369	296.4	17.4	65.2	5.6	19.2	63.8	5.5
8836	338.3174	-7.9104	2013.599	16.735	17.255	12.95	163.5	-67.7	79.7	5.6	-70.2	84.5	5.8
8837	338.3375	15.3323	2013.2325	12.563	13.09	42.277	262.9	-177.4	-114.9	5.4	-177	-111.4	5.4
8838	338.3488	12.7717	2013.563	11.178	12.12	48.145	307.4	79.9	21.9	5.1	80.3	24.5	5.1
8839	338.3629	12.8094	2013.543	11.994	13.391	48.522	23.3	61.7	-21.4	5.1	65.7	-18.6	5.1
8840	338.4254	53.5704	2013.299	17.981	18.193	36.325	230.2	64	1.3	6	61.8	-2.6	6
8841	338.4801	28.9066	2013.3375	9.764	15.597	28.953	240	69.6	-0.6	5.5	71.9	5.9	5.6
8842	338.5084	8.0253	2013.3675	14.413	15.392	12.422	-0.527	8.7	-62.8	6.5	11.6	-65.2	6.5
8843	338.5556	-4.6812	2012.887	13.057	14.105	8.671	333.2	75.6	-39.6	6	76.7	-38.7	6
8844	338.5629	-5.4123	2013.2845	13.37	15.144	22.527	42.1	69.1	19	5.9	68.9	21.5	6
8845	338.5689	31.4795	2013.193	12.016	14.415	40.742	55.9	52	-119	6.1	48.6	-118.3	6.1
8846	338.6551	-6.4426	2013.3145	12.143	17.694	10.966	1.7	23.3	-80.4	5.7	20.8	-80.3	6.3
8847	338.7560	74.8845	2013.422	9.724	12.817	9.608	261.8	54.8	71	5.9	52.7	68.2	5.9
8848	338.7561	41.0603	2013.4175	11.632	15.793	23.451	195.3	166.4	-26.3	5.5	163.3	-26.9	5.6
8849	338.7637	18.3440	2013.225	10.327	10.935	16.338	47.9	-53.3	-83.1	5.3	-50.5	-90.1	5.3
8850	338.7972	-5.2257	2012.892	13.718	14.529	5.213	333.3	69.4	-4.5	6	71.2	-2.3	7.1
8851	338.8053	-0.4098	2013.2615	11.11	16.674	22.714	31.6	103.6	33.6	6.1	100.9	33.2	6
8852	338.8215	11.9535	2013.5765	15.962	16.713	5.485	278.8	-46.3	-60.8	5.2	-47.3	-61.2	5.2
8853	338.8616	9.5748	2013.489	10.9	12.248	11.367	56.9	116.3	42.8	6.4	113.9	44.3	6.3
8854	338.8757	-6.8106	2013.4345	16.011	16.48	5.727	307.2	-115.4	-110.4	5.6	-116.8	-108.8	5.6
8855	338.8902	-2.8253	2012.883	10.069	13.331	21.057	232	86.9	31.4	6	92.7	24.9	6
8856	338.9548	54.5532	2013.372	13.084	16.972	32.546	158.3	67.6	8.8	5.6	67	6.7	5.6
8857	338.9882	24.4308	2013.2135	15.95	15.993	13.761	317.6	75.2	16.3	5.6	72.6	15.2	5.6
8858	338.9887	26.6330	2013.2875	14.956	15.893	40.46	24.6	33.1	-80.6	5.6	33.9	-80.9	5.6
8859	339.0430	4.0295	2013.369	15.184	15.19	57.152	300.2	-51.9	-67.8	6.6	-48.4	-67.1	6.6
8860	339.0672	21.8533	2013.205	12.574	15.654	46.757	293.6	112.8	-86.6	5.3	109.7	-81.6	5.3
8861	339.0895	52.9492	2013.4075	12.805	16.104	21.729	70	65.9	32.2	6.4	66.8	31.7	6.4
8862	339.1386	-2.4009	2012.8855	11.251	12.346	12.747	191.8	47.6	-64.6	6	46	-64.5	6
8863	339.2433	39.0952	2013.3725	14.449	14.807	4.474	302.7	120.6	11.7	5.5	118.4	11.8	5.5
8864	339.2502	44.0424	2013.435	12.077	15.671	11.151	331	76.4	18.9	5.5	77.8	21.7	5.5
8865	339.3320	22.4165	2013.2465	11.905	13.311	15.335	168.5	101.2	0.9	5.2	98.9	0.1	5.2
8866	339.4018	0.2520	2013.686	10.598	11.021	4.889	78.2	-40.7	-204.3	7.1	-44.9	-207.1	6.8
8867	339.5242	42.2259	2013.32	17.508	18.139	57.387	124.8	-40.7	-64.1	5.6	-31.5	-72.4	5.8
8868	339.5469	48.0118	2013.4775	9.967	14.073	9.062	47.7	-18.3	-98.1	5.6	-17	-99.4	5.5
8869	339.5501	38.7073	2013.306	14.284	17.913	11.154	23.8	70.1	11.9	5.5	68.4	11.2	5.7
8870	339.5563	61.2143	2013.3725	9.952	14.235	6.588	149.8	69	68.8	6	61.4	70.9	6
8871	339.5710	51.2527	2013.4465	12.254	14.463	40.106	265.8	29.6	67.8	5.5	29.8	66.3	5.6
8872	339.5998	26.5449	2013.139	16.836	17.845	16.685	255.5	66.6	-0.3	5.3	65	3.8	5.4
8873	339.6883	47.6758	2013.4495	17.858	18.564	25.044	307.5	63.4	25.5	6	61.2	31.1	6.2
8874	339.7246	30.4150	2013.2405	13.663	16.589	9.088	247.1	77.2	-27.3	5.4	78	-26.9	5.5
8875	339.7655	3.8964	2013.6455	14.674	14.759	6.042	247.4	97.6	-19.3	6.1	96.3	-19.7	6.4
8876	339.7729	60.1020	2013.2625	18.49	18.837	44.185	357.9	75.4	-56.5	6.4	84.8	-47.9	6.5
8877	339.7742	22.8097	2013.219	11.556	16.245	10.187	143.5	-58.6	-130	5.3	-53.2	-125.9	5.4

8878	339.8277	19.8105	2013.1175	13.432	13.932	5.563	215.2	-28.6	-83.3	5.2	-28.9	-83.7	5.3
8879	339.8292	0.8568	2013.3355	12.174	14.518	8.984	44	-107.3	-69.9	6.6	-108.9	-66.3	6.7
8880	339.8397	-3.5156	2012.8905	10.012	13.86	34.106	141.4	116.3	-20.7	5.8	115.3	-25	5.7
8881	339.8555	3.6573	2013.338	14.436	15.789	35.324	281.5	80.7	-20.1	6.5	77.6	-17.7	6.5
8882	339.9323	0.2968	2013.2125	13.755	15.452	8.272	264.2	137	58.1	6.7	136.5	58.8	6.6
8883	339.9368	71.8862	2013.344	11.445	15.776	14.624	13.8	89.1	26.8	5.9	86.9	29	6
8884	339.9664	-3.0493	2012.888	14.19	14.74	5.594	280.3	-69.4	-81.7	5.7	-67.7	-80.2	5.8
8885	340.0029	64.3868	2013.3955	11.643	15.095	50.647	112.4	148.1	-43.9	6	147.2	-44.9	6
8886	340.0036	28.6109	2013.229	11.936	16.883	23.026	23.1	106.7	-15.9	5.3	104.7	-18.2	5.4
8887	340.0038	44.2096	2013.377	14.805	17.615	46.187	6.5	80.6	8.9	6	80	6.4	6.1
8888	340.0038	53.2254	2013.391	12.401	16.485	28.198	189.3	82.8	36.2	6.4	86	37.2	6.5
8889	340.0488	0.1536	2013.336	11.666	13.427	39.027	254	131.3	4.7	6.5	129.5	5.9	6.5
8890	340.0832	31.7875	2013.606	8.086	11.334	6.478	96.3	86.4	-24.8	5.4	87.7	-28	5.3
8891	340.0839	13.7321	2013.49	10.387	11.895	13.855	323.7	116	-4	5.9	114	-6.5	5.9
8892	340.0898	76.0770	2013.3795	11.758	15.512	20.564	269.1	63.7	7.5	5.9	63.2	6.6	5.9
8893	340.0931	10.7246	2013.524	16.092	16.903	22.783	15.5	47.6	-134	6.5	47.4	-133.6	6.6
8894	340.1423	4.8130	2013.4215	14.342	15.105	12.468	187.2	98.4	-53.6	6.6	97.8	-57.1	6.6
8895	340.2368	19.1064	2013.13	11.906	17.95	16.269	334	67.1	20	5.3	69.7	22.9	5.5
8896	340.2459	44.3709	2013.463	9.916	16.979	21.993	208	63.7	-30.5	5.9	66.6	-27.3	6
8897	340.2956	41.5163	2013.3155	15.979	16.562	9.865	147.5	-1.9	-76.4	5.5	-0.3	-77.2	5.6
8898	340.3069	-9.8846	2013.599	14.124	17.889	19.907	226.9	-4.3	-139	5.5	1.6	-138.1	7.7
8899	340.3373	58.2950	2013.216	17.348	17.807	51.638	34.4	68.7	-17.3	6.2	68.4	-14.4	6.3
8900	340.4038	45.5516	2013.409	15.839	16.032	9.609	174.4	71.9	78.7	5.9	71.6	76.7	6
8901	340.4119	-2.5528	2012.886	8.74	15.88	35.204	296	94.9	-61.4	5.8	91.9	-66.8	5.8
8902	340.4520	58.2758	2013.4055	16.82	18.56	39.423	147.3	81.6	67.7	7.4	83.3	67.1	6.5
8903	340.5024	33.8625	2013.3365	12.399	17.868	26.049	121.4	124.5	28.1	5.5	129.3	21.1	5.9
8904	340.5595	43.9257	2013.5245	11.934	15.899	7.677	100.3	8.3	-130.1	5.5	4.1	-131.6	5.5
8905	340.5723	62.2549	2013.4415	10.611	10.669	13.242	262.4	68.2	18	5.9	66.1	17.2	5.9
8906	340.6296	2.8698	2013.2095	10.459	16.273	21.992	342.9	-7.4	-62.3	6.4	-9	-64.3	6.5
8907	340.6647	36.2998	2013.2805	16.331	16.517	4.865	51.8	81.5	52.9	5.6	76.9	52.9	5.8
8908	340.7044	34.4829	2013.358	10.177	17.629	26.875	153.5	38.1	-113.8	5.4	32.2	-111.3	5.6
8909	340.7447	61.5774	2013.4175	14.22	15.805	38.451	320.6	88.1	24.7	5.9	87.6	22.9	5.9
8910	340.7468	-2.7213	2012.8895	12.076	16.095	8.227	296.5	-123.4	-114	5.7	-127	-108	6
8911	340.7597	3.7933	2013.483	10.689	13.933	9.07	123.2	110	-43.6	6.5	108.2	-42.5	6.5
8912	340.7931	21.0196	2013.211	10.656	16.588	10.339	105.3	85.5	11.8	5.2	82.9	10.2	5.3
8913	340.8857	46.9729	2013.464	12.688	15.036	6.06	78.8	-48.4	-99.9	5.5	-56.2	-91.2	5.6
8914	340.8999	12.8656	2013.6335	10.545	10.937	29.837	295.3	61.4	-17.4	5.4	61.4	-18.1	5.4
8915	340.9702	-1.8373	2012.8905	11.964	13.208	5.796	222.1	-60.7	-92.3	5.7	-58.6	-90.7	5.8
8916	340.9860	42.1682	2013.3455	17.451	17.755	5.268	67.2	129.1	18.9	5.7	123.9	12.2	5.9
8917	340.9881	15.1453	2013.1645	13.934	14.887	4.573	157.9	65.4	22	5.5	65.5	25.2	5.9
8918	341.0976	14.3316	2013.4395	11.157	12.619	38.055	109.1	136.8	38.8	5.6	137.4	39.4	5.5
8919	341.1919	63.2726	2013.369	13.589	16.201	9.224	200.8	65.8	48.4	6	69.2	49.7	6
8920	341.3321	4.9063	2013.3645	13.096	15.711	18.488	116.7	86.2	-16.5	6.5	84.1	-14.4	6.5
8921	341.3560	23.0423	2013.3245	12.466	18.025	10.004	183.5	39.6	-66.9	5.2	35.6	-65.6	5.5
8922	341.3712	-5.9177	2013.4275	14.153	17.744	8.819	23.8	87	-13.1	5.8	88.3	-14.7	6.1
8923	341.4249	41.2034	2013.4815	8.779	11.862	21.064	322.9	159.6	116.4	5.5	159.4	115.7	5.6

8924	341.4286	26.4965	2013.0925	9.985	14.179	6.747	41	109.8	16.8	6.4	113.4	12.5	6.9
8925	341.4629	-1.0795	2012.883	12.082	13.296	5.388	102.1	-63.8	-83.7	5.7	-70.2	-80.5	6.9
8926	341.5837	-8.4212	2013.599	10.462	14.207	10.282	2.4	87	-2.1	5.8	90.3	-9.3	5.8
8927	341.6353	56.7100	2013.0655	18.185	18.454	50.035	200.7	-10.9	-81.7	7.5	-10.4	-81.1	6.5
8928	341.7063	27.2704	2013.16	15.008	15.149	4.08	189.9	62.1	4.8	5.8	63.5	2.6	5.9
8929	341.7070	19.0163	2013.254	8.767	12.954	25.504	89.9	-67.8	-190.6	5.4	-64.8	-186.2	5.4
8930	341.7634	51.5531	2013.4465	15.385	15.773	5.611	261.3	-58.8	-83.5	5.6	-60.2	-83.4	5.5
8931	341.8157	51.1207	2013.2565	16.969	17.755	6.405	233.6	69.1	-45.4	5.6	68.9	-39.3	5.7
8932	341.8548	1.1551	2013.164	9.197	18.16	25.894	26.5	33.6	-97	6.6	44.7	-94.1	7.2
8933	341.9095	57.2756	2013.3715	13.742	15.311	45.916	160.6	149.2	10.9	6.2	146	9	6.3
8934	341.9843	3.5962	2013.5265	10.934	11.303	23.208	334.3	74.7	3.7	6.3	75.3	4.9	6.3
8935	342.0630	27.3840	2013.32	13.531	14.677	6.483	114.7	167.3	-3.1	5.3	162.9	2.3	5.3
8936	342.1063	55.7993	2013.4845	8.768	13.704	45.938	171.4	28	-63.9	6.2	25.1	-63	6.2
8937	342.1106	29.5554	2013.1635	12.387	13.358	13.69	238.6	-68	-80.2	5.3	-67.9	-78	5.3
8938	342.1183	33.9351	2013.29	13.95	17.108	13.818	273.2	91.7	57.7	5.6	90.2	61.3	5.6
8939	342.1436	-3.4172	2012.894	14.24	17.608	10.79	336.1	66	-8.8	5.7	63.2	-7.6	6.3
8940	342.1753	45.4269	2013.444	11.585	17.74	25.647	112.2	175.7	-29.5	5.5	173.5	-24.7	5.7
8941	342.2234	13.1197	2013.48	9.807	10.85	6.161	155.1	-11.6	-82.1	5.6	-19.1	-80.1	5.5
8942	342.2259	33.4210	2013.3065	14.151	14.175	7.15	33.2	-71.5	-81.5	5.6	-73	-82.8	5.6
8943	342.2388	-8.6934	2013.599	16.6	16.811	34.116	228.2	157	-121.9	5.9	155.9	-119.6	5.9
8944	342.2490	42.3029	2013.426	14.626	16.229	5.345	208.4	74.1	42.3	5.5	74.2	45.5	5.5
8945	342.3315	49.3428	2013.4535	14.726	16.453	20.959	56	-47.4	-63.3	5.5	-45.9	-61.7	5.5
8946	342.3392	-9.5260	2013.599	14.084	16.198	44.976	63.4	69.7	-20.6	5.5	68.2	-22.2	5.6
8947	342.3577	5.2833	2013.3355	13.881	16.049	26.468	303.1	96	38.1	6.6	94	36.4	6.6
8948	342.3938	43.6973	2013.4115	13.046	16.198	9.439	336.2	97.4	24	5.5	94.2	24.5	5.5
8949	342.3946	7.8840	2013.4055	12.359	16.105	7.615	38.9	91.8	-17.3	6.4	93.2	-23.6	6.5
8950	342.5757	-2.4861	2012.8915	11.988	14.598	30.307	253.4	70.9	-130.5	5.7	71.3	-128.6	5.7
8951	342.6082	21.7026	2013.175	12.418	13.539	8.285	37.5	60.5	33	5.2	62.1	37.1	5.2
8952	342.6681	30.5488	2013.168	15.023	16.415	10.965	279.8	67.1	-7.7	5.5	66.6	-12.7	5.5
8953	342.7315	6.0864	2013.386	13.885	14.267	29.292	158.2	106.6	-70.3	6.7	105.1	-72.4	6.6
8954	342.7418	68.0448	2013.379	13.622	15.732	35.008	283.5	170.7	96.2	5.9	171	97.3	6
8955	342.7619	49.0927	2013.4245	18.154	18.518	54.503	181.2	-23.8	-64.3	5.7	-19.7	-67.9	6.1
8956	342.7710	14.3860	2013.171	15.621	17.491	10.546	76.1	47.3	-146.4	5.8	46.5	-144.1	5.9
8957	342.7863	55.2661	2013.3235	18.03	18.226	8.862	98.9	102.9	48.4	6.4	101.3	58.3	6.5
8958	342.8550	-5.0857	2013.316	12.124	16.383	7.217	235.9	67.2	41	5.7	69	43.3	5.6
8959	342.8733	-1.9357	2012.891	14.963	16.504	18.728	86.9	79.9	2.7	5.8	82.6	1.1	5.8
8960	342.8885	13.5762	2013.2685	14.934	16.217	55.446	318.6	-50.2	-60.1	5.6	-43.3	-61.3	5.7
8961	342.9462	16.9522	2013.2985	15.044	18.139	8.477	151.5	101	-46.1	6	104.8	-46.4	6.6
8962	342.9658	50.7231	2013.414	17.147	17.408	4.337	4.4	-57.7	-63.3	5.6	-51.2	-68.5	5.6
8963	343.0487	83.9126	2013.6945	10.809	13.417	5.995	59.2	70.7	57.4	6	65.8	61.1	6.2
8964	343.1839	34.9883	2013.457	14.537	14.615	5.589	127.3	144.3	119.3	5.6	143.2	117.7	5.5
8965	343.2412	57.9573	2013.234	14.652	16.97	11.442	219.5	-64.3	111.1	7.6	-56.8	117.7	7
8966	343.2652	27.4462	2013.164	15.524	16.108	12.886	52.5	180.4	-13.8	5.4	182.1	-12.7	5.4
8967	343.3512	30.7691	2013.374	10.564	13.636	33.701	265.5	-154.9	-176	5.6	-156.3	-176.2	5.7
8968	343.3689	30.3866	2013.3115	14.972	16.01	42.733	260.8	68.5	16.2	5.5	65.2	20.8	5.5
8969	343.3886	67.2300	2013.538	13.915	14.902	3.685	54.3	128.2	16.4	5.9	125	18.6	5.9

8970	343.4798	22.4070	2013.6255	8.922	15.906	11.552	217.7	-7.8	-89.7	5.2	-11.2	-90.6	5.4
8971	343.4969	33.8494	2013.296	12.465	16.525	23.359	49.1	-51.1	-89.9	5.5	-52.7	-90.7	5.5
8972	343.5045	38.0282	2013.4535	14.779	18.081	24.955	331.4	-5.7	-70.8	5.5	-8.8	-73.3	5.6
8973	343.5221	65.6230	2013.1745	10.068	14.063	5.768	3.1	112	96.5	5.2	108.8	94	5.5
8974	343.5277	57.6097	2013.398	8.847	14.688	14.249	221.5	30.2	-92.5	5.9	27.1	-91	5.9
8975	343.5373	23.2838	2013.0355	14.029	18.465	38.739	56.2	127.1	13.6	5.3	131	8.6	5.7
8976	343.5693	17.7113	2013.2675	10.627	14.67	9.548	239.7	-66.3	-65.9	5.6	-65.8	-65.1	5.6
8977	343.5699	60.8200	2013.049	18.004	19.554	25.298	62	4.7	-64.8	5.4	8.2	-62.3	6.2
8978	343.5798	-7.9369	2013.599	15.787	16.194	5.279	97.4	85.7	6.3	5.7	84.5	7	5.7
8979	343.5820	30.3717	2013.4125	10.087	10.638	14.485	233.7	140.7	-12.3	5.4	139.3	-12.5	5.5
8980	343.6193	-5.2327	2013.0235	10.874	12.939	9.933	247.5	79.1	-14.3	6.1	79.2	-11.6	6
8981	343.6404	48.5685	2013.4205	16.928	17.648	30.839	253.1	70.8	100.9	6	79.4	98.4	6.1
8982	343.6499	22.5448	2013.2545	11.613	16.332	42.545	30.1	64	14.2	5.2	62.5	11.6	5.2
8983	343.6502	15.2707	2013.244	10.318	11.361	22.549	316.5	109.3	9.4	5.6	108.8	9	5.6
8984	343.6780	64.9820	2013.415	14.388	15.23	32.744	62	-51.8	-64.2	5.2	-55.8	-65.8	5.2
8985	343.8465	10.4570	2013.297	13.04	16.257	14.05	45.4	70.2	1.8	6.4	70	2.4	6.4
8986	343.9081	25.7439	2013.1955	13.026	14.323	12.885	198.4	122.2	-14	5.3	122.9	-18.3	5.3
8987	343.9312	48.9566	2013.4525	11.334	12.603	15.455	298.3	-124.9	-78.7	5.9	-127.1	-77.6	5.9
8988	343.9412	12.2744	2013.255	15.3	16.873	19.836	194.9	66.2	-14.1	5.6	63.7	-12.3	5.7
8989	343.9481	45.8963	2013.3765	17.309	17.611	52.379	325.8	-36.1	-99.3	5.7	-33.2	-96.7	5.7
8990	343.9550	8.4164	2013.3125	13.137	13.419	19.181	14	83.6	-119.3	6.6	80.1	-118.8	6.6
8991	344.0162	60.4571	2013.415	10.072	11.909	26.579	251.8	84.7	75.3	5.2	79.9	76.5	5.2
8992	344.0767	-7.1712	2013.419	16.698	16.881	37.606	50.3	70.9	6.3	5.9	68.5	5.7	5.8
8993	344.1717	17.8233	2013.2155	10.752	17.668	14.925	17.3	110.3	-7	5.2	110.9	-10.3	5.6
8994	344.1804	46.2327	2013.4065	12.705	14.785	17.549	251.6	66.2	31.3	5.5	66.1	30.7	5.5
8995	344.1909	53.1120	2013.397	14.41	15.466	10.064	194.3	65.4	-9	5.9	61.3	-15	5.9
8996	344.2040	12.7249	2013.2835	16.421	16.792	7.842	264.9	62.2	24	5.2	60.6	22.6	5.3
8997	344.2443	13.3059	2013.3035	10.748	11.165	8.388	153.8	116.7	-38.3	5.2	113.4	-34.7	5.2
8998	344.2565	19.7473	2013.222	14.578	17.052	54.483	307.1	185.6	119.8	5.9	179.4	116.6	6
8999	344.2662	74.9635	2013.3685	12.527	16.397	6.148	1.4	74.2	109.2	6.3	73.2	107.3	6.6
9000	344.3482	16.3573	2013.306	10.827	16.59	26.477	35.5	94.7	-63.7	5.2	91.1	-59.3	5.3
9001	344.3638	9.9687	2013.232	16.412	17.137	14.377	214.1	104.6	-48.5	6.6	100.7	-50.7	6.6
9002	344.4069	48.6244	2013.38	16.269	16.475	7.294	44.2	99	10.8	5.6	94.8	11.3	5.6
9003	344.4384	17.5129	2013.1925	16.733	18.379	33.517	357.1	-18.8	-66.4	5.3	-23.2	-68.1	6.2
9004	344.4614	-2.3723	2012.8895	13.659	15.222	58.111	5.7	200.9	50.7	5.7	204.1	48.8	5.8
9005	344.4941	75.0950	2013.2765	17.003	17.789	6.961	223	108.5	67.1	6.4	110.2	69.1	6.5
9006	344.5239	28.1474	2013.2625	10.623	14.11	59.396	120.4	-94.2	-116.1	5.3	-96.3	-120.8	5.3
9007	344.5626	1.4043	2013.4025	10.84	12.138	37.024	241.7	-39.3	94.1	6.5	-35	92.8	6.5
9008	344.5896	63.9448	2013.3325	10.411	13.537	7.511	136.7	-16.5	-64.3	5.2	-16.9	-65	5.2
9009	344.7356	20.2217	2013.1085	14.553	15.317	21.178	235.6	-56.9	-120.4	6.6	-56.2	-121.4	6.6
9010	344.7729	22.2602	2013.418	12.572	12.996	4.38	283.1	-14.2	-79.7	6.5	-11.2	-76.8	6.5
9011	344.8027	23.4173	2013.2645	13.378	17.666	16.921	251.2	114.9	55.5	6.4	112.8	53.9	6.8
9012	344.8076	20.1459	2013.1425	12.961	15.973	9.404	324.3	-47	-78.1	6.5	-53.8	-72.1	6.6
9013	344.8085	-2.2284	2012.8885	16.016	16.704	7.485	111.3	103.2	-44.1	6.7	99.1	-46.8	6.7
9014	344.8562	43.7972	2013.346	12.352	16.789	17.729	332.7	53.1	65.4	5.5	53.6	62.9	5.6
9015	344.8705	72.1717	2013.0295	17.626	18.973	54.108	349.6	-5.6	-71.8	6.5	3.8	-69.7	7

9016	344.9264	34.6767	2013.389	14.998	15.896	6.228	2.9	-73.4	-61.1	5.4	-73.5	-60.6	5.5
9017	344.9598	32.6499	2013.3695	11.111	15.619	18.774	229.9	49.4	-177	5.5	44.4	-182.8	5.5
9018	344.9761	18.3629	2013.257	12.98	14.611	11.191	154.9	74.6	36.7	5.6	72.7	38.5	5.6
9019	345.0449	-4.3962	2012.8865	15.403	17.078	12.034	255.3	60.8	-3.4	6.7	60.7	-2.9	7.1
9020	345.0550	48.2480	2013.405	13.837	15.588	10.204	218.4	-55.3	-96	5.6	-55.4	-97.6	5.6
9021	345.0551	16.2250	2013.4605	14.07	15.17	5.17	326.5	75.7	-69.5	5.3	77.7	-68.5	5.4
9022	345.0994	41.1242	2013.4965	8.312	10.213	14.722	294.2	-26	-137.4	6.3	-31.8	-139.8	6.4
9023	345.1020	19.3772	2013.328	12.702	13.596	6.045	186.7	97.1	0.4	5.6	99	1.2	5.6
9024	345.1228	-2.7547	2012.8875	10.541	14.074	51.693	300.7	46	-66.1	6.6	41.1	-70.7	6.6
9025	345.1695	24.7984	2013.333	16.455	17.57	58.773	213.1	77.6	-4.5	5.3	78.8	3.8	6.1
9026	345.2074	41.7209	2013.359	17.536	18.426	16.557	81.7	16.6	-100.2	6.4	6.5	-104.8	7.2
9027	345.2339	41.8013	2013.4175	15.55	18.413	39.23	320.4	-11.2	-63.2	6.4	-13.7	-65.8	6.5
9028	345.2818	40.9384	2013.349	10.768	16.902	25.261	257.2	-43.4	-88.2	6.4	-35.3	-93.6	6.5
9029	345.2832	27.5969	2013.2525	16.255	16.938	4.186	1.9	71.5	-12.1	5.4	68.1	-14.7	5.3
9030	345.2907	19.9597	2013.2405	16.219	17.921	20.644	137.8	138.6	33.9	5.7	136.8	34.9	5.9
9031	345.2920	56.4830	2013.447	12.588	15.56	6.125	267.7	-14.2	-69.7	5.6	-14.9	-67.6	5.6
9032	345.3671	-3.7499	2012.8865	13.812	14.263	5.994	60.6	-40	-71.2	5.7	-44.7	-65.9	5.8
9033	345.3826	44.4367	2013.3535	17.748	19.219	32.163	151.3	70.9	-31.2	5.5	70	-33.9	7.6
9034	345.4420	5.1658	2013.315	9.604	10.292	48.939	19.3	-66.3	-76.1	6.7	-67.5	-72.4	6.6
9035	345.4626	-5.6656	2012.883	13.729	16.315	9.794	171.7	168.9	-17.2	5.8	173.4	-20	6.4
9036	345.5811	12.8419	2013.3985	13.767	15.681	5.182	297.8	3.8	-178.8	5.3	9.1	-175	6.1
9037	345.6547	39.3526	2013.3145	15.035	17.247	12.617	335.6	62.8	34.8	5.9	63.1	35.2	6
9038	345.7948	27.7891	2013.36	17.374	17.625	6.25	353.3	64.3	-51.8	5.4	65.6	-52.9	5.3
9039	345.8066	-5.0755	2012.883	10.912	16.352	40.153	173.7	116.3	-13.6	6	111.7	-12.5	6.6
9040	345.8140	7.5335	2013.261	11.253	17.168	14.296	230.4	148	-11.2	6.6	143.7	-14.8	6.7
9041	345.8652	43.1339	2013.3465	15.886	16.429	12.819	78.2	-77.9	-105	5.5	-78	-104.1	5.6
9042	345.8920	47.9951	2013.513	12.909	16.825	32.799	15	92	16.7	5.6	92.6	17.4	5.9
9043	345.8944	27.8402	2013.233	16.584	16.826	10.49	125.4	74.7	52.6	5.3	71	53.7	5.4
9044	345.9427	-0.4630	2012.9435	15.854	15.958	7.198	358.2	83.1	-13.4	6	80.7	-10	6.1
9045	345.9675	28.8554	2013.539	12.452	15.346	6.156	191.8	-46.7	-61	5.2	-47.1	-62	5.4
9046	346.1342	14.3476	2013.358	12.738	15.466	6.995	206.9	100.5	-27.4	5.2	100.7	-27.2	5.3
9047	346.1430	21.1950	2013.216	11.583	15.463	15.145	135	173.4	-161.2	5.4	173.6	-162.1	5.5
9048	346.1567	77.9355	2013.383	14.221	14.642	9.477	42.9	4.6	-69.1	5.9	5.4	-69.5	5.9
9049	346.1798	23.9241	2013.4025	11.388	11.576	9.676	126.1	29.3	-86.2	5.2	29.9	-86.4	5.2
9050	346.3031	57.9107	2013.5125	14.276	14.819	3.628	2.5	68.8	52.5	5.6	68.8	50.7	6.2
9051	346.3631	11.4916	2013.1855	12.564	15.513	11.21	253.2	91.7	12.2	6.5	89.7	18.4	6.5
9052	346.4615	17.4279	2013.16	17.135	18.451	18.874	97.2	62.4	-3.6	5.3	60.5	-2.6	5.6
9053	346.4668	10.1000	2013.02	14.319	17.056	15.759	71.2	67.7	-43.7	6.6	70.6	-45.3	6.7
9054	346.4816	35.4716	2013.315	12.25	17.774	21.444	159.8	82.4	62.2	6	84.5	63.6	6.2
9055	346.4881	19.8154	2013.3565	12.27	14.637	12.742	249.2	67.9	-12.1	5.2	69.1	-10.8	5.2
9056	346.5256	36.8153	2013.3465	12.938	16.035	7.523	149.8	168.3	-52.4	5.9	172.5	-56.1	6
9057	346.5481	16.7857	2013.036	10.979	14.965	7.086	263	56.2	-108.8	5.3	56.9	-114.7	5.4
9058	346.5552	48.8497	2013.4675	11.962	14.436	8.978	101.6	39.5	94.4	5.6	37.2	93.7	5.5
9059	346.5778	20.5007	2013.0335	17.649	18.184	4.517	221.7	-19.5	-94.3	5.4	-26.2	-96.4	6.5
9060	346.5929	-2.4492	2012.884	13.736	14.313	41.337	83.9	52.9	-62	5.7	56.2	-64.6	5.7
9061	346.5943	8.0232	2013.031	11.637	15.831	13.6	226.1	89.4	28.5	6.5	85.8	29.1	6.6

9062	346.5954	49.3582	2013.1865	17.115	18.441	29.107	10.9	108.6	-45.6	6.8	105.1	-53.3	6.4
9063	346.6167	48.0516	2013.421	13.105	14.965	32.893	278.8	45.3	62.3	5.6	42.7	61.3	5.6
9064	346.6750	-4.2011	2012.883	9.308	10.463	14.889	289.2	122	-30.2	5.8	118	-32	5.8
9065	346.6793	5.5667	2013.1505	11.564	12.534	45.842	41	97.6	4.7	6.6	96.7	4.3	6.6
9066	346.6832	16.2903	2013.2025	12.504	16.395	13.253	169.6	63.4	-36.2	5.2	60.7	-35	5.3
9067	346.7068	-2.6575	2012.883	17.02	17.173	9.033	202.3	3.1	-63.4	6	5.7	-61.6	6.4
9068	346.7543	18.6350	2012.9925	13.304	18.829	54.336	23.5	120.3	-34.5	6.3	115.4	-45.4	7.8
9069	346.7896	-0.5919	2012.925	14.42	16.191	14.449	168	-37.7	-69	5.7	-36.8	-69	5.8
9070	346.8085	-9.9183	2012.946	14.696	14.963	11.532	202.5	-14.5	-93.8	5.9	-13.8	-90.5	5.8
9071	346.8103	-3.4786	2012.883	11.728	16.373	9.561	291.8	69	-82.9	5.7	67.3	-89.3	6
9072	346.8323	41.2020	2013.4035	12.296	17.165	25.003	146.5	71.7	-13.1	5.9	71	-13.9	5.9
9073	346.8454	43.3761	2013.39	11.224	11.725	24.142	47.4	93.3	37.1	5.5	94.3	36.7	5.5
9074	346.8479	38.2212	2013.3425	13.058	17.98	14.899	46.5	93.9	8.8	5.9	92.2	6.2	6.1
9075	346.9034	1.9311	2012.9295	12.093	14.968	6.857	35.5	63.9	-73.7	6.6	62.6	-75.3	6.7
9076	346.9110	58.6216	2013.3915	9.574	15.521	30.729	71.1	115	70.5	5.6	119.7	68.1	5.6
9077	347.0009	66.9539	2013.008	18.249	19.053	48.765	190.8	10.3	-67	6.5	10.1	-68.6	7.6
9078	347.0094	20.8593	2013.0765	16.078	17.204	4.387	66.2	136.4	-11.1	5.5	138.4	-11.7	6.4
9079	347.0259	26.0678	2013.343	11.166	14.989	15.974	181.5	60.9	20.9	5.2	61.9	16.8	5.3
9080	347.0621	4.6619	2013.188	12.159	12.969	21.623	30.2	211.4	-69.1	7.1	215.5	-65.6	7.2
9081	347.0982	-8.6326	2012.946	10.22	11.935	35.996	270.2	40.4	-80.1	5.8	33.6	-80.6	5.7
9082	347.1385	26.3454	2013.1195	12.895	17.624	10.317	354.1	-55.5	-85.9	5.2	-54.9	-80.5	5.6
9083	347.1578	1.4627	2013.235	13.547	13.641	37.082	202	121.6	-49.3	6.7	119.2	-58.1	6.8
9084	347.2003	41.9166	2013.309	13.979	16.484	25.014	287.6	115.3	15.4	5.9	116.6	16.7	6
9085	347.2030	33.1673	2013.357	9.724	14.155	31.223	146.2	-90.8	-63	6.2	-95.7	-60.1	6.3
9086	347.2052	18.6695	2013.1935	15.187	15.389	8.667	244.7	104.6	4.9	5.3	102.9	4.5	5.3
9087	347.3160	6.1589	2012.871	16.924	17.597	25.732	222.7	62.7	-15.7	6.8	64	-20.3	7
9088	347.3862	49.9732	2013.4135	11.524	12.879	14.645	32.5	-87.2	-111.6	5.6	-83.5	-112.7	5.6
9089	347.4011	48.8952	2013.4545	15.011	15.374	45.312	309.8	68.2	-25.6	5.5	65.8	-29.1	5.5
9090	347.4436	11.8408	2013.0575	14.803	16.96	10.25	317.5	-14.3	-66.1	6.5	-14.1	-67.7	6.6
9091	347.4622	26.6290	2013.278	14.035	16.233	28.348	242.7	35.7	-68.7	5.8	40.7	-61.5	5.6
9092	347.4830	14.7258	2013.1035	12.914	13.745	9.796	115.1	72.6	0.8	5.2	72.9	-1	5.3
9093	347.4962	47.9594	2013.6505	7.252	7.61	15.68	256.2	156.4	-1	5.5	152.5	3.6	5.4
9094	347.5251	66.6977	2013.406	12.515	15.116	17.091	14	-47.3	-89.6	6.2	-49.8	-88.1	6.2
9095	347.5306	2.0471	2012.877	14.307	14.501	11.833	158.1	118.5	5.9	6.6	115.4	3.9	6.6
9096	347.5335	64.4602	2013.133	17.467	18.511	34.068	84.1	19.4	-60.8	6.2	17.4	-64.3	6.5
9097	347.5983	26.8751	2013.4445	11.227	11.435	11.511	175.2	71.6	-10.1	5.5	73.5	-13.6	5.5
9098	347.6136	45.7087	2013.5065	13.84	14.953	6.731	44.6	107.7	-18.3	5.9	107.3	-23	5.8
9099	347.6268	6.9614	2013.091	13.745	14.904	19.581	144.7	60.9	-11.2	6.6	60.2	-19.1	6.7
9100	347.6376	8.5425	2013.119	10.106	11.2	13.072	205.8	95.3	-10.8	6.5	98.7	-10.5	6.5
9101	347.7430	32.0535	2013.396	11.459	16.245	41.342	281.6	71.1	2.5	5.5	70.7	-0.6	5.5
9102	347.7761	41.7140	2013.428	11.703	12.366	5.479	21.2	87.9	15.1	5.9	88	16.3	5.9
9103	347.8318	-6.6133	2012.926	11.726	16.744	20.957	329.6	-26.4	-75.5	5.7	-28.8	-76.3	6.2
9104	347.8469	31.5410	2013.4375	10.654	11.946	23.336	220.1	87	44.8	5.5	85.6	45.7	5.5
9105	347.9739	67.0386	2013.448	12.446	15.077	6.012	326.6	110	81.3	6.2	108.1	85.2	6.2
9106	347.9914	50.4311	2013.6185	18.015	18.631	45.228	191.4	62.1	25.6	5.7	63.1	17.4	5.8
9107	348.0533	1.1823	2013.032	10.533	15.664	19.86	219.4	112	-97	6.9	107.3	-95.9	6.9

9108	348.0581	42.7346	2013.3765	16.832	18.09	20.028	356.9	70.9	48.6	5.6	73.9	42.1	5.7
9109	348.0905	32.6986	2013.3755	12.081	13.336	29.134	40.4	148.7	29.3	5.5	146.6	30.2	5.5
9110	348.1017	20.5453	2013.204	12.509	17.091	26.451	196.2	96.9	-7.5	5.2	96.7	-10	5.3
9111	348.1056	19.0411	2013.2555	11.209	11.785	6.361	7.9	76.9	20.3	5.2	74.6	19.7	5.3
9112	348.1070	1.3229	2013.0055	13.339	15.298	10.49	174.5	66.8	20.6	6.5	68.5	19	6.6
9113	348.1275	45.9903	2013.3925	14.281	16.152	12.293	47.6	65.6	25.6	5.5	65.8	25.2	5.5
9114	348.1314	12.5504	2013.049	15.045	15.649	6.673	255.9	23.7	-114	5.7	21.5	-115.5	5.8
9115	348.1336	16.1444	2013.016	13.734	18.61	16.527	17.1	70.7	0.2	5.6	69.3	4.9	6.2
9116	348.1904	15.4371	2013.2355	12.843	13.989	13.657	317.7	-38.9	-69.6	5.6	-37.5	-69.6	5.6
9117	348.1955	21.5342	2013.07	14.798	15.865	56.059	61.7	-19.5	-61	5.3	-17.5	-62.2	5.3
9118	348.1965	27.0178	2013.375	10.971	17.291	53.39	109.4	150.3	-22.1	5.6	155.4	-23.7	5.8
9119	348.2769	40.0028	2013.429	7.729	9.531	14.233	285	70.2	12.7	5.9	71.3	5.8	5.9
9120	348.3313	87.8813	2013.7595	12.633	15.644	8.408	216.5	66.4	17	5.8	67	18.8	5.8
9121	348.3393	20.1952	2013.1435	12.803	15.914	6.315	123.1	138.6	-61	5.2	140.9	-62.6	5.6
9122	348.3528	59.3039	2013.4155	11.318	13.573	4.949	249.4	135.3	21.2	6	139.9	24.2	6
9123	348.3805	52.3589	2013.4455	16.643	17.114	4.018	307.3	67.9	10.6	5.6	70.1	8	5.7
9124	348.4504	4.4117	2012.9195	15.255	16.951	17.394	332.9	36.8	-78.4	6.6	37.6	-80	6.8
9125	348.4572	25.5958	2013.389	10.445	12.161	49.385	139	72.2	20.1	5.5	71	28.5	5.5
9126	348.4870	2.1294	2012.8755	15.232	15.768	5.759	211.1	11.9	-70.1	6.6	10.6	-66.9	6.7
9127	348.5001	6.5408	2013.092	15.292	16.566	39.386	7.7	97.7	-5.6	6.6	96.7	-5.7	6.9
9128	348.5051	21.9509	2013.0285	18.207	18.53	5.912	18.1	93.1	29.2	6.1	92.1	28.3	6.5
9129	348.5072	5.1552	2012.9705	12.557	16.988	9.134	204.8	136.1	-10.5	6.6	132.3	-12.5	7.1
9130	348.5729	40.1593	2013.4025	11.425	12.987	40.731	246	173.9	36	6	178	41.2	5.9
9131	348.5934	68.7392	2013.1795	17.974	18.736	55.064	254.5	-18.4	-74.6	6.6	-30.5	-67.7	7.2
9132	348.5980	55.2501	2013.159	17.415	18.355	14.93	177.1	57.1	-65	6.4	52.2	-65.1	6.6
9133	348.6007	31.7780	2013.513	8.798	10.385	16.871	157.4	67.9	15.1	5.4	68.8	20	5.5
9134	348.6231	26.9290	2013.195	16.227	16.737	12.069	342.2	29.9	-75.5	5.6	29	-79.3	5.7
9135	348.6772	40.5936	2013.367	15.829	18.034	7.989	102.8	77.3	-10	5.9	81	-5.4	6.1
9136	348.6997	61.5375	2013.3345	10.283	14.294	6.49	347.1	82.9	6.6	5.9	82.9	9.4	6
9137	348.7078	0.9035	2012.882	15.928	17.462	12.124	113.7	63.8	-36.4	6.7	60.2	-36.7	7.3
9138	348.7846	27.3158	2013.3195	13.446	13.556	57.495	152.5	-46.4	-67.6	5.5	-44.3	-73.3	5.5
9139	348.8563	53.2323	2013.383	13.988	15.948	14.761	26.8	30.5	-88.2	5.9	31.7	-89.5	5.9
9140	348.8951	11.9318	2013.2525	13.16	14.383	6.082	59.1	90.8	-72.1	5.6	87.5	-74.2	5.7
9141	348.9648	27.4827	2013.542	8.879	11.249	17.366	233.4	-25.9	-72.2	5.4	-26.3	-75.1	5.5
9142	348.9719	45.8360	2013.4595	16.683	17.613	6.61	146.3	78.9	31.2	5.5	77.2	30.4	5.6
9143	348.9725	84.5338	2013.7265	8.98	14.041	34.271	260.2	101.1	14.4	5.8	104.3	17.8	5.8
9144	349.0102	52.2569	2013.4285	11.377	13.474	34.656	272	75.2	6	5.9	75.6	11.4	5.9
9145	349.0541	30.6593	2013.2805	17.148	17.544	18.324	162.5	66.6	14.3	6	67.7	18.7	6
9146	349.1537	-0.2634	2012.954	14.391	14.716	7.934	31.7	95.4	68.4	5.8	92.2	67.9	5.8
9147	349.2045	63.1271	2013.4185	14.478	14.873	7.61	264.2	63.8	19	5.9	63.1	15	5.9
9148	349.2261	61.1621	2013.4025	13.505	16.541	5.89	113.7	154.7	130.7	6	157.4	128.7	6.1
9149	349.3059	19.9736	2013.2985	8.692	16.038	53.829	111.5	-28.1	-87.7	5.2	-29	-83.7	5.3
9150	349.3745	44.9324	2013.491	11.026	13.933	5.409	238.6	91.3	-28.7	6.4	93.9	-23	6.3
9151	349.3851	9.6933	2013.254	9.377	14.338	20.711	6.9	122.8	-48.2	6.5	126.3	-49.7	6.7
9152	349.4084	-5.0146	2012.907	10.769	12.942	15.245	177.1	94.2	-21.4	5.7	94	-20.7	5.7
9153	349.4161	20.5403	2013.147	12.634	17.267	9.009	179	-18.6	-82	5.2	-19.9	-81.3	5.5

9154	349.4182	31.0800	2013.389	12.722	14.988	14.459	76.2	27.2	-136.6	6	25.6	-135.7	6
9155	349.4428	27.7046	2013.2415	15.543	15.601	8.942	212.4	89.8	-0.3	5.6	91.8	-0.3	5.6
9156	349.5097	5.3497	2013.04	13.99	15.758	51.794	216	130.6	79.7	6.9	128.1	77.6	7.2
9157	349.5112	29.7891	2013.435	12.335	14.829	11.133	97.1	112.7	-51.8	5.5	114.9	-57.7	5.6
9158	349.5244	-7.3410	2012.9395	15.912	16.623	17.857	150.8	-1.1	-132.8	6.3	-0.6	-135.6	6.7
9159	349.6155	-9.2706	2012.946	10.342	11.12	22.524	150.1	169.8	-75.4	5.8	171.1	-74.7	5.7
9160	349.7346	58.5642	2013.353	12.99	16.492	14.681	358.4	64.1	-25.1	5.6	62.4	-24.7	5.7
9161	349.8347	7.3320	2013.031	10.897	12.157	6.169	6.2	67.2	57.8	6.4	65.6	55.5	6.5
9162	349.8512	-1.7947	2012.883	11.008	15.652	20.749	302	88	-40.9	5.7	87.8	-40.1	5.8
9163	349.8584	15.6378	2013.1235	15.549	17.324	13.222	82	-29.3	-77.8	5.3	-25.9	-77.2	5.3
9164	349.8650	79.0038	2013.585	7.296	10.456	10.888	214.4	209.2	84.2	6.4	217	75.7	6.4
9165	349.8900	48.2923	2013.43	13.602	17.194	9.151	48.9	63.7	-22.1	5.5	64.7	-22.8	5.6
9166	349.9479	11.3391	2013.07	16.797	17.116	6.638	76.8	-82.3	-77.3	6.6	-85.3	-78.3	6.6
9167	349.9646	31.8940	2013.3505	10.239	17.265	15.508	295.3	-76.5	-116.7	5.4	-74.8	-117.4	5.8
9168	349.9716	-1.0906	2012.883	14.73	16.962	23.42	324	-42.7	-80.2	5.7	-43	-76.8	5.9
9169	350.0757	54.4879	2013.3625	11.032	16.509	12.416	200.7	170.7	82.5	5.6	168.2	81.9	5.7
9170	350.0809	41.1086	2013.3745	14.972	15.706	41.217	253.9	60.9	20.6	5.5	63.3	19.6	5.5
9171	350.0957	83.4638	2013.687	14.311	15.871	4.918	10.4	102.7	-1.7	6.2	103.2	-1.6	6.4
9172	350.1689	48.7644	2013.596	9.157	13.468	6.57	114.1	-38.3	-71	5.5	-46.3	-65.8	5.6
9173	350.1832	83.4975	2013.7475	13.821	14.454	7.096	226	79.3	-19.6	6.2	76.1	-18.8	6.2
9174	350.1942	48.7928	2013.4845	8.499	12.938	8.751	244.9	152.9	34.4	5.5	154.4	24.6	5.5
9175	350.2927	57.6622	2013.4035	12.711	14.845	6.627	59.4	75.2	-15	5.6	76.4	-16.1	5.6
9176	350.2945	1.8413	2012.8785	10.609	12.652	11.834	196.3	64.4	-8.2	6.6	63.8	-10.5	6.6
9177	350.3076	25.4126	2013.4275	11.897	13.366	4.845	344.9	-44.2	-71.6	5.5	-44.8	-71.3	5.6
9178	350.4048	24.4773	2013.1695	14.309	17.013	19.037	219.9	71.1	-6.3	5.6	72.8	-6.4	5.7
9179	350.4072	29.7400	2013.4585	12.469	14.929	16.526	53.2	107.9	17.1	5.5	108.1	16.9	5.5
9180	350.4154	2.2815	2012.878	11.836	14.338	5.831	332.4	-40.2	-99	6.6	-38	-104.8	6.7
9181	350.4554	2.2801	2012.8775	14.091	15.787	6.666	176.5	88.2	14.6	6.6	84.2	20.2	6.7
9182	350.4662	27.2395	2013.4005	12.966	16.307	13.781	350.9	62.4	5.3	5.2	60.6	4.3	5.2
9183	350.5171	45.4935	2013.609	17.164	18.977	58.515	68.7	9.2	-63.5	5.5	17	-64.8	6.8
9184	350.5386	59.3620	2013.4475	18.324	19.03	15.437	312.7	77.8	-18.1	6.2	81.3	-8.4	6.8
9185	350.5773	15.0372	2012.923	16.64	16.809	8.442	64.3	-25.2	-61.7	5.4	-27.4	-62.3	5.4
9186	350.6020	-5.7451	2012.9205	12.307	12.641	4.632	172	74.3	-30.5	5.7	72.7	-30.2	5.8
9187	350.7056	23.3744	2013.411	14.711	14.787	12.656	176.7	145.9	-26	5.2	144.7	-23.4	5.2
9188	350.7241	40.8727	2013.342	16.17	16.8	6.38	346.7	72.4	-57.3	6	70.3	-56.3	6
9189	350.7662	16.0685	2013.3715	8.436	13.903	12.226	214	-70	-101.8	5.3	-71.5	-98.8	5.3
9190	350.7869	21.6569	2012.8935	11.44	11.628	4.836	82.6	-26.4	-164.7	5.4	-23.8	-161.2	6.2
9191	350.8072	-2.2673	2012.884	10.846	18.015	21.479	342.7	-23.7	-78.8	5.7	-26.1	-79.7	6.4
9192	350.8242	10.5727	2013.0805	14.235	16.077	5.933	260.1	142.6	-29.4	6.4	141.1	-32	6.6
9193	350.8517	-7.8671	2012.946	11.491	13.811	25.825	250.2	76.9	-14.5	5.7	77.8	-15	5.7
9194	350.8711	45.7932	2013.4685	8.14	8.698	25.852	131.1	173.1	-26.3	5.5	173.5	-18.6	5.5
9195	350.9829	-0.4474	2013.065	12.76	14.965	18.85	212.1	69.1	-23.6	5.6	68.4	-21.9	5.7
9196	350.9870	-7.8472	2012.946	14.299	16.356	11.891	137.4	57.2	61.7	5.7	55.3	60.6	5.9
9197	351.0438	32.9232	2013.42	13.39	16.145	17.656	185.7	72.8	23.5	5.4	73.5	24.2	5.4
9198	351.0533	71.0241	2013.416	13.837	17.616	16.312	235.8	81.6	69.7	6.4	84.4	67.1	6.6
9199	351.1091	51.8105	2013.4565	15.742	16.9	19.841	225.6	90	7.4	5.5	86.6	6.9	5.6

9200	351.1164	18.1170	2013.378	10.073	13.456	41.321	212.8	105	27.3	5.2	102.9	26.5	5.2
9201	351.2020	66.9331	2013.093	18.883	19.103	59.568	228.5	-9.4	-72.1	7.5	-23	-71.8	7.1
9202	351.2722	80.6421	2013.6135	9.539	16.242	17.028	278	33.3	87	6.2	34.8	86.4	6.3
9203	351.3789	3.4927	2012.9025	15.093	16.217	53.83	187.3	65.9	-108.2	6.8	60.2	-111.4	6.7
9204	351.3820	4.8804	2012.968	13.828	14.726	6.515	105.9	-148.8	-121.6	6.8	-148	-124	6.8
9205	351.4158	31.9786	2013.57	14.953	15.786	4.889	348.5	99.5	-32.1	5.4	95.4	-33.8	5.4
9206	351.4516	17.5937	2013.163	16.601	18.027	5.759	324.5	181.3	14.9	5.4	184.1	8	5.7
9207	351.4702	-8.2754	2012.946	11.044	17.07	27.497	317.4	23.8	-68	5.7	20.9	-65.9	6.6
9208	351.4995	27.2393	2013.4775	14.907	16.297	7.248	355.1	85.5	-14.7	5.2	86.5	-13.5	5.2
9209	351.5097	36.8558	2013.3975	13.765	15.541	8.922	211	65.5	16.3	5.5	66.7	17.5	5.6
9210	351.5326	2.5179	2012.881	15.241	16.64	10.473	76.9	99.1	-14.1	6.7	98.1	-9.2	6.7
9211	351.5580	40.3137	2013.6205	14.47	15.364	3.925	103.4	63.5	-12.5	5.5	62.2	-12.3	6.1
9212	351.5860	42.1373	2013.3615	10.861	13.925	27.807	274.8	67	0.8	5.5	64	0.4	5.5
9213	351.6265	39.0299	2013.277	12.679	14.453	5.316	142.8	45.2	-73.3	5.6	44.4	-72.9	5.6
9214	351.6289	43.1415	2013.3825	11.303	16.868	38.048	135.9	64.2	23	5.5	64.3	25.6	5.5
9215	351.6754	30.7056	2013.512	11.012	16.447	16.824	81.6	-158.4	-99.1	5.4	-163	-97.6	5.5
9216	351.7180	38.7898	2013.432	17.241	18.497	48.258	50	11.8	-62	5.6	19.8	-62.6	6.2
9217	351.7740	6.4079	2012.991	14.617	15.216	6.16	277.9	115.1	-42.2	6.5	110.8	-42.7	6.5
9218	351.7869	85.3407	2013.7015	17.397	17.464	4.148	63.4	95.5	53.2	6.6	90.2	54.2	5.9
9219	351.8519	50.2634	2013.4705	10.757	17.703	31.403	43.7	191.4	68.4	5.6	190	72.3	5.7
9220	351.8589	4.8934	2012.997	11.908	13.143	4.555	206.9	92.6	35.4	6.5	88.9	38.6	6.7
9221	351.8727	51.5263	2013.186	17.411	18.566	40.583	353.4	26.9	-70.3	7.2	35	-69.2	7.4
9222	351.9110	17.0065	2013.099	15.448	17.763	18.248	244.8	95.8	8.4	5.3	97.9	9.3	5.4
9223	351.9348	-9.3135	2012.946	16.357	16.588	8.323	129.4	38.8	-68.2	6.5	39.4	-66.3	6.8
9224	351.9409	30.0861	2013.469	10.188	11.555	58.06	155.2	61.2	-26.9	5.9	63.1	-27.7	5.9
9225	351.9425	12.3949	2013.2365	11.864	13.961	11.571	278.1	135.9	62.5	5.3	135.7	58.4	5.4
9226	352.0860	82.2745	2013.722	12.642	16.195	46.723	7.5	109.3	48.3	6.2	107.9	45.4	6.2
9227	352.0899	47.6955	2013.4845	13.527	16.697	11.301	236.2	167	-18.6	5.5	166.5	-16.9	5.5
9228	352.1090	16.0663	2013.2225	9.122	12.222	6.344	176.4	43	-91.8	5.2	46.3	-89.6	5.4
9229	352.1735	33.6708	2013.399	13.786	14.23	8.558	183.4	132.4	-40.8	5.9	131.5	-40	5.9
9230	352.2103	7.0569	2013.096	13.341	14.07	14.898	325.5	142.2	-36	6.6	146.7	-35.7	6.6
9231	352.2266	0.9487	2012.95	11.467	17.542	21.924	120.7	-35.6	-78.4	6.5	-33.1	-78.9	6.8
9232	352.2478	-6.6399	2012.916	9.502	10.012	20.42	318.2	75.4	-20	5.8	75.2	-21.1	5.8
9233	352.3920	45.6810	2013.4215	14.682	16.508	5.854	197.3	-79.8	-70.5	6.4	-81.9	-69	6.4
9234	352.4166	26.6346	2013.4425	13.07	16.134	47.597	260.9	51.2	-81	5.2	53.1	-77.5	5.2
9235	352.4440	56.4173	2013.462	10.462	14.103	8.943	125	-4.6	-232.5	6.5	-2.5	-227.5	6.4
9236	352.4819	35.8596	2013.406	14.614	16.286	4.997	276.8	85	5.5	5.9	84.3	-1.5	6.1
9237	352.4926	3.7178	2012.9715	15.358	17.04	15.239	201.9	72.7	24.2	6.5	69.1	27.4	6.7
9238	352.5485	77.8006	2013.4865	14.248	16.143	6.541	174.2	49.7	71	6.2	46.4	71.4	6.2
9239	352.5693	68.3692	2013.413	10.711	12.332	6.232	125.7	70.7	10	5.9	68.3	11	5.9
9240	352.5729	-2.9911	2013.308	9.174	13.411	48.801	63.1	-4.8	-102	5.9	-3.1	-105.9	5.7
9241	352.6101	30.8317	2013.5535	7.396	9.682	18.794	202.7	80.1	-36.7	6	83.4	-33.7	5.8
9242	352.6320	21.5624	2013.097	16.349	16.849	3.751	53.6	98.9	-68.5	5.3	100.2	-75.6	5.9
9243	352.6350	40.3107	2013.375	16.654	18.161	22.449	241.3	67.6	-18.2	5.9	67.7	-19.7	6.2
9244	352.6546	8.3028	2012.924	12.505	16.322	21.389	294.3	76.5	-7.5	6.6	74.5	-9.5	6.7
9245	352.6774	79.5707	2013.6205	9.546	15.632	17.764	319.2	72.9	-17.4	6.1	70.5	-10.8	6.2

9246	352.6961	-9.7907	2012.946	14.333	15.679	17.197	81.7	-16.7	-82.6	5.9	-13.2	-83.3	5.8
9247	352.7012	-5.8063	2012.937	15.655	17.88	37.061	325.6	76.6	-82.9	5.8	82.5	-76.6	7.1
9248	352.7124	14.6410	2013.19	10.719	12.866	5.89	107	91.9	-4.6	5.2	90.2	-3.8	5.3
9249	352.7224	4.4789	2012.987	12.552	18.725	57.384	302.9	66.4	1.3	6.4	64.2	12.3	7.9
9250	352.7260	27.7779	2013.5445	8.863	15.018	26.691	250.7	-139.2	-131.6	5.2	-136	-130.2	5.3
9251	352.7754	8.7005	2013.057	12.847	13.108	30.766	359	228.4	86.7	7	225.1	85.6	7
9252	352.8023	13.4245	2013.0235	16.957	17.357	5.96	316.5	40.7	-89.8	5.3	39.3	-89.3	5.3
9253	352.8328	52.4106	2013.5445	7.578	10.373	17.388	243.6	107.7	-39.1	5.5	103.6	-33.9	5.5
9254	352.8345	56.9587	2013.445	11.521	11.897	34.374	265.9	90.7	31.7	6	92.1	31.3	6
9255	352.8915	35.7979	2013.42	14.331	18.283	35.298	40.2	-66.9	-80.3	5.5	-71.3	-72.6	5.8
9256	352.9055	75.3580	2013.4465	10.174	17.253	22.388	117.3	72.7	-22.9	6.1	77.2	-17.8	6.3
9257	352.9913	37.7886	2013.3755	11.606	17.484	11.813	303	122.8	11.5	5.9	125.9	11.5	6
9258	353.0429	29.8878	2013.4675	10.873	14.48	7.287	119.6	-103.2	-74.2	5.2	-105.4	-73.8	5.3
9259	353.3450	3.7616	2012.9415	14.062	16.527	43.665	126.9	94.9	10.3	6.6	97.7	8.1	6.6
9260	353.4144	15.7271	2013.0695	14.753	16.408	24.242	25.8	-77.1	-115.9	5.3	-78.4	-107.2	5.4
9261	353.4562	-5.7551	2012.914	11.432	15.224	10.117	44.5	68.4	40.6	6.6	71.8	38.3	6.9
9262	353.6015	3.5209	2013.1995	11.628	12.385	6.82	234.8	100.3	-6.5	6.5	99.5	-4.6	6.5
9263	353.6388	59.9699	2013.54	11.157	12.169	4.037	356	153.1	-5.6	5.6	155.5	-12.8	5.7
9264	353.6867	41.1354	2013.4065	14.111	14.459	7.371	122	63.2	9.7	5.9	62.3	8.3	5.9
9265	353.7143	27.2935	2013.4485	11.647	14.321	6.677	294.9	-42.3	-98	5.5	-42.7	-97	5.6
9266	353.7432	7.9284	2012.851	18.484	18.631	22.782	106.8	13.2	-61.2	7	7.1	-63.3	7.9
9267	353.7785	38.6554	2013.327	16.813	18.098	6.555	65.4	25.3	-91.3	5.9	26.4	-86.7	6.4
9268	353.8918	15.8732	2013.3385	11.733	12.709	13.339	337.9	58.6	-81.4	5.6	55.4	-84.4	5.6
9269	353.9037	11.2064	2013.1625	12.559	15.077	41.002	150	76.7	-23.9	6.5	75.7	-21.4	6.6
9270	353.9818	14.6691	2013.0335	15.747	16.645	11.757	278.2	79.9	-63.6	6.7	79.4	-62.8	6.7
9271	353.9902	27.2684	2013.545	10.388	10.836	8.266	225.3	-11.3	90.4	5.5	-11.4	88	5.5
9272	354.0307	28.1478	2013.5395	10.02	12.601	17.468	31.4	-15.5	-61.9	5.4	-16	-62.5	5.5
9273	354.0498	78.4334	2013.505	12.492	14.316	4.27	323.7	50.9	-64.8	5.9	56	-60.1	6.4
9274	354.0678	62.0319	2013.4505	9.698	11.368	11.078	112.6	-70.9	-66.2	5.9	-67.8	-69.6	5.9
9275	354.0774	50.2072	2013.4785	12.765	16.347	7.007	198.7	132.8	56.1	5.5	132.1	61	5.6
9276	354.1315	45.5146	2013.469	8.2	14.167	9.254	144.7	14	-99	5.5	12.3	-99.2	5.6
9277	354.1971	25.4952	2013.351	15.626	16.841	12.644	311.8	-34.2	70.3	5.5	-35	73.6	5.6
9278	354.3677	60.2240	2013.3625	14.454	15.384	10.68	311.1	82.3	6.1	5.9	81.3	5.5	6
9279	354.3736	6.3136	2012.9495	15.286	15.601	16.396	73.4	96.7	-33.5	6.7	100.9	-37.1	6.8
9280	354.3888	13.4898	2012.984	14.868	18.15	27.073	226.4	65.6	52.9	6.6	63.8	52.3	7.1
9281	354.4004	12.5997	2013.348	10.015	10.089	20.475	247.4	42.5	-64.7	6.5	45.1	-66.9	6.5
9282	354.4169	0.7765	2013.3025	10.273	11.695	34.172	156.8	-32.3	-65.8	6.4	-30.2	-65.5	6.5
9283	354.5016	12.1509	2012.999	14.158	18.233	58.081	20.3	-28.9	-78.7	6.6	-19.4	-77	7.3
9284	354.5024	51.9494	2013.443	18.052	18.206	36.409	280.8	69.6	-9.1	5.6	67	-4	5.6
9285	354.5467	15.5900	2012.9805	11.955	17.091	7.747	214.4	71.1	-26.8	6.6	74.4	-22.8	7.2
9286	354.5760	61.7184	2013.4235	13.499	13.599	29.757	173.4	86.1	-6.6	5.9	87.4	-3.9	5.9
9287	354.6610	20.9837	2013.323	13.592	17.153	22.613	346.6	-54.2	-66.2	6.4	-55	-67	6.6
9288	354.6786	29.9799	2013.373	12.444	16.419	11.083	76.5	-13.8	-65.4	5.9	-20.9	-64.6	6
9289	354.6866	37.4483	2013.5025	10.765	15.262	9.48	263.9	88	-8.5	5.8	84.9	-6.1	5.8
9290	354.7072	62.3892	2013.173	18.144	18.485	52.783	280.3	-23.4	-67.2	6.2	-19.4	-70.4	6.3
9291	354.7747	20.7849	2013.573	14.809	15.354	4.839	11.3	123.4	-44.1	6.4	119.8	-45.8	6.4

9292	354.7958	52.4665	2013.422	12.465	14.537	5.881	32	67.6	-1.2	5.5	66.3	0.8	5.5
9293	354.8335	37.3893	2013.404	16.201	16.562	8.773	221.2	64.8	-30.3	5.6	64.3	-36.7	5.6
9294	354.8493	31.5920	2013.3705	15.22	15.595	8.227	210.1	68.9	-9.4	6	70.8	-7.4	6
9295	354.9147	36.9382	2013.426	16.394	16.616	13.548	297.9	114.5	-6.6	5.6	113.8	-8.6	5.6
9296	354.9331	3.5126	2013.5315	9.753	10.644	7.703	287.1	77.6	-10.6	6.3	76.2	-10.6	6.3
9297	354.9838	-1.4500	2012.8615	16.245	17.761	15.31	164.7	-104.2	-190.7	5.8	-100.6	-190	6.3
9298	355.0117	37.6523	2013.4685	6.806	10.519	15.003	162.5	0.8	-82.8	5.6	-1.4	-83	5.5
9299	355.2399	55.4725	2013.5285	14.598	17.314	30.095	115.6	69	-22.4	5.8	67.7	-22.3	5.9
9300	355.2935	0.7903	2013.399	9.638	10.689	28.296	124.6	-3.3	66.3	6.3	0.1	68.1	6.4
9301	355.2981	6.2627	2013.381	7.717	8.318	14.177	310.4	88.7	-6.5	7.1	93.1	-7.6	6.5
9302	355.3009	56.3232	2013.514	15.278	17.132	29.183	53.9	22.3	-83.6	5.8	23	-84.3	5.9
9303	355.3222	11.7852	2013.113	14.336	15.118	12.648	294.5	-35	-86.8	6.6	-36.7	-85.5	6.6
9304	355.3231	14.9148	2013.166	11.731	12.945	15.142	83.3	137.7	82.4	5.3	143.4	80.4	5.3
9305	355.3693	42.8082	2013.415	13.933	16.582	7.68	37.5	66.5	39.2	5.5	66.1	38.7	5.5
9306	355.4194	13.4245	2013.3015	13.209	14.381	4.865	185.6	77.3	-17.9	5.3	75.7	-8.5	5.4
9307	355.4220	1.0868	2012.973	14.749	17.341	7.859	24.7	-53.2	-75.3	6.5	-54.7	-79.5	7.1
9308	355.5137	52.2575	2013.4485	14.358	16.806	18.236	231.8	93.5	28.7	5.5	92.7	29.4	5.6
9309	355.5329	20.3957	2013.17	16.047	16.383	12.837	159.1	65.3	-25.8	5.6	68.1	-24.7	5.7
9310	355.6366	67.8256	2013.4205	9.834	16.254	15.504	241.2	69.3	5.6	6.4	66.8	6.3	6.5
9311	355.6567	24.7021	2013.51	12.111	16.23	6.622	245.3	12.3	-65.2	5.2	17.6	-65.1	5.2
9312	355.6709	9.4075	2013.19	10.359	11.552	15.227	222.1	86.9	-36.4	6.5	89.7	-34.5	6.6
9313	355.6871	1.3026	2013.031	14.316	14.776	5.089	274.3	87.6	20.7	6.5	87.9	21.2	6.6
9314	355.7727	-6.9197	2012.9025	15.156	15.765	6.458	280.2	67.9	-20.7	5.9	68.1	-17.2	6
9315	355.7850	-2.5814	2013.138	11.429	13.859	5.961	233.3	84.4	-88.3	6.1	84.2	-89.1	5.9
9316	355.8186	-0.8240	2012.9735	13.292	15.573	6.578	19.9	198	-64.5	5.9	196.7	-68.5	6.1
9317	355.8568	11.2959	2012.875	14.527	17.694	25.043	193	86.8	14	6.6	85.1	12.1	6.9
9318	355.8912	45.5449	2013.5055	10.955	13.828	10.447	163.5	13.9	-163.7	5.5	17.8	-161.5	5.5
9319	355.8953	67.0615	2013.404	14.286	15.698	8.152	194.6	3.4	-95	6.2	6	-93.8	6.2
9320	355.9010	35.6827	2013.479	11.65	12.058	20.014	22.3	89.6	40.8	5.9	89.4	39.8	5.9
9321	355.9332	-7.8084	2012.903	14.753	15.663	5.559	22.5	-35.7	-155.5	5.8	-41.2	-154	6
9322	355.9778	1.8823	2012.9125	15.663	16.061	4.737	334	134.1	-61.3	6.8	131.9	-61.7	6.8
9323	355.9786	24.9804	2013.2535	11.949	12.572	5.376	328.4	146.2	0.6	5.3	146.5	5.2	5.5
9324	355.9942	-2.7306	2012.949	13.055	17.172	25.685	171.6	69	-23.2	5.7	68.4	-17.4	6.2
9325	356.0169	9.0511	2012.9935	11.942	15.519	13.099	66.6	74.3	11.2	6.6	73.2	10.3	6.6
9326	356.1020	51.8688	2013.598	17.18	18.397	58.444	165.6	62.7	-54.3	5.5	73.1	-44	6.5
9327	356.1304	8.5039	2013.2985	8.459	18.616	57.662	329.1	64.2	-40.6	6.2	67.9	-31.7	7.3
9328	356.1487	27.0083	2013.455	10.47	10.922	56.17	62.3	141.9	-23.8	5.2	141.5	-24.7	5.2
9329	356.1602	44.0045	2013.434	14.801	15.277	4.299	4.8	68.7	-22.3	5.5	68.1	-22.3	5.8
9330	356.2058	1.5583	2013.4675	14.596	14.801	4.174	243.2	-0.5	-95.5	6.7	2.1	-99.8	6.4
9331	356.2225	17.2998	2013.0405	15.142	15.951	8.759	252.2	64.2	-22	5.3	62.8	-24.8	5.3
9332	356.2284	28.2342	2013.4225	13.116	14.351	31.608	240.5	108.6	-42.7	5.2	109.2	-43	5.2
9333	356.2289	55.0819	2013.504	16.634	17.693	37.528	0.5	66.6	4.4	5.9	67.6	-5.8	5.9
9334	356.2368	48.5105	2013.637	8.603	10.36	16.947	238	151.4	-57.4	5.4	151.5	-66.7	5.5
9335	356.2721	21.1018	2013.0565	16.598	17.566	22.736	209.6	-69.5	-117.4	5.9	-68.1	-112.6	5.9
9336	356.3088	27.3251	2013.435	12.894	13.811	13.745	17.7	-11.5	-138.9	5.2	-8.5	-140	5.2
9337	356.3090	8.2308	2013.4405	8.429	10.478	28.161	310.5	178.9	-0.6	6.3	178.1	-6.3	6.6

9338	356.3748	41.3063	2013.4565	11.01	15.863	16.468	25	62.4	-38.2	5.9	63.7	-34.6	5.9
9339	356.4404	58.2211	2013.416	12.678	18.377	52.205	294.7	83.1	-29	5.9	87.3	-23.7	6.3
9340	356.4616	-0.4487	2013.1105	12.365	14.935	8.286	61.6	-76	-66.8	6.6	-79.5	-68.4	6.6
9341	356.4754	30.6449	2013.496	8.704	10.831	6.537	290.8	107.1	-63.6	6	101.2	-62.3	6
9342	356.4819	-7.9331	2012.946	14.44	15.391	11.337	216.2	73.3	-18.5	5.8	72.8	-17	5.9
9343	356.4884	12.3709	2013.03	13.002	15.138	42.87	129.3	133.1	-13.3	5.3	132.7	-11.7	5.4
9344	356.4934	52.9593	2013.4645	12.266	16.306	10.709	141.1	83.5	-5.1	5.9	82.7	-3.4	5.9
9345	356.5849	58.4711	2013.3795	16.327	17.29	5.785	327.8	98.1	-10.5	5.9	102.4	-15	6
9346	356.6516	11.1115	2013.018	11.165	16.028	48.279	267.2	108.6	5	6.6	108.6	3.4	6.8
9347	356.6873	4.5744	2013.065	11.912	14.516	28.55	292.8	8.5	-62.4	6.6	8.5	-60.6	6.7
9348	356.6987	9.8692	2012.87	15.654	17.153	15.529	204	-7.5	-70.6	6.6	-12.9	-72.5	6.7
9349	356.7044	48.7995	2013.482	17.044	17.139	24.326	168.6	34.2	-61.5	5.9	32.6	-65	5.9
9350	356.7982	-6.0090	2012.9205	11.112	16.493	44.676	109.7	79.8	-36.5	5.7	79.7	-40.5	6.1
9351	356.7993	35.2799	2013.4855	9.483	13.401	14.407	358.3	114.3	-0.1	5.9	111.8	-5.5	5.9
9352	356.8292	-7.4787	2012.909	12.054	17.005	41.652	65	66.6	4.2	5.7	67.1	-4.5	6.8
9353	356.9027	-1.1815	2012.8505	15.623	16.895	29.603	331.3	74.9	-4	6.6	72.4	-0.5	7
9354	356.9117	-3.5426	2013.0265	10.418	16.051	28.043	90.1	71.4	7.9	6.5	71.1	9.3	6.7
9355	356.9460	45.6097	2013.5295	9.434	15.415	15.041	229.9	88.6	-3.9	5.5	88.2	-5.9	5.5
9356	356.9508	35.9635	2013.2955	13.748	16.557	5.514	19	88.3	-17.2	5.9	93	-7.8	6.6
9357	356.9663	50.2320	2013.5095	12.225	16.904	11.984	15.5	65.6	-13.4	5.9	64.1	-11.4	5.9
9358	356.9825	-9.6224	2012.946	15.648	15.8	12.607	160.2	-15.2	-130.2	5.9	-15.2	-132.7	6
9359	357.0000	29.0671	2013.2215	17.588	18.798	17.077	197.3	94.1	46	5.8	102.7	37.4	7.4
9360	357.0253	4.2016	2012.9455	12.187	16.711	14.301	47.9	120.9	30.1	6.7	120.1	29.5	6.8
9361	357.1065	22.8952	2013.517	13.486	13.996	4.847	18	-99.2	-124.9	6.5	-101.1	-126	6.2
9362	357.1151	-2.5879	2012.8825	15.958	16.975	39.087	223.5	89.2	10.2	6.8	91.8	0.4	6.9
9363	357.1278	-3.7005	2012.9005	15.958	17.456	36.517	73.5	13.2	-68.2	6.7	22	-63.1	7.1
9364	357.2043	12.6778	2013.0005	14.232	15.336	18.838	85.7	108.1	45.9	5.7	110.1	45.2	5.7
9365	357.2324	21.3696	2013.3325	13.392	18.337	10.188	182.3	81.5	-84.4	6.2	81.2	-92.7	7
9366	357.2443	43.3657	2013.3845	10.416	17.529	43.125	305.5	76.3	-10.4	5.5	75	-6.4	5.6
9367	357.3068	21.4743	2013.2965	15.326	17.704	54.883	206.8	-11.3	-66	6.2	-1.1	-69.3	6.5
9368	357.3155	16.7072	2013.2055	10.383	15.992	18.055	138.1	34.5	-112.1	5.5	35	-114.5	5.7
9369	357.3240	11.1470	2012.959	12.772	16.858	17.008	45.5	115.2	-168	6.8	114.3	-164.4	6.7
9370	357.3313	25.8731	2013.313	15.036	15.069	18.534	343.1	176.2	25.5	5.6	176.7	24.8	5.6
9371	357.3518	36.9005	2013.511	9.836	13.463	20.131	210.8	116.7	-87.2	5.9	116.4	-83.9	5.9
9372	357.3831	2.9895	2013.3135	9.531	12.832	28.649	231.5	68.9	-58.5	6.4	71.7	-56.5	6.5
9373	357.4132	10.4625	2012.953	15.703	16.23	17.218	314.6	133.2	-67.9	6.7	133.9	-66.7	6.9
9374	357.4290	41.9874	2013.475	10.024	15.999	26.778	218.2	108.3	-64.6	5.5	107	-67.2	5.5
9375	357.5146	5.5106	2013.275	10.31	11.426	11.074	305.8	117.8	-112.7	6.6	127.8	-105.2	6.7
9376	357.5166	48.3544	2013.448	11.579	17.589	43.5	119.9	84.7	6	5.9	87.8	0.4	6
9377	357.6909	49.2289	2013.5265	14.106	17.055	11.87	195.4	64.6	20.4	5.8	62.8	23.1	5.9
9378	357.8066	16.5069	2013.263	11.486	11.929	12.056	318.7	148.4	-36.7	5.6	149.2	-33.5	5.6
9379	357.8099	-6.9548	2012.924	15.871	17.398	38.245	41.4	68.9	0	5.8	69.7	11.6	6.5
9380	357.8446	23.5130	2013.199	16.613	17	7.228	140.2	75.4	1.6	5.7	75.1	-3.7	5.7
9381	357.8619	53.6232	2013.587	17.362	18.419	34.176	141.6	-15.5	69.6	5.9	-15.2	72.2	6.4
9382	357.8657	58.9950	2013.556	17.765	18.412	24.546	91.8	61	-8.3	6.2	63.7	-1.8	6
9383	357.9839	79.0191	2013.443	14.439	16.005	27.066	275.5	83.9	16.1	6.2	79.5	17.3	6.2

9384	358.0636	46.4313	2013.53	15.374	15.975	56.414	347.1	154.7	31.9	5.5	155.7	32.1	5.5
9385	358.0709	4.4656	2013.0825	11.3	16.719	37.818	0.5	94.3	39.7	6.5	93.5	37.5	6.7
9386	358.0796	36.1312	2013.381	12.766	18.136	13.34	75	70.7	22.4	6.2	70.5	28.6	6.7
9387	358.0923	14.0425	2012.9005	14.169	17.143	8.823	237.2	67.4	-20.7	5.7	64.6	-21.3	5.8
9388	358.1046	-0.4414	2012.8435	16.2	16.665	6.108	335.7	-30.7	-66.5	5.7	-29.9	-63.7	5.8
9389	358.1214	60.4827	2013.4395	11.823	13.304	6.297	88	101.3	25.2	5.9	100.5	22.9	5.9
9390	358.1436	62.5817	2013.5225	13.257	15.444	4.88	180	79.5	16.1	5.9	78.5	14.5	6.4
9391	358.1566	27.3348	2013.3535	13.521	15.978	5.992	134.3	-153.3	-96.7	5.7	-152.9	-100.7	5.9
9392	358.1590	-3.4447	2012.8935	11.401	13.533	5.937	0.9	-47.2	-148.6	5.7	-44.1	-155.8	5.9
9393	358.2052	39.9621	2013.4775	15.108	17.101	34.594	60.9	67.2	-2.2	6.2	64.8	0.5	6.2
9394	358.2072	20.8498	2013.3525	14.465	14.975	8.05	230.3	63.9	-20.5	5.5	62.9	-22.6	5.5
9395	358.2082	70.1173	2013.351	18.313	18.546	42.977	57.4	65.9	-13.9	6.6	68.9	-1.4	6.5
9396	358.2374	22.9449	2013.3005	16.11	16.727	7.385	184	61.6	-59.6	5.6	61.1	-61.1	5.6
9397	358.2568	74.3840	2013.4935	13.872	13.951	7.041	162.6	78.5	11.1	6.2	80.4	7.6	6.2
9398	358.3227	41.7052	2013.4	17	17.049	43.336	171.9	68.8	-6.1	6.3	68.7	-1.8	6.2
9399	358.3339	14.2589	2012.838	16.815	18.074	53.845	245.7	65	-37.9	6	66.8	-39.3	5.8
9400	358.3456	72.7527	2013.435	14.021	16.558	7.485	123.5	-39.9	-78.7	6.2	-44.5	-77.5	6.3
9401	358.3739	4.1758	2013.023	11.675	17.43	30.867	280	67.3	51.1	6.5	72.3	51	6.8
9402	358.3885	11.1038	2013.4835	13.299	13.882	5.395	143.3	99.1	3.2	6.7	99.8	1.7	7.3
9403	358.3982	12.1058	2013.569	9.849	10.691	5.847	165.3	39.4	-111.6	5.8	43.2	-112.3	5.6
9404	358.4002	-5.7907	2012.908	13.041	17.806	15.189	10.8	116.7	-33	5.7	119.9	-36.7	7.5
9405	358.4214	49.7276	2013.542	14.687	17	5.104	341.6	103.1	-44.6	5.8	103.5	-47.8	5.9
9406	358.4627	6.2386	2013.035	12.258	14.939	15.822	124.9	7.3	-65.2	6.5	10.6	-63.5	6.6
9407	358.4717	12.1734	2012.934	14.701	17.18	6.575	161.4	70.5	12.5	5.7	71.7	13.2	5.8
9408	358.4756	6.2089	2012.9325	14.731	16.716	12.301	109.3	-4.6	-67.8	6.6	-4	-68.8	6.7
9409	358.4785	-3.4186	2013.0015	14.001	17.756	28.033	162.5	65.9	-37.9	5.7	67.4	-42.1	6.3
9410	358.4899	13.2708	2013.0765	12.05	13.653	7.694	123.7	74.4	-20.4	5.7	74	-19.3	5.7
9411	358.5457	2.3138	2013.451	9.422	10.189	14.862	185.8	42.2	-66.8	6.4	46.2	-60.8	6.4
9412	358.5468	1.4858	2013.047	11.718	12.985	5.195	149.6	36	-95.7	6.5	37.9	-92.7	6.7
9413	358.5887	29.1629	2013.395	10.962	16.183	14.581	335.5	80.8	0.2	5.9	80.1	-2.9	6
9414	358.6237	29.6375	2013.4885	8.235	14.783	22.47	201.7	11.4	-191	5.9	5	-195	6
9415	358.6470	24.4626	2013.3305	14.179	15.216	14.38	124	91.4	-32.9	6	90.6	-32.6	6
9416	358.7217	84.6093	2013.765	11.566	14.946	8.898	282.1	100.4	23.8	5.8	102.1	23.1	5.8
9417	358.7379	82.6258	2013.7385	12.924	15.038	7.272	357.4	98.4	113.5	6.1	95.4	108.4	6.1
9418	358.7402	46.8775	2013.4595	13.215	18.125	10.092	33	60.3	9.6	5.5	62	12.1	6.1
9419	358.7585	-4.6239	2013.1135	11.272	13.126	9.706	15.6	63.7	-34.4	5.6	63	-34.7	5.7
9420	358.7867	14.8123	2013.319	14.243	15.285	6.45	289.7	124.4	-68.4	5.3	126.6	-67.1	5.3
9421	358.9801	-8.2960	2012.946	11.758	13.461	7.595	212.3	66	-43.6	5.8	65.7	-40.5	6.5
9422	358.9835	21.2562	2013.3495	14.613	15.369	7.608	328.9	64.2	-28.9	5.2	63.6	-29	5.2
9423	359.0729	6.6633	2013.046	12.675	13.298	10.668	355.4	-51.4	-62.9	6.6	-52.6	-61.2	6.6
9424	359.0862	-7.9095	2012.946	14.647	14.727	8.55	302.3	-26.2	-87.9	5.8	-31.9	-86.7	5.8
9425	359.1510	25.6225	2013.3885	13.491	16.474	9.128	155.8	85.1	9.7	6.4	83.6	9.4	6.7
9426	359.2016	45.8795	2013.542	12.476	13.981	5.184	220.6	142.5	-37.7	5.5	143.3	-41	5.5
9427	359.2037	67.3020	2013.428	11.865	12.053	9.23	229.4	133.5	-11.1	6.2	135.6	-6.3	6.2
9428	359.2682	30.5770	2013.435	10.798	11.599	12.813	296.3	41.5	-96.9	5.5	42.9	-95.7	5.5
9429	359.2807	14.2213	2013.1505	15.423	16.767	5.344	110.7	62.7	-21	5.2	63	-12.8	5.3

9430	359.3095	29.9882	2013.305	16.315	17.268	11.975	203.9	41.2	-67.8	5.6	45	-66.5	5.3
9431	359.3573	71.2903	2013.424	14.941	16.853	5.363	216	34.9	-71.8	6.2	38.2	-72	6.3
9432	359.4034	69.8786	2013.232	18.126	18.587	15.251	200.2	78	29.3	6.5	75.1	37	6.5
9433	359.4594	33.7302	2013.3305	10.668	17.735	24.61	174	108.8	10.6	5.5	109.6	13.4	5.8
9434	359.4612	52.6839	2013.5515	15.957	16.898	4.335	29	112	13.8	6.2	110.3	10.5	6.5
9435	359.4820	64.9310	2013.5435	15.207	15.562	3.79	196.9	99.1	49.4	6.2	105.2	48.9	6.3
9436	359.5190	61.6389	2013.452	12.551	12.933	9.206	333.2	70.7	19.7	6.2	70.9	18.9	6.2
9437	359.5420	30.9672	2013.3835	14.557	14.908	34.865	62.7	146.3	-6.7	5.6	144.4	-6.3	5.5
9438	359.5505	64.1349	2013.427	16.048	16.449	6.494	173.9	81.4	6.9	6.2	78.2	4.7	6.2
9439	359.5666	2.5055	2013.0275	14.627	16.21	5.63	47.1	84.3	-4.9	6.5	86.3	-4.2	6.8
9440	359.5872	27.2458	2013.4235	13.747	14.522	7.7	192.9	66.4	16.4	5.3	64.8	14.4	5.3
9441	359.6257	24.2005	2013.2705	10.717	12.988	13.265	257.2	-65.3	-186.3	5.7	-65.4	-191.3	5.7
9442	359.6395	-5.5678	2012.882	14.429	15.833	12.093	23.2	-23.9	-95.6	6.7	-18.6	-98.6	6.9
9443	359.6564	51.0265	2013.4775	11.027	16.104	20.837	187.1	83	53.3	6.2	87.2	51.7	6.2
9444	359.7183	10.5728	2013.06	13.436	15.982	5.422	314.8	111.9	-27.1	6.5	115.1	-31.5	7.1
9445	359.7227	1.8124	2013.2155	13.151	13.583	37.851	143.2	83	-33.5	6.4	81.2	-29.2	6.5
9446	359.7450	9.1085	2012.8695	15.741	18.768	7.251	61.9	-28.9	-78.9	6.6	-41.4	-71	7.7
9447	359.7615	54.7113	2013.629	8.198	13.827	7.577	260.3	61.4	-19.4	5.8	65.7	-7.2	6.6
9448	359.7976	4.7965	2013.0835	15.093	15.866	6.763	206.6	96.9	1.4	6.6	101.3	7.7	6.9
9449	359.8214	60.5387	2013.2015	17.234	18.829	29.157	327.2	12.3	68.1	6.7	13.5	66.8	6.8
9450	359.8828	56.1039	2013.4745	10.642	14.495	52.302	98.6	76.7	-63.5	5.9	74.3	-62.5	5.9

BY THE SAME AUTHOR

All are available from Amazon.com and from Amazon.co.uk

1800 new double stars for amateur observers

3600 celestial asterisms for amateur astronomers

Discover your own variable star

Identifying Common Proper Motion Binary Star Systems

Identifying Identical Twin Star Systems from the SDSS Data Release 10